The Future Role of Dwarf Honeybees in Natural and Agricultural Systems

The Future Role of Dwarf Honeybees in Natural and Agricultural Systems

Edited by
Dharam P. Abrol
Faculty of Agriculture, Sher-e-Kashmir University of Agricultural Sciences and Technology
Jammu, India

CRC Press
Taylor & Francis Group
Boca Raton London New York

CRC Press is an imprint of the
Taylor & Francis Group, an **informa** business

First edition published 2020
by CRC Press
6000 Broken Sound Parkway NW, Suite 300, Boca Raton, FL 33487-2742

and by CRC Press
2 Park Square, Milton Park, Abingdon, Oxon, OX14 4RN

CRC Press is an imprint of Taylor & Francis Group, LLC

Library of Congress Cataloging-in-Publication Data

Names: Abrol, D. P., editor.
Title: The future role of dwarf honeybees in natural and agricultural systems / Dharam P. Abrol.
Description: First edition. | Boca Raton, FL : CRC Press, 2020. | Includes bibliographical references and index.
Identifiers: LCCN 2020009463 | ISBN 9781138335820 (hardback) | ISBN 9781003033936 (ebook)
Subjects: LCSH: Honeybee--South Asia. | Honeybee--Variation--South Asia. | Pollination by bees--South Asia.
Classification: LCC QL568.A6 F888 2020 | DDC 595.79/90954--dc23
LC record available at https://lccn.loc.gov/2020009463

ISBN: 9781138335820 (hbk)
ISBN: 9781003033936 (ebk)

Typeset in Times
by Lumina Datamatics Limited

Contents

Foreword

The major challenge before agricultural scientists in the twenty-first century is to sustain higher productivity in agriculture without compromising the future in terms of resource degradation and depletion. The world is facing food deficit coupled with instability of climatic cycles. The growing population pressure has hastened the environmental degradation, ultimately posing a threat to natural resources and fast approaching famine. In the next 50 years, the global population is expected to reach 9 billion, doubling the food, feed and crop demand. The situation has further been aggravated due to pollinator decline worldwide resulting in a pollination crisis adversely affecting crop productivity and due to lack of prophylactic progress in the conservation of biodiversity.

Insect pollinators play an important role in producing crops in global agriculture. Pollinator-dependent crops contribute to maintaining a healthy variety in the human diet and often have a high market value, beneficial for local or regional economies.

We dwell in the midst of yet another world food shortage that is exacerbated by escalating the prices world over. Coupled with the apparent instability of climate cycles in recent years, one-fourth of our growing human population is fast approaching famine. Concomitantly, this situation includes rampant declines in honeybee populations across three continents, for as yet incompletely resolved reasons, and with no remedy or end in sight. These problems have been further aided and abetted by a lack of prophylactic progress in the conservation of biodiversity and increased agricultural production. Pollinators and pollination are crucial in the functioning of almost all terrestrial ecosystems including those dominated by agriculture because they are in the front line of sustainable productivity through plant reproduction.

A growing and increasingly human population is expected to double the demands for food production by 2050. Food production is also a central driver of global environmental change, contributing with negative effects on the climate, water resources, soils and the rich biodiversity the world harbours. Humanity stands in front of a great challenge to tackle these issues and needs to develop the food production system as a part of the solution for a sustainable world.

Insect pollination is important for global food nutrition. One often-highlighted ecosystem service is insect pollination. The loss of biodiversity among pollinators has raised questions about whether the pollination services they provide are at risk. Pollination and food production that pollinators provide are threatened by land-use change, agricultural intensification, climate change, pesticide use, pathogens, genetically modified organisms and invasive species. The economic market value of pollination in crop production globally is estimated to be $235 billion to $577 billion annually. Among the various pollinating agents, honeybees play a very important role in pollinating various crops. The honeybee pollination not only results in higher yields, it also gives a better quality of produce, and the efficient pollination of flowers also serves to protect the crops against pests.

Of the nine species of *Apis*, the dwarf honeybee *Apis florea* — generally confined to warm climates – is one of the most important pollinators of agricultural, horticultural and thousands of plants growing in natural landscapes. One of the most outstanding traits of this honeybee species is its ability to survive in a very hot and dry climate.

The greatest advantage of the dwarf honeybee *A. florea* is its capability to work under extremely hot, arid weather conditions, which is highly beneficial for crops blooming at higher temperatures. The dwarf bee is able to survive in climates with ambient temperatures reaching 50°C or more. Another interesting feature of *Apis florea* is its small foraging range, which makes it a suitable pollinator of small crops providing less amounts of nectar mostly abandoned by large-sized bees. *Apis florea* foragers have a small foraging range and visit a vast array of plants compared to larger ones. The small foraging range helps in the production of unifloral honey, which is normally difficult to obtain from bees with wider foraging ranges.

The role of dwarf honeybee *Apis florea* assumes utmost importance in the context of pollinator decline throughout the world, threatening stability of ecosystems and global food security. *Apis florea* is a lowland species of South Asia, extending more to the west than other Asiatic *Apis* species. It is distributed in India, Pakistan, Sri Lanka, Thailand, Indochina Malaysia and parts of Indonesia in altitudes up to 500 m but extends up to 1300 m in Yunnan, China, Iran, Oman, India and Thailand. It lives sympatrically with other native honeybee species *A. dorsata*, *A. cerana* and introduced *A. mellifera*. It nests in the open on slender branches of trees, shrubs under dense foliage, sheltered buildings and rock cliffs. It is well known for its swarming and absconding behaviour. Absconding is used to defend against predators, and the bees abandon their nest even at the slightest disturbance. The increase in the number of colonies occurs through swarming. The dwarf bee is an important pollinator of crops in hot and dry agricultural plains of India.

This book, *The Future Role of Dwarf Honeybees in Natural and Agricultural Systems* by Dharam P. Abrol is the first of its kind that deals in details on varied aspects of *Apis florea* biology, management, conservation strategies for protecting biodiversity and enhancing crop productivity. It provides a multidisciplinary perspective about the different facets of dwarf honeybees. The book aims to promote a large, diverse, sustainable and dependable bee pollinator workforce that can meet the challenge for optimizing food production well into the twenty-first century.

I congratulate the author for his efforts to provide this timely contribution.

M.S. Swaminathan
Founder Chairman
M.S. Swaminathan Research Foundation
Ex-Member of Parliament (Rajya Sabha)

Preface

Of the nine species of *Apis*, the dwarf honeybee *Apis florea* Fabricius is one of two small wild honeybee species confined to warm climates where it does very well. It is known to have a wider distribution in many tropical and subtropical parts of Asia, especially in Southeast China, India, Burma, Laos, Vietnam, Malaysia, Indonesia and the Philippines, and it is rarely found at altitudes above 1500 m above sea level.

The dwarf bee is able to survive in very hot and dry climates with ambient temperatures reaching 50°C or more. The greatest advantage of the dwarf honeybee *A. florea* is its capability to work under extremely hot, arid weather conditions, which is highly beneficial for crops blooming at higher temperatures. Another interesting feature of *Apis florea* is its small foraging range, which makes it a suitable pollinator of small crops that provide less amounts of nectar such as ber, coriander, cumin, fennel, carrot, phalsa, jamun, neem, grape etc. The small foraging range helps in the production of unifloral honey, which is normally difficult to obtain from bees with wider foraging ranges.

The research on *Apis florea* has often been ignored by beekeepers and scientists because of the species' small size and very low honey production capacity. In view of the above, *Apis florea* is considered a vital component of the natural ecosystem and needs conservation as a sustainable option for rural development, crop production and bio-diversity conservation. Keeping the above facts in mind, there is an urgent need to generate information on this important species. The aim of this book is to fill the gap by providing detailed information on different aspects of *Apis florea* leading to food security sustainability and environmental protection.

Agricultural scientists and graduate students working in the field pollination research, development and teaching will be the primary audience as well as principal purchasers of the book. The book will be used in the pollination-related courses at the graduate and post-graduate levels as supplemental reading at all the agricultural colleges/universities in the developed and developing countries. It is an excellent source of advanced study material for academics, researchers, students and programme planners. The information provided in this book will be useful for pollination biologists, honeybee biologists in entomology departments, students, teachers, scientists of agriculture, animal behaviour, botany, conservation, biology, ecology, entomology, environmental biology, forestry, genetics, plant breeding, horticulture, toxicology, zoology, seed growers and seed agencies.

This book provides a multidisciplinary perspective about the different facets of dwarf honeybees. The book aims to promote a large, diverse, sustainable, and dependable bee pollinator workforce that can meet the challenge for optimizing food production well into the twenty-first century and shall serve as a reference book for students, teachers, researchers, extension functionaries and policy planners. This book is the outcome of my personal experiences and the contributions of several workers, which have been incorporated. All the contributors deserve special appreciation for writing chapters in their respective fields in great depth with dedication. I thank university authorities for the excellent working atmosphere and needed encouragement for compiling such a voluminous book. I thank all our contributing authors who have worked tirelessly to provide their timely contributions.

I express my special thanks and appreciation for the editorial guidance of Renu Upadhyay (Commissioning Editor, CRC Press) and Jyotsna Jangra (Editorial Assistant, CRC Press) and Preethi Sekar (Project Management Executive, Lumina Datamatics) Ms Shikha Garg (Editorial Assistant, CRC Press) whose efforts have helped in shaping the text to bring out the best to the readers.

A final word of gratitude to all those who did not find mention above but have helped in the publication of this book.

Dharam P. Abrol
Jammu, India

Editor

Dr Dharam P. Abrol, Dean, Faculty of Agriculture, Sher-e-Kashmir University of Agricultural Sciences, has served university in various capacities. He has served as Head of the Division of Entomology from August 2008 to February 2015 and Controller of Examinations during 2012. He has specialized in pest management, honeybee management and pollination biology. He has authored 18 books, 12 manuals and published over 250 original research papers in various national and international journals. He has completed several externally funded collaborative research projects with international organizations in Poland and Switzerland. Prof Abrol has visited South Korea, Malaysia, Saudi Arabia and several other countries as a special invitee to these countries. In addition, he has received letters of appreciation from different organizations. Prof Abrol is a Fellow of the National Academy of Agricultural Sciences, India, and a Fellow of the Royal Entomological Society London, UK. He is a recipient of the Young Scientist Award (1992) — conferred by the Jammu and Kashmir State Council for Science and Technology — a prestigious state award for his outstanding contributions in the field of agricultural sciences. He is also the recipient of Pran Vohra Award — 1993 — a prestigious Young Scientist Award conferred by the Indian Science Congress Association, Calcutta, for his outstanding and innovative research in the field of agricultural sciences. He was also conferred the Prof T. N. Ananthakrishnan Award 1997–1998 — a prestigious national award for his outstanding contributions in the field of entomology by the T. N. Ananthakrishnan Foundation, G.S. Gill Research Institute, Chennai. Prof Abrol won the Dr Rajendra Prasad Puruskar 1999–2000 award — a prestigious national award from the Indian Council of Agricultural Research, New Delhi, for his Hindi book on beekeeping entitled 'Madhmakhi Palan-Sidhant Evam Vidhian'. He received the 11th Apicultural Association Award (2010) for outstanding contributions in apiculture. King Saud University conferred on him a gold medal for development of apiculture in Asia.

Contributors

Dharam P. Abrol
Faculty of Agriculture
Sher-e-Kashmir University of Agricultural Sciences and Technology
Jammu, India

R. Aruna
Department of Agricultural Entomology
Tamil Nadu Agricultural University
Coimbatore, India

N. S. Bhat
Department of Apiculture
College of Agriculture
University of Agricultural Sciences
Bengaluru, India

Michael Burgett
Department of Horticulture
Oregon State University
Corvallis, Oregon

Avinash Chauhan
Department of Entomology
School of Agricultural Sciences and Rural Development
Nagaland University
Dimapur, India

Bajaree Chuttong
Science and Technology Research Institute
Chiang Mai University
Chiang Mai, Thailand

S. Devanesan
College of Agriculture
Kerala Agricultural University
Thiruvananthapuram, India

Orawan Duangphakdee
King Mongkut's University of Technology Thonburi
Bangkok, Thailand

B. Fakrudin
Department of Biotechnology and Crop Improvement
Post-Graduate Centre
University of Horticultural Sciences
Bengaluru, India

Sampat Ghosh
Department of Biological Sciences
Sardar Patel University
Balaghat, India

Mohammed M. Ibrahim
Department of Plant Protection
Faculty of Natural Resources and Environmental Studies
University of Kordofan
El Obeid, Sudan

Jasmi Jambak
Program of Biological Education
College of Teacher Training and Education PGRI West Sumatra
Padang, Indonesia

Chuleui Jung
Department of Plant Medicals
Agriculture Science and Technology Research Institute
Andong National University
Andong, South Korea

İrfan Kandemir
Department of Biology
Faculty of Science
Ankara University
Ankara, Turkey

Prakash Karmakar
Department of Botany and Forestry
Vidyasagar University
Midnapore, India

Jameel Khan
Department of Biotechnology and Crop Improvement
Post-Graduate Centre
University of Horticultural Sciences
Bengaluru, India

Ujjwal Layek
Department of Botany and Forestry
Vidyasagar University
Midnapore, India

S. Mohankumar
Department of Plant Biotechnology
Tamil Nadu Agricultural University
Coimbatore, India

N. Nagaraja
UGC-Human Resource Development Centre
Bangalore University
Bengaluru, India

Saeed Mohamadzade Namin
Department of Plant Protection
Faculty of Agriculture
Islamic Azad University
Varamin, Iran

K. Nandini
Department of Biotechnology and Crop Improvement
Post-Graduate Centre
University of Horticultural Sciences
Bengaluru, India

T. Neethu
Department of Apiculture
College of Agriculture
University of Agricultural Sciences
Bengaluru, India

Muhammad Noor-ul-Ane
Department of Entomology
Faculty of Agricultural Sciences
Bahauddin Zakariya University
Multan, Pakistan

K. Omkar
Department of Biotechnology and Crop Improvement
Post-Graduate Centre
University of Horticultural Sciences
Bengaluru, India

Ayça Özkan Koca
Department of Gastronomy and Culinary Arts
Faculty of Fine Arts
Maltepe University
Istanbul, Turkey

Christian Pirk
Social Insect Research Group
Department of Zoology and Entomology
University of Pretoria
Pretoria, South Africa

M. Praghadeesh
Department of Plant Biotechnology
Tamil Nadu Agricultural University
Coimbatore, India

Rika Raffiudin
Department of Biology
Bogor Agricultural University
Bogor, Indonesia

T. Sonai Rajan
Department of Agricultural Entomology
Tamil Nadu Agricultural University
Coimbatore, India

A. J. Solomon Raju
Department of Environmental Sciences
Andhra University
Visakhapatnam, India

Preecha Rod-im
King Mongkut's University of Technology Thonburi
Bangkok, Thailand

Siti Salmah
Department of Biology
Andalas University
Padang, Indonesia

P. A. Saravanan
Department of Agricultural Entomology
Tamil Nadu Agricultural University
Coimbatore, India

B. V. Shwetha
Department of Apiculture
College of Agriculture
University of Agricultural Sciences
Bengaluru, India

H. K. Singh
Department of Entomology
School of Agricultural Sciences and Rural Development
Nagaland University
Dimapur, India

M. R. Srinivasan
Department of Agricultural Entomology
Tamil Nadu Agricultural University
Coimbatore, India

Ratna Thapa
Zoology Department (Entomology Section)
Kathmandu, India

1

Introduction

Dharam P. Abrol

CONTENTS

Introduction

Of the nine species of *Apis*, the dwarf honeybee *Apis florea* Fabricius is one of two small wild honeybee species confined to warm climates where it does very well. Molecular analysis indicates that *A. dorsata* is a member of the most ancient lineage within the genus and progenitor of *Apis florea* and *A. andreniformis*. The line giving rise to *Apis florea* and *A. andreniformis* separated from the *dorsata* line approximately 11 million years later and remained monophyletic until approximately 500,000 years ago (Wills 1991). It is known to have a wider distribution in many tropical and subtropical parts of Asia, especially in Southeast China, India, Burma, Thailand, Laos, Vietnam, Malaysia, Indonesia and Philippines, and it is rarely found at altitudes above 1500 m above sea level. In the west, the species is present in the warmer parts of Oman, Iran, coasts of the Persian Gulf and Pakistan, through the Indian subcontinent and Sri Lanka. It is found in tropical forests, in woody areas and even in farming areas. Finding nests of *Apis florea* is very common in South East Asia. The dwarf honeybee *Apis florea*, as its name implies, is the smallest species of honeybee, both in terms of its body size as well as its nest. The nest of *A. florea* consists of a single comb, the upper part of which expands to form a crest that is attached with the branch or other object from which it is suspended. The nests are generally covered with branches of trees or dense foliage. They may be found usually from 1 to 8 metres above the ground. In Oman, *A. florea* nests are most frequently found in caves, and the combs are without crests.

Comb Structure and Castes

Apis florea, the little bee, builds a single vertical comb nest. The comb is constructed around the branch or stem of a bush and hedges in the shaded places. The nesting locations of *A. florea* are unique, not easily accessible to humans and other animals. *A. florea* avails itself of various plant species including human-built structures for nesting under shady places on the twigs/branches. Nesting in shady places helps the colony members to protect themselves from the bright light, strong winds and inclement weather conditions. The height of *A. florea* nests may be from ground level to 50 ft, but 15 ft has been reported as a preferred level (Woyke et al. 2001; Manjunath 2008; Vaudo et al. 2012). The nest encircles the strong stem at the top so as to give it good strength. The portion of the comb that encircles around the stem is thick, but as it is built further down, it becomes thin in depth. The colony members make short-distance movements to unshaded nesting sites during autumn. The adult bees hang around the nest in clusters to control the nest conditions. The brood is present in the lower section and honey in the upper section of the comb. The comb is broad at the top and serves as a landing platform for the foragers.

The comb itself is less than 1 ft in length and contains about 14,000 cells. The sizes of all the three types of cells vary with geography and become smaller as we go from north to south. In a colony, there are about 6000 bees, and the queen lays 350 eggs per day in a brood area of 600 sq. cm. The worker brood cells are made on the flat surface in the middle of the comb. The comb is always covered by a curtain of bees. Drone cells are 1.5 times larger in diameter than worker cells and are found on the lower part of the comb. Queen cells are raised in spring and autumn at the bottom edge of the comb and are quite large. Compared to the other species of honeybees, its workers have a long life of 61.2 days, that is, 2.5 times that of an *A. mellifera* worker. Since this honeybee shifts a lot, it would appear its biology is adapted to multiplication, swarming and dissolution of the colony with the purpose of reassembling under a new situation. During a short span of 2–5 months, the nest is built, the brood is reared, the honey is stored and 12–16 queen cells are built for further propagation. When the first virgin queen emerges, half the workers leave along with the mother queen. Subsequently, the virgin queens mate and smaller swarms leave every few days. In the end, a handful of bees remain on the comb, and eventually they also swarm away or simply disperse, leaving the comb bare and deserted.

The distribution of *Apis florea* is much wider compared to its sister species, *Apis andreniformis*. Both the species are very primitive, which is evident from their small colonies and simple method of construction of their nests. The exposed combs are built on the branches of shrubs and trees, and the foragers do not perform a waggle dance as in *Apis mellifera*; rather, the dance is in the form of straight runs pointing directly to the direction of the food source. In the other bee species, the dance is vertical and not directed towards the food source. Except for the small size of combs, exposed nests and simple dance language, the life cycle is fairly similar to other *Apis* species. This species is mostly confined to plains up to 500 m; however, it may perform seasonal migrations up to 1500 m or more (Muttoo 1956). In spite of its small size, it competes well with other *Apis* species (Koeniger 1976). The bees with larger sizes are found in the north, and size decreases moving towards the south. Phenotypic variation among *A. florea* has not been fully understood. The other honeybee species, *A. andreniformis*, is darker in colour and found mostly in sub-mountain regions of Malay Peninsula, Sri Lanka, Sumatra, Java and Borneo and Thailand, but not recorded in India so far.

The dwarf honey *A. florea* can survive in very hot and dry climates with ambient temperatures reaching 50°C or more. Its capability to work under extremely hot arid weather conditions is highly beneficial for crops blooming at higher temperatures (Sihag 1984, 1994, 2000). Another interesting feature of *Apis florea* is its small foraging range, which makes it a suitable pollinator of small crops, providing less amounts of nectar such as ber, coriander, cumin, fennel, carrot, phalsa, jamun, neem, grape etc. The small foraging range helps in the production of unifloral honey, which is normally difficult to obtain from bees with wider foraging ranges. The research on *Apis florea* has often been ignored by beekeepers and scientists because of the species' small size and very low honey production capacity. In view of the above, *Apis florea* is considered a vital component of the natural ecosystem and needs conservation as a sustainable option for rural development, crop production and biodiversity conservation. Keeping the above facts in mind, there is an urgent need to generate information on this important species (Oldroyd and Nanork 2009).

The combs of *A. florea* are well covered with three to four layers of bees clinging with each other forming a protective cover. About three-fourths of the total population of bees are involved in forming the protective cover. When disturbed, this protective cover shows a shimmering movement, with individual bees shaking their abdomens side to side in a synchronus manner and releasing hissing sounds simultaneously. If the disturbance continues, the worker bees take off from the curtain and start stinging the intruder.

The section of the comb surrounding the support consists of honey storage that forms the crest from whose cells the inroad curved surface is used by the bees to take off and land. The scout bees convey the discovery of food source from this platform. Adjoining honey storage cells in this part of the comb are used for storing pollen.

The worker brood is reared immediately below the pollen storage cells. Below the worker brood cells, the drone brood is added prior to swarming. If a colony loses its queen, emergency queen cells are built from the young larvae in normal cells. The comb support at both ends is covered with sticky propolis or plant gums 2.5 to 4 cm wide to ward off ants. Among all the honeybee species, this type of defensive technique is used by *A. florea* only. When a good supply of nectar and honey is available, the populous colonies issue multiple reproductive swarms. The mobility of the colonies increases considerably. Absconding of colonies, on other hand, results due to shortage of food, inclement weather conditions and attack of diseases and enemies.

Honey storage in *A. florea* colonies is very small, not exceeding more than a few hundred grams per colony. However, in some parts of Asia, rural people have devised techniques to collect honey by shifting the combs to the village from their natural sites and clamping with a small branch using two twigs. The honey is removed by cutting the upper part of the comb and squeezed out from it. After a period of six to eight weeks when the bees have replenished the comb, honey is harvested again. This method is not always successful, as the bees may abscond after one harvest. Though in some areas where *A. florea* colonies are located, more may help the local populations to generate money through honey harvest, but this practise is ecologically highly destructive, resulting in reduction of the population of valuable natural pollinators. The workers and the laying queen are able to respond to nest predation very quickly. In the event of attack from predators, the whole colony accompanied by a laying queen flies several meters away to regroup and abscond. Some absconding colonies are able to survive and build their combs in new area. Another species of dwarf bee is *A. andreniformis*, which is darker and a little larger than *A. florea*, mostly found in tropical and semievergreen forests. The abodemen is black in colour as compared to *A. florea*, which has white stripes alternating with orange. The species has been observed in cultivated and inhabited areas. All activities including building of nests is similar to *A. florea*. According to Ruttner (1988), *A. florea* can be categorized into three geographic types that include a group found in Sri Lanka and South India; a second one in Iran, Oman and Pakistan; and a third in Thailand. It is expected that all three types are found in India.

The dwarf bee can tolerate temperatures as high as 50°C or more. The comb architecture of *A. florea* is similar to all other *Apis* species, except for the honey storage portion, which is distinct in the *A. florea* comb. In *Apis florea*, the comb shows a very distinct honey portion situated at the top. The honey cells are constructed around the support where the support is free from above. The cells surrounding the twig are 3 cm deep and extend more or less horizontally from the support on both sides, thus forming a cylindrical comb with the twig passing through its middle. The comb is used to store honey and is called the honey cap or crest. The dwarf bee shows a great adaptability in its nesting behaviour. When the nests are constructed in wall niches or under the eaves of houses, the crest is not in the form of a cylinder. The nature of the support determines the shape of the crest. If it is vertical as in the case of a wall niche, the crest is attached to the wall laterally. If on a horizontal ceiling, the crest is positioned adjacent to the vertical portion of the comb. The comb abruptly reduces in thickness beneath the crest to about 16 mm and extends vertically with cells 2.7 to 3.1 mm in width. When the flowers are available in good abundance, a conspicuously large celled drone brood portion is seen below the worker brood, and the colony reproduces by issuing swarms. The queen cells are raised along the lower edge of the comb. The worker bees are much smaller than the drone and queen bees. The number of worker cells per dm2 was 1190 in north India, 1240 in central India and 1560 in south India (Muttoo 1956), indicating that size of bees increases from south to north. The number of bees present in a colony may range from 5000 to 10,000 depending upon the condition of the colony. The queen normally lays 350 to 700 eggs per day (Kshirsagar et al. 1980; Ruttner 1988). The bees migrate most frequently when the forage is limited. The bees' stay at a place may range from 2 to 5 months, and the bees may stay even for one year if the forage potential of the area is good. Muttoo (1956) reports that the bees undertake migrations in summer from the hot plains to the hills involving long distances of several kilometres. Moritz et al. (2010) reported that migrations of *A. florea* may pose a threat to *A. mellifera*, as they also compete for the same resources. The bees have a short flight range, often hardly travelling 100 m from the nest.

Honey Quantity and Quality

The storage capacity of the honey cap in the nest is 500–1000 g or more (Muttoo 1956; Ruttner 1988). Large colonies in areas rich in bee forage are known to produce up to 4 kg of honey each. Phadke (1968) reports that honey of the dwarf bee is similar to that of the dammar bees and has a high dextrin content and low tendency to granulate. Dextrins are four times that in other *Apis* honeys. Levulose content is also high. The reason for the differences in the composition of the dwarf bee honey is supposed to be the sources of forage (Abrol and Kapil 1987). These usually include weeds and small flowered plants like those in the families Apiaceae, Asteraceae and Euphorbiaceae. The honey is valued in Ayurveda for its high medicinal properties. Handling *A. florea* colonies is easy. The twig supporting their nest is cut loose. The bees cling securely to the brood nest and not even a single bee flies out during transportation from one site to another. The bees are mild and rather than drive away the enemy by stinging, the bees resort to absconding and constructing a new nest in a sheltered and protected place. Given its open nesting and small size, which makes it difficult for the bees to defend their nest effectively from enemies, particularly humans, this behaviour perhaps ensures the continuation of the species. Because of this behaviour, rural and tribal people, including children, often plunder *A. florea* nests for the sweet honey and nutritious brood that is eaten raw.

Domestication of *Apis florea*

Several attempts at managing and hiving of *A. florea* have been recorded from India, but none of them met with complete success (Bharadwaj and Kapil 1980). Ghatge (1949) reported that these bees were amenable to gentle handling. After cutting the honey cap, the comb could be tied securely in a brood frame of *A. cerana* and the bees immediately join the comb to the top of the frame. A super frame of *A. cerana* would be held above the brood frame. According to Ghatge, the bees would construct a honeycomb and fill it with honey. However, validation of this technology has not been reported so far. Pundir (1971) succeeded in keeping an *A. florea* colony for about two months in Ranipokhari, near Dehradun, Uttar Pradesh, in a 4-frame nucleus box of *A. cerana*. Dutton and Simpson (1977) were the first to successfully manage *A. florea* colonies in Oman. Dutton and Free (1979) and Williams (1979) made further observations on the practises of the beekeeper. The farmers collected wild colonies from nearby mountains and installed them in artificial caves and cavities in old buildings and trees. Due to the bees often absconding, more and more colonies need to be discovered and collected in order for colonies to be maintained. According to FAO (1987), because of their typical open nesting behaviour, this prevents the bees from being kept in human-made structures, which is essential to manipulate them (Akratanakul 1987).

Commercial Beekeeping with *A. florea*

Commercial beekeeping with *Apis florea* is not very common due to very low production of honey. However, Dutton and Free (1979) reported that *florea* colonies are collected by experienced beekeepers in Oman from the wild and established in a split stalk of a date palm leaf and located in wall niches or close to buildings. Honey is harvested without destroying the brood. Sometimes during winter, colonies are transported to sunny areas. Such efforts are extremely necessary to develop *florea* beekeeping in excessively hot climates, where *A. mellifera* cannot normally be kept (Ruttner 1988).

One of the most remarkable qualities of *Apis florea* is that it is an industrious field bee; it is very easy to handle and when it migrates, the flight range is quite short (Ruttner 1988). Rearing of *A. florea* does not require any sophisticated technology, and bees do not even require the box needed by *A. mellifera* and *A. cerana*. Though honey production is very low, the pollination potential of *A. florea* is unparalleled for agricultural and horticultural crops. It is an important pollinator in dry areas where temperatures go very high and other bees abandon their activities. Atwal (1970), Sihag (1982) and Abrol (2006) consider *A. florea* as one of the most efficient pollinators of sarson, toria and raya varieties of mustard, lucerne, onion and cultivated Apiaceae that include ajwan, anise, carrot, cumin, dill seed and fennel. *Apis florea* are very small bees, and their nest is small, too, often no larger than a man's hand. A colony might contain 20,000 bees.

Pollination

Farmers value *A. florea* colonies for their pollination service. To ensure pollination and good yields, they collect *colonies* from wild shrubs and trees, take the nests to their farmlands and locate them, generally, along hedges having small trees that can provide adequate shelter to the nests. While collecting the colony, the honey portion is cut and consumed. The brood comb is secured between two twigs or a split bamboo piece. The twig with the nest is tied to a thin branch at the new site.

Apis andreniformis Frederick Smith, 1858

Apis andreniformis is a species of honeybee whose native habitat is in the tropical and subtropical regions of Asia. *A. andreniformis* It was the second honeybee species to be recognized, and its biology, geographic distribution and its specific status were recognized by many authors. However, it must be mentioned that the species was only recently separated from *Apis florea* since there are sites where both *A. andreniformis* and *A. florea* live conspecifically. They are distributed in tropical and subtropical areas, including Southeast China, India, Burma, Laos, Vietnam, Malaysia, Indonesia (Java and Borneo) and the Philippines (Palawan).

The most significant morphological characteristic of the species is the black-striped legs, specifically on the tibia and on the dorsolateral (back and side) surface of the basitarsus. *A. andreniformis* has blackish pigmentation, while that of *A. florea* has yellowish pigmentation. The other distinguishing traits of the two species include different respective cubital indexes: *A. andreniformis* has an index of 6.37, and *A. florea* has one of 2.86. Also, the proboscis of *A. andreniformis* has a length of 2.80 mm, while that of *A. florea* is 3.27 mm. This physical difference contributes to a division in the distribution of naturally occurring nectar between the two species. Finally, there are differences in the barbs of the stinger, and in the basitarsus of the drones.

Nests of both *Apis florea* and *Apis andreniformis* are small, made in shrubs and in bushes. They may be made at 1 to 15 m from the ground, with an average 2.5 m above ground. The honeycomb ranges from 7 to 90 mm, however, need further confirmation. *A. andreniformis* has been reported to be more defensive than *A. florea*. It is known to attack when there are disturbances 3 to 4 m from the hive.

Taxonomy and Phylogeny

Apis florea was first identified by Danish zoologist Johan Christian Fabricius in the late eighteenth century in 1787. *A. florea* is a member of the *Apis* genus. The name florea refers to Romanian origin. *A. florea* is considered to be phylogenetically the most basal bee that is native to South East Asia. It is suggested that honeybees originated in this region and *A. mellifera* evolved from *A. florea*, which later migrated to Europe. The other honeybee species, called *Apis andreniformis*, is a sister species of *A. florea*. These species represent the most ancient honeybee lineage.

Description and Identification

Both the species are the most primitive of the living species. This is evident from their simple nest construction and small colony size. Both the species belong to subgenus *Micrapis*. The combs are built on branches of shrubs or small trees. The bees do not perform a gravity-based dance on the vertical surface as occurs in other honeybee species. They perform a dance on the horizontal upper surface. The dance is composed of straight runs pointing directly to a food source such as nectar and pollen. Distinctiveness of two honeybee species, *A. florea* and *A. andreniformis*, was clearly defined in 1990. The colour of *A. florea* workers is redder whereas *A. andreniformis* is darker in colour and the first abdominal segment is black (Wu and Kuang 1987).

Colony Cycle

The annual colony cycle of honeybees involves migration, swarming and absconding. *A. florea* migrate seasonally from one habitat to another. This might increase colony fitness, as the honeybees search for new territories, resources or a reduction in parasites. Once a colony has outgrown its hive space, it will reproduce via swarming.

During warmer seasons, the prevailing temperatures are favourable enough to allow foraging and reproduction of honeybees. During the colder seasons of autumn and winter, colonies diminish in size because they depend on food stores. Before swarming, the colony will build queen cells in order for virgin queens to rear young queens. The workers start searching for nesting sites before of the emergence of queens. Afterwards, the bees will choose who stays and who goes. Similar to absconding behaviour, the old queen will swarm while the new queens prepare for emergence (Suwannapong et al. 2011).

Apis florea Fabricius, commonly known as the dwarf or small honeybee, is one of two species of dwarf honeybees found in South East Asia. It builds an exposed single comb nest most frequently in a shaded location (Akratanakul 1987; Seeley et al. 1982). In northern Thailand during the dry winter season, the average *A. florea* colony occupies a comb area of 490 cm^2, contains 5000 adult workers, 35 adult drones, 215 immature drone stadia, a single female reproductive and 7 reproductive queen cells (Burgett and Titayavan 2004).

Breeding and Lifespan

In a single patriline, more workers mean the colony will produce more drones (Palmer and Oldroyd 2001). High mating frequency in *A. florea* is an evolutionary phenomenon as observed in other members of the genus. The average lifespan of a drone is 15.6 days, ranging 6–41 days. The queen has a much longer life span that continues up to the life of the colony (Breed et al. 2007). *A. florea* has workers using oophagy to ensure that workers' eggs are destroyed at twice the rate of the queens' egg-laying speed (Halling et al. 2001). The queens' eggs are marked so as to avoid destruction by worker bees. Since the worker policing is also observed in *A. mellifera*, this clearly indicates that this mechanism evolved before the divergence of the genus *Apis* (Halling et al. 2001).

Foraging

Foraging behaviour of *A. florea* is very similar to other honeybee species. It is generally a top worker and shows floral fidelity. Abrol (2010) reported dwarf honeybee *A. florea* as the most important pollinator of onion flowers. It was most abundant and composed of more than 94% of the total flower visitors. Flight activities in the morning were initiated when a minimum threshold of environmental variables was attained, while the cessation was controlled by a decline in light intensity and solar radiation. The honeybees were not observed to forage during night as in the case of *A. dorsata*, which continues to visit flowers at night (Abrol 2004). They prefer to forage on crops with small flowers. *A. florea* tends to forage for longer durations as compared to other honeybees. Because of their small size, they usually take more time to forage for pollen or nectar (Dyer and Seeley 1991).

Competition with Other Species

A. florea demonstrates aggressive behaviour when competing for territory (Koiger and Vorwohl 1979). As the honeybees' flight range is very small, they defend flowers in a small territory (Koiger and Vorwohl 1979).

Honey Collection

A. florea is a poor honey yielder, and a comb may yield 200–900 g of honey. The honey is thin in consistency. Honey hunters remove the whole of the combs to collect honey and thereby destroy a large number of colonies. *A. florea* honey is considered to be of very high medicinal qualities. In some countries such as Oman, *A. florea* honey has been reported to fetch very high prices. *A. florea*, even though quite small in body size, is quite aggressive and resorts to robbing the nests of *A. mellifera*, which is much bigger in body size and strength of its colonies.

Swarming

Even though they are gentle in temperament, the bees do sting when irritated. The colonies can be shifted to crops requiring pollination. They coexist with other bees such as *A. cerana* and *A. dorsata*. One of the special traits of *A. florea* is that it can live in deserts and tolerate temperatures up to 50°C or more. The heat tolerance ability is further demonstrated by the fact that a worker bee's daily activity starts at a temperature above 18°C and continues up to 40°C. It is no wonder that this bee is a relentless visitor of flowers of crops, trees, shrubs and annuals. Because of their very short flight range, the bees exploit the flowers intensively. The swarms forming the new colonies generally settle at a distance of less than 100 m. Some swarms do go far but not more than a few hundred metres. It has been observed that if a queen is removed or it dies naturally, the workers build a queen cell by modifying a worker cell somewhere in the middle of the comb.

Mating Systems

Unlike *A. mellifera*, females of *A. florea* produce no mucus and after mating the sperms are transferred into the spermatheca directly (Koeniger et al. 1989). The males of *A. florea* use a clasper organ at the hindleg to hold the queen's leg during mating. Evidently, *A. florea* mates through direct sperm transfer (Koeniger et al. 1989). The *Apis florea* queen mates with a few drones (3–4) as compared with *A. mellifera* (25) drones, which simply deposit spermatozoa in the oviducts of the queen, where only 10% of spermatozoa reach the spermatheca. This shows that *A. florea* uses energetically a more efficient method of sperm transfer (Koeniger et al. 1989).

Defensive Behaviour

A. florea shows highly specific social defence behaviour when it encounters predators nearby. Besides showing hissing and shimmering behaviour, the bees nest amidst dense foliage to camouflage themselves from potential predators (Narayanaswamy and Basavarajappa 2013). One of the best examples is the use of sticky material to obstruct the attack of the weaver ant, *O. smaragdina.* The guard bees take cover behind the sticky barrier and alert other bees through hissing sounds to prevent a large-scale attack by weaver ants. The recruitment increases over time to this sticky zone barrier, which reinforces the bees' defence (Sciences and Sciences 2005). Simultaneous action of many individuals requires a strong communication between workers, which is accomplished by warning signals emitted by one individual, called piping, which is followed by 0.3–0.7 seconds later by a large number of bees through hissing (Sarma et al. 2002). Hissing is a noisy, broadband signal produced by slight movements of the bees' wings audible to the human ear. Individuals close to the piping bee are the first to hiss, which spreads rapidly to neighbours until an impressive coordinated action is taken by the entire colony. Sarma et al. (2002) reported that piping and hissing are associated with the workers ceasing their foraging dancing and departures from the colony.

Communication Behaviour

Apis florea requires a mechanism of communication to convey information about foraging and distance and direction (Narayanaswamy and Basavarajappa 2013). The side-to-side distance to a site is conveyed by *Apis florea* through a waggle dance wherein a worker moves forward while shaking the abdomen. The longer the duration of the waggle dance, the more the distance to a site and vice versa. The direction is indicated by orienting the waggle phase in the direction of the site (Narayanaswamy and Basavarajappa 2013). *Apis florea* does not change the precision of the waggle dance to indicate a nest site or food patch, unlike *A. mellifera* (Narayanaswamy and Basavarajappa 2013).

Genetic Diversity

Strong genetic diversity exists even in single *Apis florea* colonies, as the queens are polygamous. This variation is responsible for *Apis florea* workers performing different tasks. For instance, Dyer and Seeley (1991) reported that a specific set of workers perform fanning when a specific threshold of temperature is attained.

Division of Labour

Young workers of *A. florea* perform indoor duties whereas old workers protect the nest and undertake foraging activities (Beshers et al. 2001). Different sets of workers perform specific activities that are genetically controlled, such as fanning of the nest to maintain temperature is done by a specific set of workers. The different patrilines respond to specific temperature thresholds (Beshers et al. 2001). Genotypically, a diverse group of workers is produced as a consequence of multiple matings of the queen with unrelated males, which are genetically more fit than uniform colonies and can tolerate a wider range of environmental variations (Crozier and Page 1985). This results in task specialization in which highly related daughters of a haploid male undertake certain tasks (Robinson and Page 1988; Oldroyd et al. 1994). This is why polyandry has evolved and is being maintained in honeybee colonies (Wongvilas et al. 2010). In a multiple mated queen, eggs laid by workers are removed; this is called worker policing (Dawkins and Carlisle 1976). The worker policing helps to control the production of males by other workers in favour of reproduction by the queen.

The workers compromise by raising the eggs laid by the queen only and removing any eggs laid by other workers as an evolutionarily stable strategy (ESS) (Dawkins and Carlisle 1976). Worker policing is therefore beneficial to workers than rearing their own eggs, as they are better able to ensure the survival of their genetic material. This explains why workers have inactive ovaries in queenright colonies of this multiply mated species (Dawkins and Carlisle 1976). Workers of queenless colonies activate their ovaries just within four days and lay eggs in a last effort to ensure gene survival before the colony disappears through absconding (Dawkins and Carlisle 1976). In queenless colonies, worker policing decreases as all workers lay male eggs (Dawkins and Carlisle 1976). The problem occurs when other workers parasitize a queenless colony and policing may be reinstalled (Dawkins and Carlisle 1976) thereby diminishing reproductive success and the benefits of remaining within that colony (Dawkins and Carlisle 1976). This results in absconding of queenless colonies within three weeks (Dawkins and Carlisle 1976). This clearly establishes that *Apis florea* prefers to have a queenright colony and use worker policing as an ESS (evolutionarily stable strategy). Chapman et al. (2009) reported that in colonies of the red dwarf honeybee, *Apis florea*, when the colonies become queenless, the proportion of nonnatal workers increases significantly. These workers develop ovaries and lay eggs. The queenless colonies are, therefore, heavily parasitized with the eggs of nonnatal workers. Chapman et al. further found that workers of queenless *A. florea* colonies are far more likely to leave their colony and join another colony compared with workers from queenright colonies.

Parasites and Predators

Since the economic significance of *A. florea* is not much acknowledged, enemies and diseases of this bee have scarcely been investigated. *Euvarroa* is the main parasitic mite that attacks both the species. *Euvarroa wongsirii* preys on *A. andreniformis* and *Euvarroa sinhai* on *A. florea*. The two species of *Euvarroa* differ morphologically; *Euvarroa wongsirii* has a triangular body and a length of 47–54 μm whereas *E. sinhai* has a circular shape and a length of 39–40 μm (Delfinado-Baker et al. 1989). The most important predators of *Apis florea* are ants and other arthropods. Since *Apis florea* colonies are open nesting bees, they employ two mechanisms to defend their nest, pushing their head against falling objects or leaves and using a sticky barrier against ants and other arthropods (Koeniger et al. 1989).

Interaction with Humans

Two types of interactions of *A. florea* are observed with humans beings, which include hunting and tourism (Oldroyd and Wongsiri 2009). In Asia, there has been regular hunting of these bees because they are harmless and also have social and cultural significance to both Hindus and Buddhists. Hindus use honey for various religious ceremonies (Oldroyd and Nanork, 2009). Similarly, in Buddhism, gifting honey is considered one of the most expressive ways that one can reward a good deed. In Nepal, honey-hunting sites are considered tourist places (Oldroyd and Nanork, 2009). Increased deforestation due to industrialisation has enhanced bee migration to areas that are less populated, thus limiting the human interaction to a greater extent (Oldroyd and Nanork, 2009).

REFERENCES

Abrol, D. P. 2004. Path coefficient analysis of environmental factors influencing flight activity *Apis florea* F. and seed yield in carrot (*Dacus carota* L.). *International Journal of Horticultural Science* 10(1): 77–82.

Abrol, D. P. 2006. Factors influencing flight activity of *Apis florea* F., an important pollinator of *Daucus carota* L. *Journal of Apicultural Research and Bee World* 45(2): 2–6.

Abrol, D. P. 2010. Foraging behaviour of *Apis florea* F., an important pollinator of *Allium cepa* L. *Journal of Apicultural Research* 49(4): 318–325.

Abrol, D. P. and Kapil, R. P. 1987. Nectar dilution pattern of bees in semi-arid environments. *Current Science* 56(13): 681pp.

Akratanakul, P. 1987. Beekeeping in Asia FAO Agricultural Services Bulletin 68/4 Food and Agriculture Organisation of the United Nations FAO 1987 Rome, 1990.

Atwal, A. S. 1970. *Insect pollinators of crops*. Punjab Agricultural University Press, Ludhiana, Punjab India, p. 142.

Beshers, S. N. and Jennifer, H. F. 2001. Models of division of labor in social insects. *Annual Review of Entomology* 46: 413.

Bharadwaj, S. C. and Kapil, R. P. 1980. Domestication and training of *Apis florea* F. To visit and artificially feeding station. *Indian Journal of Entomology* 42: 290–292.

Breed, M. D., Deng, X.-B., and Buchwald, R. 2007. Comparative nestmate recognition in Asian Honey Bees, *Apis florea, Apis andreniformis, Apis dorsata*, and *Apis cerana. Comparative and General Pharmacology* 38: 411–418.

Burgett, M. and Tityavan, M. 2004. Apis florea F colony biometrics in northern Thailand. *Proceedings of the 8th IBRA conference on tropical bees and VI Encontro sobre Abelhas Ribieirao Preto Brazil*, 6–10 September 2004, pp. 46–54.

Chapman, N. C., Nanork, P., Gloag, R. S., Wattanachaiyingcharoen, W., Beekman, M., and Oldroyd, B. P. 2009. Queenless colonies of the Asian red dwarf honey bee (*Apis florea*) are infiltrated byworkers from other queenless colonies. *Behavioral Ecology* 20(4): 817–820.

Crozier, R. H. and Page, R. E. 1985. On being the right size: Male contributions and multiple mating in social Hymenoptera. *Behavioral Ecology and Sociobiology* 18(2): 105–115.

Dawkins, R. and Carlisle, T. R. 1976. Parental investment, mate desertion and a fallacy. *Nature* 262: 131–133.

Delfinado-Baker, M., Baker, E. W., and Phoon, A. C. G. 1989. Mites (Acari) associated with bees (Apidae) in Asia, with description of a new species. *American Bee Journal* 129(9): 609–613.

Dutton, R. W. and Free, J. B. 1979. The present status of beekeeping in Oman. *Bee World* 60: 176–185.

Dutton, R. W. and Simpson, J. 1977. Producing honey with Apis florea in Oman. *Bee World* 58(2): 71–76.

Dyer, F. C. and Seeley, T. D. 1991. Dance dialects and foraging range in 3 Asian honey-bee species. *Behavioral Ecology and Sociobiology* 28: 227–233.

Ghatge, A. 1949. Some research work on Apis florea. *Indian Bee Joural* 11(1/2): 5–6, 9.

Halling, L. et al. 2001. Worker policing in the Dwarf Bee *Apis florea. Behavioral Ecology and Sociobiology* 49: 509–513.

Koeniger, N. 1976. Interspecific competition between Apis florea and Apis mellifera in tropics. *Bee World* 57, 110–112.

Koeniger, N., Koeniger, G., and Wongsiri, S. 1989. Mating and sperm transfer in *Apis florea. Apidologie* 20: 413–418.

Koiger, N. and Vorwohl, G. 1979. Competition for food among four sympatric species of Apini in Sri Lanka (*Apis dorsata, Apis cerana, Apis florea* and *Trigona iridipennis*). *Journal of Apicultural Research* 182: 95–109.

Kshirsagar, K. K., Salvi, S. R., Mahindre, D. B., and Chauhan, R. M. 1980. Menthol as an effective acaricidal fumigant. *Proceedings of the 2nd international conference on apiculture in tropical climates*, New Delhi, India.

Manjunath, M. G. 2008. Studies on nesting behaviour of giant honeybee, *Apis dorsata* F. in Mysore, Karnataka. M. Phil, Dissertation, pp. 1–103.

Moritz, R. F. et al. 2010. Invasion of the Dwarf Honeybee *Apis Florea* into the Near East. *Biological Invasions* 12: 1093–1099.

Muttoo, R. N. 1956. Bee diseases in Indian honeybee. *Indian Bee Journal* 18: 14–18.

Narayanaswamy, R. and Basavarajappa, S. 2013. Nesting Attributes of Dwarf Bee, *Apis florea* F. under Urban Ecosystem of Manasagangotri Campus, Mysore, India. 2: 6–15.

Oldroyd, B. P. and Nanork, P. 2009. Conservation of Asian Honey Bees. *Apidologie* 40: 296–312.

Oldroyd, B. P., Sylvester, H. A., Wongsiri, S., Rinderer, T. E. 1994. Task specialization in a wild bee, *Apis florea* (Hymenoptera: Apidae), revealed by RFLP banding. *Behavioral Ecology and Sociobiology* 34(1): 25–30.

Oldroyd, B. P. and Wongsiri, S. 2009. *Asian honey bees: Biology, conservation, and human interactions.* Cambridge, MA: Harvard University Press.

Palmer, K. A. and Oldroyd, B. P. 2001. Mating frequency in *Apis florea* revisited (Hymenoptera, Apidae). *Insectes Sociaux* 48: 40–43.
Phadke, R. P. 1968. Studies on Indian honeys: Proximax composition and physic chemical characteristics of Indian honeys from wild honeybees *Apis dorsata A. florea* and *Trigona*. *Indian Bee Journal* 30(1): 3–8.
Pundir, R. S. 1971. Hiving *Apis florae*. *Indian Bee Journal* 33(3/4), 63.
Robinson, G. E. and Page, R. E. 1988. Genetic determination of guarding and undertaking in honey-bee colonies. *Nature* 333(6171): 356–358.
Ruttner, F. 1988. *Biogeography and taxonomy of Honeybees*. Springer-Verlag, Berlin, Heidelberg and New York. http://dx.doi.org/10.1007/978-3-642-72649-1.
Sarma, M. S. et al. 2002. Worker piping triggers hissing for coordinated colony defence in the dwarf honeybee *Apis florea*. *Zoology* 105: 215–223.
Sciences, E. D. P. and Sciences, E. D. P. 2005. Reinforcing a barrier – A specific social defense of the Dwarf Honeybee (*Apis*). *Apidologie* 36: 505–511.
Seeley, T. D., Seeley, R. H., and Akratanakul, P. 1982. Colony defense strategies of honeybees in Thailand. *Ecological Monographs* 52: 43–63.
Sihag, R. C. 1984. Influence of environmental factors on the pollination activity of bees. *Environment & Ecology* 2(2): 149–152.
Sihag, R. C. 1982. Environmental regulation of pollination process in cultivated crops. *Indian Bee Journal* 44(2): 42–45.
Sihag, R. C. 1994. Pollination of Asian crops. *J. Entomol. Zool.* 8: 158–160.
Sihag, R. C. 2000. Eco-biology of the little honeybee (*Apis florea* F.) in semi arid-sub-tropical climates of India. In: *Asian Bees and Beekeeping: Progress of Research and Development* (Eds.: M. Matsuka et al.). Oxford & IBH Publ. Co. Pvt Ltd, New Delhi, pp. 46–49.
Suwannapong, G., Eric, M., and Nieh, J. C. 2011. Biology of Thai Honeybees: Natural History and Threats. N.p.
Vaudo, A. D., Ellis, J. D., Cambray, G. A., and Hills, M. 2012. Honey bee (*Apis mellifera capensis/A. M. scutellata* hybrid) nesting behavior in the Eastern Cape, South Africa. *Journal of South Africa* 59(3): 323–331.
Williams, I. H. 1979. Bee-keeping in the Sultanate of Oman. *American Bee Jouranl* 510–511, 532–533, 535.
Wills, L. G. 1991. The phylogenetic relationships in the genus Apis as determined by the sequence of the cytochrome oxidase II region of mitochondria DNA. M.Sc thesis Simon Fraser University, pp. 55.
Wongvilas, S. et al. 2010. Interspecific and conspecific colony mergers in the dwarf honey bees *Apis andreniformis* and *A. florea*. *Insectes Sociaux* 57: 251–255.
Woyke, J., Wilde, J., and Wilde, M. 2001. Swarming, migration and absconding of *Apis dorsata* colonies. In: *Proceedings of the 7th international conference on apiculture in tropical climates*, Chiang Mai, pp. 183–188.
Wu, Y. R. and Kuang, B. 1987. Two species of small honeybee – A study of the genus Micrapis. *Bee World* 68: 153–155.

2

Biology of Dwarf Honeybee, Apis florea *Fabricius (Hymenoptera: Apidae)*

N. Nagaraja

CONTENTS

Introduction

Honeybees, the eusocial and beneficial insects, play a crucial role in pollination of crop plants across the globe and produce various hive/bee products including nature's golden wonder, the honey to humankind. The dwarf honeybees, *Apis florea* Fabricius, *Apis andreniformis* Smith and giant honeybees, *Apis dorsata* Fabricius, *Apis laboriosa* Smith are open-nesting species that build medium to large-sized single exposed combs. They exist wild in nature. Furthermore, the hive honeybee species, *Apis cerana* Fabricius, *Apis koschevnikovi* Enderlein, *Apis nuluensis* Tingek, Koeniger and Koeniger, *Apis nigrocincta* Smith and *Apis mellifera* Linnaeus construct nests generally in cavities with 10–15 multiple parallel combs. They are domesticated in human-made hives.

Apis florea, the smallest bee among honeybee species, apparently occupies a primitive position in the genus *Apis*. *A. florea* is a most peculiar wild bee in several ways (Free, 1981). Its ecological niche is found in the stratum of dense bushes and small trees of tropical Asia. The ability of these species to survive in dry climates is one of the outstanding traits (Ruttner, 1988). They secure shade and safety of the nest for living in nature (Kshirsagar, 1980). Pratt (2004) opined that the comb not only provides shelter to bees, but it also gives protection for developing young ones. Nagaraja and Rajagopal (1999) recorded a greater brood area in *A. florea* colonies during winter and summer seasons. The workers protect their colonies against predators by warning them through shimmering and hissing behaviours (Pirk et al., 2002).

Biogeography

Apis florea occurs in tropical and subtropical regions of the Asian continent. It is widely distributed by extending up to 7000 km from Oman through southern Iran and tropical mainland Asia and to Palawan and the islands of Indonesia (Crane, 1999). The main habitat of this species is known to be in India, Sri Lanka, Pakistan, Thailand, Indonesia, Malaysia and Indochina (Ruttner, 1988). However, *A. florea* is continuously expanding its distribution regions. It was recorded to be found in Saudi Arabia, Sudan and Ethiopia (Bezabih et al., 2014) and recently in Suez, northern Egypt and the Egyptian delta (Shebl, 2017). *A. florea* occupies a large area covering rainforests, savannas, subtropical grasslands and semideserts (Hepburn et al., 2005). It occurs in plains and rarely in places higher than 1500 m above mean sea level with summer records up to 900 m in Iran (Ruttner et al., 1995) and 1900 m in Oman (Whitecombe, 1984). *A. florea* is quite common in lowlands below 1000 m in the regions of Iran, Yemen, Thailand and China.

Morphological Features

Apis florea is the smallest among true honeybee species and shows striking differences in size among its castes. The workers of these species are smaller in size, and the small size could be an adaptation for nesting structure and foraging activity. They have a distinct bicoloured abdomen, with the reddish interior contrasting sharply with the grey posterior. However, the first abdominal segment of the old worker bees is red whereas in young workers, it is paler in colour. The body length of worker bees ranges from 7 to 10 mm. Halling et al. (2001) recorded the mean body length of 9 mm in *A. florea* distributed in Indian conditions. Similarly, Ruttner (1988) found the forewing lengths of 6–7 mm from the bee samples collected from different Asian regions with a cubital index of 2.82. The studies on biometrics of *A. florea* revealed that the worker bees of hilly regions have larger hind wings than those of the plains and were varied from 4.25 to 5.4 mm (Ozkan et al., 2009). The long hairs on the margin of the hind tibia were whitish and have less black pigment (Wongsiri et al., 1997). The drones of *A. florea* show grey abdominal segments with white hairs. They have a thumb-like bifurcation called a basitarsus, which is quite long and more than two-thirds of the length of the tibia. A drone's endophallus has a pair of bursal cornua, and the fimbriate lobe has three protrusions with a strong and curved terminal (Wongsiri et al., 1997). Queens have orange-yellow abdominal segments.

The compound eyes of honeybees are similar in appearance, oval in shape and have a large number of sensory hairs between the ommatidia (Somanathan et al., 2009). These sensory hairs differ in length between the species and are shortest in *A. florea* (50 μm). The anatomical spatial resolution and absolute sensitivity of *A. florea* shows least sensitive eyes with smallest acceptance angles. However, the eyes of these species are potential for higher spatial resolution than *A. mellifera* and *A. dorsata*. *A. florea* shows caste-specific variation in structure and function of compound eyes (Streinzer et al., 2013). The drones have disproportionately enlarged eyes in relation to body size and have more ommatidia than those of *A. dorsata* and almost as many ommatidia as those of *A. mellifera*. The eyes of queens have relatively enlarged facets than worker bees. The microscopic studies on the tongue of *A. florea* show that it exhibits many sensory structures such as sensilla chaetica, sensilla basiconica, sensilla trichodea, branched processes, hair membrane plate sensilla and cuticular plate sensilla on the mouth parts (Kumar and Kumar, 2015). Sensilla chaetica is effective in the flow of liquid food to the buccal cavity, and sensilla basiconica act as thermoreceptors and hygroreceptors. The cuticular scales are functional in adjusting the segments of the organ and serve as mechanoreceptors.

Nesting Biology

Nest-Site Selection

The nest sites are crucial and ecologically critical for social insects for growth and development of their colonies. Nest-site selection is a process in which a reproductive swarm selects a suitable new nest site. Due to availability of abundant potential nest sites, individual behaviour of scout bees was simpler than *A. mellifera* scouts engaged in nest-site selection, and regular inspection and re-evaluation of a potential nest site does not occur in *A. florea* (Makinson et al., 2011). During swarming, *A. florea* scouts search the suitable environment for nest sites and report their location if found in the form of recruitment dances. Other bees are recruited towards those sites, as the number of bees dancing for a particular site increase. They select the nest site within a few hours whereas *A. mellifera* takes many days (Villa, 2004). The shorter time required for nest-site selection in *A. florea* is probably due to availability of many nesting sites with equal quality (Oldroyd et al., 2008). Furthermore, Nasanov pheromones are helpful for scouts to mark a focal twig or tree. If the selected site proves unsuitable, then the colony moves again in search of a new nest site.

Nest Building

Apis florea builds nests with single exposed combs vertically often suspended on branches of hedges, house chimneys, empty caves and piles of dried sticks. The bees prefer selective nest sites that the colonies are not easily accessible to pests, predators and parasites. They select shady places that are quite effective in protecting colonies from bright sunlight, strong wind speed, heavy rains etc. (Nagaraja and Rajagopal, 2009). They were found on shrubs and small trees in larger numbers throughout south India (Nagaraja and Rajagopal, 1999). They preferred trees in greater numbers (42.60%) than the shrubs (39.40) for nest building. In Oman, constructed nests in rock cliffs were also found. The nests of *A. florea* were generally occurring singly rather than in aggregations on a nesting site. However, occasionally, 2–5 nests were also recorded on a single tree. Distribution of multiple numbers of nests on a single tree was attributed to availability of plenty of suitable nesting sites and rich availability of forage in the vicinity. It could be also be due to a lack of competition for pollen and nectar between the open-nesting bee species. However, the nests of *A. florea* and *A. dorsata* in the same tree were found at the heights of 6 and 13 m, respectively. Though *A. florea* colonies build nests at the heights between a few centimetres to many meters above ground, a maximum number of colonies were recorded at the heights between 2 and 6 m (Figure 2.1). The number of colonies declined as the height of the nesting structure was increased. It is predicted that the nest-height selection could be decided on the basis of the height of the host plants from which they can easily gather floral resources by spending the least energy.

Comb Architecture

Honeybees build the combs with hexagonal cells. The growth rates of the combs in respect to width and length were almost similar; the comb is circular in shape with the exception of the protruded drone brood area at the bottom of the nest. These species commence construction of the comb by laying a foundation of hexagonal cells on the surface around the substratum. The first row of cells helps to attach the comb to the substratum, and the comb encircles the branch completely. The combs of *A. florea* have a midrib only in the brood area below the supporting branch (Rinderer et al., 1996). The surface of the brood area of the nest is parallel to the branch direction. The large cells are very deep and are meant for honey storage, and the comb bulges out on either side and at the top. The small worker cells are located below the supporting branch, which is used for worker brood rearing. The worker brood area is arranged in specific concentric rings in the plane of the comb (Figure 2.2).

The larger drone cells are mostly found in the lower margin of the comb. The pear- shaped queen cells, which are the largest of the cells, are found protruding vertically from the lower edge of the brood comb. They have a midrib through the brood and pollen storage area with worker cells from

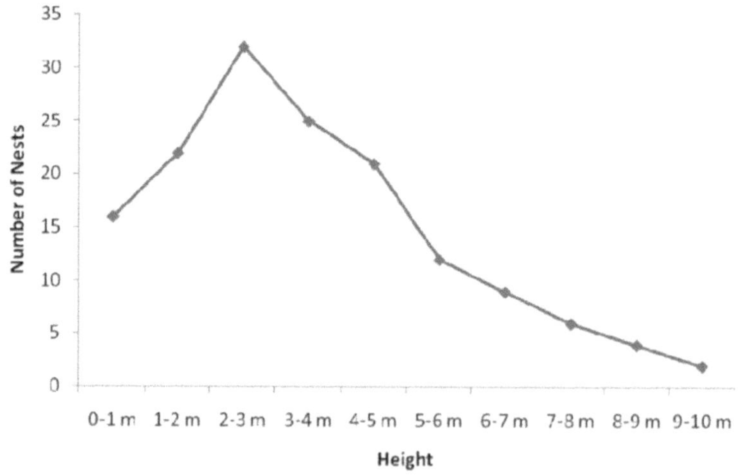

FIGURE 2.1 Distribution of *Apis florea* colonies in relation to height in and around Bengaluru, India ($N = 30$).

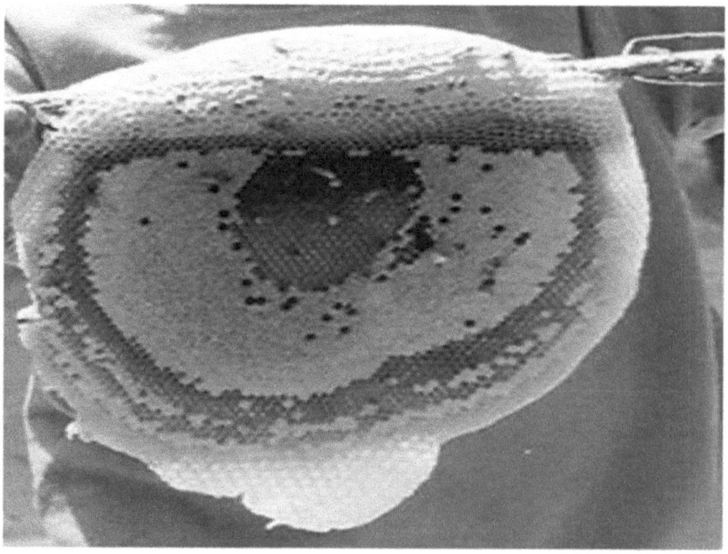

FIGURE 2.2 The comb of *Apis florea* exhibiting arrangement of a worker brood in concentric rings.

outside meeting worker cells from the other side (Rinderer et al., 1996). The honey portion of the comb is usually attached to a substratum and provides a platform for communication of foragers, and they perform dances on the flat upper surface of the comb (Figure 2.3). The dance is short-run, pointing directly to the source of pollen and nectar; dances on the vertical surface of the combs have also been recorded (Gould et al., 1985). This horizontal comb surface provides a platform for orienting foragers towards celestial cues rather than to gravity or towards landmarks if the celestial cues are unavailable (Dyer, 2002).

The comb-building activity of *A. florea* was slow at the beginning of colony establishment and increased steadily for 45–120 days (Table 2.1). There was a comb size of 23.70 cm² on 15 days of settlement of the colony and reached a maximum size of 365.70 cm² in 120 days. The faster comb-building activity of *A. florea* in the beginning of colony establishment was attributed to immediate need of the

FIGURE 2.3 The comb of *Apis florea* showing honey portion on the upper surface of the comb by attaching to substratum.

TABLE 2.1

Mean Comb Growth of Dwarf Honeybee, *Apis florea* Colonies in and Around Bengaluru, India

Duration (days)	Comb Height (cm)	Comb Width (cm)	Comb Size (cm^2)
15	4.80	4.95	23.70
30	5.60	7.30	40.90
45	7.80	9.00	70.20
60	8.68	11.80	102.40
75	10.90	12.50	136.30
90	13.10	16.10	210.90
105	15.20	10.00	304.00
120	16.40	22.30	365.70

N = 25.

shelter for the newly established colony, and the young worker bees in the swarm initiate comb construction by producing required quantities of beeswax. Whitecombe (1984) recorded a large-sized comb of 370–740 mm in Oman. The studies on comb dimensions of *A. florea* revealed the mean width, height and thickness of the combs were 17.07 ± 1.97, 14.50 ± 1.80 and 4.25 ± 0.70, respectively (Table 2.2).

Colony Development

The size of the eggs laid by *A. florea* queens is almost similar to the size of *A. cerana* eggs but smaller than eggs of *A. dorsata* (Woyke, 1975). The total development period of *A. florea* workers from egg to adult is about 20–22 days. Similarly, the drones develop in 24–26 days and queen bees in 16–17 days. The sizes of the worker and drone brood cells were found to be 2.7–3.1 and 4.2–4.8 mm, respectively. The cells maintain a temperature between 33° and 38° in the central area of the brood comb. Unlike worker brood cells that are used several times, the drone cells are used only once. The colonies of *A. florea* are found to be rear 12–16 queens during the swarming season (Kshirsagar et al., 1983; Soman and Chawda, 1996).

TABLE 2.2

Comb Dimensions of Dwarf Honeybee, *Apis florea* Colonies in and Around Bengaluru, India

Comb Dimension	Maximum	Minimum	Mean
Width (cm)	25.24	8.90	17.07 ± 1.95
Height (cm)	22.80	6.20	14.50 ± 1.80
Thickness (cm)	7.10	1.40	4.25 ± 0.70
Size (cm^2)	502.40	85.70	294.00 ± 81.50

$N = 97$.

Akratanakul (1977) found an average production of eight queen cells/colony. Colonies that have lost their queen may produce up to 19 emergency queen cells by extending worker brood cells (Akratanakul, 1977). Similarly, if the queen is not replaced, laying workers usually develop within the colony.

The colony establishment of *A. florea* in respect to brood rearing, pollen stores, honey stores and adult bee population was recorded (Nagaraja and Rajagopal, 1999; Rod-im et al., 2015). The area of worker brood shows geographic variability in tropical regions. The maximum and minimum brood area recorded in the colonies was 525.67 ± 0.86 and 220.32 ± 0.89 cm^2, respectively. The brood-rearing activity of *A. florea* colonies was greater during September to November. The maximum pollen and honey stores recorded were 53.41 ± 0.86 and 114.53 ± 0.70 cm^2, respectively. Similarly, the minimum pollen and honey stores were 19.46 ± 0.80 and 36.24 ± 0.91 cm^2, respectively. Further, there was a seasonal production of drones and queen bees that was found in greater numbers during the summer season (March and April). The high brood-rearing activity was attributed to blooming of agro-forest crops in the area during honey flow seasons. It was also due to maximum egg- laying capacity of the queen on availability of plenty of pollen and nectar for bees. The large number of larvae triggered pollen-gathering activity of bee colonies. In an *A. florea* colony, there were about 6000 worker bees, and the queen laid bout 100–500 eggs/day depending on the availability of floral resources. Comparatively, the workers of *A. florea* have a long life span of 61 days, which is almost 2.5 times more than those of *A. mellifera*.

The honey stores were greater in the *A. florea* colonies from October to December and were attributed to the availability and blooming of nectar flora in the region. The greater bee population generally coincides with the high brood-rearing activity. *A. florea* is a poor honey yielder, and a colony yields 500–1000 g of honey/annum. The honey produced by this species is known to have special medicinal properties.

Colony Cycle

It is well known that the worker bees of *A. florea* perform colony activities in relation to age. The young worker bees take care of indoor duties, and older bees perform outdoor activities similar to other *Apis* species. Worker bees initiate foraging activity by an age of 15–25 days. Dyer and Seeley (1991) showed that the difference in worker maturation is a consequence of the bee curtain, which inhibits worker bees from performing other tasks of the colony.

Colony Migration

The colonies of *A. florea* generally nest on a suitable nesting site for 3–6 months. If there is a shortage of floral resources, they migrate for short distances to nearby hills by deserting old combs seasonally. However, seasonal migration of *A. florea* varied from place to place depending on the climatic conditions and availability of floral resources. In south Karnataka, migration of these colonies takes place from May to July (Nagaraja, 2012). At the time of migration, these bees are known to take along the pollen and honey stores and even the wax to new sites. The scarcity of food due to prevailing high temperatures and incidence of pests and predators, such as ants and wax moths, causes seasonal migration.

Swarming Behaviour

Swarming, the process of colony division, needs a queen and an adequate number of worker bees to form a new colony (Rangel and Seeley, 2008). It is a complex, multistep process that is mediated by multiple factors such as environmental, social, physiological and molecular mechanisms. Generally, the colonies initiate rearing new queens when the colonies have huge adult bee populations with plenty of pollen and honey stores (Winston et al., 1981). The initiation of queen rearing is the first visible step in a colony's preparation for swarming. During the swarming, the colonies exhibit general restlessness and repeated movement of workers. In *A. florea* when the first virgin queen emerges, about half the workers leave along with the mother queen. Subsequently, the virgin queen mates and the smaller swarms leave every few days. In the end, only a handful of worker bees remain on the comb. Later, they also swarm away, leaving the empty comb and deserted. After departing from the parent colony, the workers in the swarm form a temporary cluster and initiate a search for a new nest site.

Absconding Behaviour

Absconding is a defensive mechanism of survival from pests and predators in bee colonies. This behaviour is quite common in the colonies of tropical honeybees (Hepburn and Radloff, 1998; Oldroyd and Wongsiri, 2006). Absconding behaviour is one of the most frequently expressed behaviours in *A. florea*. These colonies would abscond at any time of the season with a peak tendency during warmer periods with poor supply of pollen and nectar. Disturbance by natural enemies coupled with unfavourable environmental conditions causes colonies to abscond. However, the bees require navigational skills for successful absconding. After absconding from the old nest, *A. florea* workers return to collect wax from the old comb. They usually collect the wax from the crown of the nest and sometimes from old queen cells.

Colony Defence

Colony defence is one of the survival strategies that *Apis florea* uses to defend its colonies against pests and diseases in honeybees. *Apis florea* generally constructs nests on the nest sites almost free from pests, predators and also natural hazards such as direct sunlight, rainfall, wind etc. (Reddy and Reddy, 1993). It tends to select dense bushes to protect the colonies especially from birds and mammals. These bees elicit a hissing sound against any disturbance caused in the colony. If an enemy tries to land on the nest forcibly, it is normally attacked by a group of worker bees that clump around it. If the disturbance continues, the colony would rather abscond than defend itself. The worker bees coat both ends of the nest support with sticky bands of propolis or plant resins up to a width of 2.5–4 cm. to prevent the ants and other insect pests from invading the nest (Wongsiri et al., 1997). When members of the weaver ant, *Oecophylla smaragdina*, are close to proximity of the nest, the bees deposit sticky barriers to obstruct their route of entry. The guard bees cover the sticky bands and alert other bees of the colony through a hissing sound. These sticky bands are taken care of by worker bees who keep them sticky by adding their secretions.

Apis florea has evolved many behavioural adaptations against environmental stress and risk of predation. The workers of *A. florea* arrange themselves to get protection from overheating (Dyer and Seeley, 1991). They perform vigorous body shaking to avoid insect pests from landing on the comb (Seeley et al., 1982). Mammals and birds may be threatened by stinging and also warned away by the production of hissing sounds. *A. florea* removes accumulated debris from the comb to break contact between the comb and objects. The workers bring their heads underneath the objects and push them upwards (Sen Sarma et al., 2000). Head-pushing behaviour is a distinct behavioural task in *A. florea* and is comparable to nest-cleaning behaviour in *A. mellifera* (Visscher, 1983).

Apis florea employs a defensive strategy called 'piping' by emitting an initial warning signal produced by the shortest movements of the bee's wings. The bees close to the piping bee initiate hissing, which spreads rapidly to their nest-mates until an impressive coordinated situation is attained by the colony. Piping and hissing are accompanied by ceasing of foraging dances and departure from the colony by the worker bees (Sen Sarma et al., 2002). *A. florea* removes 89% of introduced the freeze-killed brood in 24 h and 95% of pin-killed brood in 48 h after their introduction through hygienic behaviour (Woyke et al., 2012).

Massed Flight Activity

The massed flight activity is an important feature of social behaviour in honeybees. A striking form of massed flight with nearly half the bee colony involved has been reported (Seeley, 1985) and was associated with defecation and temperature regulation in the nest (Mardan and Kevan, 1989; Woyke et al., 2005). Similarly, the influence of massed flight in colony defence, seasonal migration (Koeniger and Koeniger, 1980) and drone flight (Koeniger et al., 1994) have been reported in *A. dorsata* colonies. In a well populous colony, the combs of *A. florea* are covered with two to three layers of worker bees clinging together (Oldroyd and Wongsiri, 2006; Phiancharoen et al., 2011). This bee curtain functions as a protective shield against unfavourable environmental conditions and predators (Dyer and Seeley, 1991). In open-nesting species such as *A. florea* and *A. dorsata*, the bee curtain is composed of several layers of bees connected in a mesh-like structure (Anderson et al., 2002; Kastberger et al., 2011). In *A. florea* colonies, the mass flight activities occurred between 11.30 and 15.00 h, and the workers flew up to a height of 15–20 m and hovered in front of the nest until they landed on the comb back (Bhagavan et al., 2016).

Olfactory Learning

Honeybees regulate most of the colony activities through bouquets of semiochemicals that are actively produced and transmitted among colony members (Thom et al., 2007). The primer pheromones induce a long-term effect on physiology and behaviour in contrast to the releaser pheromones, which trigger short-term behavioural responses in honeybees. The mandibular pheromones released by the queen bee are crucial for performing different tasks in bee colonies (Le Conte and Hefetz, 2008). Similarly, the drones of *A. florea* learn and recall olfactory memory of queen mandibular odorants, i.e., 9-ODA, 10-HDA and 9-HDA. However, they showed better learning and memory recall towards 10-HDA. It would be due to greater receptivity of drones sensory cells towards 10-HDA coupled with synthesis of large quantities of 10-HDA by queen bees in the colonies (Keeling et al., 2000). The drones of *A. florea* recall most of the odorants at 15 min of post-olfactory conditioning, and the level of memory gradually decreased with increasing duration of conditioning (Nagaraja, 2014).

Sex Pheromones

Queens of *A. florea* and unmated queens of *A. mellifera* synthesize higher quantities of (2E)-10-hydroxydecenoic acid (10-HDA) than 9-ODA (Plettner et al., 1997). Sannasi et al. (1971) reported that, drones of *A. florea* are attracted to 9-ODA. The queen mandibular gland component (2E)-9-oxodecenoic acid (9-ODA) has been suggested to function as the major sex pheromone component in all honeybee species. However, chemical analyses showed that in *A. florea*, a different decenoic acid, (2E)-10-hydroxydecenoic acid (10-HDA), is the major component in the mandibular gland secretion. *A. florea* drones are attracted to 9-ODA as well as to 10-HDA. However, 10-HDA attracted greater numbers of drones even at lower dosages than 9-ODA (Nagaraja and Brockmann, 2009).

Worker Policing

Honeybee workers have the potential to produce unfertilized eggs from which viable drones are developed. This behaviour is quite common in queenless colonies, where many workers lay eggs and drones are reared (Page and Erickson, 1988). Workers of *A. florea* are known to produce viable drones in the absence of a queen (Woyke and Wongsiri, 1992), but in queenright colonies, workers are predicted to be sterile. The rapid removal of worker-laid eggs by *A. florea* workers shows that this species can discriminate between worker- and queen-laid eggs. This suggests that *A. florea* queens use pheromones to mark their eggs, and workers respond to their signal by selectively rearing these marked eggs. *A. florea* is known to control worker reproduction by mutual policing mechanisms to keep worker reproduction at a low level. Policing mechanisms involve the destruction of worker eggs or offspring. The worker-laid eggs are generally eliminated as fast as queen-laid eggs, which shows that, *A. florea* uses oophagy of worker-laid eggs as a mechanism of worker policing (Halling et al., 2001).

Conclusion

Apis florea, the basal species among true honeybees, exhibits unique biological and behavioural characteristics that eventually differ from other honeybee species. The uniqueness in decisions taken by swarms during nest-site selection, differential morphometrics among its castes, ability to survive in warmer climates and evolution of specialized defensive strategies against pests and diseases ensured *A. florea's* successful establishment in Asia in general and South East Asia in particular. Furthermore, there is a greater scope of research on the detailed biology and behaviour of *A. florea* for effective utilization of its services in pollination of crops and honey production.

ACKNOWLEDGEMENTS

I am thankful to Prof. C. Srinivas the director of the UGC-Human Resource Development Centre, Bangalore University, Bengaluru, for his constant encouragement. I am also grateful to Bangalore University, Bengaluru, for providing research facilities.

REFERENCES

Akratanakul, P. The natural history of the dwarf honeybee, *Apis florea* F. in Thailand, Thesis, Cornell University, Ithaca, NY, pp. 1–91, 1977.

Anderson, C., Theraulaz, G. and Deneubourg, J.L. Self assemblages in insect societies. *Insectes Soc.* 49: 99–110, 2002.

Bezabih, G., Adgaba, N., Hepburn, H.R. et al. The territorial invasion of *Apis florea* in Africa. *Afr. Entomol.* 22(4): 888–890, 2014.

Bhagavan, H., Muthmann, O. and Brockmann, A. Structural and temporal dynamics of the bee curtain in the open-nesting honeybee species, *Apis florea. Apidologie* 47: 749–758, 2016.

Crane, E. *The World History of Beekeeping and Honey Hunting*, Taylor & Francis Group, New York, p. 675, 1999.

Dyer, F.C. The biology of the dance language. *Annu. Rev. Entomol.* 47: 917–949, 2002.

Dyer, F.C. and Seeley, T.D. Nesting behaviour and the evolution of worker tempo in four honeybee species. *Ecology* 72: 156–170, 1991.

Free, J.B. Biology and behaviour of the honeybee *Apis florea* and possibilities for beekeeping. *Bee World* 62(2): 46–59, 1981.

Gould, J.L., Dyer, F.C. and Towne, W.F. Recent progress in the study of the dance language. *Fortschr. Zool.* 31: 140–161, 1985.

Halling, L.A., Oldroyd, B.P., Wattanachaiyingchareon, W. et al. Worker policing in the *Apis florea. Behav. Ecol. Sociobiol.* 49: 509–513, 2001.

Hepburn, H.R. and Radloff, S.E. *Honeybees of Africa*. Springer-Verlag, Berlin, Germany, p. 365, 1998.

Hepburn, H.R., Radloff, S.E., Otis, G.W. et al. *Apis florea*: Morphometrics, classification and biogeography. *Apidologie* 36: 359–376, 2005.

Kastberger, G., Maurer, M., Weihmann, F. et al. Stereoscopic motion analysis in densely packed clusters: 3D analysis of the shimmering behaviour in giant honeybees. *Front. Zool.* 8: 3, 2011.

Keeling, C.L., Slessor, K.N., Koeniger, N. et al. Quantitative analysis of the mandibular gland components of the dwarf honeybee (*Apis florea* Fabricius). *Apidologie* 31: 293–299, 2000.

Koeniger, N. and Koeniger, G. Observations and experiments on migration and dance communication of *Apis dorsata* in Sri Lanka. *J. Apic. Res.* 19(1): 21–34, 1980.
Koeniger, N., Koeniger, G., Tingek, S. et al. Drones of *Apis dorsata* (Fabricius 1793) congregate under the canopy of tall emergent trees in Borneo. *Apidologie* 25: 249–264, 1994.
Kshirsagar, K.K. Morphometric studies on *Apis cerana indica*. F. Worker. III. Effect of geographical position on morphometric characters. *Indian Bee J.* 43: 1–5, 1980.
Kshirsagar, K.K., Muvel, K.S., Mittal, M.C. et al. Some observations on the behaviour of *Apis florea* F. In: *Proceeding of 2nd International Conference of Apiculture in Tropical Climates*, New Delhi, India, pp. 356–366, 1983.
Kumar, R. and Kumar, N.R. Scanning electron microscopic study on the mouthparts of *Apis florea* (Hymenoptera: Apidae). *Apidologie* 47: 717–727, 2015.
Le Conte, Y. and Hefetz, A. Primer pheromones in Social Hymenoptera. *Annu. Rev. Entomol.* 53: 523–542, 2008.
Makinson, J.C., Oldroyd, B.P. and Schaerf, T.M. Moving home nest-site selection in the red dwarf honeybee (*Apis florea*). *Behav. Ecol. Sociobiol.* 65: 945–958, 2011.
Mardan, M. and Kevan, P.G. Honeybees and yellow rain. *Nature* 341: 191, 1989.
Nagaraja, N. Asian honeybees: Biology, threats and their conservation. In: R.M. Florio (Ed.), *Biology, Threats and Colonies*, pp. 99–123. Nova Publishers, Hauppauge, NY, 2012.
Nagaraja, N. Olfactory learning and memory in drones of *Apis cerana* Fab. and *Apis florea* Fab. towards queen mandibular pheromone odorants. *J. Entomol. Res.* 38(3): 165–168, 2014.
Nagaraja, N. and Brockmann, A. Drones of red dwarf honeybee, *Apis florea* are attracted by two components from the mandibular gland, 10-HDA and 9-ODA. *J. Chem. Ecol.* 35: 653–655, 2009.
Nagaraja, N. and Rajagopal, D. Colony establishment, nesting and foraging activity of little honeybee, *Apis florea* F. (Hymenoptera: Apidae). *J. Entomol. Res.* 23(4): 331–338, 1999.
Nagaraja, N. and Rajagopal, D. *Honeybees: Diseases, Parasites, Pests, Predators and Their Management.* MJP Publishers, Chennai, India, p. 210, 2009.
Oldroyd, B.P., Gloag, R.S., Even, N. et al. Nest-site selection in the open nesting honeybee, *Apis florea*. *Behav. Ecol. Sociobiol.* 62: 1643–1653, 2008.
Oldroyd, B.P. and Wongsiri, S. *Asian Honeybees, Biology, Conservation, and Human Interactions*, Harvard University Press, Cambridge, MA, 2006.
Ozkan, A.K., Gharleko, M.M., Oezden, B. et al. Multivariate morphometric study on *Apis florea* distributed in Iran. *Turk. J. Zool.* 33(1): 93–102, 2009.
Page, R.E. and Erickson, E.H. Reproduction by worker honeybees (*Apis mellifera*). *Behav. Ecol. Sociobiol.* 23: 117–126, 1988.
Phiancharoen, M., Duangphakdee, O. and Hepburn, H.R. Biology of nesting. In: H.R. Hepburn and S.E. Radloff (Eds.), *Honeybees of Asia*, pp. 109–131. Springer-Verlag, Berlin, Germany, 2011.
Pirk, C.W.W., Hepburn, R., Radloff, S.E. et al. Defense posture in dwarf honeybee, *Apis florea*. *Apidologie* 33: 289–294, 2002.
Plettner, E., Otis, G.W., Wimalaratne, P.D.C. et al. Species and caste determined mandibular gland signals in honeybees (Apis). *J. Chem. Ecol.* 23: 363–377, 1997.
Pratt, S.C. Collective control of the timing and type of construction by honeybees (*Apis mellifera*). *Apidologie* 35: 193–205, 2004.
Rangel, J. and Seeley T.D. The signals initiating the mass exodus of a honey bee swarm from its nest. *Anim. Behav.* 76: 1943–1952, 2008.
Reddy, M.S. and Reddy, C.C. Studies on the distribution of nests of giant honeybee (*Apis dorsata* F.). *Indian Bee J.* 55: 36–39, 1993.
Rinderer, T.E., Wongsiri, S., Kuang, B. et al. Comparative nest architecture of the dwarf honeybees. *J. Apic. Res.* 35: 19–26, 1996.
Rod-im, P., Hepburn, R. and Duangphakdee, O. Developmental cycles of the red dwarf honeybee, *Apis florea*. In: *Proceedings of International Conference on Agricultural, Ecological and Medical Sciences*, Phuket, Thailand, pp. 72–75, 2015.
Ruttner, F. *Biogeography and Taxonomy of Honeybees*, Springer-Verlag, Berlin, Germany, p. 284, 1988.
Sannasi, A., Rajulu, G.S. and Sundara, G. 9-Oxodec-trans-e-enoic acid in the Indian honeybees. *Life Sci.* 10: 195–201, 1971.
Seeley, T.D. *Labour Specialization by Workers.* Honeybee Ecology, Princeton University Press, Princeton, NJ, pp. 31–35, 1985.

Seeley, T.D., Seeley, R.H. and Akratanakul, P. Colony defense strategies of the honeybees in Thailand. *Ecol. Monogr.* 52(1): 43–63, 1982.

Sen Sarma, M., Fuchs, S. and Tautz, J. Debris removal by head pushing in *A. florea* F. honeybees. *Naturwissenschaften* 87: 241–243, 2000.

Sen Sarma, M., Fuchs, S., Werber, C. et al. Worker piping triggers hissing for coordinated colony defense in dwarf honeybee, *Apis florea. Zoologia* 105: 215–223, 2002.

Shebl, M.A. Discovery of *Apis florea* colonies in north-eastern Egypt. *Afr. Entomol.* 25(1): 248–249, 2017.

Soman, A.G. and Chawda, S.S. A contribution to the biology and behaviour of the dwarf bee, *Apis florea* F. and its economic importance in Kutch, Gujarat, India. *Indian Bee J.* 58: 81–88, 1996.

Somanathan, H., Warrant, E.J., Borges, R.M. et al. Resolution and sensitivity of the eyes of the Asian honeybees *Apis florea, Apis cerana* and *Apis dorsata. J. Exp. Biol.* 212: 2448–2453, 2009.

Streinzer, M., Brockmann, A., Nagaraja, N. et al. Sex and caste-specific variation in compound eye morphology of five honeybee species. *Plos One* 8(2): e57702, 2013.

Thom, C., Gilley, D.C., Hooper, J. et al. The scent of the waggle dance. *Plos Biol.* 5(9): e228, 2007.

Villa, J.D. Swarming behaviour of honeybees (Hymenoptera: Apidae) in south-eastern Lousiana. *Ann. Entomol. Soc. Am.* 97(1): 111–116, 2004.

Visscher, P.K. The honeybee way of death: Necrophoric behaviour in Apis mellifera colonies. *Anim. Behav.* 31(4): 1070–1076, 1983.

Whitecombe, R.P. Beekeeping with *Apis florea* in Oman. *Aramco World Mag.* 35(2): 34–40, 1984.

Winston, M.L., Dropkin, J.A. and Taylor, O.R. Demography and life history of two honeybee races (*Apis mellifera*). *Oecologia* 48(3): 407–413, 1981.

Wongsiri, S., Lekprayoon, C., Thapa, R. et al. Comparative biology of *Apis andreniformis* and *Apis florea* in Thailand. *Bee World* 78: 23–35, 1997.

Woyke, J. The eggs in the comb cells of three honeybee species in India. *Proc. Int. Beekeep. Congr.* 25: 340, 1975.

Woyke, J., Wilde, J. and Phaincharoen, M. First evidence of hygienic behaviour in the dwarf honeybee, *Apis florea. J. Apic. Res.* 51(4): 359–361, 2012.

Woyke, J., Wilde, J., Reddy, C.C. et al. Periodic mass flights of the giant honeybee *Apis dorsata* in successive days at two nesting sites in different environmental conditions. *J. Apic. Res.* 44(4): 180–189, 2005.

Woyke, J. and Wongsiri, S. Occurrence and size of laying worker eggs in *Apis florea* colonies. *J. Apic. Res.* 31: 124–127, 1992.

3

Reproductive Biology of Apis florea

N. Nagaraja and Dharam P. Abrol

CONTENTS

Introduction

Honeybee colonies normally consist of a queen bee, tens of thousands of worker bees and a few hundred drone bees especially during pollen and nectar flow seasons. The queens and drones are fertile and have well-developed reproductive systems (Winston, 1987). Honeybee colonies exhibit reproductive division of labour between the female castes. The queen is the reproductive female whilst the workers are functionally sterile. However, the workers have poorly developed ovaries and vestigial spermatheca (Gotoh et al., 2013). The worker sterility is a characteristic of eusociality that continues in worker bees (Ronai et al., 2017). The worker bees perform all the activities of colonies assigned by the queen bee except reproduction. The workers and queens are developed by diploid eggs in contrast to drones by haploid eggs laid by the queen bee. However, the workers of Cape honeybee, *Apis mellifera capensis*, produce diploid female offspring from unfertilized eggs (Anderson, 1963). This process is called thelytoky (Crozier, 1975). The female offspring can develop into either workers or queens (Ruttner, 1988). The queen propagates the colony through reproduction with selected drones in the beginning of her life (Ruttner, 1956).

The Queen and Drone Bees

The queens are developed from fertilized eggs laid by the mother queen and are reared in specially built conical cells in the bottom of the bee comb. The queen bee is fed with royal jelly, a hypopharyngeal glandular secretion of young worker bees, throughout her life. The queen inspects the colony for other queens, if any, that are nearly ready to emerge. She shows the signs of sexual maturity by opening her sting chamber and also by vibrating the tip of her abdomen and dipping it into worker cells after a few days. Queens make a few short orientation flights before a longer flight for mating with drones in drone congregation areas. The colony's long-term survival depends on the reproductive female's ability to produce a large number of offspring. The queen monitors the colony activities through her mandibular pheromones. If the laying queen is removed from the colony, emergency queen cells are constructed by modifying worker cells (Kshirsagar et al., 1980). The emergency queen cells of *A. florea* mainly occurring in the centre of the comb are slightly shrunk beneath the comb surface. In queenless colonies, the ovaries are developed in worker bees followed by egg laying (Akratanakul, 1977). The queens lay around 200,000 eggs a year (Snodgrass, 1984) for 3–4 years and may survive for up to 8 years (Bozina, 1961).

The drones develop from the unfertilized eggs laid by the queen in drone cells. They are fertile males and are essential for propagation of the bee colony. Each colony produces several hundred drones during queen-rearing season. The peak drone brood rearing activity was recorded during March to April in the colonies of *A. florea* in Bengaluru (Figure 3.1). Similarly, the rearing activity also coincides with drone production. The drones feed on royal jelly for the first two days of their larval stage and honey and pollen during the late larval stage and adult life. The young drones learn flight activity from 5 to 10 days after emergence. The drones have a clasper at the basitarsus of their hind leg that is used to hold the queen's leg during mating. They have the lowest number and density of olfactory sensilla and much smaller sex pheromone-processing macroglomeruli than those of *A. mellifera*. The longevity and maturity of *A. florea* are shorter than *A. mellifera* (Buawangpang et al., 2009).

Reproductive System of the Queen

The reproductive system of a honeybee queen consists of a pair of ovaries and associated accessory organs. Each ovary consists of a bundle of around 150 ovarioles. At the one end of each ovariole, there is a germinal tissue from which the nurse cells, follicle cells and true egg cells are formed. The ovarioles open in to a pair of lateral oviducts that join the median oviduct. The eggs are moved from the ovarioles into the oviducts and then to the spermatheca. The queens store the spermatozoa in their spermathecae and are viable for many years (Koeniger, 1986). When the queen lays an egg in to

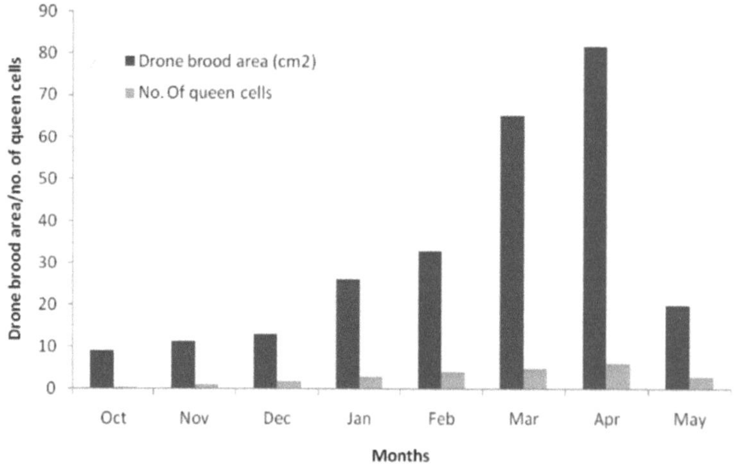

FIGURE 3.1 Drone and queen brood-rearing activity of *Apis florea* in Bengaluru, India.

a worker cell, a few spermatozoa are released out of the spermatheca and fertilize the egg. However, when she lays an egg in to a drone cell, no spermatozoa are released and the egg is not fertilized. The unfertilized eggs laid by the queen are developed into drones and fertilized eggs into queen and worker bees (Winston, 1987).

Reproductive System of the Drone

The drones of honeybees become sexually mature at the age of about 12 days. The testes are composed of bundles of tubules where the spermatozoa are produced and matured. The testes are large at the time of emergence from the cell, and the contents of each testis pass through the vas deferens to the seminal vesicle. The seminal vesicles are large, as they contain semen suspended with spermatozoa and become shrunk later. The endophallus is the copulatory organ and is everted by strong muscular contraction and compression of the abdomen. The endophallus of *A. florea* is an isolated, specialized structure among the *Apis* species (Ruttner, 1988). It has a pair of bursal cornua at the sides of the vestibulum, each of which is composed of three joined protrusions; a continuous non-indented hairy fold on the ventral surface extending from the vestibulum, a small protrusion proximal to a small dorsal hairy patch and a fimbriate lobe with three protrusions (Wongsiri et al., 1990). During development, the spermatozoa do not undergo a complete meiosis in drones, and they inherit only the mother's genes to the next generation (Laidlaw and Page, 1986). The drones of *A. florea* have rudimentary mucus glands.

Sex Attractants

Sex pheromones are odours produced by either male or females. They induce physiological effects in males and females, bringing the sexes together for mating (Ayasse et al., 2001). Honeybees are known to use sex pheromones for attraction of mates. The major active chemical component of the *A. mellifera* queen's mandibular gland was identified as (2E)-9-oxodecenoic acid (9-ODA by Callow and Johnston (1960). Among several other important biological functions, 9-ODA was found to be the main component of the *A. mellifera* queen's sex attractant (Gary, 1962). Later, it was found that extracts of queens of major Asian honeybee species (*A. dorsata*, *A. cerana* and *A. florea*) also attract *A. mellifera* drones, which contained 9-ODA. Similarly, the dead extracted queens or black dummies of similar size impregnated with 9-ODA successfully attracted drones of *A. cerana*, *A. dorsata* (Koeniger et al., 1994) and *A. koschevnikovi* (Koeniger et al., 1998).

Chemical analyses of the mandibular glands of virgin queens of *A. florea* showed relative amounts of the three major components: (2E)-9-oxodecenoic acid (9-ODA), (2E)-9-hydroxydecenoic acid (9-HDA) and (2E)-10-hydroxydecenoic acid (10-HDA). However, the virgin and mated *A. florea* queens synthesize 10-HDA in greater amounts. Sannasi et al. (1971) showed that *A. florea* drones are attracted to 9-ODA, but attempts to attract its drones with 9-ODA lures failed (Plettner et al., 1997). Brockmann et al. (2006) demonstrated that 10-HDA and 9-HDA are part of the sex-pheromone blend in *Apis mellifera*. These components do not attract drones as single components but increase the attractiveness of 9-ODA. Recently, it was found that the drones of *A. florea* were attracted to both 9-ODA and 10-HDA but towards 10-HDA in greater numbers, showing the latter as a major sex pheromone component in queens of *A. florea* (Nagaraja and Brockmann, 2009).

Drone Congregation Areas (DCAs)

A drone congregation area (DCA) is an identified location where drones of honeybees gather and mate with the virgin queens. This is different from most other mating places because it is high above the ground. During mating season, the drones of honeybees are known to fly to drone congregation areas where they mate with visiting queens (Koeniger et al., 2005). The drones assemble in congregation at 10–40 m above ground with a diameter of up to 30 to 200 m (Koeniger and Koeniger, 2004). They emit

FIGURE 3.2 Drone congregation area (DCA) of *Apis florea* in Bengaluru, India.

a pheromone to attract other drones, thereby increasing the size of congregation. The drone congregations may contain more than 10,000 drones from up to 240 different colonies (Koeniger et al., 2005) in the locations where colonies are kept in large numbers. Furthermore, many more numbers of drones are attracted, owing to the virgin queen's olfactory signals and visual cues at shorter range (Gries and Koeniger, 1996). The virgin queens fly to the vicinity of drone congregation after a few minutes of the drones' arrival (Koeniger and Koeniger, 2004), and mating takes place in these areas.

The drones of honeybees congregate close to vegetation and tree canopies, and there are species-specific differences in the height at which drones fly (Otis et al., 2000). The drones of *A. mellifera* congregate in open air, while *A. cerana* gathers in close proximity to the trees (Punchihewa et al., 1990). Similarly, the drones of *A. koschevnikovi* assemble under the canopy of lower trees (Koeniger et al., 2000). The drones of *A. dorsata* are known to fly to large prominent trees as a landmark for orientation and assemble near or under the branches (Koenoger et al., 1994). However, the drones of *A. florea* gather in DCAs that are close to small trees with dense leafage. The DCA of these species is an open space with a diameter of 10–12 m illuminated with bright sunlight (Figure 3.2). Selection of mating yards with limited open space would be an adaptive strategy to deter insect and bird predators in the vicinity. The drones of *A. florea* are attracted and assembled in the DCA at the heights between 2 and 4 m. This congregation area was reconfirmed by visiting drones of *A. florea* even after three years. The drone congregation areas are constant in a location for many years, and some are reported to exist even for over decades (Ruttner, 1988).

Mating Behaviour

Mating Flights

The drones of honeybees start performing mating flights at the age of about 10–12 days. In the colonies of *A. florea*, the mean age for first flight was 5.6 days; the mean age for final flight was 15.6 days (Buawangpang et al., 2009). The drones perform mating flights from 11.30 to 14.00 hrs, with a peak between 12.30 to 13.00 h in Bengaluru, India (Figure 3.3). The preferential mating flights of drones during midday hours of the day is possibly owing to no or least occurrence of bee predators in the vicinity. Obviously, the bright sunlight is also helpful in detection of the predators, if any, by the drones and

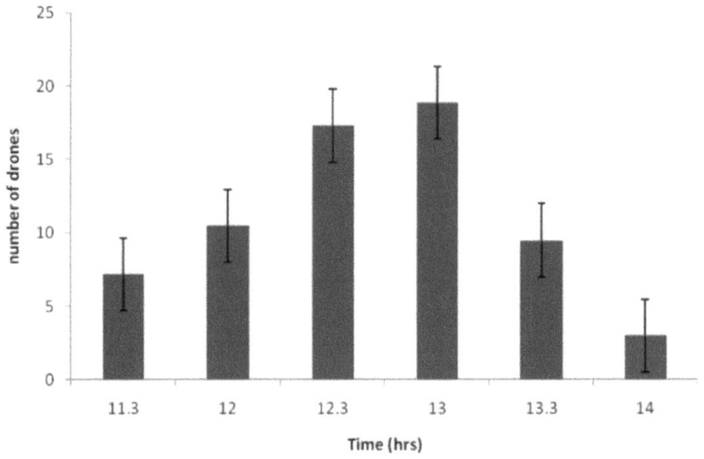

FIGURE 3.3 Mating flights of drones of *Apis florea* during different hours of the day in Bengaluru, India (error bars with Stdev).

queens in the mating sites. Besides, the presence of drones of other *Apis* species in the vicinity facilitates selection for pre-mating isolation through temporal separation (Oldroyd and Wongsiri, 2006). Whenever honeybee species co-occur, the smaller species mates first. The mating flights in honeybees last for 10–30 min. However, in *A. florea* drones, the mating flights lasted for 20–30 min, which is on par with the flights performed by *A. florea* in Thailand (Buawangpang et al., 2009). The short duration of mating flights would be an adaptive strategy against the possible harm by the insect and bird bee predators. Mating flights in honeybees is influenced by several factors. Among them are age of the queen, environmental factors such as temperature, wind, clouds etc. Similarly, availability of the number of drones within the flight range of a queen affects the duration of individual nuptial flights. However, the radio frequency identification was found to be an effective tool in recording the duration and frequency of nuptial flights of honeybee queens (Heidinger et al., 2014).

Mating Strategies

The bees ranging from solitary to eusocial display a spectrum of mating behaviours (Danforth et al., 2006). Social insects have developed a polyandrous mating system. Furthermore, the mating behaviour of honeybee species is striking. The virgin queens of honeybees perform a few short orientation flights before they mate with drones (Winston, 1987). However, a considerable number of queens do not return to the colony after their flights, showing mating risk (Ruttner, 1988). Interestingly, the mating takes place at drone congregation areas that are species specific. The mating system in honeybees is of a male aggregation type (Boomsma et al., 2005), as the males establish the congregation areas before arrival of females for mating (Winston, 1991). However, the female mate choice is also possible because the successful copulation depends on queens opening their sting chamber to allow a male to insert his genitalia (Strassmann, 2001). Furthermore, the queens need to contract their bursa copulatrix to press the endophallus in order to transfer sperm into their lateral oviducts.

During mating flight, a comet of drones follows the queen and the drones compete to copulate with her. The virgin queen in flight responds to drones in the vicinity of the drone congregation area by holding open the entrance to her sting chamber. One of the drones flying in the drone congregation areas succeeds in seizing her by clasping the virgin queen between his legs. The first drone inserts his endophallus into the bursa copulatrix of the queen and ejaculates his spermatozoa into the oviducts of the queen. After the ejaculation, the male separates from the queen but leaves a mating sign consisting of several gland secretions and a chitinous plate (Koeniger et al., 1979). Each consecutive drone inserts his endophallus beneath the mating sign of his predecessor and removes it by a special organ on the endophallus (Koeniger, 1986).

The tip of the endophallus of *A. florea* is pointed and smaller than that of *A. mellifera* and can probably deposit the sperm into or near the orifice of the spermathecal duct. Therefore, the drones of *A. florea* ejaculate their sperm directly into the spermathecal duct from where it is moved to the spermatheca of the queen (Koeniger and Koeniger, 1991). It was recorded sperm in their spermathecae of the returning queens but not in their lateral oviducts. The drones of *A. florea* have rudimentary mucus glands, and no mucus for a mating sign is produced (Koeniger and Koeniger, 1991). However, in *A. mellifera*, the drone is attached during the copulation to the queen by the endophallus filled with mucus (Koeniger, 1986). After the semen from the succession of drones has filled the queen's oviducts, it moves to oviducts as the contractions of the oviduct muscles push the semen into the spermatheca. The drones of *Apis* species die during their first and only copulation (Baer, 2005), as the eversion of the endophallus is irreversible and results in the paralysis of the drone (Koeniger and Koeniger, 1991). As a consequence, drones can copulate only once and mating is lethal, so a drone commits his entire reproductive effort to a single queen. After one or two nuptial flights, the queen returns to the colony and starts laying eggs after a few days.

Mating Frequency

The queens of honeybees usually mate with more than 10 drones (Kryger and Moritz, 1997; Neumann et al., 2000; Woyke et al., 2001). Similarly, Woyke (1975) found that the queens of *A. cerana* are mated by an average of 10 drones. After initiation of egg laying, the queens never remate even if they are run out of sperm (Winston, 1987). Thus, the sperm number in the spermatheca sets a limit to the fitness of the queen. The mating frequency of *A. florea* and *A. andreniformis* has the lowest of all *Apis* bees. The queens of *A. florea* and *A. andreniformis* mate with three to four males in a single flight (Koeniger et al., 2000). However, Palmer and Oldroyd (2001) found that the queens of *A. florea* typically mate with more than 13 drones. Furthermore, the mating frequency of *A. florea* queens was known to be varied from 13 to 19, which is similar to other honeybee species. Sometimes, queens of *Apis* species are found to mate with more than 44 drones (Moritz et al., 1996).

Sperm Storage

Queens of honeybees efficiently keep the stored sperm alive for a long time (Boomsma et al. 2005) and use them meticulously during fertilization of eggs. They store around 3%–5% of the sperm they acquired during copulations (Baer, 2005). The queens of *A. andreniformis* mated with 11 drones received about 1.43 million sperm, which is very close to 1.26 million actually found in spermatheca. However, in *A. florea*, the male storage success seems lower given the 3.44 million sperm transferred during copulations and the typical storage of 1.05 million sperm in the spermatheca. Polyandry in these species may have been maintained because males might also benefit from genetic diversity among worker offspring or because they do not have enough sperm to fully inseminate a female (Kraus et al., 2005). A single *A. florea* drone has an average of 0.44 million spermatozoa, and the spermatozoa counted from the spermatheca of two laying queens were 0.38 and 0.022 million (Ruttner, 1988). However, the maximum number of spermatozoa found in *A. florea* was 1.35 million (Koeniger et al., 1989). In *A. mellifera* and *A. cerana*, about 90% of the spermatozoa deposited by the drones in the oviducts are lost and only 10% of sperms reach the spermatheca. In *A. mellifera*, about 6–8 million spermatozoa are found in a fully mated young queen, while in *A. cerana*, the queen has only 1.3–3.5 million spermatozoa (Ruttner et al., 1973).

Reproductive Swarming

Swarming is a complex process that is mediated by many intrinsic and extrinsic factors. It generally occurs when the bee colonies are quite populous and the resources are available aplenty. The bee colonies rear the new queens especially during brood rearing season (Winston et al., 1981). Similarly, the reproductive success of the colony is closely associated with the fertility of the queen.

Initiation of Swarming

Initiation of mass queen rearing is the first visible step of swarming preparation in bee colonies. The colonies of *A. florea* construct up to 12–16 queen cells at the lower rim of the comb. They stop further construction of the comb and cease the brood-rearing activity. During the swarming period, the workers are coordinated by both chemical and non-chemical signals. The bee colonies exhibit a general restlessness and show repeated runs of workers on the surface of the bee curtain. The first virgins seem to hatch before or shortly after the first swarm leaves probably together with the mother queen (Whitecombe, 1984). During this period, another swarm issues as young bees and queens hatch successively. While the first swarm takes about half of the colony with it, the following swarms become smaller and smaller until a little group of bees is left on the comb, which leaves with the last queen or disperses. Therefore, the swarming period ends with dissolution of the colony in *A. florea*. Lindauer (1957) observed five swarms from an individual colony in Sri Lanka within 12 days and found only 400–600 bees left with the mother comb.

Nest-Site Selection

The success of the swarm depends upon its size and the number of worker bees present in it. Generally, larger swarms are most successful in locating new nest sites (Seeley and Visscher, 1985). *A. florea* does not seem to select a particular twig or branch prior to the swarm flying off as is evident by the wide divergence in dances prior to departure (Oldroyd et al., 2008). Instead, the swarms of these species fly in the general direction indicated by the average direction of the dances performed in the last half hour or so. They are also known to make rapid decisions as every swarm observed took only a few hours to move to a new site (Oldroyd et al., 2008). The decisions taken by scouts of *A. florea* are fast and inaccurate because they did not reach a decision on specific location before departure of the swarm. It seems that *A. florea* swarms make a quick decision about the new nest site and leave again if they find the location is inaccurate. Dancing of more than half of all bees suggests that regular inspection and re-evaluation of a potential nest site does not occur in *A. florea*. The absence of dance decay suggests that nest-site quality is not reflected in the bees' dance behaviour (Makinson et al., 2011). It is probable that all the bees that dance prior to departure attempt to guide the swarm in their preferred direction. The swarms of *A. florea* are mostly guided by the scouts that are active prior to departure, which represents a limited number of bees in the swarm. The distance to the new nesting site of a swarm may be only a few up to several hundred m. Sometimes, the swarm changes its site several times until comb building is initiated.

Reproductive Isolation

Apis florea shows distinctive reproductive isolation similar to other honeybee species. The drones of *A. florea* found in Sri Lanka perform mating flights about one hour earlier than their counterparts in Southeast Thailand (Koeniger et al., 1996). However, in Bengaluru, India, the mating flights peaked between 12.30 to 13.00 h. Similarly, the successful mating flight of queens in Bangkok coincides with the period of drone mating flights (Koeniger et al., 1989; Rinderer et al., 1993). The temporal sequence of mating periods is correlated with the size of bee species, as it starts with the smallest (*A. andreniformis*) and ends with the largest honeybees (*A. dorsata*). The dwarf honeybees have tiny mucus glands, but in *A. florea*, no mucus secretion was found in the sting chamber of queens returning from mating flights (Koeniger et al., 1989). They cannot produce enough mucus to strengthen the connection between the copulatory pair. However, they have a forceps-like appendix at the metatarsus of the hind leg, which is shorter in *A. andreniformis* than in *A. florea*. The spermatheca of *A. florea* is comparatively smaller and stores a low number of sperms.

Reproductive Conflict

The production of drones is a major conflict in honeybee colonies, as workers can lay abundant unfertilized male eggs (Ratnieks, 1993). This conflict is resolved by worker policing in which workers eat eggs laid by them. Worker policing is a behaviour in which workers inhibit the reproduction by other workers

(Ratnieks, 1988). This behaviour is quite common in queenless colonies, where many workers lay eggs from which drones are reared (Page and Erickson, 1988). Similarly, the workers of *A. florea* produce viable drones in the absence of the queen (Woyke and Wongsiri, 1992), but in queenright colonies, workers are predicted to be sterile. The rapid removal of worker-laid eggs by *A. florea* workers shows that this species can discriminate between worker- and queen-laid eggs. *A. florea* queens use pheromones to mark their eggs, and workers respond to their signal by selectively rearing these marked eggs. *A. florea* is known to control worker reproduction by mutual policing mechanisms also. The worker-laid eggs are generally eliminated fast, showing that *A. florea* uses oophagy of worker-laid eggs as a mechanism of worker policing (Halling et al., 2001). Recently, it was found that the rebel workers, which usually develop after swarming, have a higher reproduction potential and produce male offspring under queenright conditions (Kuszewska et al., 2018).

Instrumental Insemination

Instrumental insemination is a technique used in bee breeding and stock improvement. This technique is followed in most parts of the world for selective breeding of *A. mellifera* colonies (Page et al., 2002). Utilization of this technique in *A. cerana* queens was successful in laboratory conditions (Woyke, 1975). However, it could not succeed in field conditions. Instrumental insemination is of special interest in *A. florea*, as the number of spermatozoa per drone is much lower than other honeybee species. Similarly, in queens of these species, spermatozoa are ejaculated directly into spermatheca (Koeniger and Koeniger, 2000). Phiancharoen et al. (2011) successfully attempted an instrumental insemination technique in *A. florea*, but only 50% of the queens began to lay eggs after insemination. However, it is a difficult task in identifying the correct depth at which to insert the syringe into the vagina or median oviduct without damaging tissues. Similarly, it proved to be a challenging endeavour. However, there is greater scope of improving the instrumental insemination technique for genetical stock improvement.

Conclusion

Despite the similarities in reproductive biology and mating behaviour of the *Apis* species and *A. florea*, there are a few differences. The drones of *A. florea* perform mating flights almost 2 and 4 h earlier than *A. cerana* and *A. dorsata*, respectively. It is also apparent that whenever many honeybee species co-occur, the smaller species is known to mate first. Selection of mating yards with limited open space would be an adaptive strategy to deter insect and bird predators in *A. florea*. Furthermore, the ejaculation of sperms directly into the spermathecal duct of the queen by the drones and non-production of mating signs are found to be unique characteristics of *A. florea*. In addition, use of 10-HDA in attraction of drones by the queen coupled with fast decisions taken by scouts in nest-site selection during swarming is surprising. However, further research on reproductive biology and mating behaviour of *A. florea* in its distribution regions is suggested for better understanding its reproductive biology.

ACKNOWLEDGEMENTS

The senior author would like to thank Prof. C. Srinivas the director at the UGC-Human Resource Development Centre, Bangalore University, Bengaluru, for his constant encouragement and Bangalore University, Bengaluru, for providing research facilities.

REFERENCES

Akratanakul, P. The natural history of the dwarf honeybee, *Apis florea* F. in Thailand, Thesis, Cornell University, Ithaca, NY, pp. 1–91, 1977.

Anderson, R.H. The laying worker in the Cape honeybee, *Apis mellifera capensis*. *J. Apic. Res.* 2(2): 85–92, 1963.

Ayasse, M. Paxton, R.J. and Tengo, J. Mating behaviour and chemical communication in the order Hymenoptera. *Annu. Rev. Entomol.* 46: 31–78, 2001.

Baer, B. Sexual selection in *Apis* bees. *Apidologie* 36: 187–200, 2005.

Boomsma, J.J., Baer, B. and Heinze, J. The evolution of male traits in social insects. *Annu. Rev. Entomol.* 50: 395–420, 2005.

Bozina, K.D. How long does the queen live? *Pchelovodsivo* 38: 13 (In Russian), 1961.

Brockmann, A., Dietz, D., Spaethe, J. et al. Beyond 9-ODA: Sex pheromone communication in the European honey bee *Apis mellifera* L. *J. Chem. Ecol.* 32: 657–667, 2006.

Buawangpang, N., Sukumalanand, P., Burgett, M. et al. *Apis florea* drone flight: Longevity and flight performance. *Apidologie* 40: 20–25, 2009.

Callow, R.K. and Johnston, N.C. The chemical constitution and synthesis of queen substance of honey bees (*Apis mellifera*). *Bee World* 41: 152–53, 1960.

Crozier, R. *Animal Cytogenetics* Vol. III, Insecta 7, Hymenoptera, Bortraeger, Berlin, Germany, 1975.

Danforth, B.N., Sipes, S., Fang, J. et al. The history of early bee diversification based on five genes plus morphology. *Proceedings of the National Academy of Sciences of the United States of America* 103: 15118–15123, 2006.

Gary, N.E. Chemical mating attractants in the queen honey bee. *Science* 136: 773–774, 1962.

Gotoh, A., Ito, F. and Billen, J. Vestigial spermatheca morphology in honeybee workers, *Apis cerana* and *Apis mellifera* from Japan. *Apidologie* 44: 133–143, 2013.

Gries, M. and Koeniger, N. Straight forward to the queen: Pursuing honeybee drones (*Apis mellifera* L.) adjust their body axis to the direction of the queen. *J. Comp. Physiol. A* 179: 539–544, 1996.

Halling, L., Oldroyd, B.P., Wattanachaiyingcharoen, W. et al. Worker policing in the bee *Apis florea*. *Behav. Ecol. Sociobiol.* 49: 509–513, 2001.

Heidinger, I.M., Meixner, M.D., Berg, S. et al. Observation of the mating behaviour of honeybee (*Apis mellifera* L.) queens using Radio-Frequency Identification (RFID): Factors influencing the duration and frequency of nuptial flights. *Insects* 5(3): 513–527, 2014.

Koeniger, G. Mating sign and multiple mating in the honeybee. *Bee World* 67: 141–150, 1986.

Koeniger, N. and Koeniger, G. An evolutionary approach in mating behaviour and drone copulatory organs in *Apis*. *Apidologie* 22: 581–590, 1991.

Koeniger, N. and Koeniger, G. Reproductive isolation among species of the genus *Apis*. *Apidologie* 31: 313–339, 2000.

Koeniger, N. and Koeniger, G. Mating behavior in honeybees (Genus *Apis*). *TARE* 7: 13–28, 2004.

Koeniger, G., Koeniger, N. and Fabritius, M. Some detailed observations of mating in the honeybee. *Bee World* 60, 53–57, 1979.

Koeniger, G., Koeniger, N. and Tingek S. Mating flights, number of spermatozoa, sperm transfer and degree of polyandry in *Apis koschevnikovi* (Buttel-Reepen, 1906). *Apidologie* 25: 224–238, 1994.

Koeniger, G., Koeniger, N. and Tingek S. Interspecific instrumental insemination of *Apis cerana* Fabricius 1793, *Apis koschevnikovi* Buttel-Reepen, 1906 and *Apis dorsata* Fabricius 1793. *Apidologie* 27: 303–304, 1996.

Koeniger, N., Koeniger, G. and Wongsiri, S. Mating and sperm transfer in *Apis florea*. *Apidologie* 21: 413–418, 1989.

Koeniger, G., Koeniger, N., Tingek, S. et al. Mating flights and sperm transfer in the dwarf honey bee *Apis andreniformis* (Smith 1858). *Apidologie* 31: 301–312, 2000.

Koeniger, N., Koeniger, G., Gries, M. et al. Drone competition at drone congregation areas in four *Apis* species. *Apidologie* 36: 211–221, 2005.

Koeniger, N., Tingek, S., Koeniger, G. et al. Exploring the biodiversity of honeybees. *Borneo* 4: 18–33, 1998.

Kraus, F.B., Koeniger, N., Tingek, S. et al. Temporal genetic structure of a drone congregation area of the giant Asian honeybee (*Apis dorsata*). *Naturwissenschaften* 92: 578–581, 2005.

Kryger, P. and Moritz, R.F.A. Lack of kin recognition in swarming honeybees (*Apis mellifera*). *Behav. Ecol. Sociobiol.* 40: 271–276, 1997.

Kshirsagar, K.K. Morphometric studies on *Apis cerana indica*. F. worker. III. Effect of geographical position on morphometric characters. *Indian Bee J.* 43: 1–5, 1980.

Kuszewska, K., Waclawska, A. and Woyciechowski, M. Reproduction of rebel workers in honeybee (*Apis mellifera*) colonies. *Apidologie* 49(2): 162–171, 2018.

Laidlaw, H. and Page, R. Mating designs. In T. Rinderer (Ed.), *Bee Genetics and Breeding*, pp. 323–341. Academic Press, Orlando, FL, 1986.

Lindauer, M. Communication among the honeybees and stingless bees of India. *Bee World* 38: 3–14, 1957.

Makinson, J.C., Oldroyd, B.P. and Schaerf, T.M. Moving home nest-site selection in the red dwarf honeybee (*Apis florea*). *Behav. Ecol. Sociobiol.* 65: 945–958, 2011.
Moritz, R.F.A., Kryger, P. and Allsopp, M.H. Competition of royalty in bees. *Nature* 384: 31, 1996.
Nagaraja, N. and Brockmann, A. Drones of red dwarf honeybee, *Apis florea* are attracted by two components from the mandibular gland, 10-HDA and 9-ODA. *J. Chem. Ecol.* 35: 653–655, 2009.
Neumann, P., Moritz, R.F.A. and Mautz, D. Colony evaluation is not affected by drifting of drone and worker honeybees (*Apis mellifera*) at a performance testing apiary. *Apidologie* 31: 67–79, 2000.
Oldroyd, B.P., Allsopp, M.H. and Gloag, R.S. Thelytokous parthenogenesis in unmated queen honey bees (*Apis mellifera capensis*): Central fusion and high recombination rates. *Genetics* 180: 359–366, 2008.
Oldroyd, B.P. and Wongsiri, S. *Asian Honey Bees Biology, Conservation and Human Interactions.* Harvard University Press, Cambridge, 2006.
Otis, G., Koeniger, N., Rinderer, T.E. et al. Comparative mating flight times of Asian honey bees. *Proceedings of the 7th International Conference on Tropical Bees*, Chiang Mai, Thailand, International Bee Research Association, 2000.
Page, R.E. and Erickson, E.H. Reproduction by worker honeybees (*Apis mellifera*). *Behav. Ecol. Sociobiol.* 23: 117–126, 1988.
Page, R.E. Jr., Gadau, J., and Beye, M. The emergence of hymenopteran genetics. *Genetics* 160: 375–379, 2002.
Palmer, K.A. and Oldroyd, B.P. Mating frequency in *Apis florea* revisited (Hymenoptera; Apidae). *Insectes Soc.* 48: 40–43, 2001.
Phiancharoen, M., Wongsiri, S. and Hepburn, H.R. Queen production and instrumental insemination of *Apis florea* queens. *Apidologie* 42: 307–311, 2011.
Plettner, E., Otis, G.W., Wimalaratne, P.D.C. et al. Species and caste determined mandibular gland signals in honeybees (*Apis*). *J. Chem. Ecol.* 23: 363–377, 1997.
Punchihewa, R.W.K., Koeniger, N. and Koeniger, G. Congregation of *Apis cerana indica* Fabricius 1798 drones in the canopy of trees in Sri Lanka. *Apidologie* 21: 201–208, 1990.
Ratnieks, F.L.W. Reproductive harmony via mutual policing by workers in eusocial Hymenoptera. *Am. Nat.* 132: 21, 1988.
Ratnieks, F.L.W. Egg-laying, egg-removal, and ovary development by workers in queen-right honey bee colonies. *Behav. Ecol. Sociobiol.* 32: 191–198, 1993.
Rinderer, T.E., Oldroyd, B.P., Wongsiri, S. et al. Time of drone flight in four honeybee species in southeastern Thailand. *J. Apic. Res.* 32: 27–33, 1993.
Ronai, I., Allsopp, M.H., Tan, K. et al. The dynamic association between ovariole loss and sterility in adult honeybee workers. *Proceedings of the Royal Society B: Biological Sciences* 284: 20162693, 2017.
Ruttner, F. The mating of the honeybee. *Bee World* 57: 3–15, 23–24, 1956.
Ruttner, F. *Biogeography and Taxonomy of Honeybees*, Springer Verlag, Berlin, Germany, p. 284, 1988.
Ruttner, F., Woyke, J. and Koeniger, N. Reproduction in *Apis cerana*. 2. Reproductive organs and artificial insemination. *J. Apic. Res.* 12: 21–34, 1973.
Sannasi, A., Rajulu, G.S. and Sundara, G. 9-Oxodec-trans-e-enoic acid in the Indian honeybees. *Life Sci.* 10: 195–201, 1971.
Seeley, T.D. and Visscher, P.K. Survival of honeybees in cold climates: The critical timing of colony growth and reproduction. *Ecol. Entomol.* 10: 81–88, 1985.
Snodgrass, R.E. *Anatomy of the Honeybee*, Comstock Publishing Associates, Ithaca, NY, 1984.
Strassmann, J.E. The rarity of multiple mating by females in the social Hymenoptera. *Insectes Soc.* 48: 1–13, 2001.
Whitecombe, R.P. Beekeeping with *Apis florea* in Oman. *Aramco World Mag.* 35 (2): 34–40, 1984.
Winston, M.L. *The Biology of the Honeybee*, Harward University Press, Cambridge, MA, 1987.
Winston, M.L. The role of queen mandibular pheromone and colony congestion in honeybee (*Apis mellifera*) reproductive swarming (Hymenoptera: Apidae). *J. Insect Behav.* 45: 649–660, 1991.
Winston, M.L., Dropkin, J.A. and Taylor, O.R. Demography and life history of two honeybee races (*Apis mellifera*). *Oecologia* 48(3): 407–413, 1981.

Wongsiri, S., Limbipichai, K., Tangkanasing, P. et al. Evidence of reproductive isolation confirms that *Apis andreniformis* (Smith, 1858) is a separate species from sympatric *Apis florea* (Fabricius, 1787). *Apidologie* 21: 47–52, 1990.

Woyke, J. The eggs in the comb cells of three honeybee species in India. *Proceedings of the International Beekeeping Congress* 25: 340, 1975.

Woyke, J., Wilde, J. and Wilde, M. *Apis dorsata* drone flights, collection of semen from everted endophalli and instrumental insemination of queens. *Apidologie* 32: 407–416, 2001.

Woyke, J. and Wongsiri, S. Occurrence and size of laying worker eggs in *Apis florea* colonies. *J. Apic. Res.* 31: 124–127, 1992.

4

The Genetics and Genomics of the Dwarf Honeybee Apis florea *Fabricius*

S. Mohankumar, T. Sonai Rajan, M. Praghadeesh, P. A. Saravanan and M. R. Srinivasan

CONTENTS

Introduction

The past decade has seen profound growth in the field of genetics and genomics, finding roots in many branches of science. In a broader perspective, genetics relates to the study of individual genes or a gene family, their pattern of inheritance and their phenotype. Genomics has outgrown the field of genetics with the series of discoveries in sequencing technologies and high-throughput gene mapping with the study of whole sets of genes present in an organism. It may be interpreted as studying the entire genome atlas of any target species with an ability to correlate the dynamics of effect of many individual genes and considering the impact of expression that one gene has on the other. With the advancement of next-generation sequencing technologies and data analytic capabilities for decoding the genome and transcriptome of a target organism, researchers have gained the freedom to understand the genetic architecture of the organism and to specifically study the role of numerous individual contributing genes in the architecture towards a specific outcome/phenotype of interest.

Although the genetic and genomics-based approach has proved to be a highly effective tool in understanding the biology of a target organism, the application of this holistic approach is still in its infancy in studying the honeybee species. Further, among the different honeybee species, the dwarf honeybee, *Apis florea* Fabricius, has received very little attention in decoding the genetic architecture and dynamics of gene functions. Arguably, the most profound impact has been in *Apis mellifera* (Italian bee), where the explosion of genomics sequence and gene profiling data has illustrated the complexity in the genetic make of honeybees. Owing to the complex ploidy level in the different castes of bee species, dozens of different genes may be aberrant in structure, arrangements, copy number, and a significant number of genes may be differentially expressed. Studies on comparative genomics that take into account the four major bee species may aid the search for the more elusive genes that play a major role in biotic and abiotic threats of bee species. Insights into the genomics and molecular genetic studies of *Apis mellifera* have provided new strategies for defining the framework of genetic architecture of other honeybee species. Further, concurrent efforts are warranted to study the individual species of honeybees such as *Apis florea* (little bee) to translate any such novel outcome in terms of gene function specific to royal jelly genes, fox, sting properties and body colour, which can be translated and applied to other species as well.

Apart from the benefits of a comparative approach, the *Apis florea* species is considered unique from the rest of the bee species owing to its focused demographic presence, stinging mode and properties, composition of sting venom and colony structure formation properties. These unique *A. florea* specific phenotypic properties may have deep roots in their genome architecture which can be effectively leveraged to study the genome of bee species holistically and exploit them to manage various biotic and abiotic stress factors that continue to threaten the existence of honeybee species and decelerate the bee decline scenario. In this chapter we briefly discuss the genetics and genomics of *A. florea*.

Karyotype of *A. florea*

The chromosome details of the honeybee species have been studied since 1901, when the haploid chromosome number was established, n = 16 (Fahrenhorst, 1977), and the chromosome numbers of haplodiploid were confirmed by Hoshiba and Kusanagi (1978) and Hoshiba et al. (1981). Based on the number of chromosomes and primitive behaviour of honeybees, Deodikar et al. (1958) reported that the domesticated tetraploid species *A. cerana* and *A. mellifera* may have evolved from the wild diploid of *A. florea*. The study on karyotype and C-banding analyses of *A. florea* (Asadi et al., 2010) demonstrated that the total number of chromosomes in *A. florea* (n = 16) was divided into two groups: 4 chromosomes in metacentric (no. 1, 4, 7 and 11), and 12 chromosomes (included in submetacentric, subtelocentric and acrocentric chromosomes) along with a distal C band were also reported to be present in chromosome 1. Karyotyping analyses of the total chromosome length showed that the longest metaphase chromosome of *A. florea* was about 2.76 μm and the smallest 0.86 μm. The karyotype observed in *A. florea* was reported to be similar with *A. mellifera* (Hoshiba, 1984) and *A. cerana*.

Genome Information of *A. florea*

The genomic revolution and attempts to decode the genetic architecture of honeybee species started with the publication of the honeybee (*A. mollifiers*) genome (Honeybee Genome Sequencing Consortium, 2006). Consequently, over time, three gene annotation versions were developed and made available in NCBI. Similarly, distributed efforts were made to sequence the genome of other honeybee species. The whole genome of *A. florea* was sequenced in 2010 by the Baylor College of Medicine, Human Genome Sequencing Center (BCM-HGSC) using a 454 platform (whole genome shotgun assembly). Male haploid drones have been used to generate the whole genome sequence of *A. florea*. The total genome size of *A. florea* was reported to be 230.47 Mb with the GC content of 34.9%. Based on comparative annotation, it was estimated that *A. florea* genome consists of contigs (19,341), gene (12,134), pseudogene (73) and protein (17,663) (NCBI: www.ncbi.nlm.nih.gov).

Gene Discovery

Although genomic information and whole genome sequence data are available for honeybee species, very few attempts were made to functionally annotate and validate the genes of interest for various traits. With increasing data on honeybee transcriptome and proteome available in the public domain, this paved the way for gene discovery that governs crucial traits including body colour, stinging nature, colony construction and royal jelly. In this context, many genes have been identified and analyzed in *A. florea*.

For example, major royal jelly protein gene (*MRJP*) (Helbing et al., 2017); gene (*Fl*) for body colour of *A. florea* (Woyke, 1997); olfactory receptor (*OR*) gene (Karpe et al., 2016); bee venom *MELT* gene (UniProtKB-P01504 (MEL_APIFL) (Kriel, 1973, 1975); and forkhead box (FOX) gene (Nie et al., 2018).

Major Royal Jelly Protein Gene (*MJPR*) in *A. florea*

Royal jelly is a secretion of the hypopharyngeal glands of worker bees composed of protein (10%–15%), sugar (10%–16%), lipids (3%–6%), water (60%–70%) and trace amounts of vitamins, salts and amino acids (von Planta, 1888; Rembold, 1983). It's mainly used to rear female larvae to queens and also plays a major role in growth and development of larvae. Among the nutrients in royal jelly, proteins play a major role in the development of honeybees. The first royal jelly protein was identified and termed as the major royal jelly protein 1 (*MJPR1*) (Hanes and Šimúth, 1992). Subsequently, the nine most abundant proteins of royal jelly were also termed major royal jelly protein, and the encoding genes (*mrjp1-mrjp9*) have been identified in the genome of the honeybee (Honeybee Genome Sequencing Consortium, 2006). In *A. florea*, *mrjp3, 1, 4, 6* and *5* were reported to show an identical location at the 5' end of the cluster, *mrjp8* and *9* were found at the 3' end and the phylogenetic reconstruction of *mrjp* amino acid sequences reported a new member of the *mrjp* family (*mrjp10*) in *A. florea* (Helbing et al., 2017). The expression and function of *mrjp10* was not reported; thus, the characterization and comparative analysis of the *mrjp* expression pattern in *A. florea* would be helpful to understand the extent of gene expression diversity.

Gene Responsible for Body Colour in *A. florea*

Body colour is considered the most distinct character of the honeybees, and the colouring pattern (light and black) of bee species varies between and within the species as well as between the three different castes in the same nests or colony. Roberts and Mackensen (1951) reported that at least seven different loci of the genes may alter the colouring pattern in honeybee species. In *A. florea*, Woyke (1997) demonstrated that the gene responsible for body colour was *Fl*, which causes drones to be black and queen and worker bees to be of the yellow type. He also reported that the expression of body colour in three castes of *A. florea* depends upon the sex (males black, females yellow), though the pattern of yellow and black colour is different in worker and queen bees.

Olfactory Receptor (*OR*) Genes in *A. florea*

Olfaction plays an important role in honeybee species for searching the food materials and communication between the honeybees (von Frisch, 1965). In general, honeybees detect the olfactory signals through olfactory sensory neurons located on the antenna, and the olfactory sensory neurons express the odorant receptor proteins on the cell membrane (Vosshall et al., 2000). Studies reported that the insect genomes (Hallem et al., 2004; Hallem and Carlson, 2006) are endowed with the range of functional olfactory genes, and each of them was demonstrated to be encoded with a unique receptor protein regulated to a specific odour. Thus, the identification of olfactory receptor genes plays a major role for better understanding olfactory perception of honeybee species. Earlier, around 100 *OR* genes were identified in *A. florea* using an automated annotation method and included in the NCBI genome. Recently, Karpe et al. (2016) identified the complete repertoire of *A. florea OR* genes (180 Nos) through a genome-wide survey in the *A. florea* genome sequence (Aflo_1.0 from NCBI Genbank – The Honey Bee Genome Sequencing Consortium (HBGC)). Most of the identified *AfOr* genes were reported to be similar to the *OR* genes identified in the *A. mellifera*, and the characteristics of *AfOr* genes *viz.*, gene structure, synteny of tandemly repeated *ORs* and phylogenetic clustering were highly conserved. Further analyses of transcription within male and female antennae of *A. florea* detected a total number of 145 and 162 *OR* transcripts in male and female antennae, respectively, and most of the *OR* genes were demonstrated to be highly expressed on the female antennae. Five distinct male-biased *OR* genes were also identified, of which three genes (*AfOr 11*, *AfOr 18* and *AfOr 170p*) have been identified to be male biased in *A. mellifera* (Karpe et al., 2016).

Mitochondrial Genome of *A. florea*

The complete mitochondrial genome of *A. florea* was described by Wang et al. (2013) and reported that the 17,694 bp long mitogenome was larger than any other honeybee species in the Apidae family. Wang et al. (2013) demonstrated that the gene order and arrangement of the *A. florea* genome was identical with the mitogenome of *A. mellifera* (Crozier and Crozier, 1993), though, *A. florea* mitogenome contained three $tRNA^{Ser}$ (AGN) located between $tRNA^{Glu}$ and $tRNA^{Met}$, each of which was described to be preceded by a 44-bp-long repeat unit (RA1-RA3) and followed by a 64-bp-long repeat unit (RB1-RB3). In addition, the fourth RA (TA4) was reported to be adjacent to $tRNA^{Met}$ (Wang et al., 2013). The A+T-rich region of the mitogenome of *A. florea* was 1987 bp long, and that size was remarkably longer than that of other honeybee species from the Apidae family (Tan et al., 2011). This A+T-rich region demonstrated to be composed of three non-repeat regions (216, 109 and 52 bp, respectively) that includes two types of tandem repeat units, and each repeat unit was composed of 141–219 bp long 5 tandem repeats (RC1–RC5) and 50–52 bp long 12 tandem repeats (RD1-RD12), respectively. DNA sequencing of $tRNA^{leu}$ and *Cytochrome oxidase II* (COII) of five honeybee species (*A. cerana, A. mellifera, A. dorsata, A. andreniformis* and *A. florea*) demonstrated that there was 7%–11% sequence divergence among the species, and their phylogenetic analyses provided evidence that *A. florea* showed early divergence in contrast to cavity nesting bees. Similarly, Yang et al. (2017) sequenced the complete mitochondrial genome of wild *A. florea* collected from Southwest China and demonstrated that the circular mitochondrial genome was 15,933 bp length and also reported that the genome has similar codon usage and gene organization to those of mitogenome reported from other hymenopteran. The complete mitochondrial genome of *A. florea* was reported to contain the typical insect mitogenome set of 13 protein coding genes, two ribosomal RNA genes, 22 transfer RNA genes and the noncoding regions with very high AT-base content (Yang et al., 2017). Phylogenetic analyses of the same study demonstrated that *A. florea* and *A. cerana* + *A. mellifera* were sister taxa (Raffiudin and Crozier, 2007) and more original. Recently, Takahashi et al. (2018) discovered the complete mitochondrial genome of *A. florea* (Thailand populations) using next-generation sequencing (NextSeq 500) and demonstrated that the mitochondrial genome of *A. florea* from Thailand was 17,693 bp in length. In addition, they reported that the heavy (H) strand of *A. florea* mitochondrial genome encoded nine protein coding genes and 14 tRNA genes, and the light (L) strand coded four protein coding genes, 8 tRNA genes and 2 rRNA genes along with the A+T-rich region (with the average AT content of 86.28%), besides 3 tRNA-*Ser* (*AGN*) repeats. Furthermore, the presence of cloverleaf secondary structures were also reported across the tRNA genes except for tRNA-Ser (S1) and tRNA Glu (E) of *A. florea* mitogenome.

Linkage Map and Recombination Rate in *A. florea*

The genetic and linkage mapping of an organism is highly useful for identifying the genome region controlling a trait of interest, cloning genes for which only a phenotype is known by map position, detecting the pattern of chromosome evolution and identifying the landing sites of transgene (Goldsmith et al., 2005). In addition, the linkage map can also provide the information on the recombination frequency and genome organization of a species. Measuring the frequency and location of recombination is an extremely useful tool in genetic research. In this context, DNA-based markers, *viz*., RAPD, AFLP, SSR and SNP have been widely used to develop the linkage map in honeybee species. Earlier, Meznar et al. (2010) investigated the meiotic recombination in *A. florea* through genotyping using *A. mellifera* species-specific microsatellite markers in the natural mapping population of *A. florea*, and the recombination rate was calculated using the physical distance from *A. mellifera*. The results showed high genome-wide recombination rate in *A. florea* and demonstrated that the open nesting *A. florea* frequently exposed to several biotic (pests and pathogens) and abiotic (environmental variations) factors (Seeley et al., 1982), which may cause higher genetic diversity and thus a higher recombination rate (Meznar et al., 2010). Recently, the genomic linkage map was generated for *A. florea* (Rueppell et al., 2016) through SNP genotyping of haploid male offspring of a single female, and the results revealed that the map spans 4782 cM with 1279 markers in 16 linkage groups and this map size resulted in an average recombination rate of 20.8 Cm/Mb. Synteny analyses indicated that the regular intrachromosomal rearrangement but no translocations across chromosomes accompany the high rate of recombination during the evolution of honeybee lineage (Rueppell et al., 2016). This recombination rate of *A. florea* adds another record of a social insect species with very high rate, strengthening the argument for the exceptional evolution of genomic recombination rates in social insects.

Sex Determination in *A. florea*

The haplo-diploid nature of sex is present in honeybee (*Apis*) species, in which males are developed from unfertilized eggs with one set of chromosome (haploid) and the females are developed from fertilized eggs with two sets of chromosome (diploid) (Dzierzon, 1845). However, the inbreeding of honeybees resulted in the appearance of diploid males, and the presence of these diploid males suggested that neither the fertilization process nor the number of chromosome (haploid or diploid) of eggs provides the primary signal for gender determination in honeybees (Mackensen, 1951). Whiting (1933, 1943) proposed a hypothesis of complementary sex determination, in which a single sex determination locus determines the sexual fate in hymenopteran and other species. In honeybees, sex is determined by a series of alleles (19 alleles) at a single locus (possibly located on chromosome 8). The studies on the isolation of locus pertinent to the sex determination in honeybees led to the identification of the complementary sex determiner (*csd*) gene (Beye et al., 2003), which is an ancestral mode of gender determination and reported to be present in four super families of Hymenoptera (Heimpel and De Boer, 2008).

The gene *csd* encodes an SR-type protein (arginine/serine-rich), which contains an RS domain in the middle and a proline-rich region at its C terminus. Apart from these domains, *csd* is characterized by the hypervariable region, forming repeated structures of species-specific amino acid motifs that are highly variable in length (Hasselmann et al., 2008; Biewer et al., 2016). Studies reported that the haploids containing a single *csd* allele will develop into normal males (drones), while diploids, heterozygous for *csd* become females. The *csd* gene product is essential for the development of female because inactivation of the *csd* gene product in the embryos of females causes a full switch into the development of male (Beye et al., 2003). Subsequently, the target of the *csd* gene has been identified by Hasselmann et al. (2008) and reported as a feminizer (*fem*) gene. The feminizer gene is reported to have two types of splice variants, and their occurrence is specific to the sex of the bee caste. The female cells alone have a functional *fem* gene product. Whereas in males, splicing introducces a stop codon into the *fem* coding sequences, resulting in a truncated version of the *fem* gene specific to males (Gempe and Beye, 2009).

The complementary sex determination gene (*csd*) has also been reported in *A. florea* with a higher level of nucleotide diversity than in other *Apis* species (*A. dorsata*, *A. cerana* and *A. mellifera*). Liu et al. (2011) cloned the genomic region 3 of the *csd* gene from 60 workers of *A. florea* and obtained 54 sequences belonging to 37 sequence variants and identified 12 *csd* alleles with high nucleotide diversity of exon, intron and coding region. The amino acid sequence analyses of *A. florea* reported two kinds of repetitive sequences ((N) $_{1\text{-}4}$Y and (KHYN) $_{1\text{-}4}$) in the hypervariable region of the *A. florea csd* region 3 (Liu et al., 2011). Among the two repetitive sequences, the (N) $_{1\text{-}4}$Y sequence was described as a basic motif and was reported to present in all alleles with varied levels. Studies also indicated the presence of the same repeats ((N) $_{1\text{-}4}$Y) in other species of *Apis* (Cho et al., 2006; Hasselmann et al., 2008). The second repetitive sequence of KHYN$_{1\text{-}4}$ was reported to be present in parts of the *A. florea csd* alleles (Liu et al., 2011), and it was considered as a species-specific allele. Recently, Biewer et al. (2016) studied the evolutionary history of paralogous sex-determining genes (*csd* and *fem*) in *A. florea* through reverse transcription PCR analysis using RNA samples from a pool of early embryonic stage eggs, and the results demonstrated that the genomic regions covered by exons encoding the *fem* and *csd* genes were 7.0 and 8.7 kb in length, respectively, and researchers speculated that the *csd* gene was 3.5 kb downstream of the *fem* gene. Further, the female splice variant form of *Af-fem* consists of 10 exons and encodes an open reading frame of 401 amino acids. While, *Af-csd* consists of nine exons, and the deduced open reading frame of 408 amino acids has also been isolated and reported in *A. florea*. Biewer et al. (2016) reported that the sex-determining function of *csd* in *A. florea* follows the same principle as that proposed in other species of *A. mellifera*, whereas the molecular mechanism of *csd* appears to be encoded by low amino acid differences, a less complex hypervariable region and more modified coiled-coil motifs than that described for other *Apis* species.

Genetic Diversity and Phylogeography of *Apis florea*

The little or dwarf honeybee, *Apis florea*, is an open nesting species of honeybee, commonly using the ground-level branches of small bushes as nesting sites; the nest is constructed from a single comb. Even in a single nest, a high level of genetic diversity was reported among *A. florea* and was attributed to the polyandrous nature of queen bees (Jones et al., 2004). Studies demonstrated that the polyandrous nature of queen bees produced a genetically diverse nature of worker bees (Jones et al., 2004). Increased genetic diversity within a honeybee colony increases variations of the frequency of tasks performed by workers (Oldroyd et al., 1994). Thus, the determination of genetic diversity of honeybee species is crucial for the identification and selection of desired traits (i.e., pests and disease resistance). In this context, different molecular markers including mitochondrial DNA (*COI* and *COII*) and nuclear (SSR) DNA markers have been widely used to assess the genetic diversity and phylogeography of honeybee species including *A. florea* (Jones et al., 2004; Lazar et al., 2013; Azadi et al., 2016). Earlier, the genetic diversity of worker bees of *A. florea* was correlated with their preliminary behaviour of fanning in nest cooling of hives using nuclear DNA markers (two sets of microsatellite markers) and demonstrated that the genetically varied workers have different thresholds for performing the regulatory task of fanning to maintain a homeostatic environment in the hives, and the study demonstrated that the genetic variation in task threshold is a common process in *Apis* species (Jones et al., 2007).

Lazar et al. (2013) studied the genetic diversity and phylogeography of *A. florea* collected from different geographical regions of India using nuclear (SSR) and mitochondrial (*COII*) DNA markers. The results of the microsatellite analyses revealed a significant amount of observed and expected heterozygosity and demonstrated that the presence of significant genetic diversity across the seven sampled locations in India (including Tamil Nadu, Kerala, Andhra Pradesh, Karnataka, Maharashtra, Assam, Jammu and Kashmir) covering various demographic and geographical origins. The combined analyses of neighbour-joining tree, structure analyses and principal coordinate analyses exhibited significant clustering across the sampled populations and also indicated the presence of six population clusters (K = 6). The populations collected from the Assam and Jammu regions showed close genetic relatedness, whereas populations from Karnataka formed a cluster with immigrants from other regions. Similarly, the populations from Kerala, Andhra Pradesh and Tamil Nadu formed clusters with shared immigrants among them whereas the populations of Tamil Nadu and Andhra Pradesh had maximum admixture populations. Populations from the Maharashtra region formed a separate cluster with negligible immigrants. The analyses of molecular variance across the populations indicated the significant genetic clustering among the populations with a mean F_{ST} value of 0.1917. The haplotype analyses of *A. florea* demonstrated the presence of new haplogroups across the sampled locations, and a total of 49 new haplotypes have been reported. The maximum likelihood tree from the *COII* sequences revealed that there is a significant gene flow between the populations from southern and northern parts of India. The populations from the southern parts were widespread across the sampling regions. The comparative analyses of newly identified haplogroups from India with that of *A. florea* haplotypes reported from China resulted in identification of many new haplotypes that are exclusive to the Indian region. Only one haplotype from India, KFG7.2_Hap4, was identical to previously identified haplotypes X8(3) from Honghe (JN107723) and X11 from Nanning (JN107726) in China and indicated the possible population flux between India and China. The genetic relationship and population differentiation of *A. florea* from the western part of Iran was investigated by Terzi et al. (2014) using random amplified polymorphic DNA (RAPD) analyses, and the result showed a total of 115 polymorphic fragments produced from 10 primers in the studied populations. The populations included in this study from Ilam, Khuzestan and Bushehr of western Iran showed a high level of genetic variation (H = 0.21), and researchers concluded that the geographical variations may influence the genetic variation of *A. florea* across the western part of Iran.

Asadi et al. (2016) investigated the genetic diversity and differentiation of four different populations of *A. florea* from southern Iran using six polymorphic microsatellite loci, and the results revealed that the samples from the Minap region showed high levels of polymorphism with an average of 3.3 alleles per locus; the samples from Qeshm Island recorded the lowest average allele of 2.2 per locus. The heterozygosity was reported to be high in most of the samples, and loci and the data on observed and expected heterozygosity showed significant differences among the populations of *A. florea*. Distance matrices analyses also demonstrated a high level of genetic distance between the populations from Minap and the Qeshm Island regions. This high level of genetic variation between Minap and Qeshm Island has been explained as a result of surrounding climate and altitude of the locations. The F_{ST} values of *A. florea* indicate inbreeding in a population level in a range from −1 as lowest to +1 as highest inbreeding. Finally, researchers concluded that the random mating of queens with several males from other colonies or apiaries, a high level of inbreeding within the colony and frequent migration of colonies to the different environmental conditions may be the possible reasons for the presence of high levels of genetically diverse workers in *A. florea*.

Conclusion

The utilization of genetic and genomic approaches is crucial for understanding the historical events that led to the present distribution of the species. The genomic mapping considerably increases the knowledge of genome organization, which resulted in major gene discoveries that may play a major role in conservation of the bee population. Many aspects of genetics and genomics are to be explored in *A. florea*. The advent of next-generation technologies in insect has resulted in ever-increasing genomic information and a public database. It is necessary that concerted efforts should be made to dissect the voluminous data generated and to derive meaningful conclusions on individual genes. Compared to other species of honeybees, very limited work has been reported in *A. florea*. The research on global gene expression profiling and SNP genotyping will also be helpful in understanding the molecular genetics of *A. florea*. The emerging field of genetics and genomics has the potential to serve as an indispensable tool in conservative genetics of *A. florea*.

REFERENCES

Asadi, N., Ghafari, S.M., Gharahdaghi, A.A., Tahmasebi, G.H.H., & Khederzadeh, S. "Karyotype and C-banding analysis of haploid male chromosome of *Apis florea* F." *African Journal of Biotechnology* 9 (27) (2010): 4471–4474.

Asadi, N., Rahimi, A., Ghaheri, M., Kahrizi, D., Dehbaghi, M.B., Khederzadeh, S., Banabazi, M.H. et al. "Genetic diversity of the dwarf honeybee (*Apis florea* Fabricius, 1787) population based on microsatellite markers." *Cellular and Molecular Biology* 62 (12) (2016): 51–55.

Beye, M., Hasselmann, M., Fondrk, M.K., Page, R.E., & Omholt, S.W. "The gene *csd* is the primary signal for sexual development in the honeybee and encodes an SR-type protein." *Cell* 114 (4) (2003): 419–429.

Biewer, M., Lechner, S., & Hasselmann, M. "Similar but not the same: Insights into the evolutionary history of paralogus sex determining genes of the dwarf honey bee *Apis florea*." *Heredity* 116 (2016): 12–22.

Cho, S., Huang, Z.Y., Green, D.R., Smith, D.R., & Zhang, J. "Evolution of the complementary sex-determination gene of honey bees: Balancing selection and trans-species polymorphisms." *Genome Research* 16 (11) (2006): 1366–1375.

Crozier, R.H., & Crozier, Y.C. "The mitochondrial genome of the honeybee *Apis mellifera*: Complete sequence and genome organization." *Genetics* 133 (1993): 97–117.

Deodikar, G.B., Thakar, C.V., & Tonapi, K.V. "Evolution in the genus *Apis* and its bearing on breeding better strains of Indian bee." *Proceedings of International Beekeeping Congress* 17 (1958): 245–250.

Dzierzon, J. "Gutachten über die von Herrn Direktor Stöhr im ersten und zweiten Kapitel des General-Gutachtens aufgestellten Fragen." *Eichstädter Bienenzeitung* 1 (1845): 109–113, 119–121.

Fahrenhorst, H. "Nachweis übereinstimminder chromosomen-Zahlen (n=16) bei allen 4 Apis-Arten." *Apidiologie* 8 (1977): 89–100.
Gempe, T., & Beye, M. "Sex determination in honeybees." *Nature Education* 2 (2) (2009): 1.
Goldsmith, M.R., Shimada, T., & Abi, H. "The genetics and genomics of the silkworm *Bombyx mori*." *Annual Review of Entomology* 50 (2005): 71–100.
Hallem, E.A., & Carlson, J.R. "Coding of odors by a receptor repertoire." *Cell* 125 (2006): 143–160.
Hallem, E.A., Ho, M.G., & Carlson, J.R. "The molecular basis of odor coding in the Drosophila antenna." *Cell* 117 (2004): 965–979.
Hanes, J., & Šimúth, J. "Identification and partial characterization of the major royal jelly protein of the honey bee (*Apis mellifera* L.)." *Journal of Apicultural Research* 31 (1992): 22–26.
Hasselmann, M., Gempe, T., Schiott, M., Nunes-Silva, C.G., Otte, M., & Beye M. "Evidence for the evolutionary nascence of a novel sex determination pathway in honeybees." *Nature* 454 (2008): 519–522.
Heimpel, G.E., & De Boer, J.G. 2008. "Sex determination in the Hymenoptera." *Annual Review of Entomology* 53 (2008): 209–230.
Helbing, S., Lattorff, H.M.G., Moritz, R.F.A., & Buttstedt, A. "Comparative analyses of the major royal jelly protein gene cluster in three *Apis* species with long amplicon sequencing." *DNA Research* 24 (3) (2017): 279–287.
Honeybee Genome Sequencing Consortium. "Insights into social insects from the genome of the honeybee *Apis mellifera*." *Nature* 443 (2006): 931–949. doi:10.1038/nature05260.
Hoshiba, H. "The C-banding analysis of the diploid male and female honeybee (*Apis mellifera*)." *Proceedings of Japan Academy* 60 (B) (1984): 238–240.
Hoshiba, H., & Kusanagi, A. "Karyological study of honey bee." *Journal of Apicultural Research* 17 (1978): 105–109.
Hoshiba, H., Okada, I., & Kusanagi, A. "The diploid drone of *Apis cerana japonica* and its chromosomes." *Journal of Apicultural Research* 20 (1981): 143–147.
Jones, J.C., Nanork, P., & Oldroyd, B.P. "The role of genetic diversity in nest cooling in a wild honeybee, *Apis florea*." *Journal of Comparative Physiology A* 193 (2007): 159–165.
Karpe, S.D., Jain, R., Brockmann, A., & Sowdhamini, R. Identification of complete repertoire of the *Apis florea* odorant receptors reveals complex orthologus relationship with *Apis mellifera*." *Genome Biology and Evolution* 8 (9) (2016): 2879–2895.
Kriel, G. "Structure of melittin isolated from two species of honeybees." *FEBS Letter* 33 (1) (1973): 241–244.
Kriel, G. "The structure of *Apis dorsata* melittin: Phylogenetic relationships between honeybees as deduced from sequence data." *FEBS Letter* 54 (1) (1975): 100–102.
Lazar, K.V., Fakrudin, B., Omkar Babu, K., Mohankumar, S., Parthiban, T.P., Chapalkar, S., Viraktamath, S., & Vastrad, A.S. "Genetic diversity and phylogeography of Apis florea Fabricius." In Viraktamath, S., Kakrudin, B., Vastrad A.S., & Mohankumar, S. (Eds.), *Morphometry and Phylogeography of Honeybees and Stingless Bees in India*, Publication Centre, Directorate of Extension, University of Agricultural Sciences, Dharwad, India (2013). pp. 101–123.
Liu, Z.Y., Wang, Z.L., Wu, X.B., Yan, W.Y., & Zang Z.J. "*csd* alleles in the red dwarf honey bee (*Apis florea*, Hymenoptera: Apidae) Show exceptionally high nucleotide diversity." *Insect Science*18 (2011): 645–651.
Mackensen, O. "Viability and sex determination in the honey bee (*Apis mellifera* L.)." *Genetics* 36 (1951): 500–509.
Meznar, E.R., Gadau, J., Koeniger, N., & Rueppel, O. "Comparative linkage mapping suggests a high recombination rate in all honeybee." *Journal of Heredity* 101 (1) (2010): 118–126.
Nie, H., Geng, H., Lin, Y., Xu, S., Li, Z., Zhao, Y., & Su, S. "Genome wide identification and characterization of Fox genes in the honeybee *Apis cerana* and comparative analysis with other bee Fox genes." *International Journal of Genomics* 2 (2018): 1–12.
Oldroyd, B.P., Smolenski, A.J., Cornuet, J.M., & Crozier, R.H. "Anarchy in the beehive." *Nature* 371 (1994): 749.
Raffiudin, R., & Crozier, R.H. "Phylogenetic analysis of honey bee behavioral evolution." *Molecular Phylogenetic Evolution* 43 (2007): 543–552.
Rembold, H. "Royal jelly." In Ruttner, F. (Ed.), *Queen Rearing: Biological Basis and Technical Instruction Chapter 2*, Apimondia Publishing House, Bucharest (1983). pp. 35–42.
Roberts, W.C., & Mackenson, O. "Breeding improved honeybees." *American Bee Journal* 91 (1951): 292–294, 328–330.

Rueppell, O., Kuster, R., Miller, K., Fouks, B., Correa, S.R., Collazo, J., Phaincharoen, M., Tingek, S., & Koeniger, N. "A new metazoan recombination rate record and consistently high recombination rates in the honeybee genus *Apis* accompanied by frequent inversions but not translocations." *Genome Biology and Evolution* 8 (12) (2016): 3653–3660.

Seeley, T.D., Seeley, R.H., & Akratanakui, P. "Colony defense strategies of the honeybees in Thailand." *Ecological Monographs* 52 (1982): 43–63.

Takahashi, J., Deowanish, S., & Okuyama, H. "Analysis of the complete mitochondrial genomes of two dwarf honeybee species, *Apis florea* and *Apis andreniformis* (Insecta: Hymenoptera: Apidae) in Thailand." *Mitochondrial DNA Part B* 3 (1) (2018): 350–353.

Tan, H.W., Liu, G.H., Dong, X., Lin, R.Q., Song, H.Q., Huang, S.Y., Yuan, Z.G., Zhao, G.H., & Zhu, Z.Q. "The complete mitochondrial genome of the Asiatic cavity-nesting honeybee *Apis cerana* (Hymenoptera: Apidae)." *PLoS One* 6 (2011): e23008.

Terzi, B., Koca, A., Gharkheloo, M.M., Sozen, M., & Kandemir I. "Genetic variation of dwarf honeybee (*Apis florea* Fabricius) populations distributed in the western parts of Iran based on RAPD analysis." *Uludag Bee Journal* 14 (1) (2014): 4–14.

von Frisch, K. *Tanzsprache und Orientierung der Bienen* (1965). Berlin, Germany: SpringerVerlag GmbH.

von Planta, A. "Ueber den Futtersaft der Bienen. Hoppe-Seyl Zeitschr." *Physiological Chemistry* 12 (1888): 327–354.

Vosshall, L.B., Wong, A.M., & Axel, R. "An olfactory sensory map in the fly brain." *Cell* 102 (2000): 147–159.

Wang, A.R., Kim, M.J., Park, J.S., Choi, Y.S., Thapa, R., Lee, K.Y., & Kim, I. "Complete mitochondrial genome of the dwarf honeybee, *Apis florea* (Hymenoptera: Apidae)." *Mitochondrial DNA* 24 (3) (2013): 208–210.

Whiting, P.W. "Selective fertilization and sex-determination in Hymenoptera." *Science* 78 (1933): 537–538.

Whiting, P.W. "Multiple alleles in complementary sex determination of *Habrobracon*." *Genetics* 28 (1943): 365–382.

Woyke, J. "Expression of body colour patterns in three castes of four Asian honey bees." In Mardan, M. (Ed.), *Proceedings of the International conference on Tropical Bees and the Environment*, Beenet Asia, Serdang (1997). pp. 85–95.

Yang, J., Xu, J., He, S., & Wi, J. "The complete mitochondrial genome of wild honeybee *Apis florea* (Hymenoptera: Apidae) in south western China." *Mitochondrial DNA Part B* 2 (2) (2017): 845–846.

5

Foraging in Apis florea

Dharam P. Abrol

CONTENTS

Survival of any living organism depends upon the constant supply of food that is acquired from the environment. The habitat of most animals in which they live may not provide sufficient resources essential for their survival close by, and they have to forage to locate these resources to bridge this gap on temporal or spatial distance (Schoener, 1971). The various tasks performed by any animal for acquisition of these resources can be defined as foraging. The information on foraging strategies is essential to understand community structure as well as to develop conservation strategies. The foraging behaviour of social insects is interesting, as the individuals do not forage to meet their own nutritional needs; rather, they forage to meet the needs of the colony. Interestingly, honeybees have quite advanced and sophisticated foraging coordination and communication mechanisms that help them in collection of resources necessary for survival (von Frisch, 1971; Suwannaong, 2000). Foraging in honeybees is a social enterprise (Seeley, 1985) in which honeybees collect nectar, pollen water and propolis from plants. This activity is performed by foraging bees or foragers that collect nectar and pollen from flowers, the former being a source of proteins and the latter of carbohydrates. The plants providing these two substances are collectively called bee pasturage. The other materials collected include water and propolis as per the needs of the colony. The foraging behaviour is an important aspect in the life of bees that helps them to adapt to available vegetation and climatic conditions. Once initiated, foraging continues to depend upon climatic conditions and availability of forage, and each trip may last for an hour or more. The foraging efficiency depends upon quantity and quality of available forage, weather conditions and distance from resources.

Introduction

Dwarf honeybees, especially *Apis florea* F., are very small in body size and valued more as pollinators than honey producers (Muttoo, 1965; Dutton and Free, 1979; Free, 1981; Singh, 1982; Gubartalla, 1997) as compared to other honeybee species *Apis dorsata, A. mellifera* and *A. cerana*. They are economically important in agriculture, medicine and the environment, and they are excellent pollinators of crops and other plants (Free, 1981; Wongsiri et al., 1991; Wongsiri et al., 2002). The dwarf honeybee is confined to warm climates where it does very well in extremely hot, arid environments (Suwannapong et al., 2011). The honeybees have a wide foraging range and are excellent pollinators of various economic crops and forest plants (Wongsiri and Tangkanasing, 2000; Wongsiri et al., 2000; Tahmasebi et al., 2002). *Apis florea* is known to have a wider distribution in many tropical and subtropical parts of Asia, especially in Southeast China, India, Burma, Laos, Vietnam, Malaysia, Indonesia and the Philippines (Suwannapong et al., 2011). *Apis florea* is generally found in plains or lowlands.

The dwarf honeybee is able to survive in very hot, dry and humid climates. The dwarf bees perform a short forage flight that is often up to a distance of 750 m so the honey stored by it is generally unifloral. The bees have successfully exploited a diversity of habitats thereby avoiding direct competition with other insects or predators. Morphologically, dwarf honeybees are smaller as compared to other bee species but have proven to have unexpected potential in maintaining ecological balance.

Foraging Behaviour

Apis florea has been recorded as versatile pollinators of a wide variety of field and fruit crops *that* varies in time and space. Abrol (1992a) reported that bees with higher energy requirements do not forage on the flowers providing low caloric rewards. The flowers of *F. vulgare, C. sativum, D. carota, A. cepa, Trigonella foenumgraecum* var Kasuri and *Mangifera indica* were highly attractive to *A. florea*, which was observed in large members. The bee with its smaller size and body weight was morphologically better adapted to extract maximum caloric reward from the flowers. Sihag and Kapil (1983) reported that *Apis florea* visited *Brassica* crops in larger numbers than *A. dorsata*. They attributed this difference to the different energy demands of the bees. *A. florea* visited flowers with low caloric rewards whereas *A. dorsata* preferred those with high rewards. Generally, the smaller flowers with little nectar are unattractive to large hovering animals such as hummingbirds and the sphinx moth, which probably cannot suffice their energetic demand (Heinrich, 1972a). Because the energy expended in walking from one flower to another is 100 times less than an equivalent period of flight and because the flowers can be visited in rapid succession (Heinrich and Raven, 1972 a,b,c).

Abrol (1998b) studied the foraging behaviour of *Apis florea* and *A. dorsata* in relation to the energy production pattern of four plant species. The energy produced ranged from 0.35 joules/flower/day in *Lantana indica* to 1031 joules in *Adhatoda vasica*. However, flowers of *A. vasica*, which are structurally more complex, were less preferred than those of the other three species. The bees preferred to forage on *Brassica juncea*, followed by *B. campestris* var. *toria*, from which they could harvest more energy per unit time. Abrol (2006) observed the foraging behaviour of four honeybee species (*Apis mellifera A. cerana, A. dorsata* and *A. florea*) in relation to energy production rates of *Prunus persica* and a simultaneously blooming weed, *Lepidagathus incurva*. The weed having higher nectar sugar concentration and high flowering density attracted more numbers of bees as compared to peach. The foraging rates of bees were much higher on the weed, and they could harvest more energy per unit time from the weed flowers. The results indicate that food acquisition efficiency and quality of food determine the foraging decision of bees. Bees are generally attracted to flowers that are available in large numbers and are more rewarding in terms of energy (Abrol, 1993). Sihag and Rathi (1992) reported that *Apis florea* populations were more attracted to *Brassica compestris* flowers that provided fewer caloric rewards than *Cajanas cajan*, which had high caloric rewards. The differential foraging resulted from the fact that bees could easily harvest more energy in a unit time from simple flowers of *Brassica compestris* compared to *Cajanas cajan* in which flowers are more complex, tunnelled and papilionaceous, making the visitor approach much more complex. Earlier, Sihag (1990) observed similar behaviour of two honeybees species, *Apis florea* and *Apis dorsata*, on *Brassica chinensis* and *Eruca sativa* wherein the flowers of *Brassica chinensis* were preferred.

Foraging Preference of Bees in Relation to Nectar-Sugar Components

Sihag and Kapil (1983) found that *Apis dorsata* with higher energy requirements visited sucrose-dominated flowers more frequently than did *Apis florea*, which mostly relied on glucose-dominated ones. Wyke (1952) reported that honeybees preferred equiproportioned sugars. Waller (1972) and Abrol (1985), on the contrary, stated that bees prefer nectar with one dominant sugar than the equiproportioned sugars. Comparative studies on foraging behaviour of honeybees and solitary bees on 13 crops (Abrol and Kapil, 1991) in relation to nectar sugar components revealed that honeybee *Apis dorsata* and solitary bees *Megaclule lanata* Lepel, *M. cephalotes* Smith and *Xylocopa fenstrata* F. visited sucrose-dominated nectars more frequently than did *Apis florea* F. and *Pithitis smargdula* F., which relied heavily on glucose-dominant nectars. Preference of sucrose or glucose-dominant nectar indicates that total caloric reward is an important factor determining the attractiveness of a foraging population on a particular crop (Abrol and Kapil, 1991).

Foraging Is Genetically Controlled

Oldroyd et al. (1994) found that different tasks performed by *Apis florea* such as liquid foraging, pollen foraging, guarding, stinging, fanning and wagging abdomen are genetically based as revealed by RFLP banding. They found that patterns of fingerprints were dissimilar in all groups observed for different behaviours.

Competition between *Apis florea* and *Apis mellifera* on Sugar Syrup

Mogga (1994) found that *A. florea* in Sudan foraged actively between 09.00 h and 14.00 h while *A. mellifera* did that between 05.00 to 08.00 h and 15.00 to 18.00 h. Mohamed (2000) stated that *A. florea* started foraging in Sudan between 08.00 and 09.00 h while *A. mellifera* between 05.00 and 06.00 h. Gubartalla (1997) showed that *A. florea* started at 05.00 h and 07.00 h in summer and winter, respectively. Mohamed (2000) found *A. florea* more abundant on flowers than *A. mellifera*. According to Mossadegh (1993), *A. florea* workers can forage all day long even during the hot months because they are adapted to fly at higher temperatures and can imbibe large amounts of concentrated nectar and can collect pollen grains that *A. mellifera* cannot harvest. They can also forage on food sources such as honeydew or pollen grains, which are not attractive to *A. mellifera*. These studies indicate that there is a spatial and temporal partitioning between the two species thereby reducing the competition.

Morphologically, the two species are different and within the same area *A. florea can* exploit flowers that larger bees cannot because of their body size (Punchihewa et al., 1985). Because of the small size of flowers, *A. florea* was more abundant on carrot (*Dacus carota)* compared to *A. mellifera* (Ahmed and Aslam, 2002) and *A. mellifera sudanensis* (El Shafie et al., 2002). *A. florea* is a very active competitor against *A. mellifera* on food sources. This is because of the fact that *A. florea* has fewer food requirements compared with *A. mellifera.*

Aggressive Behaviour of Apis florea *towards* Apis mellifera

Chahal et al. (1986) observed robbing of *Apis mellifera* colonies by *Apis florea* bees in Punjab, India. They reported that relatively weak colonies of *Apis mellifera* were robbed and disorganized. Observations under simulated conditions indicated that *A. florea* bees behaved aggressively towards *A. mellifera.* Although casualties occurred in both species during robbing, *A. florea* bees stung *A. mellifera* bees but avoided being stung by the latter. *A. florea* is less aggressive as compared with *A. mellifera* and other bee species (Dutton and Free, 1979; Ali, 1996; Gubartalla, 1997; Pirk et al., 2002). However, this docile behaviour of *A. florea* could be a disadvantage because the bees will be more vulnerable to attack by honey hunters.

Defensive Behaviour of Apis florea

Honeybees are known to defend their colonies by showing several reactions such as wing shimmering, hissing and unprovoked massive attacks, but these responses vary from species to species with alarming stimulus (Casey, 1976). Ruttner (1988) reported that if the sheltering leaves were removed, artificially or during the dry season, an *Apis florea* colony would abscond. Pirk et al. (2002) reported that colonies of *A. florea* were very 'gentle' compared to other honeybee species but exhibited clear-cut changes of behaviour when disturbed or alarmed. *Apis florea* colonies are so sensitive that even the close presence of an observer and or intruder induced the outermost layer of the bee curtain to react by flexing their wings, shimmering and hissing (Akratanakul, 1977). Fuchs et al. (2001) stated that *A. florea* shimmered and hissed when disturbed. Pirk et al. (2002) reported that stimulus significant changes in body posture and orientation occur, as the bees were found to expose their abdomens and pack more closely together on the comb.

Effect of Repellents on Foraging Behaviour

Indiscriminate use of pesticides has been a potential danger to the ecological system including the survival of bees. Mass-scale mortality of honeybees due to the widespread use of insecticides seriously affects the pollination of crops and other plants (Shaw, 1941; Stute, 1957; Anderson and Atkins, 1958; Palmer-Jones, 1958; Bye et al., 1959; Chamberlain, 1959; King, 1959; Wafa et al., 1960; Morse, 1964; Johansen, 1977; Mayer et al., 1980; Gupta, 1982; Sanford, 1983; Gupta, 1986a, 1986b). Protection of bees can be obtained through the use of a repellent individually or in combination with pesticides (Bhardwaj, 1974; Kumari, 1976; Goyal, 1977; Gupta, 1982, 1983; Gupta and Kapil, 1984; Gupta, 1985; Gupta, 1986a, 1986b). Gupta (1987) tested three ketones (methyl propyl ketone, benzyl methyl ketone and methyl pentyl ketone) on onion (*Allium cepa*) and three (methyl pentyl ketone, ethyl propyl ketone and ethyl pentyl ketone) on carrot (*Daucus carota*) under field conditions for their repellent action against *Apis florea* F. Field experiments clearly demonstrated that the duration of stability of ketones is greater on *D. carota* than on *A. cepa*. The effect of repellents decreased with the passage of time, clearly indicating that temperature is the main factor influencing repellent effect whereas relative humidity acts as a balancing factor.

Foraging in Relation to Environmental Factors

The physical environment influences the flower visiting, foraging behaviour and effectiveness of pollination in complex ways. Foraging activities of pollinating insects are under the key control of different environmental variables. The European honeybee (*A. mellifera*) and the Himalayan species (*Apis cerana*) have done better at exploiting temperate climates than such other species as the Asian little honeybee (*Apis florea*) and giant honeybee (*Apis dorsata*).

The capability of *A. florea* to work under extreme climatic conditions and to survive in very hot, dry climates is one of the species' outstanding features (Ruttner, 1988). It tolerates temperatures over 50°C in the Persian Gulf (Qinglan and Corlett, 2016). Both the species differed in their responses to climatic conditions, *A. florea* abundance had a positive correlation, while the presence of *A. andreniformis* was positively related to temperature and humidity.

Whitcombe (1984) compared the foraging behaviour of *A. florea* with *A. mellifera* and found that temperature tolerance of the former species was much higher than the latter. However, *A. florea* started morning activities late as compared to *A. mellifera* and needed a minimum threshold of 18°C as compared to *A. mellifera*, which commenced at 10°C. The foraging activity peaked at 30°C–32°C in *A. florea* and 22°C in the case of *A. mellifera*. This behaviour pattern helped the beees to avoid possible competition at floral resources. During the winter, brood rearing in *A. florea* colonies was significantly reduced, possibly due to the colonies' poor thermoregulatory capabilities, and brood rearing in summer continued despite floral dearth periods (Dutton and Simpson, 1977; Dutton and Free, 1979; Williams, 1979). *A. florea* forages even at 43°C when all other honeybee species either shun foraging or shift to collection of water (Sihag, 1984, 1991). Foraging in *A. florea* did not commence till the temperature threshold of 18°C was attained; the foragers were active at 34°C. The peak of activity of *Apis florea* was obtained during the later part of the day thus reducing competition with *A. mellifera* for floral sources. Jadhav (1982) reported that on the day of the solar eclipse both *A. florea* and *A. dorsata* stopped foraging at 1540 h and did not resume activities even when the eclipse was over at 1650 h. The activity of *A. florea* was much influenced by temperature while as that of *A. dorsata* light intensity. Sihag and Abrol (1986) reported that the flight activity of *A. florea* was positively correlated with air temperature, light intensity, solar radiation and nectar sugar concentration and negatively with relative humidity. Of all the factors studied, relative humidity and solar radiation were found to influence flight activity directly.

Absconding and Swarming

The tendency of *Apis florea* to abscond when disturbed and the low yield of honey per colony are the two properties that unlike *mellifera*, *cerana* and *dorsata*, do not qualify it as a major honey source though its honey is widely esteemed (Muttoo, 1956; Crane, 1975). Little is known about the swarming process in *Apis florea* (Free, 1981). Though in Oman, large colonies have been reported to produce up to eight swarms a year (Dutton and Free, 1979). Akratanakul (1977) reported that they form an 'interim' reproductive swarm 20 m from the parental nest. Absconding behaviour of *Apis florea* is also poorly understood. It is not yet clear whether swarming and absconding by *Apis florea* extend to long-distance migration as in *dorsata* (Koeniger and Koeniger, 1980) or are purely a local phenomenon. Thakar and Tonapi (1962) reported that *Apis florea* is highly migratory and periodically migrates between plains and adjacent low hills according to seasonal variations in forage availability. The bees are essentially adapted to plains up to 1500 or 2000 ft and may be rarely seen migrating up to 4000 to 5000 ft in hills. The other authors reported that local movements of less than 500 m do occur (Tirgari et al., 1969; Tirgari, 1971). The possible causes for absconding include dearth of available forage, attack of enemies and temperature changes (Free, 1981b).

A. florea is highly migratory, but long-distance migrations are unknown. This honeybee also has a habit of nest shifting within short distances. The bees may shift within two months in the dry season and within five months in the rainy season. When they shift, they carry away all the provisions, including wax. When disturbances cause them to desert the comb, then they leave behind honey, brood and pollen stores. But the bees continue to return to old combs to take away comb wax, honey and pollen.

Lindauer (1957) and Tirgari (1971) also recorded frequent swarming behaviour of *A. florea* throughout the year. The swarming during May and June may be due to floral scarcity that triggers the absconding by colonies in order to secure rich food sources. Similar observations were reported by Tirgari (1971). *A. florea* continues to abscond throughout the year for different reasons, which may be due to nonavailability of food resources, attack of pests and diseases and/or disturbance caused by honey hunters (Morse and Benton, 1967). Several other investigators have also recorded absconding behaviour of *A. florea* (Sihag 1982) such as Ruttner et al. (1985a), Free (1994) and Gubartalla (1997). Swarming of *A. florea* occurs when the mean depth of the comb is 25 cm and the mean width is 17.7 cm. Absconding occurs at comb depth of 28.5 cm and comb width of 21.1 cm. This shows that florea comb dimensions (depth and width) at absconding were partially larger than at swarming. Thus, beekeepers can predict the occurrence of swarming and absconding by measuring the depth and width of the combs.

Commencement and Cessation of Foraging Activity

The data on foraging activity of different bee species showed that commencement and cessation of foraging activities of different honeybee species were related to their specific ecological threshold to environmental factors in relation to their physiology and physiology of host plants.

Each bee species was found to have its specific ecological threshold below which activity does not occur (Osgood, 1974; Szabo and Smith, 1972; Reddy, 1979; Lerer et al., 1982; Butler, 1954). The time of commencement of bee activity varies from one day to another, which depends upon attainment of minimum threshold conditions for their foraging activity. Abrol (2006) studied the foraging behaviour of four honeybee species – *A. dorsata F*, *A. mellifera* L, *A. cerana* F. and *A. florea* F. – on litchi flowers (*Litchi chinensis* Sonn.) in relation to weather factors and found that 16°C temperature for *A. dorsata* and *A. mellifera*, 15.5°C for *A. cerana* and 18.5°C for *A. florea* appeared to be the minimum threshold conditions for initiation of flight activity. Cessation of activities in all the honeybee species was governed mainly by a decline in values of light intensity and solar radiation irrespective of other factors. Abrol (2006) studied the foraging behaviour of honeybee *Apis. florea* F. on carrot (*D. carota*) flowers and found that flight activities of *Apis florea* commenced between 07.26 and 07.44 hrs in the morning when temperature ranged between 19.4°C and 25°C. Cessation of activities was governed mainly by a decline in values of light intensity (500–1300 lx) and solar radiation (10–43 mW/cm^2), which were appreciably low at cessation than at the commencement. Abrol further found that temperature, light intensity and solar radiation were the three most important factors that exerted a pronounced and positive direct effect on flight activities of *Apis florea* while relative humidity had a negative direct effect. Direct effect of nectar sugar concentration was positive and negligible.

Abrol (1985) found that *A. dorsata* was physiologically more tolerant to large fluctuations in ambient temperature than *A. florea*, which restricts its activity during unfavourable weather conditions. Cold weather conditions and wind seriously constrain the foraging activity of *A. florea*. It has been observed that when a small shower of rain in winter results in cooling of the atmosphere, the foraging activity of *A. florea* was not resumed for many days even when the temperature was above the threshold level of 15°C. This may be due to poor thermoregulatory capabilities of this species. Lindauer (1957) reported that habits of *A. florea* are limited by climate conditions, especially by cold. The bee worker is active at a higher temperature where it has to expend less energy to raise thoraces temperature to the level required for flight. Abrol and Kapil (1987) found that *Apis florea* can forage at extremely high temperatures at 43°C and utilize highly concentrated nectar of 68.00%. Abrol (1985) found that *Apis florea* were found to be very sensitive to variations in ambient temperature than *A. dorsata*. This implies that *A. florea* is adapted to work at relatively higher temperatures (43°C) and is limited by fluctuations in lower ambient temperatures. It is probably advantageous for the small bees to work at extremely higher temperatures because they can efficiently radiate enough heat to the environment by their larger surface area. Similar observations were made by Willmer (1983) who reported that small lightly coloured species visited fully isolated flowers and were highly correlated with solar radiations.

Abrol (2006b) studied the foraging ecology of the dwarf honeybee *Apis florea* visiting carrot flowers (*Daucus carota*) in relation to five environmental variables and found that the direct effect of temperature was high and positive followed by light intensity and solar radiation. The direct effect of relative humidity was high and negative. However, the direct effect of nectar-sugar concentration was negative and negligible. Gubartalla (1997) found that *A. florea* was more abundant on flowers of broad bean than other pollinators but was less on sunflower. Ahmed and Aslam (2002) found that among the bees, *A. florea* was more abundant compared to *A. mellifera* on carrot (*Dacus carota*) flowers. Gubartalla (1997) observed that the water collection activity of *A. florea* started early in the morning, then increased gradually till midday and then steadily declined thereafter.

Thermoregulation in *A. florea*

A. florea as an open-air nesting species is more directly exposed to different ambient and seasonally fluctuating conditions than the hive bee *A. mellifera*, the colonies of which are well protected within the hives or tree trunks etc. Ruttner et al. (1985b) reported finding colonies of *A. florea* up to an elevation of 850 m above sea level in Iran. The colonies could survive at temperatures that fell to below +5°C. Mossadegh (75) reported that in Khuzestan, southwest Iran, *A. florea* colonies maintain a central temperature of 34.5°C–36°C while rearing broods in spring, during the summer (ambient temperature of up to 55°C) and up to mid-autumn.

The smallest species, *A. florea*, shows the smallest T_{th}–T_a, but the intermediate-sized *A. cerana* and *A. mellifera* both show a consistently higher T_{th}–T_a than the largest species, *A. dorsata*. The workers of the Asian honeybees range in mass from 118mg for *A. dorsata*, 44 mg for *A. cerana*, 23 mg for *A. florea*. *Apis cerana*, like *A. mellifera*, builds nests consisting of 5–7 parallel wax combs suspended from the roof of an enclosed cavity in a tree or in rocks. *Apis florea* and *A. dorsata*, however, build nests in the open, each one consisting of a single comb that is protected from predators and fluctuations in climatic conditions by a thick curtain of interlinked bees. The small nests of *Apis florea* are hidden in dense vegetation, while enormous nests of *A. dorsata* (1–1.5 m^2) are suspended from heavy branches, overhanging rocks or buildings.

Thermoregulatory capabilities of *Apis florea* are strongly dependent upon prevailing ambient temperature. *Apis florea* shows a strong dependence of Tth upon Ta, and this species' Tth–Ta of 4°C–5°C at Ta = 25°C–30°C would mean that it is more affected than *A. dorsata* by low temperature. *A. florea* colonies in the morning indicated that foragers do not leave the nest until Ta~20°. However, some foragers could show departures at cooler Ta values, at Ta as low as 19°C. All three Asian species forage at temperatures as high as 40°C, but the lethal body temperature is 46°C both in the case of *A. cerana* and *A. dorsata*, but not in *A. florea*, which would either stop flying to cool off, as bumblebees do (Heinrich, 1975), or, more likely, to dump heat in flight, as *A. mellifera* does (Heinrich, 1979a, 1980a, 1980b).

Diurnal Trends in Foraging of *Apis florea*

Rao and Suryanarayana (1990) studied the foraging activity of different honeybee species. Foraging activities of *Apis florea* generally started at late hours as compared with other honeybee species such as *A. cerana indica*, *A. dorsata* and *A. c. indica* (Panda et al., 1993). *Apis florea* continued nectar and pollen collection from 0700 to 1800 hours. Mogga (1994) studied the comparative foraging behaviour of *A. mellifera yemenitica* and *A. florea*. *A. m. yemenitica* was found to forage much before the initiation of flight activities of *A. florea*. It started activities before dawn, reaching a peak at 500 hours; however, activities of *A. florea* were lower at 500 hours and increased thereafter reaching a peak at 800 hours. Palanichamy et al. (1995) found that the little bee *A. florea* was more active at about 1000 h and less at about 1600 h. Similarly, Sihag and Khatkar (1999) found that foraging activities of *A. florea* were very low in the morning and subsequently increased, reaching a peak at 1100 and 1300 hours.

Chandrasekhar and Sattigi (2006) found that the foraging activity of *A. florea* commenced at 0800 h and continued throughout the day; however, peak was observed at 10:00 a.m. onwards up to 14.00 p.m. Earlier, Verma and Phogat (1994) observed the peak foraging activity of *A. florea* between 8000 to 1000 h. Gubartalla (1997) also reported that *A. florea* scored its maximum morning activity at 10.00 h, while *A. mellifera* was at 08.00 h. The maximum summer activities were recorded at 31°C and 35°C for *A. mellifera* and *A. florea*, respectively. However, Whitcombe (1984) obtained maximum foraging activities at 22°C and 31°C for *A. mellifera* and *A. florea*, respectively, in Oman. It could be concluded that *A. florea* tolerates higher temperatures than *A. mellifera*. Similar results were obtained by Whitcombe (1984) and Ruttner (1988). Midday activities for both species were almost devoted for water collection.

Mogga (1994) and Whitecombe (1988) reported that during winter *A. mellifera* started foraging at 17°C (at 08.00 h), although *A. florea* couldn't forage below 20°C. Thus, *A. mellifera* starts foraging earlier and tolerates lower temperatures than *A. florea*. Winter peaks of activity were recorded at 10.00 a.m. and 14.00 p.m. for *A. mellifera* and *A. florea*, respectively. Summer peaks of activity occurred at 08.00 h for *A. mellifera* and at 10.00 h for *A. florea*. This means that there were two hours between *A. mellifera* and *A. florea* activities during summer and four hours during winter. The two bee species stopped foraging at sunset (6:00 p.m.), which could be due to reduced light intensity. Foraging activity of *A. florea* increased with an increase in temperature during May. However, an increase in temperature leads to a decrease in foraging activity of *A. mellifera*. In contrast, an increase in temperature during January led to an increase in activities of both bee species.

Whitcombe (1984) observed that *A. mellifera* tolerates higher wind speed and higher relative humidity than *A. florea*. However, both species may stop foraging at high wind speed (more than 15 km). Temperature could be considered the limiting factor for bee foraging activity. *A. florea* forages for water throughout the day in summer. Similar results were obtained by Gubartalla (1997) and Mohamed (2000).

Foraging in Relation to Different Crops

Apis florea prefers certain crops (cruciferae) having small flowers that provide low energy, which are unsuitable for large bees such as *A. mellifera* (Whitcombe, 1984). *A. florea* has been reported to show less preference for sunflower as compared to *A. mellifera* (Punchihewa et al., 1985; Mossadegh, 1993; El Shafie et al., 2002). However, *A. florea* was dominant on rocket and broad beans. It could be concluded that bee preference to some flowers is related to the size of flower and size of bees. *A. florea* dominated over other pollinators on onion flowers throughout the day. Similar findings were made by Sihag (1982) who also described *A. florea* as the most important pollinator of *Allium cepa*. Ahmed and Aslam (2002) also recorded a maximum number for *A. florea* as compared to *A. mellifera* and *A. cerana* on carrot flowers. El Shafie et al. (2002) and Mogga (1994) also obtained similar results. However, Chandel et al. (2003) considered *A. florea* less active than *A. dorsata* on onion flowers.

Apis florea bees are the most important pollinators of field crops in the plains of India and Pakistan. Atwal (1970) reported that among all the insects visiting *Brassica campestris*, *B. juncea* and *Medicago sativa* flowers, *A. florea* was the dominant visitor and comprised of 73%–84%. Sihag (1986) reported *A. florea* among the most efficient pollinators of Umbelliferae, *Allium* and *Brassica* sp. *A. florea* is an excellent pollinator of orchard and field crops, and its colonies are frequently harvested by bee hunters for their brood as well as honey. The dwarf bee is an important pollinator of crops growing in hot and dry agricultural areas of India. The species has a very short flight range that usually does not extend 500 m. *Apis florea* foragers visit a broader spectrum of plant species than the larger ones. Mohana Rao and Suryanarayana (1990) reported that on niger *A. florea*, *A. dorsata*, and *A. cerana* constituted 51.5%, 29.5% and 10.3%, respectively. Panda et al. (1995) reported that on niger *A. mellifera* started foraging at 17°C (at 08.00 h), although *A. florea* couldn't forage below 20°C. Thus, *A. mellifera* starts foraging earlier and tolerates lower temperatures than *A. florea*. In winter, both species have only one peak of foraging activity as compared to two peaks in summer. Winter peaks of activity were recorded at 10.00 h and 14.00 h for *A. mellifera* and *A. florea*, respectively. Summer peaks of activity occurred at 08.00 h for *A. mellifera* and at 10.00 h for *A. florea*.

Foraging in Relation to Nectar Collection

Nectar is the most important floral reward offered by plants to attract pollinators. Nectar variation can determine the preference of pollinators for certain types of nectars over others (Van Riper, 1958; Hainsworth and Wolf, 1976; Stiles, 1976; Pyke and Waser, 1981; Tamm and Gass, 1986; Alm et al., 1990; Erhardt and Rusterholz, 1998). Panda et al. (1993) in Orissa found that the niger crop was visited by six species of bees. *A. florea* was found to be the major visitor, predominant nectar gatherer and most efficient pollinator among the three *Apis* species observed. Maximum bee activity in all bee species was recorded between 11.00 and 12.00 hr. Foraging in *A. dorsata* differed from other bee species, as it spent the least time per flower compared to other *Apis* species.

Foraging Preference of Bees in Relation to Nectar-sugar Components

Sihag and Kapil (1983) found that *Apis dorsata* with higher energy requirements visited sucrose dominated flowers more frequently than did *Apis florea* which mostly relied on glucose dominated ones. Wyke (1952) reported that honeybees preferred equiproportioned sugars. Waller (1972) and Abrol (1985) on the contrary stated that bees prefer nectar with one dominant sugar than the equiproportioned sugars. Abrol and Kapil (1991) studied the foraging strategies of honeybees and solitary bees in relation to nectar sugar components of 13 crops plants and found that honeybee *Apis dorsata* and solitary bees *Megaclule lanata* Lepel, *M. cephalotes* Smith and *Xylocopa fenstrata* F. visited sucrose dominated nectars more frequently than did *Apis florea* F. and *Pithitis smargdula* F. which heavily relied on glucose dominant nectars. Preference of sucrose or glucose dominant nectar indicates that total caloric reward is an important factor determining the attractiveness of foraging population on a particular crop (Abrol 1989; Abrol and Kapil, 1991).

Foraging in Relation to Pollen Collection

Foraging patterns showed an inverse relationship of pollen-gathering activity with nectar gathering and non–foraging activity (Reddy, 1980a). At Pune, *A. cerana* and *A. florea* commenced foraging activity on onion between 0900–0930 h and 0815–0840 h, respectively, and continued till 1830 h (Rao and Lazar, 1980). Dhaliwal and Atwal (1986) reported that the foragers of *A. florea*, *A. dorsata*, and *A. mellifera* foraging on alfalfa (*Medicago sativa* L.) carried 60.0%, 60.0% and 66.6% pure pollen loads. A change from pollen to nectar foraging was more common than vice versa. Dhaliwal and Atwal (1986) also observed this type of behaviour in *A. mellifera, A. dorsata* and *A. florea* foraging on alfalfa *M. sativa*. Foraging patterns showed an inverse relationship of pollen-gathering activity with nectar-gathering and non–foraging activity (Reddy, 1980a). Pollen-gathering activity of *A. cerana indica* in Delhi was reported by Bisht and Pant (1968). The bees collected pollen throughout the year, and the maximum collection was from January to March. But in the Shimla Hills of the north Himalayas, this bee collected pollen and nectar throughout the year (Mattu and Verma, 1985). Foraging activity was greater in summer and autumn than in winter and in monsoon periods. The time of greater flight activity varied from season to season. Similar observations were reported by Singh (1981) in Saharanpur, who also found that foraging hours varied with the season of the year. Verma (1983) found that in *A. cerana indica*, the peak of pollen collection in the Jeolikote area occurred between 0800 and 1100 h in February-March.

Proportions of bees in a colony that foraged for pollen, the pollen stores in hive and amount of brood reared followed a similar pattern (Reddy, 1980b). While studying comparative foraging behaviour, Thakur et al. (1982) reported that *A. c. indica* on mustard had considerable activity in the morning, and *Apis mellifera* picked up by 1000 to 1030 h and remained active till late in the evening. At Pune, *A. cerana* and *A. florea* commenced foraging activity on onion between 09:00–09:30 h and 08:15–08:40 h, respectively, and continued till 1830 h (Rao and Lazar, 1980). The peak foraging of *A. cerana* was observed in the afternoon between 1500 and 1600 h. *Apis cerana indica* foragers started foraging on mustard under mid-hill conditions between 09:00 and 10:00 h and stopped foraging beyond 1500 h (Bhalla et al., 1983a; Mattu et al., 1994). But its activity on stone fruits continued up to 1600 h (Bhalla et al., 1983b). Choudary (1978) analyzed 5200 pollen loads; only 56 contained pollen from more than one plant species, which showed high floral fidelity of *A. indica*. Similar observations were reported by Chaturvedi (1973, 1977),

Sharma (1970a, 1970b), and Jhajj and Goyal (1979b). Dhaliwal and Atwal (1986) reported that the foragers of *A. florea*, *A. dorsata* and *A. mellifera* foraging on alfalfa (*Medicago sativa* L.) carried 60.0%, 60.0% and 66.6% pure pollen loads. Jhajj and Goyal (1979c, 1986) also reported that in *A. cerana indica* and *A. mellifera*, the flower constancy decreased with time; after five days only 20% of the original *A. cerana indica* and 28% of *A. mellifera* remained flower constant. A change from pollen to nectar foraging was more common than vice versa. In caged pollination experiments, Mohan (1973) found that bees did not pollinate sunhemp flowers and collected pollen without touching the stigma. Dhaliwal and Atwal (1986a) also observed this type of behaviour in *A. mellifera*, *A. dorsata* and *A. florea* foraging on alfalfa *M. sativa*. The Indian bee *A. cerana indica* collected an average pollen load of 0.019 g from *Brassica juncea*, *Pyrus malus* and Zea mays (Punjabi et al., 1969). Little pollen was collected from maize. The average pollen load from mustard was 8 mg (Naim and Bisht, 1979). Dhaliwal (1970) compared the pollen collection by *A. mellifera* reared in *A. cerana indica* combs or in *A. mellifera* combs. The comb cell sizes were found to considerably affect the pollen-carrying capacities of bee foragers. Jhajj and Goyal (1979a) have also reported reversal of pollen foraging to hive guarding by carbon dioxide anaesthesia.

Devika et al. (2017) found that on squash flowers, *A. dorsata* initiated pollen collection early in the morning from 6:00 h of the day, *A. mellifera* from 0630 h while *A. cerana* and *A. florea* from 0700 h of the day. Pollen collection peaked at 06:00–08:00 h (*A. dorsata*), 08:00–10:00 h (*A. mellifera* and *A. cerana*) and 10:00–12:00 h (*A. florea*) of the day. In earlier studies, a number of investigators have reported that *A. florea* works at relatively higher temperatures (18°C) compared with other honeybee species and initiates forging activities later than other honeybees (Abrol, 2007). Other honeybee species such as *A. dorsata*, *A. cerana* and *A. mellifera* collect pollen during the early hours of the day when nectars are diluted due to high humidity and dewdrops on flowers. In contrast, *A. florea*'s activities start during the later part of the day to collect nectar and pollen simultaneously. Interestingly, *A. florea* continues to collect nectars from small exposed flowers with very small and concentrated nectars.

Water Collection

Apis florea forages for water throughout the day in summer (Gubartalla 1997; Mohamed, 2000). Gubartalla (1997) stated that the water collection activity of *A. florea* started early in the morning (at sunrise), then increased gradually till midday and then steadily declined towards the end of the day. In general, two peaks were observed, first at noon and the second at 16.00 h. Shwetha (2007) studied the water collection activity of *Apis florea* workers during May at Bangalore and obtained similar results.

Foraging Speed and Foraging Rate

Dhaliwal and Atwal (1986a) observed that on alfalfa (*M. sativa*) the foraging behaviour of honeybees was related with their body weights. But foraging rates were in the order of *A. mellifera* < *A. florae* < *A. dorsata*. On onion blooms, *A. cerana* indica was a faster flier between the umbels than *A. florea*, visiting 1.93 and 1.34 umbels per min, respectively (Rao and Lazer, 1980). But on an umbel, *A. florea* showed a better foraging rate than *A. cerana* visiting 6.53 and 5.93 flowers per min, respectively. Dhaliwal and Atwal (1986a) compared the tripping efficiency of different bees visiting alfalfa (*M. sativa*), where the order of efficiency was *A. dorsata* > *A. mellifera* > *A. florea*. Abrol (2006b) found that *A. florea* visited fewer numbers of flowers/umbels during the earlier and later parts of the day. Maximum flowers/umbel/min were visited between 1200–1400 hours. On average, flowers visited/min ranged between 17.13 and 27.31 during different hours of the day with an overall average of 22.78 flowers/min. Similarly, the number of umbels visited/min ranged between 0.98 and 1.29 with an average of 1.14 umbels/min.

Impact of Solar Eclipse on Foraging

Jadhav (1982) studied the impact of the solar eclipse on foraging behaviour of *A. florea* and *A. dorsata*. He found that *A. florea* and *A. dorsata* stopped foraging activities when the solar eclipse was at its peak and did not resume even after the eclipse was over. Diwan (1980), on the other hand, found that returning foragers of *Trigona iridipennnis* had difficulty in locating the hive.

Foraging in Relation to Chemicals

The dwarf honeybee (*Apis florea*) is widely distributed in southern Asia (Oldroyd and Wongsiri, 2006). Members of this species are very successful colonizers in tropical, subtropical and even in semi-arid and less favourable hostile conditions (Oldroyd and Wongsiri, 2006). *Apis florea* has successfully established in new areas and coexisted with *Apis mellifera* (Lord and Nagi, 1987). It is capable of nesting and foraging in new habitats (Mogga and Ruttner, 1988). It has a very good learning capability as is evidenced by modification of its foraging behaviour by different chemicals (Gupta, 1987c, 1988, 1989b), semi-field (Gupta, 1987a; Mishra et al., 1987a; Naik et al., 2010) and field (Gupta, 1987b; Mishra et al., 1987b), which were found to repel under laboratory conditions.

Species-Specific Variations in Foraging Activity

Abrol (1992a) studied the foraging behaviour of six bee species in relation to 25 crops. *Apis dorsata* with its larger size and higher energetic demands preferred *Cajanus cajan, Parkinsonia aculeata, Pongamia glabra* and *Luffa cylindrica* than flowers of *F. vulgare, C. sativum D. carota, A. cepa, T. foenumgraecum* and *M. indica*. The latter group of flowers was visited by *A. florea* through the day. The latter bee species with its small size and body weight is physiologically better adapted to collect nectar from these flowers. In the case of *Brassica* crops, *Apis dorsata* visited flowers during early hours of the day when nectar production was at its peak. But from11.00 hours onward, foraging population shifted to *T. alexandrium* due to reduction in quantity of rewards available from the *Medicago sativa* and *Brassica* crops. *Apis florea*, which commenced activity between 10:00 and 11:00 a.m., dominated throughout the day. In the case of sunflower, all types of pollinators continued foraging throughout the day. Though the caloric reward per flower is low, each bee species was able to maintain energy balance since flower heads with a platform provide no barrier for landing of the foragers. The energetic cost is reduced due to temporary suspension of hovering flight and a large number of flowers visited.

Foraging Distance

Foraging Range

The foraging range of pollinators provides information regarding their intrinsic capabilities and limitations in gathering nectar and pollen resources. The distances up to which the maximum number of bees can orient from their domiciles and effect pollination are of utmost significance. The foraging range of bees depends largely upon their body size and energy needs, yet it is also determined by the quantity of reserve fuel carried when in flight. The smaller bees with relatively low-energy requirements and little reserve fuel may hardly risk foraging long distances. Therefore, the orienting range of bees from their domiciles makes an important component for investigation Abrol (1988) reported that counts of marked *A. florea* bees were considerably higher up to 150 m from the nesting site but thereafter the numbers declined gradually, none being observed at more than 500 m.

von Frisch (1967) reported that the foraging range of honeybees was up to 13.5 km. Visscher and Seeley (1982) showed that honeybees (*A. mellifera*) regularly forage several kilometres from the nest. The most common distance was 600–800 m. The mean was 2.3 km, and the circle enclosing 95% of the colony's foraging activity had a radius of 6 km. The maximum distance that *A. florea* workers foraged was 101.8 and 100.6 m, and the minimum was 13.6 and 13.4 m; as the food was available near the nesting site, the bees did not forage at longer distances.

Speed of Flight

Apart from various other factors such as abundance of pollinating insects, rate of flower visitation and pollen-carrying capacity, the pollinating efficiency of a bee depends upon its speed of flight and the rate at which it flies from one flower to another. Since wingbeat frequencies express the flight intensity of fliers, its evaluation may help in assessing their pollinating efficiency. In honeybees, the speed of flight has been found to vary from 20.9 to 25.7 km per hour (average 24 km per hour) for loaded bees and from 10.9 to 29.0 km per hour (average 20 km per hour) for empty bees. The speed of flight of honeybees varies from 4.81 ms^{-1} to a maximum of 12.61ms^{-1} and wing loading from 7.28 Nm^{-1} to a maximum of 14.21 Nm^{-1}. Both *A. cerana* and *A. mellifera* exceed *A. dorsata*'s speed in flight. Because a higher flight speed in a smaller bee implies a disproportionately high rate of power output – i.e., a greater mass-specific metabolic rate in flight – it follows that an elevated rate of heat production could account for the higher Tth of *A. cerana* and *A. mellifera* in flight. Wing loading is another functional correlate of power output (Casey, 1976). The two intermediate-sized species, *A. cerana* and *A. mellifera*, have similar high rates of heat production, whereas the smallest and largest species, *A. florea* and *A. dorsata*, have similar low rates.

Size-Related Foraging in Relation to Weather Conditions

The large-sized *A. laboriosa* workers are able to forage at ambient temperatures at least 5°C–6°C lower than the minimum ambient temperature at which *A. dorsata* workers can fly (Dyer and Seeley, 1987, 1991; Underwood, 1991). This has apparently been accomplished through an increase in body size, without a disproportionate increase in thoracic mass, as is the case with *A. cerana*. Activities of *A. florea* normally remain suspended during inclement weather conditions, as the bees are unable to maintain their body temperature due to small size (Abrol, 2006).

Factor-Compensating Mechanisms

Abrol (1985) found that resultant foraging activity in bees may be a consequence of several interacting factors in which value of one factor with higher values may compensate for the lower values of the other factors. He found that *Apis florea* commenced foraging activities where a combination of 19.5°C temperature, 87% relative humidity, 1700 lx light intensity and 64 mW/cm² was attained; however, on another day activity was initiated at higher values temperature when other factors were comparatively less in intensity. This implies that higher temperature can compensate for the lower values of others and this triggers the activity, while in the former case the higher values of light intensity and solar radiation could compensate for the lower values of temperature and result in the initiation of the activity. Abrol and Kapil (1986) reported similar compensating mechanisms operating at the commencement of pollination activity of *Megachile lanata* in which lower values of temperature were compensated jointly by higher values of light intensity and solar radiation and vice versa.

Ecological Threshold of Nectar Secretion and Foraging Activity of Honeybees

Coevolutionary adaptations have brought a close correlation between plants and their pollen vectors in a manner that functional activity of both more often synchronizes. Each bee has its ecological threshold for initiation of field activity until surpassed activity is limited (Abrol and Kapil, 1986). In plants, a characteristic rhythm of nectar secretion occurs with peak periods of nectar production usually synchronizing when the conditions are optimum for bee activity (Corbet, 1978). Abrol (1985) reported that in *Apis* species such as *A. dorsata* and *A. florea* the onset of flight activities occurred at 15°C under subtropical conditions in Hisar during winter months. Abrol (1986b) found that in flowers of *Brassica juncea* the nectar secretion occurred between 12°C and 15°C, thereby indicating 12°C–15°C as a minimum threshold temperature for activation of enzymatic machinery to secrete nectar. This overlap of activity threshold of two different groups makes them interdependent for energy reward and for reproduction and perpetuation of their species.

Sequence and Timing of Bee Visits

The visits of the bees to the flowers depend largely upon the energy requirement reward system and occur in a very predictable and ordered manner (Schlising, 1970; Heinrich, 1975a; Murrell and Nash, 1981). Abrol (1987a, 1987b) found that larger bees such as *A. dorsata, A. mellifera* and *A. cerana* with higher energetic demands and superior thermoregulatory capabilities visited flowers early in the morning when the caloric reward was more and competition with other pollinators was minimal. Whereas, the smaller ones such as *A. florea* with relatively low energy requirements and poor thermoregulatory capabilities visited flowers in the late morning hours, when energy expenditure due to prevailing higher temperature was minimized.

Foraging as a Function of Species-Specific Adaptation

Abrol (1998) found that under similar set of climatic conditions the responses of *Apis cerana* and *Apis mellifera* were absolutely different. *A. cerana* started flying earlier in the morning than *A. mellifera* did, and also at a lower temperature, light intensity and solar radiation and also worked for longer periods. Solar radiation directly affected the activity of *A. cerana*, and solar radiation and light intensity directly affected that of *A. mellifera.* Forbes and Cervancia (1994) also reported similar findings in the case of *Apis cerana* and *A. mellifera* in Majayjay, Philippines. Abrol (2006a) studied the foraging behaviour of four honeybee species – *A. dorsata F*; *A. mellifera* L; *A. cerana* F. and *A. florea* F. – visiting litchi flowers *(Litchi chinensis* Sonn.) and found that all four honeybee species differed in their responses to environmental factors prevailing under a similar set of conditions depending upon physiological adaptation of each honeybee species. Of all the factors studied, temperature, light intensity and solar radiation were the three important factors whose influence on the foraging population was more pronounced.

Competition for Floral Resources

Very recently, information has been obtained as evidence of competition between bee species in natural situations. Morse (1977) noted that workers of two bumblebee species were partitioned along sprays of goldenrod, *Solidago canadensis.* Thus, workers of *Bombus ternarius* avoided competition with the larger *B. terricola* workers (Morse, 1977). Morse (1978) recorded interspecific displacement of bumblebee workers on roses as well. From his data on nectar depletion and resource partitioning in Maine, Heinrich (1976) inferred that four bumblebee species competed for nectar. Chahal et al. (1986) found that *Apis florea* showed aggressiveness towards *A. mellifera* both at hives and on flowers. Interactions among different species revealed that small individuals generally attacked larger ones. *A. dorsata*, however, was attacked only by *A. cerana* workers and not by other species. In the case of *A. florea*, the bees' small size does not deter them from showing aggression towards other *Apis* species at a food source. When foraging at dishes containing sugar syrup, *Apis florea* was as successful (or more so) than *A. cerana* and *A. dorsata* in maintaining its station there and inducing the departure of bees of other species. However, *A. florea* and *A. cerana* workers sometimes forage together at an artificial food source without showing aggression.

Resource Partitioning

In a mutualistic relationship with flowering plants, honeybees gather nectar and pollen from blossoms. All honeybee species do not exploit the same floral resources. This is partly due to their geographical variation because honeybee species or subspecies often live in allopatric ranges that contain different types of vegetation (Schneider and McNally, 1992, 1993). In cases of sympatry, resources may be partitioned in other ways. For example, resources may be partitioned based on size. Sharma et al. (2000) studied resource partitioning in *Apis mellifera* and *A. cerana* and found that of the 23 plant species studied, *Apis mellifera* and *A cerana* avoided competition by visiting the plants in differing intensities. During spring, *A. mellifera* preferred *Rubus ellipticus* (7.85 bees) and *Prunus armeniaca* (7.6 bees) whereas *A. cerana* was more active on *Malus domestica* (9.91 bees). Sihag (2000) reported that *A. florea* starts foraging when the ambient temperature surpasses 18°C and continues foraging until the ambient temperature approaches 43°C. Maximum foraging activity is shown at 30°C–40°C. These ranges are highest than those shown by *A. mellifera* and *A. dorsata.* Body size and tongue length play a predominant role in determining the resource partitioning of the bees. For example, Oldroyd et al. (1992a, 1992b) found evidence suggesting that pollen resources were partitioned among *A. dorsata, A. cerana, A. florea*, and *A. andreniformis* based on bee size, with the larger two species using the richest pollen resources.

El Shafie et al. (2002) studied 21 pollen sources for the honeybee *Apis mellifera sudanensis* and the dwarf honeybee *Apis florea* between early December and late March, and they found some hints of competition between the two species. However, further studies showed that *A. mellifera sudanensis* and *A. florea* do seem to coexist. This coexistence is based on different daily rhythms of pollen collection. *A. mellifera* collected pollen of *Acacia seyal*, date palm and onions early in the morning (and partly in the late afternoon), while *A. florea* started pollen collection mostly later in the morning and ended it earlier in the afternoon. In contrast to *A. mellifera, A. florea* collected pollen all day without interruption, even at very high air temperatures. Niche overlap (concerning the times of visits to flowers) between the two bee species was very low in date palms and of medium importance in *Acacia seyal*. It is remarkable that in total, *A. florea* is always present in higher numbers than *A. mellifera sudanensis* on flowers. In a study of bee behaviour and pollination in onion, Rao and Lazar (1983) revealed that *A. dorsata* never visited the flowers of onion during the entire period of its blooming, while *A. cerana* and *A. florea* were observed collecting nectar.

Foraging in Relation to Flavour, Taste and Colour of Flowers

The recognition of food or nectar sources is influenced by various factors such as colour, flavour and taste. For instance, bees can distinguish green, blue, white, yellow and violet flowers but are insensitive to red flowers. The bees can discriminate between different concentrations of sweetness and varying salt content in flowers. In the case of simultaneously blooming flowers of different concentrations, the bees will prefer those that have a higher sugar content. Once the bees are conditioned to a particular source of nectar or pollen, they continue to visit it till it is exhausted. Abrol (2003) studied foraging behaviour of *Apis florea* on *Portulaca grandiflora* flowers and found that maximum foraging populations were observed on straw yellow flowers followed by dark yellow, purple and red flowers, respectively. Earlier, Faegri and van der Pijl (1971) reported that *A. mellifera* preferred yellow, blue and purple flowers. Sharma et al. (1999) found that pink, red and blue colours attracted more honeybees and bumblebee pollinators; black and green were less effective than the control (white).

Nocturnal Foraging

The truly diurnal bees, *A. florea* and *A. cerana*, have less sensitive eyes but the potential for a higher spatial resolution than both *A. mellifera* and *A. dorsata*. Dyer (1985) observed that in the case of the Asian honeybee, *A. dorsata*, although the moon's illumination is essential for nocturnal flight, the moon itself is ignored for orienting the dances. Rather, bees probably use the sun's position as a reference point for their dances, even though the sun is below the horizon. Rao et al. (2008) found that *Pterocarpus santalinus* blooms massively in the dry season. The flowers open in the middle of the night and are bright yellow in colour. They are homogamous and are visited by the rockbee *A. dorsata* on moonlit nights or otherwise at dawn and up to 0730 h. The natural habitat of *P. santalinus* has a hot, dry climate, and the nocturnal flowering and foraging of the rockbee on moonlit nights appears to be an adaptation to avoid the adverse effects of daytime high temperatures. The honeybee *A. dorsata* was reported to employ the mechanism to find the sun on overcast days.

Flight Pattern

No information is available on the flight pattern of *Apis florea*. The flight of *A. cerana* is very much similar to that of a fly, which makes rapid, hasty and unpredictably zigzag movements compared to the steady, clumsy flight pattern of European honeybee *mellifera* bees (Kellog, 1941; Lindauer, 1956; Sakagami, 1959). This behaviour helps the bees in escaping from flying predators, such as hornets and bee-eating birds. The bee behaviour is species specific. When dislocated 50 m, the speed of homing (Atwal and Dhaliwal, 1969) was found to be shorter for *A. cerana indica* (192 s) than for *A.m. ligustica* (295 s).

General Flight Activity and Foraging Range

Bees are the most important pollinators, and understanding their foraging behaviour is most important component for their conservation (Axelrod, 1960; Bawa, 1990). Bees are the most predominant pollinators for most ecological regions of the world. Flight activity of honeybees is an index of their foraging potential for pollination and honey production. Bees recognize their hive and surroundings and can return to their hive after completing the field activities. If the hive is shifted by more than 3 feet from its original position, the bees returning from their foraging trips get confused and fail to identify the new location. They will stay on the old spot, and some of them may eventually die. The bees have a tendency to fly in one or two major lines of flight called a beeline. Bees generally prefer foraging near the hive. The longer flights are generally avoided because the food required for such an effort will be more than the nectar gathered. The bees' foraging distance strongly influences the sexual reproduction of most flowering plants and can determine the genetic structure of plant populations (Campbell, 1985; Waser et al., 1996). Bee foraging distance also affects agricultural production. Many bees that pollinate crops nest in natural habitats and forage on crops within their daily travel distance (Ricketts, 2004). Foraging distance therefore determines the spatial scale at which bees can provide pollination services to crops (Kremen, 2005). The foraging range of pollinators provides information regarding their intrinsic capabilities in gathering nectar and pollen resources (Cherian and Mahadeven, 1945a, 1945b; Cherian et al., 1947; Eckert, 1955; Dhaliwal and Sharma, 1972, 1973, 1974; Naim and Phadke,1972; Kapyla, 1978; Naim, 1984; Abrol and Kapil, 1986; Abrol, 1988; Abrol, 1993, 2008; Dyer and Seeley, 1991). The foraging range of bees depends largely on their body size and energy needs (Abrol, 1986a; Abrol and Kapil, 1986), yet it is also determined by the quality of reserve fuel carried when bees are in flight. The smaller bees with low-energy requirements and little reserve fuel may hardly risk foraging long distances (Hocking, 1953). Abrol (1988) reported *Apis florea* workersforage up to 150 metres from their nesting sites compared with *Megachile flavipes* and *Megachile nana*, which forage up to 250 and 150 m, respectively. Bees generally visit flowers within a radius of 1 to 3 km. *Apis cerana* normally forages 1 to 1.5 km whereas *Apis mellifera* may forage up to 3 km. But if the food sources are limited and the competition is more, the foragers may go long distances. The foragers of *Apis cerana* have been reported to fly 600–1040 m while as that of *A. mellifera* from 3 to 14 km.

Normally, the bees prefer to forage near their nesting sites. Decrease in return from increased distances from nesting sites may be related to partial familiarity with an area. Bacon et al. (1965) reported that bees extend their foraging range when the flowers become sparse. Evidently, the reasonable flight range of bees may be less than the reported values in literature.

In a flight room, an *A. cerana* colony from Poona (India) with permanent light and feeding within the hive showed 5.5 times as much flight activity relative to the number of bees of a colony of *A. mellifera carnica* (Preuhs,1971). Lindauer (1956) was not able to train *cerana* workers in Sri Lanka farther than 750 m from the hive. During these experiments the flight distance of bees collecting on natural sources did not exceed 300 m. A main foraging distance on cauliflower (75% of the bees) of up to 400 m, on barberry 600–700 m, was found in Punjab, India (max. distance 4900 m, or 1100; Dhaliwal and Sharma, 1973). The uphill foraging distance was much shorter (300 m) than in flat country (Dhaliwal and Sharma, 1974). In Kashmir, *cerana* bees were observed to collect pollen from *Crocus sativus* at a distance of 3.75 km (Shah and Shah, 1980). *A. florea* bees are the most important pollinators of field crops in the plains of India and Pakistan. Atwal (1970) and Sihag (1986) observed 73%–84% *florea* bees among the insects visiting a broader array of plant species (Koeniger and Vorwohl, 1979; Whitcombe, 1981, 1984) than larger species, thus making up for the short flight range, which usually does not exceed 500 m (Lindauer, 1957; Tirgari, 1971).

Dhaliwal and Sharma (1973) found that the maximum foraging range of *Apis cerana* on barberry and on cauliflower was 900 and 1100 m, respectively. However, maximum distribution of foragers was within 400 m in cauliflower and 600–700 m in barberry. Sharma (1972) found that under similar conditions *A. mellifera* worked within 1200 m on mustard flowers; however, few individuals could be found up to 5 km. Few workers have found that effective foraging range from the hive is within 800–2000 m and in exceptional cases foragers have been recovered up to 8–10 km from the hive (Knaffl, 1953; Eckert, 1955;

Peer, 1955; Kettner, 1967). The relatively small flight range of *Apis cerana* may be related to its body size and is perhaps responsible for low honey yields and high incidence of absconding. Ichikawa et al. (2000) reported that the foraging range of *A. mellifera* varied during seasons on different crops depending upon the availability of forage from 0.9 to 2.5 km. In the central district of Tokyo, Japan, however, maximum foragers were available within 100 m when resources were abundant.

It is one of the most common assumptions that honeybees only forage close to their hive. However, in certain instances, they have been reported to forage up to 12 kilometres or more from the hive. Karl von Frisch (1967), the discoverer of the honeybee dance language, was able to calibrate the dance language at distances of up to 9.5 km. The greatest estimate came from Eckert who studied bees in the desert in the western United States and reported that they flew up to 13.5 km from the hive.

Foraging distances will vary with environmental conditions, such as the density and distribution of floral resources and the general physical resistance of the different habitats to flight (e.g., Ricketts, 2001). Cresswell et al. (2000) provided evidence that the quantity and quality of available floral resources also affect foraging distance. For instance, *A. mellifera* and *Megachile rotundata* have been observed to extend their foraging range as the distance to high-reward resources increased (Beekman and Ratnieks, 2000; Bacon et al., 1965) and with resource scarcity (Steffan-Dewenter and Kuhn, 2003). Honeybees will fly farther to get some resources than to acquire other resources (Gary et al., 1972), and their foraging distance varies as a function of landscape context (Steffan-Dewenter and Kuhn, 2003). The dwarf bee *Apis florea* has a short flight range, often hardly reaching 100 metres from the nest. The maximum distance it can fly from the nest for foraging is often less than 750 m.

Flight Intensity of Foraging Bees

Wingbeat frequency is an expression of flight intensity and is an important parameter not only from the aerodynamics and bioenergetics points of view, but it also helps in assessing the pollinating efficiency of bees (Puranik et al., 1977; Heinrich, 1975). The wingbeat frequency is generally species specific. Smaller insects have been reported to have higher wingbeat frequency than the larger ones (Sotavalta, 1952). The smaller the surface area of the wings, the more rapidly the wings must beat to keep the animal in flight (Kammer and Heinrich, 1978). Wingbeat frequency is a function of body parameters. Besides the body parameters, wingbeat frequency is also influenced by environmental temperature in all the insects except hymenoptera (Reed et al., 1942).

Abrol (1985) found that wingbeat frequency of *Apis florea* weighing between 12 and 19 mg ranged between 92 to 106 cps with an average of 101 cps, and that of *A. dorsata* weighing between 106 to 129 mg was between 111 to 116 with an average of 113.95 cps. Interestingly, smaller bees on the basis of per g body weight were found to have higher values of wingbeat frequency than the larger ones. It was found that *Apis florea* weighing 12–19 $\times$ 10^{-3} g had a wingbeat frequency of 101 cps in contrast to *A. dorsata* weighing 106–127 $\times$ 10^{-3} g, which had 113.95 + 0.60 cps. *Apis florea*'s per g body weight exhibited 6688.79 cps compared with *A. dorsata* where the number was found to be 1162.16 cps/g/body weight – a difference of more than fivefold. These results suggested that smaller bees exhibited higher wingbeat frequency than the larger ones.

A. cerana from the plains of northern India (Punjab) showed higher frequencies than yellow *A. mellifera* (presumably *A. m. ligustica*) from Europe: 306 wingbeats/s versus 235 for workers and 283 versus 225 for drones (Goyal and Atwal, 1977). These frequency differences could well be a mere consequence of wing-size differences.

Foraging for Other Nutritive Plant Materials

Bees in several genera of Anthophorinae and Mellitidae collect oils from special floral glands called elaiophors (Buchmann and Hurley, 1978; Vogel, 1974), which occur in plants belonging to at least five dicot and two monocot families. Oil-collecting bees have morphological specializations, such as modified setal combs on the tarsi and thickened scopal setae (Simpson et al., 1977). The oil is mixed with pollen, and possibly nectar, and used in larval provisions. Baker et al. (1973) suggested that stigmatic exudates serve as food for flower visitors. These substances are high in lipid content and may contain amino acids and other nutrients. They sometimes contain substances that are toxic to bees, however, such as galactose in tulip exudates (Barker and Lehner, 1976). Abrol (1985) observed *Apis florea* foraging on extra floral nectaries of peach located at the base of leaves.

Foraging in Relation to Biochemical Characters of Host Plants

Insects require appreciable amounts of protein, carbohydrate lipid and nitrogen for growth and development (Bertsch, 1983; Murugan and George, 1992). A direct correlation is expected to exist between the honeybee visits and biochemical nature of flower parts. Nathan et al. (1999) monitored the foraging activity of *Apis florea* on *Ipomea staphylina* and *Eucalyptus tereticornis* flowers and found that in both these plants the nectar secretion activity is minimal, thus forcing the bees to make repeated visits to a large number of flowers to satisfy their nutritional demands. Interestingly, they found that *A. florea* stayed for a longer time in the flowers of *Ipomea staphylina* than *Eucalyptus tereticornis.* The increased rate of visits by *Apis florea* to the flowers of *E. tereticornis* compared with *Ipomea staphylina* was attributed to higher amounts of protein, carbohydrate and optimum levels of lipid and nitrogen in *E. tereticornis.* Furthermore, *A. florea* preferred to visit *E. tereticornis* than *Ipomea staphylina* since *E. tereticornis* had higher flower density. The percentage of *A. florea* population was highest in the months of November, December, January and February. The foraging population was greater in the morning (6–10 h) with moderate peak at 9 h, decreasing gradually in the afternoon (12–14 h) and suddenly dropping to 18 h under Coimbatore conditions (Western Ghats: 10°58′N,82°18′E). Nathan et al. further observed that *A. florea* confined its visits to *Ipomea staphylina* and *E. tereticornis* and did not make interplant movements.

Influence of Honeybees on Other Bee Species

Introduction of the honeybee species has been reported (Pearson, 1933) to influence native bees in a profound manner. Several authors have since presented experimental evidence of competition between honeybees and wild bees for flower resources. Roubik (1978) reported that the hives of Africanized honeybees introduced near patches of various flower species in French Guiana appeared to displace stingless bees from some of the flower species. Displacement and even aggression between honeybees at artificial nectar sources have been reported. *Apis mellifera* aggressively displaced *A. cerana* from artificial sugar feeders in Japan (Sakagami, 1959). Abrol (2009) reported that introduction of *Apis mellifera ligustica* has severely marginalized the indigenous honeybee *Apis cerana* in India. Interactions suggesting competition between honeybees and wild bees in more natural situations have also been reported. Wratt (1968) found that the number of *Bombus ruderatus* foragers carrying pollen on plots of red clover in New Zealand decreased with increasing temperature. Taken together, these studies provide strong evidence that honeybees do influence the foraging patterns of native bees by competition at resource sites.

Nectar Robbing

Nectar robbing in cauliflower bloom by many honeybee foragers has been found in Solan India by Kumar et al. (1994). Side foraging of cauliflower nectar by *A. cerana* and *A. mellifera* (Kapoor and Dhaliwal, 1989) and that of okra by *A. cerana* (Mishra et al. (1987) has been reported. Kapil and Kumar (1975) reported that few foragers of *A. dorsata* collected nectar from *Brassica juncea* flowers without actually entering the flowers. Sharma et al. (2001) found that all the bees – *A. florea, A. dorsata, A. cerana and A. mellifera* – foraged as top workers on flowers of *Brassica campestris* var. *sarson, Allium cepa, Daucus carota, Trifolium alexandrinum* and *Helianthus annuus*, except that *A. florea* foraged as a side worker on *Brassica* flowers.

Communication and Orientation in Honeybees

Lindauer (1957) reported that the communication pattern in *Apis cerana* indica, *A. dorsata* and *Apis florea* was similar to *A. mellifera*, with slight differences. *A. cerana indica* and *A. dorsata* communicated the direction in a wagtail dance on a vertical comb surface, but *A. florea* performed the dance on the top of the horizontal surface. In all the Apis species, the communication tempo was correlated with the distance. Atwal and Goyal (1969) and Goyal and Atwal (1976) confirmed the finding of Lindauer and found that *A. cerana* indica performed a round dance up to 0.3–0.7 m foraging distance; workers moved clockwise for 2–6 times, then anticlockwise several times. But *A. mellifera* performed a round dance for 0.5–0.27 m distance food source; workers alternated in single clockwise and anticlockwise circles with an increase in distance. Dixit (1956) in his studies on visual activity found that bees preferred the black figures to grey background, and *A. cerana indica* preferred more subdivided figures. He further reported that *A. florea* also behaved similarly.

Effect of Pathogens and Diseases on Foraging Activity of Bees

Anderson and Giacon (1992) reported that *Apis mellifera* colonies fed preparations of 50% of sucrose containing sac brood virus particles and *Nosema apis* Zander spores collected significantly less pollen than colonies fed only on the sucrose solution. Bailey and Fernando (1972) found that worker bees infected with sacbrood virus have a foraging behaviour different from noninfected bees in that they collect mainly nectar rather than pollen. Evidently, infection of either sacbrood or nosema in honeybees may reduce their pollinating efficiency to pollinate plant crops, especially those plants that flower during spring or early summer or that produce only pollen, such as kiwifruit (*Actinida deliciosa*). Abrol (1995b) found that *Apis cerana* and *A. mellifera* infested with ectoparasitic mites had reduced pollen collection activity as compared to healthy ones. Similar information is not available in the case of *Apis florea*.

Foraging in Relation to Body Size

Abrol (2006b) studied the foraging behaviour of honeybee *Apis florea* F. on carrot (*D. carota*) flowers. Of all these insects, the dwarf honeybee *Apis florea* F. was the most predominant and comprised more than 94% of the total flower visiting insects. This may be due to the fact that *A. florea* had much suitability of its tongue length to the corolla length of carrot bloom. While collecting nectar or pollen, *Apis florea* lay across the umbel, pollinating large numbers of flowers. This behaviour of *A. florea* makes it a most efficient pollinator of carrot blossoms. The bees are probably less selective than other *Apis* species, and their small size enables them to forage successfully on many plant species. A rewarding system associated with the flowers enables pollinators to discriminate between the closely related plant species or ecotypes. This has resulted in a co-partnership between the flowers and their pollen vectors. Co-evolution has brought a close correlation between pollinator needs and floral energy expenditures (Heinrich, 1975a).

Foraging in Relation to Pheromones

There are several pheromones to attract bees for the crops that need cross-pollination. (Free, 1962, 1968, 1981, 1982, 1987; Free et al., 1989; Waller, 1970; Woyke, 1981, 2001a, 2001b; Ohe and Praagh, 1983; Currie et al., 1992) and to repel them to save from hazardous effects of pesticides (Atkins Jr. et al., 1975a, 1975b; Attri and Singh, 1978; David and Somasundaran, 1985). Use of synthetic queen pheromones to stimulate foraging (especially pollen collection) and Nasanov pheromone whose two components (citral and geraniol) have been reported to increase honeybee foraging in onion (Woyke, 1981) and yield of apple (Ohe and Praagh, 1983) hold much promise.

Use of synthetic alarm pheromones a few minutes before insecticide application may help in reducing bee loss by repelling them from the crop. *Apis florea* foragers were observed to leave pheromone at the source of forage that was attractive to other foragers (Butler, 1954). However, they were not seen to expose their Nasonov gland when foraging on flowers, or at dishes of sugar syrup (Free and Williams, 1979). Naik et al. (2002) found 2-heptanone as a repellent for *Apis florea* and Ethylthio cyclopentane for *Apis cerana* (Naik et al., 1999).

Control of Foraging

A major goal in crop production is increased crop production through enhanced pollination, which can be achieved through control of foraging bees through various type of chemicals that may be used to attract bees or repel them from areas where there is danger from insecticides. Work with other insects – both social and non-social – indicates that this is possible by chemical and physical means. It has been found that different plant species produce varying attractant compounds associated with their nectar and pollen. Bees are highly attracted to the scent of honey being extracted or heated. The pollination efficiency can be enhanced by incorporating scents collected from flowers in their food. Isolation and identification of compounds attractive to bees, and application of these compounds to attract bees, and breeding crops with compounds attractive to bees are an area of research of potential importance to crop pollination.

Swarming Behaviour

Lindauer (1957) recorded as many as five swarms from an individual colony of *A. florea* within 12 days in Sri Lanka. Wenning (2000) observed that less honey was stored by colonies that swarmed because of their reduced population. The higher honey yield from the nuclei was basically due to their non-swarming because they had a young queen, whereas the colonies from swarms had older ones. Seeley and Morse (1977) observed that the *A. florea* swarms preferred a site near the parent colony. They frequently moved beyond 300 metres. Caron (1979) recorded that the swarming in *A. florea* occurred before first July. Over half the swarms clustered on trees, and nearly three-quarters formed at a height of less than three metres. Different numbers of queen cells at the time of swarming have been reported: 11–13 queen cells (Kshirsagar et al., 1983), 15–20 queen cells (Ruttner et al., 1985a), 12–15 queen cells (Soman and Chawda, 1996) and 10–11 queen cells (Malathi, 2005).

Absconding Behaviour

Seeley et al. (1982) found that absconding in *A. florea* may occur at any time of the season, but the maximum is in a dry and hot period when there is a poor supply of food and water. Whitcombe (1984) observed a mean residence time of 101 days with his experimental colonies; however, one of them was found staying at the same site for more than one year. He further reported that absconding in *A. florea* colonies increased in summer due to scarcity of food resources and predation from enemies. Siriwat et al. (2004) observed that even after absconding, some *A. florea* bees may come back to collect old wax from discarded combs. Malathi (2005) also observed similar behaviour of wax collection from the abandoned combs.

Conclusions

Bees serve as excellent test animals for exploring optimal foraging theories, as the energy gained from nectar can be accurately measured. Most energetics research is on bumblebees and honeybees, primarily because of their high-energy requirements and because they can be easily monitored while foraging. More work is needed on the energetics of foraging in bees and on the importance of factors other than energy in determining bee foraging patterns such as nutritive values of different pollens, floral cues, floral scents, shape of flowers etc. Most studies of competition for floral resources among bee species, especially between honeybees and wild bees, have been inconclusive because competition is difficult to prove in the field. Future studies should include manipulation of forager and resource levels and behavioural observations of the interactions between bees on flowers.

BIBLIOGRAPHY

Abrol, D.P. 1985. Analysis of biophysical interactions in causing foraging behaviour of some bees: A study in bioenergetics. PhD thesis, CCS Haryana Agricultural University Hisar, India, p. 286.

Abrol, D.P. 1986a. Time and energy budgets of alfalfa pollinating bee, *Megachile nana* Bingh and *M. flavipes* Spinola (Hymenoptera: Megachilidae). *Proc. Indian Acad. Sci. (Anim. Sci.)* 95:579–586.

Abrol, D.P. 1986b. Ecophysiological adaptations between pollinating bees and their flowers. *Environ. Ecol.* 4:161.

Abrol, D.P. 1986c. Metabolic expenditures of *Megachile nana* Bingh and *M. flavipes* Spinola (Hymenoptera: Megachilidae). *J. Anim. Morphol. Physiol.* 33:107–112.

Abrol, D.P. 1987. Activity time budgets and pollination potential of *Apis cerana indica* workers. *Environ. Ecol.* 5(4):707–709.

Abrol, D.P. 1988. Foraging range of subtropical bees, *Megachile Jlavipes, Megachile nana* (Hymenoptera: Megachilidae) and *Apis florea* (Hymenoptera: Apidae). *Indian Inst. Sci.* 68:43–47.

Abrol, D.P. 1989. Energy intake and expenditure in *Andrena ilerda* and *A. leaena* (Hymenoptera: Andrenidae). *Korean J. Apic.* 4:41–47.

Abrol, D.P. 1990a. Energetics of nectar production in some apple cultivars as a predictor of floral choice by honeybees. *Trop. Ecol.* 31(1):116–122.

Abrol, D.P. 1990b. Energy intake and expenditure in *Xylocopa fenestrata* F. and *X. pubescens* Spinola (Hymenoptera: Anthophoridae). In: G.K. Veeresh, B. Malik and C.A. Viraktamath (Eds.), *Social Insects and the Environment. Proceedings of the 11th International Congress of IUSSI*, Bangalore, India. Oxford and IBH Publishing, New Delhi, India, p. 574.

Abrol, D.P. 1991a. Foraging strategies of honeybees in pollinating apple flowers. *J. Anim. Morphol. Physiol.* 38(1/2):109–113.

Abrol, D.P. 1991b. Path analysis of environmental factors influencing daily flight activity of *Apis dorsata* F. *Acta Oecol.* 12(6):819–824.

Abrol, D.P. 1992e. Diel pattern of *Apis cerana indica* in *Asparagus officinalis* L. *Sci. Cult.* 58(1/2):25.

Abrol, D.P. 1992a. Bioenergetics in beeflower interrelationship and analysis of foraging behaviour. *Korean J. Apic.* 7(1):39–66.

Abrol, D.P. 1992b. Foraging in honeybees *Apis cerana indica* F. and *A. dorsata* F. (Hymenoptera: Apidae activity and weather conditions. *J. Indian Inst. Sci.* 72(5):395–340.

Abrol, D.P. 1992c. Energetics of nectar production in some strawberry cultivars as a predictor of floral choice by honeybees. *J. Biosci.* 17(1):41–44.

Abrol, D.P. 1992d. Oxygen consumption of foraging bees. *J. Anim. Morphol. Physiol.* 39:27–42.

Abrol, D.P. 1993a. Intervarietal differences in nectar production affecting pollinator attractiveness and fruit set in almond. In: G.K. Veeresh, R. Uma Shankar and K.N. Ganeshaiah (Eds.), *Pollination in Tropics.* IUSSI Indian Chapter, Bangalore, India, pp. 120–122.

Abrol, D.P. 1993b. Energy intake and expenditure in *Megachile femorata* Smith (Hymenoptera: Megachilidae). *Trop. Ecol.* 34(2):173–180.

Abrol, D.P. 1995a. Energetics of nectar production in some almond cultivars as a predictor of floral choice by honeybees *Apis cerana indica* F. and *A. mellifera* L. (Hymenoptera: Apidae). *Proc. Indian Natl. Sci. Acad. B* 57(2):127–132.

Abrol, D.P. 1995b. Haemocytes of honeybees *Apis mellifera* L. and *Apis cerana* F. (Hymenoptera: Apidae) and their alteration by mite parasitosis. *Indian Bee J.* 57(1):6–7.

Abrol, D.P. 1998b. Environmental factors influencing flight activity in honey bees, *Apis cerana* Fab. and *Apis mellifera* L. (Hymenoptera: Apidae). *Indian Bee J.* 60(2):71–75.

Abrol, D.P. 1998c. Foraging in honeybees *Apis cerana indica* F. and *A. dorsata* F. (Hymenoptera: Apidae) activity and weather conditions. *J. Indian Inst. Sci.* 72(5):395–401.

Abrol, D.P. 1998a. Environmental factors influencing flight activity in honeybees *Apis cerana* Fab. and *Apis mellifera* L. (Hymenoptera: Apidae). *Indian Bee J.* 60:71–75.

Abrol, D.P. 1998d. Food acquisition efficiency as a determinant of honey bee foraging decisions. *Indian Bee J.* 60(3):133–136.
Abrol, D.P. 2003. Foraging behaviour of *Apis florea* F. on *Portulaca grandiflora* L. *Sci. Cult.*, 69:289–290.
Abrol, D.P. 2006a. Factors influencing flight activity of *Apis florea* F, an important pollinator of *Daucus carota* L. *J. Apicult. Res. Bee World* 45(2):2–6.
Abrol, D.P. 2006b. Diversity of pollinating insects visiting litchi flowers (*Litchi chinensis Sonn.*) and path analysis of environmental factors influencing foraging behaviour of four honeybee species. *J. Apicult. Res. Bee World* 45(4):180–187.
Abrol, D.P. 2007. Differences in sugar metabolism among honey bees, *Apis mellifera* and *Apis cerana* potentially impact floral attractiveness. *J. Apicult. Res.* 46(3):181–190.
Abrol, D.P. 2009. *Bees and Beekeeping in India.* 2nd ed., Kalyani Publishers, Ludhiana, India, 720p.
Abrol, D.P. and Kapil, R.P. 1991. Foraging strategies of honeybees and solitary bees as determined by nectar sugar components. *Proc. Indian Natl. Sci. Acad. B* 57(2):127–132.
Abrol, D.P. and Kapil, R.P. 1986. Factors affecting pollination activity of *Megachile lanata* Lepel. *Proc. Indian Acad. Sci. (Anim. Sci.)*, 95:757–769.
Abrol, D.P. and Kapil R.P. 1987. Nectar dilution pattern of bees in semi-arid environments. *Curr. Sci.*, 56(13): 681.
Abrol, D.P., Kapil, R.P. and Sihag, R.C. 1988. Physiological adaptations of subtropical bees in relation to energy intakes. *Res. Dev. Reporter*, 5:68–79.
Abrol, D.P. Bhat, A.A. and Khan, A.R. 1989. Influence of behaviour of *Apis cerana indica* Fabr. on cross pollination in apple. *Indian Bee J.*, 51(3):103–104.
Ahmed, M. and Aslam, M. 2002. Pollinators visiting carrot (*Dacus carota*). *J. Res. Sci.* 13(12) 31–35.
Akratanakul, P. 1977. The natural history of the dwarf honey bee, *Apis florea* F. in Thailand. Cornell University, Ithaca, NY, pp. 1–91. PhD Thesis.
Ali, A.M. 1996. *Beekeeping in Sudan.* University of Khartoum, Sudan, 223 pp.
Alm, J., Ohnmeiss, T.E., Lanza, J. and Vriesenga, L. 1990. Preference of cabbage white butterflies and honey bees for nectar that contains amino acids. *Oecologia* 84:53–57.
Anderson, D.L. and Giacon, H. 1992. Reduced pollen collection by honeybee (Hymenoptera: Apidae) colonies infected with *Nosema apis* and sacbrood virus. *J. Econ. Entomol.* 85(1):47–51.
Anderson, L.D. and Atkins, E.L. Jr. 1958. Toxicity of pesticides to honey bees in laboratory and field tests. *J. Econ. Entomol.* 51:103–108.
Atkins Jr., E.L., Mcdonald, R.L. and Mcgouern, T.D. et al. 1975b. Repellent additives to reduce pesticide hazards to honey bees: Laboratory testing. *J. Apic. Res.* 14:85–97.
Atkins Jr., E.L., Mcdonald, R.L. and Greywood-Hale, E.A. 1975a. Repellent additives to reduce pesticide hazards to honey bees: Field tests. *J. Environ. Entomol.* 4(2):207–210.
Attri, B.S. and Singh, R.P. 1978. A note on the biological activity of the oil of *Lantana camara* L. *Indian J. Entomol.* 39(4):384–385.
Atwal, A.S. 1970. Insect pollinators of crops. Punjab Agricultural University Press, Ludhiana, India.
Atwal, A.S. and Dhaliwal, G.S. 1969. Some behavioural characteristics of *Apis Indica* F. and *A. mellifera* L. *Indian Bee J.* 31(1):1–8.
Atwal, A.S. and Goyal, N.P. 1969. Communication in the indian honeybee *Apis indica* F. and *A. mellifera* L. *22nd Intl. Beekeeping Congr. Summ.* 107.
Axelrod, D.I. 1960. The evolution of flowering plants. In: S. Tafl (Ed.), *Evolution After Darwin: The Evolution of Life*, Volume 1. University of Chicago Press, Chicago, IL.
Bacon, O.G., Burton, V.E., McSwain, J.W., Marble, V.L., Stanger, W. and Thorp, R.W. 1965. Pollinating alfalfa with leaf-cutting bees (AFLT 160). University of California Agricultural Extension Service, Berkeley, CA.
Bailey, L. and Fernando, E.F.W. 1972. Effects of sacbrood virus on adult honey bees in Brittain. *Ann. Appl. Biol.* 97:109–118.
Baker, H.G., Baker, I. and Opler, P.A. 1973. Stigmatic exudates and pollination. In: N.B.M. Brantjes, H.F. Linskens (Eds.), *Pollination and Dispersal*, pp. 47–60. Univesity Nijmegen, Nijmegen, the Netherlands, 125 p.
Barker, R.J. and Lehner, Y. 1976. Galactose, a sugar toxic to honey bees, found in exudate. *Apidologie* 7:109–111.
Bawa, K.S. 1990. Plant–pollinator interactions in tropical rain forests. *Annu. Rev. Ecol. Syst.* 21:299–422.

Beekman, M. and Ratnieks, F.L.W. 2000. Long range foraging in the honey bee. *Funct. Ecol.* 14:490–496.
Bertsch, A. 1983. Nectar production of *Epilobium angustifolium* L. at different air humidities. Nectar sugar in individual flowers and the optimal – foraging theory. *Oecologia* 91:187–193.
Bhalla, O.P., Verma, A.K. and Dhaliwal, H.S. 1983a. Foraging activity of insect pollinators visiting stone-fruits. *J. Entomol. Res.* 7(2):91–94.
Bhalla, O.P., Verma, A.K. and Dhaliwal, H.S. 1983b. Insect visitors of mustard bloom (*Brassica campestris* var. *sarson*), their number and foraging behaviour under midhill conditions. *J. Entomol. Res.* 7(1):15–17.
Bhardwaj, S.C. 1974. Evaluation of some chemicals as repellents to *Apis florea* F. Ph.D. Thesis, Haryana Agricultural University, Hissar, India, 125 p.
Bisht, D.S. and Pant, N.C. 1968. Studies on the pollen gathering activity of the Indian honeybee, *Apis cerana indica* F. under Delhi conditions. *Indian J. Entomol.* 44:368–372.
Buchmann, S.L. and Hurley, J.P. 1978. A biophysical model for buzz pollination in angiosperms. *J. Theor. Biol.* 72:639–657.
Burill, R.M. and Dietz, A. 1981. The response of honeybees to variations in solar radiation. *Apidologie* 12:319–328.
Butler, C.G. 1954. *The World of Honeybees.* Collins, London, UK, pp. 16, 32, 41, 140–141, 154, 211–212.
Bye, E.F., Kaeser, W. und Buchner R. 1959. Zur Wirkung von Aktivsubstanzen verchiedener. *Anzeiger für Schädlingskunde* 32:121–123.
Campbell, D.R. 1985. Pollen and gene dispersal—the influences of competition for pollination. *Evolution* 39:418–431.
Caron, D. 1979. Swarm emergence date and cluster location in honeybees. *Am. Bee J.* 119(1):24–25.
Casey, T.M. 1976. Flight energetics of sphinx moths: Power input during hovering flight. *J. Exp. Biol.* 64:529–543.
Chahal, B.S., Brar, H.S., Gatoria, G.S. and Jhajj, H.S. 1986. Aggressive behaviour of *Apis Florea* towards *Apis Mellifera* in Hive Robbing and in Foraging. *J. Apic. Res.* 25(3):134–138.
Chamberlain W. F. 1959. The behavior of agricultural insects toward olfactory repellents in the olfactometer and in split-arena tests. *J. Econom. Entomol.* 52(2):286–289.
Chandel, R.S., Thakur, R.K., Bhardaj, N.R. and Pathania, N. 2003. Onion seed crop pollination: A missing dimension in mountain horticulture. *ISHS Acta Horticult.* 631:XXVI. International Horticultural Congress: Issues and Advances in Transplant Production and Stand Establishment Research. 1st ed., Cornell University Press.
Chandrasekhar, G.S. and Sattigi, H.N. 2006. Pollinator fauna and foraging activity in radish of bees and other pollinators. *Karnataka J. Agric. Sci.* 19(3):719–721.
Chaturvedi, M. 1973. An analysis of honeybee pollen loads from Banthra, Lucknow, India. *Grana* 13(3):139–144.
Chaturvedi, M. 1977. Further investigation on the pollen analysis of bee loads from Banthra, India. *New Bot.* 4(1–4):41–47.
Cherian, M.C. and Mahadevan, V. 1945a. The longevity of Indian bee (*Apis indica*). *Indian Bee J.* 8(3/4):41–43.
Cherian, M.C. and Mahadevan, V. 1945b. Preliminary studies on the pollen carrying capacity of the Indian bee (*Apis indica* F). *Indian Bee J.* 8(7/8):109–111.
Cherian, M.C., Ramachandran, S. and Mahadevan, V. 1947. Studies in bee behaviour. *Indian Bee J.* 9:116–124.
Choudary, R.K. 1978. Floral fidelity in the Indian honey bee (*Apis cerana indica*). *Indian Bee J.* 40(2):33–35.
Corbet, S.A. 1978a. A bee's view of nectar. *Bee World* 59:25–32.
Corbet, S.A. 1978b. Bees and nectar of *Echium vulgare.* In: A.J. Richards (Ed.), *The Pollination of Flowers by Insects.* Academic Press, London, UK, pp. 21–30.
Crane, E. 1975. *Honey, a Comprehensive Survey.* Russak & Co., Inc., New York.
Crane, E. 1979. *Honey: A Comparative Survey.* Heinemann in co-operation with IBRA, London, UK, pp. 412–425.
Cresswell, J.E., Osborne, J.L., Goulson, D. 2000. An economic model of the limits to foraging range in central place foragers with numerical solutions for bumblebees. *Ecol. Entomol.* 25:249–255.
Currie, R.W., Winston, M.L. and Slessor, K.N. 1992. Effect of synthetic queen mandibular pheromone sprays on honey bee (Hymenoptera: Apidae) pollination of berry crops. *J. Econ. Entomol.* 85:1300–1306.
David, B.V. and Somasundaran, L. 1985. Syntetic pyrethroids – an evaluation of their potential effects nontarget organisms. *Pesticides* 19(1):9–12.

Devika, R.D, Triveni, B., Nandini and Narsa Reddy, G. 2017. Study of evaluation of foraging behaviour of major insect pollinators on summer squash (*Cucurbita Pepo* L.). *J. Entomol. Zool. Stud.* 5(5):28–30.

Dhaliwal, G.S. 1970. Observations on pollen carrying capacity of *Apis indica* F. and *Apis mellifera* L. *Indian J. Entomol.* 32(2):173–174.

Dhaliwal, H.S. and Sharma, P.L. 1972. Homing ability and the foraging range of the Indian honeybee, *Apis indica* F. *Indian J. Entomol.* 34(3):225–227.

Dhaliwal, H.S. and Sharma, P.L. 1973. The foraging range of Indian honeybee on two crops. *J. Apic Res.* 12(2):131–134.

Dhaliwal, H.S. and Sharma, P.L. 1974. Foraging range of the Indian honeybee. *J. Apic Res.* 13(2):137–141.

Dhaliwal, G.S. and Atwal, A.S. 1986a. Effect of age of crop, plant spacing, soil moisture, and phosphate fertilizerson bee activity on *Brassica* crop. In: R.P. Kapil (Ed.), *Pollination Biology: An Analysis.* Inter India Publications, New Delhi, India, pp. 91–102.

Dhaliwal, G.S. and Atwal, A.S. 1986b. Factors influencing tripping efficiency of bees on alfalfa (Medicago sativa L.) In: R.P. Kapil (Ed.), *Pollination Biology: An Analysis*, Inter India Publications, New Delhi, India, pp. 79–80.

Diwan, V.V. 1980. Honey bee behaviour during total solar eclipse. *Indian Bee J.* 42(4):97–99.

Diwan, V.V. and Salvi, S.R. 1965. Some interesting behavioural features of Apis dorsata Fab. *Indian Bee J.* 27(1):52.

Dixit, K.K. 1956. Differential response to form and pattern in two species of Indian honeybees. *J. Bombay Nat. Hist. Soc.* 54(1):107–117.

Dutton, R. and Simpson, J. 1977. Producing honey with *Apis florea* in Oman. *Bee World* 58(2):71–76.

Dutton, R.W. and Free, J.B. 1979. The present status of beekeeping in Oman. *Bee World* 60(4):176–186.

Dyer, F. and Seeley, T.D. 1987. Interspecific comparisons of endothermy in honey-bees (*Apis*): Deviations from the expected size-related patterns. *J. Exp. Biol.* 127:1–26.

Dyer, F.C. 1985. Nocturnal orientation by the Asian honey bee. *Apis dorsata. Anim. Behav.* 33:769–774.

Dyer, F.C. and Seeley, T.D. 1991. Dance dialects and foraging range in three Asian honey bee species. *Behav. Ecology Sociobiol.* 28(4):227–233.

Eckert, J.E. 1955. The flight of the honeybee. *Am. Bee J.* 95:395–401.

El Shafie, H.A.F., Mogga, J.B.B. and Basedow, Th. 2002. Studies on the possible competition for pollen between the honey bee, *Apis mellifera sudanensis*, and the imported dwarf honey bee *Apis florea* (Hym., Apidae) in North-Khartoum (Sudan). *J Appl. Entomol.* 126(10):557–562.

Erhardt, A. and Rusterholz, H.P. 1998. Do peacock butterflies (*Inachis io* L.) detect and prefer nectar amino acids and other nitrogenous compounds? *Oecologia* 117:536–542.

Faegri, K. and van der Pijl, L. 1971. *The Principles of Pollination Ecology*, 2nd ed. Pergamon, Oxford, 291 p.

Forbes, M.F. and Cervancia, C.R. 1994. Foraging behaviour of *Apis cerana* F. and *Apis mellifera* L. (Apidae: Hymenoptera) in Majayjay. *Philipp J. Sci.* 123(1):21–27.

Free, J.B. 1962. The attractiveness of geraniol to foraging honey bees. *J. Apic. Res.* 1:52–54.

Free, J.B. 1968. The conditions under which foraging honey bees expose their Nasonov glands. *J. Apic. Res.* 7:139–145.

Free, J.B. 1970. *Insect Pollination of Craps.* Academic Press, London, UK.

Free, J.B. 1981b. Biology and behaviour of the honeybee *Apis florea*, and possibilities for beekeeping. *Bee World* 62(2):46–59.

Free, J.B. 1981a. Ecology and behaviour of the honeybee *Apis florea* and possibilities for beekeeping. *Bee World* 62(1):49–59.

Free, J.B. 1982. The biology and behaviour of the honeybee *Apis florea.* In: P. Jaisson (Ed.), *Social Insects in the Tropics*, Vol. 1. Presses de;'Universitě Paris-Nord, Paris, France, pp. 181–187.

Free, J.B. 1987. *Pheromones of Social Bees.* Chapman and Hall, London, UK.

Free, J.B. (1994). The biology and behaviour of the honeybee *Apis florea. Bee World.* 75(2):181–186.

Free, J.B. and Williams, I.H. 1979. Communication by pheromones and other means in *Apis florea* colonies. *J. Apic. Res.* 18:16–25.

Free, J.B., Ferguson, A.W. and Simpkins, J.R. 1989. Honeybee responses to chemical components from the worker sting apparatus mandibular glands in field tests. *J. Apic. Res.* 7–21.

Fuchs, S., Sen Sarma, M., Werber, C. and Tautz, J. 2001. The cry of the honeybee *Apis florea. Proceedings of the 2001 meeting of the European sections of the International Union for the Study of Social Insects (IUSSI)*, September 25–29, Berlin, Germany, p. 164.

Gary, N.E., Witherell, P.C. and Marstonm, J. 1972. Foraging range and distribution of honey bees used for carrot and onion pollination. *Environ. Entomol.* 1:71–78.
Goyal, K. 1977. Olfacto-gustatory response of *Apis florea* F. to some ketones. MSc Thesis, Haryana Agricultural University Hisar, India, p. 64.
Goyal, N.P. 1974. *Apis Indica* indica F. and *A. mellifera* L as complimentary to each other for the development of apiculture. *Bee World* 55(3):98–101.
Goyal, N.P. and Atwal, A.S. 1976. Comparative communication behaviour among *Apis Indica indica* F. and *A. mellifera* L. dancing behaviour. *Indian Bee J.* 38(1–4):1–2.
Goyal, N.P. and Atwal, A.S. 1977. Wing beat frequency of *Apis Indica* indica F. and *A. mellifera* L. *J. Apic Res.* 16(1):47–48.
Gubartalla, A.E. 1997. Studies on some aspects of the biology of the dwarf honeybee *Apis florea* F. in Khartoum State. M.Sc., Faculty of Agriculture, University of Khartoum, Sudan.
Gupta, M. 1982. Olfactory physiology of *Apis florea* F. with reference to repellents. Ph.D. Thesis, Haryana Agricultural University, Hisar, India, p. 168.
Gupta, M. 1983. Honey bees: Pesticidal hazards. *Haryana Review*, 17:19.
Gupta, M. 1985. Olfactory response of *Apis florea* F. to some compounds of essential oils. *Proc. 72nd Indian Science Congress*, A 267, Lucknow, India.
Gupta, M. 1986a. A quantitative study and ultrastructure of flagellar sensillae of *Apis florea* F (Hymenoptera: Apidae). *Proc. Indian Acad. Sci. Anim. Sci.* 95:595–603.
Gupta, M. 1986b. Conditioning of *Apis florea* F as a function of duration of exposure to chemical vapours at Rc50 and Rc60 dilutions. *Beitr. Entomol.* 36:225–229.
Gupta, M. 1987a. Behaviour of *Apis florea* Fabr. to some ketones and essential oils under semifield conditions. *Indian Bee J.* 49:29–32.
Gupta, M. 1987b. Efficacy of ketones on the foraging behaviour of *Apis florea* F in field conditions. *Apidologie* 18:121–128.
Gupta, M. 1987c. Essential oils: A new source of bee repellents. *Chem. Ind.* 162–163.
Gupta, M. 1988a. Olfacto-gustatory response of *Apis florea* F to some halogen substituted ketones. *Indian J. Anim. Res.* 22:55–59.
Gupta, M. 1988b. Olfactory response of *Apis florea* F to some phenolic ketones. *Comp. Physiol. Ecol.* 13:64–68.
Gupta, M. 1989a. Conditioning of *Apis florea* F worker bees as a function of exposure to chemical vapours at 0.05 N concentration. *Zool. Jahrb. Abt. Allg. Zool. Physiol. Tiere.* 93:215–218.
Gupta, M. 1989b. Effect of repellents on the foraging behaviour of *Apis florea* F under field conditions. *J. Res. Haryana Agric. Univ.* 19:6–15.
Gupta, M. and Kapil, R.P. 1984. Olfactory response of workers of *Apis florea* F. to some repellents in semi-field conditions. *Proceedings of the 5th International Symposium Pollin.*, Versailles, pp. 65–78.
Gupta, M. and Kwrt, R.P. 1984. Olfactory response of workers of *Apis florea* F. to some repellents in semi-field conditions. *Proc. 5th Intern. Symp. Pollin.*, Versailles, pp. 65–78.
Hainsworth, F.R. and Wolf, L.L. 1976. Nectar characteristics and food selection by hummingbirds. *Oecologia* 25:101–113.
Heinrich, B. 1972a. Energetics of temperature regulation and foraging in a bumble bee, *Bornbus terricola* Kirby. *J. Comp. Physiol.* 77:49–65.
Heinrich, B. 1972b. Patterns of endothermy in bumblebee queens, drones and workers. *J. Comp. Physiol.* 77:65–79.
Heinrich, B. 1972c. Energetics of temperature regulation and foraging in a bumblebee, *Bombus terricola* Kirby. *J. Comp. Physiol.* 77:49–64.
Heinrieh, B. 1975a. Energeties of pollination. *Ann. Rev. Ecol. Syst.* 6:139–170.
Heinrich, B. 1975b. Bee flowers: A hypothesis on flower variety and blooming times. *Evolution* 29:325–334.
Heinrich, B. 1975c. The role of energetic in bumblebee-flower interrelationships. *Coevolution of Animals and Plants*, pp. 141–158.
Heinrich, B. 1975d. Thermoregulation in bumble bees. II. Energetics of warm-up and free flight. *J. Comp. Physiol.* 96:155–166.
Heinrich, B. 1975e. The role of energetics in bumblebee-flower interrelationships. In: L.E. Gilbert and P.H. Raven (Eds.), *Coevolution of Animals and Plants*. University Texas Press, Austin, TX, pp. 141–158.
Heinrich, B. 1976a. Heat exchange in relation to blood flow between thorax and abdomen in bumblebees. *J. Exp. Biol.* 64:561–585.

Heinrich, B. 1976b. Resource partitioning among some eusocial insects: Bumblebees. *Ecology* 57:874–889.
Heinrich, B. 1976c. The foraging specializations of individual bumblebees. *Ecol. Monogr.* 46:105–128.
Heinrich, B. 1976d. Flowering phenologies: Bog, woodland, and disturbed habitats. *Ecology* 57:891–899.
Heinrich, B. 1979c. "Majoring" and "minoring" by foraging bumblebees, *Bombus vagans*: An experimental analysis. *Ecology*. 60(2):245–255.
Heinrich, B. 1979d. Thermoregulation of African and European honeybees during foraging, attack, and hive exits and returns. *J. Exp. Biol.* 80:217–229.
Heinrich, B. 1979a. *Bumblebee Economics*. Harvard Universiy Press, Cambridge, MA.
Heinrich, B. 1979b. Keeping a cool head: Honeybee (*Apis mellifera*) thermal regulation. *Science* 205:1269–1271.
Heinrich, B. and Raven, P.H. 1972. Energetics and pollination ecology. *Science* 176:597–602.
Hocking, B. 1953. The intrinsic range and speed of insect flight. *Trans. R Entomol. Soc. Lond.* 104:223–345.
Hocking, B. 1968. Insect flower association in the high Arctic with special reference to nectar. *Oikos* 19:359–388.
Ichikawa, N., Sasaaki, M. and Obata, H. 2000. Monitoring foraging range of Apis mellifera in the central district of Tokyo through dance communication. *Proceedings of the Fourth Asian Apicultural Association Conference*. Asian Apicultural Association, Kathmandu, Nepal, pp. 37–38.
Jadhav, L.D. 1982. Effect of solar eclipse on the foraging activity of honey bees. *Biovigyanam* 8(2):157–161.
Jhajj, H.S. and Goyal, N.P. 1979a. Comparative behaviour of pollen foragers of *Apis cerana indica* and *Apis mellifera*. *J. Apic. Res.* 18(4):279–284.
Jhajj, H.S. and Goyal, N.P. 1979b. Comparative studies on the flower constancy of *Apis cerana indica* F. and *Apis mellifera* L. *Proceedings of the Fourth International Symposium on Pollination*, Maryland, pp. 333–339.
Jhajj, H.S. and Goyal, N.P. 1979c. Reversal in the normal behavioural sequence of honeybee foragers with carbon dioxide anaesthesia. Anonymous. *Indian Bee J.* 41(3/4):17–18.
Jhajj, H.S. and Goyal, N.P. 1986. Comparative flower constancy of *Apis cerana indica* F and *Apis mellifera* L. on removing their colonies to a new site. In: R.P. Kapil (Ed.), *Pollination Biology: An Analysis*. Inter–India publications, New Delhi, India, pp. 73–77.
Johansen, C.A. 1977. Pesticides and pollinators. *Annu. Rev. Entomol.* 22:177–191.
Kammer, A.E. and Heinrich, B. 1974. Metabolic rates related to muscle activity in bumble bees. *J. Exp. Biol.* 61:219–227.
Kammer, A.E. and Heinrich, B. 1978. Insect flight metabolism. *Adv. Insect. Physiol.* 13:113–228.
Kapil, R.P. and Kumar, S. 1975. Foraging activity of *Apis dorsata* F. on *Brassica juncea* Czern and Coss. *J. Bombay Nat. Hist. Soc.* 71(2):327–331.
Kapoor, K.S. and Dhaliwal, H.S. 1989. Comparative energetics of *Apis cerana* indica Fabr. and A. *mellifera* L. on *Brassica oleracea* L. var. botrytis. *Indian Bee J.* 51(3):99–101.
Kapyla, M. 1978. Foraging distance of a small solitary bee, *Chelostoma maflillosum* (Hymenoptera, Megachilidae). *Ann. Entomologici Fenn.* 35:63–64.
Kellog, C.R. 1941. Some characteristics of the oriental honeybee, *Apis cerana* F. in China. *J. Econ. Entomol.* 34:717–779.
Kettner, H. 1967. Experiment with honeybees for pollinating cucumbers in greenhouses. *Garten U. Kleintierz* 6(17):12–13.
King, C.C. 1959. The effect of fungicides. *Glean. Bee Cult.* 87:678–681.
Knaffl, H. 1953. uber die flug weete und entfernungs meldung der. bienen z. bienen forsch 7:131–140.
Koeniger, N. and Koeniger, G. 1980. Observations and experiments on migration and dance communication of Apsis dorsata in Sri Lanka. J. Apic. Res. 19(1):21–34.
Koeniger, N. and Vorwohl, G. 1979.Competition for food among four sympatric species of Apini in Sri Lanka (*Apis dorsata, Apis cerana, Apis florea* and *Trigona iridipennis*). *J. Apic. Res.* 18(2):95–109.
Kremen, C. 2005a. Managing ecosystem services: What do we need to know about their ecology? *Ecol Lett* 8:468–479.
Kremen, C. 2005b. The ecology and management of crop pollination services. In: M.J. Groom, G.K. Meffe, C.R. Carroll (Eds.), *Principles of Conservation Biology, 3rd ed.* Sinauer and Associates, Sunderland, MA, p. 54.
Kshirsagar, K.K., Muvel, K.S., Mittal, M.C. and Phadke, R.P. 1983a. Unusual nesting choice by dwarf honeybee, *Apis florea* Fab. *Indian Bee J.* 60(4):217–218.

Kshirsagar, K.K., Muvel, K.S., Mittal, M.C. and Phadke, R.P. 1983b. Some observations on the behaviour of *Apis florea* F. In: *Proc. Second Int. Conf. on Apiculture in Tropical Climates*, New Delhi, India, pp. 356–366.

Kumar, J., Rao, K.V.K. and Gupta, J.K. 1994. Pollination efficiency of bees visiting blossom of *Brassica campestris* L. var. toria in mid hills of Himachal Pradesh, India. *Indian Bee J.* 56(3–4):202–206.

Kumari, R. 1976. Olfacto-gustatory response of *Apis florea* F. to some compounds of essential oils. MSc Thesis, Haryana Agricultural University, Hisar, India, p. 54.

Lerer, H., Bailey, W.G., Mills, P.F. and Pankiw, M. 1982. Pollination activity of *megachile rotundata* (Hymenoptera: Apidae). *Environ. Entomol.* 11:997–1000.

Lindauer, M. 1956. Uber die Verstandigung bei indischen. *Bienen. Z vgl. Physiol.* 38:521–557.

Lindauer, M. 1957. Communication among the honeybees and the stingless bees of India. *Bee World* 38:3–14, 34–39.

Lord, W.S. and Nagi, S.K.A. 1987. *Apis florea* in Africa. *Bee World* 68:39–40.

Malathi, K. 2005. Bioecology of little bee, *Apis florea* Fab. M.Sc. thesis, University of Agricultural Sciences, Bangalore, India.

Mattu, V.K. and Verma, L.R. 1985. Studies on the annual foraging cycle of *Apis cerana indica* F. in Simla hills of northwest Himalayas. *Apidologie* 16(1):1–18.

Mattu, V.K., Chaudhary, D.K. and Kumar, L. 1994. Foraging ecology of *Apis cerana* F. and *A. mellifera* L. in pollinating stone fruit crops. *Pest Manage. Econ. Zool.* 2(1):35–39.

Mayer, D.F., Jonsien, C.A., Eves, J.D., Britt, R. and Bergin, D. 1980. Monitoring of honey bee mortality from insecticides. *Am. Bee J.* 120:352–355.

Mishra, R.C., Gupta, J.K. and Kumar, J. 1987a. Nectar secretion, amount and type of nectar sugars and insect foraging in *Woodfordia floribunda Salisb. Apidologie* 18:173–178.

Mishra, R.C., Kumar, J. and Gupta, J.K. 1987b. Effect of mode of pollination on fruit characteristics of okra, *Abelmoschus esculentus* (L) Moench. *Proc. Indian Natl. Sci. Acad.* B 53:157–160.

Mogga, J. 1994. Comparison between the daily flight activity of *Apis mellifera yemenitica* R. and *Apis florea* F. in Khartorum. *Apidologie* 25(5):490–491.

Mogga, J. and Ruttner, F. 1988. *Apis florea* in Africa: Source of the founder population. *Bee World* 69:3, 100–103.

Mohamed, N.O. 2000. Studies on some aspects of the biology of honeybees: *Apis mellifera* L. and *Apis florea* F. in Khartoum State. M.Sc. Thesis, Faculty of Agriculture, University of Khartoum, Sudan.

Mohan, J.K.V. 1973. Some observations on the bee pollination in *sunhemp* (*Crotalaria juncea*) and their importance in breeding work. *Jute Bull.* 25:214–216.

Morse, D.H. 1977. Foraging of bumble bees: The effect of other individuals. *J N Y Entomol Soc.* 85:240–248.

Morse, D.H. 1978. Interactions among bumble bees on roses. *Insectes Soc.* 25:365–371.

Mossadegh, M.S. 1993. Plants visited by Apis florea F. in Iran. In: Connor, L. J, Rinderer, T., Sylvester, H. A. & Wongsiri S. (Editors). *Asian Apiculture: Proceedings of the International Conference on the Asian Honey Bees and bee Mites.* Wicwas Press, Cheshire, CT. pp. 369–371.

Murrell, D.C. and Nash, W.T. 1981.Nectar secretion by toria (*Brassica campestris* L. v. *toria*) and foraging behaviour of three Apis species on toria in Bangladesh. *J. Apic. Res.* 20(1):34–38.

Murugan, K. and George, A. Sr. 1992. Feeding and nutritional influence on growth and reproduction of *Daphnis nerii* (Linn.) (Lepidoptera: Sphingidae). *J. Insect Physiol.* 38:961–968.

Muttoo, R.N. 1956. *Beekeeping for Beginners.* Farming Bulletin Indian Council of Agricultural Research, New Delhi, pp. 1–48.

Muttoo, R.N. 1965. Facts about beekeeping in India. *Bee World* 37:125–133.

Naik, D.G., Banhatti, P., Chawda, S.S. and Thomas, D. 1999. Ethylthio cyclopentane as a repellent to Apis cerana. *Indian Bee J.* 61(1–4):43–48.

Naik, D.G., Banhatti, P., Chadawa, S.S. and Thomas, D. 2002. 2-heptanone as a repellent for *Apis florea. J. Apic. Res.* 40(1–2):59–61.

Naik, D., Puntambekar, H.M. and Anantpure, P. 2010. Essential oil of *Terminalia chebula* fruits as a repellent for the Indian honeybee *Apis florea. Chem. Biodivers.* 7(5):1303–1310.

Naim, M. 1984. Field behaviour of honeybee foragers. *Indian J. Entomol.* 46(3):335–339.

Naim, M. and Bisht, D.S. 1979. Pollen gathering capacity of the Indian honeybee, *Apis cerana indica* Fabr. *Indian J. Entomol.* 41(3):295–296.

Naim, M. and Bisht, D.S. 1989. Fidelity and time spent on flowers by *Apis cerana* F. foragers. *Indian J. Entomol.* 51(2):222–223.

Naim, M. and Phadke, K.G. 1972. Flight range of *Apis cerana indica* when enticed to a feeding dish. *J. Apic. Res.* 11(3):167–169.

Nathan, S.S., Murugan, K., Kumar, N.S., Jeyabalan, D. and Muthuraman, S. 1999. Pollination potential of honeybee *Apis florea* Linn. In relation to biochemical properties of host plants. *Zoos Print* 14(4):3–4.

Ohe, W. and Praagh, J.P. 1983. Versuche zur Duftlenkung in Obstanlagen. *Nordwetsdeutsche Imkerzeitung* 35:100.

Oldroyd, B.P., Allen Sylvester, H., Wongsiri, S. and Rinderere, T.E. 1994. Task specialization in wild bee *Apis florea* (Hymenoptera: Apidae), revealed by RFLP banding. *Behav. Ecol. Sociobiol.* 34:25–30.

Oldroyd, B.P., Rinderer, T.E. and Buco, S.M. 1992b. Intra-colonial foraging specialism by honey bees *(Apis mellifera)* (Hymenoptera: Apidae). *Behav. Ecol. Sociobiol.* 30:291–295.

Oldroyd, B.P., Rinderer, T.E. and Wongsiri, S. 1992a. Pollen resource partitioning by *Apis dorsata, A. cerana, A. andreniformis* and *A. florea* in Thailand. *J. Apic. Res.* 31(1):3–7.

Oldroyd, B.P. and Wongsiri, S. 2006. *Asian Honey Bees.* Biology, Conservation and Human Interactions, Harvard University Press, Cambridge, MA, p. 340.

Osgood, C.E. 1974. Relocation of nesting populations of *Megichile rotundata* as an important pollinator of alfalfa. *J. Apic. Res.* 13:63–73.

Palanichamy, P.S., Baskaran, S. and Mohandoss, A. 1995. Insect pollination of the moringa plant *Moringa concanensis* Nimmo Linn. *Environ. Ecol.* 13(1):47–51.

Palmer-Jones, T. 1958. Laboratory methods for measuring the toxicy of pesticides to honey bees. *New Zeal J. Agric. Res.* 1:290–300.

Palmer-Jones, T., Foster, I.W. and Jeffery, G.L. 1959. Effect on honey bees of rogor and endothion applied from the air as spray to Brassica trials of M.G.K. repellent 874. *New Zeal J. Agric. Res.* 2:475–480.

Panda, P., Rath, L.L., Padhi, J. and Panigrahi, J. 1995, Relative abundance and foraging behaviour of common bee species on niger in Parbhani district, Orissa. *Indian Bee J.* 57:10–14.

Panda, P., Sontakke, B.K. and Panda, B. 1993. Foraging behaviour of honey bee species on different varieties of niger, *Guizotia abyssinica* Cass. in *Orissa. J. Insect Sci.* 6(1):104–106.

Pearson, J.F.W. 1933. Studies on the ecological relations of bees in the Chicago region. *Ecol. Monogr.* 3:375–441.

Peer, D.F. 1955. The foraging range of the honey bee. Part 1, Ph.D. Thesis, University of Wisconsin, Madison, WI.

Pirk, C.W.W., Hepburn, R., Radloff, S.E. and Erlandsson, J. 2002. Defense posture in the dwarf honeybee, *Apis florea. Apidologie* 33:289–294.

Preuhs, E.1971. Institute Oberursel, unpublished. Thesis.

Punchihewa, R.W., Koeniger, N., Keven, P.G. and Gawaski, R.M.G. 1985. Observations on dance communication and natural foraging rates of *Apis cerana, Apis dorsata* and *Apis florea* in Sri Lanka. *J. Apic. Res.* 24(3):168–175.

Punjabi, A.A., Saraf, S.K. and Singh, G. 1969. Studies on the pollen carrying capacity of the hill strain of the Indian honeybee, *Apis indica* F. in Kashmir in relation to its body weight. *Indian Bee J.* 31(1):9–11.

Puranik, P.G., Krishna, G.G., Ahmad, A. and Chari, N.C. 1977. Wing beat frequency of a flier – mass flow theory. *Proc. Indian Acad. Sci.* 85:327–339.

Pyke, G.H. and Waser, N.M. 1981. The production of dilute nectars by hummingbird and honeyeater flowers. *Biotropica* 13:260–270.

Quinglan, C. and Corlett, R.T. 2016. Seasonal and diurnal patterns of activity in honeybees (*Apis* spp.) on the northern edge of the Asian tropics; their implications for the climate-change resilience of pollination. *Trop. Conserv. Sci.* 9(3): 1–9. doi:10.1177/1940082916667142.

Rao, G.M. and Lazar, M. 1980. Studies on the behaviour and pollination on onion (Allium cepa L.). *Proc. 2nd Conf. Apic. Trop. Climates*, New Delhi, India, pp. 580–589.

Rao, G.M. and Suryanarayana, M.C. 1990. Studies on the foraging behavior of honey bees and its effect on the seed yield in niger. *Indian Bee J.* 52(1–4):31–33.

Rao, S.P., Atluri, J.B. and Reddi, C.S. 2008. Intermittent mass blooming, midnight anthesis and rockbee pollination in *Pterocarpus santalinus* (Fabaceae). *Nord. J. Bot.* 21(3):271–276.

Reddy, C.C. 1979. The influence of various meteorological factors on the foraging behavior of *Apis cerana indica.* In: *Proceedings of the Fourth International Symposium on Pollination*, Maryland, pp. 341–345.

Reddy, C.C. 1980a. Observations on the annual cycle of foraging and brood rearing by *Apis cerana indica* colonies. *J. Apic. Res.* 19(1):17–20.

Reddy, C.C. 1980b. Studies on the nesting behaviour of Apis dorsata F. *2nd Int. Conf. Apic. Trop. Climates* 2:391–397, Indian Agricultural Research Institute, New Delhi, India.
Reed, S.C., William, C.M. and Chadwick, L.E. 1942. Frequency of wing beat as a character as a character for separating species, races and geographic varieties of *Drosophila. Genetics* 27:349–361.
Ricketts, T.H. 2001. The matrix matters: Effective isolation in fragmented landscapes. *Am Nat* 158:87–99.
Ricketts, T.H. 2004. Tropical forest fragments enhance pollinator activity in nearby coffee crops. *Conserv. Biol.* 18:1262–1271.
Roubik, D.W. 1978. Competitive interactions between neotropical pollinators and Africanized honey bees. *Science* 201:1030–1032.
Ruttner, F., Pourasgher, D. and Kauhausenn, D. 1985a. Die Honigbienen des Iran. II. *Apis mellifera* Meda skorikov. *Apidologie* 16(3):241–264.
Ruttner, F., Pourasgher, D. and Kauhausenn, D. 1985b. Die Honigbienen des Iran. I. *Apis florea* Fabricus. *Apidologie* 16(2):119–138.
Ruttner, F. 1988. *Biogeography and Taxonomy of Honeybees*. Springer-Verlag, Berlin, Germany.
Sakagami, S.F. 1959. Some interspecific relations between Japanese and European honey bees. *J. Anim. Ecol.* 28:51–68.
Schlising, R.A. 1970. Sequence and timing of bee foraging on flowers of *Ipomoea* and *Aniseia* (convolvalacae). *Ecology* 51:1061–1067.
Schneider, S.S. and McNally, L.C. 1992. Seasonal patterns of foraging activity in colonies of the African honey bee, *Apis mellifera scutellata*, in Africa. *Insectes Soc.* 39:181–193.
Schneider, S.S. and Mc Nally, L.C. 1993. Spatial foraging patterns and colony energy status in the African honey bee, *Apis mellifera scutellata. J. Insect Behav.* 6:195–210.
Schoener, T.W. 1971. Theory of feeding strategies. *Ann. Rev. Ecol. Syst.* 2:369–404.
Seeley, T.D. 1985. *Honeybee Ecology. A Study of Adaptation in Social Life*. Princeton University Press, Princeton, NJ.
Seeley, T. D. and Morse, R. A. 1977. Dispersal behavior of honey bee swarms. *Psyche* 84:199–209
Seeley, T.D., Seeley, R.H. and Akratanakul, P. 1982. Colony defense strategies of the honeybees in Thailand. *Ecol. Monogr.* 52, 43–63.
Shah, F.A. and Shah, T.A. 1980. Flight range of *Apis cerana* from Kashmir. Anonymous. *Indian Bee J.* 42(2):48.
Sharma, M. 1970a. An analysis of pollen loads of honeybees from Kangra, India. *Grana* 10:35–42.
Sharma, M. 1970b. Studies on the pollen loads of honeybees from Kangra, India. *J. Palynol.* 6:104–110.
Sharma, P.L. 1972. Homing ability and the foraging range of the Indian honey bee *Apis indica* Fabricius. *Indian J. Entomol.* 34(3):225–227.
Sharma, H.K., Gupta, J.K. and Rana, B.S. 2000. Diurnal activity of *Apis cerana* F. and *A. mellifera* L. on different flora during spring and honey flow period. *Pest Manage. Econ. Zool.* 8:151–154.
Sharma, M.K., Shashikala, K. and Kapil, A. 1999. Studies on the effect of different colour attractants on cross pollination in *Boswellia serrata* Roxb. ex Colebr. *Indian For.* 125(12):1244–1247.
Sharma, S.K., Singh, J.R. and Mahla, J.C. 2001. Foraging behaviour of *Apis* spp. in semi-arid sub-tropical climate. 3. On flowers of mustard, onion, carrot, berseem and sunflower. *Crop Res. (Hisar)* 21(3):332–334.
Shaw, F.R. 1941. Bee poisoning review of more important literature. *J. Econ. Entomol.* 34:16–21.
Shwetha, B.V. 2007. Studies on some behavioural aspects of little bee, *Apis florea* Fab. (Hymenoptera: Apidae), Msc Thesis, University of Agricultural Sciences Bangalore, India, pp. 115.
Sihag, R.C. 1982. Foraging behaviour and pollination ecology of honeybees in relation to agricultural crops. *Proc. Int. IUSSI Congr.* p. 39.
Sihag, R.C. 1983. Honey bee forage for grape (*Vitis vinifera* L.) juices during dearth period. *Indian Bee J.* 45:111–112.
Sihag, R.C. 1984. Influence of environmental factors on pollination activity of bees. *Environ. Ecol.* 2:149–152.
Sihag, R.C. 1986. Insect pollination increases seed production in cruciferous and umbelliferous crops. *J. Apic Res.* 25(2):121–126.
Sihag, R.C. 1990. Differential attractiveness of two cruciferous species to honeybees. In: G.K. Veeresh, B. Malik, C.A. Virktamath (Eds.), *Social Insects and the Environment Proceedings of 11th International Congress of IUSSI (International Union for the Study of Social Insects)*, India, pp. 434–436.
Sihag, R.C. 1991. Ecology of European honey bee (*Apis mellifera* L.) in semi-arid sub-tropicalclimates. 2. Seasonal incidence of diseases, pests, predators and enemies. *Korean J. Apic.* 6(1):16–26.

Sihag, R.C. 2000. Ecobiology of the little honeybee (*Apis florea*) in semi and subtropical climates of India. In: *Proceedings of the Fourth Asian Apicultural Association Conference*. Asian Apicultural Association, Kathmandu, pp. 46–49.

Sihag, R.C. and Abrol, D.P. 1986. Correlation and path-coefficient analysis of environmental factors influencing flight activity of *Apis florea* F. *J. Apic. Res.* 25(4):202–208.

Sihag, R.C. and Kapil, R.P. 1983. Foraging strategies of honeybees as determined by quality and quantity of nectar. *Proc. 5th Int Sympo. Polln.* Versailles, pp. 51–59.

Sihag, R.C. and Khatkar, S. 1999. Foraging pattern of three honeybee species on eight cultivars ofoilseed crops. 2. Foraging during the entire blooming period of the crops. *Intl. J. Trop. Agri.* 17: 253–261.

Sihag, R.C. and Rathi, A. 1992. Foraging modes and foraging rates of different bee pollination of pigeon pea *Cajanus cajan* (L.) MILLSP. In: G.K. Veeresh, R. Umashankar, K.N. Ganeshiah (Eds.), *Pollination in Tropics: Proc. Intl. Symp. Pollin. Trop.* 1993. IUSSI Indian Chapter, India, pp. 93–95.

Simpson, B.B., Neff, J.L. and Seigler, D. 1977. Krameria, free fatty acids and oil collecting bees. *Nature* 267:150–151.

Singh, S. 1982. *Beekeeping in India*. Indian Council of Agricultural Research, New Delhi, India, 214 pp.

Singh, Y.P. 1981. Studies on Pollen gathering capacity of Ind. Honey bee (*Apis cerana irrdica* F.) under Saharanpur conditions. *Prog. Horf.* 12:31–38.

Siriwat, W.R., Ratna, T., Thomas, E.R. and Benjamin, P.O. 2004. Nesting behaviour and nest architecture of *Apis florea* F. and *Apis andreniformis* in Thailand. In: I. Bernardinelli, N Milani (Eds.), *Proceedings of the First European Conference of Apidology "EurBee" Udine (Italy),* 19–24 September 2004, pp. 19–13.

Social Insects and the Environment. 1990. In: G.K. Veeresh, A.K. Mallick and C.A. Viraktamath (Eds.), *XI Congress IUSSI*. IUSSI, Bangalore, India, pp. 434–436.

Soman, A.G. and Chawda, S.S. 1996. A contribution to the biology and behaviour of the dwarf bee, *Apis florea* F. and its economic importance in Kutch, Gujarat, India. *Indian Bee J.* 58(2):81–88.

Sotavalta, O. 1952. Flight-tone and wing-stroke frequency of insects and the dynamics of insect flight. *Nature* 170:1057–1058.

Steffan-Dewenter, I. and Kuhn, A. 2003. Honeybee foraging in differentially structured landscapes. *Proc. R. Soc. Lond. Ser. B Biol. Sci.* 270:569–575.

Stiles, F.G. 1971. Time, energy, and territorially of the Anna Hummingbird (*Calypte anna*). *Science* 173:818–821.

Stiles, F.G. 1976. Taste preferences, color preferences, and flower choice in hummingbirds. *The Condor* 78:10–26.

Stute, K. 1957. Bisherige Ergebnisse von Versuchen mit « repellents » bei der Honigbiene. *Anz. Schdlingskd.* 30(3):38–43.

Suwannaong, G. 2000. Ultrastructure and Pheromones of the Mandibular Glands of Honeybee Foragers in Thailand. PhD Thesis, Chulalongkorn University, Thailand. p. 177.

Suwannapong, G., Yemor, T., Boonpakdee, C. and Benbow, M.E. 2011. *Nosema ceranae*, a new parasite in Thai honeybees. *J. Invertebr. Pathol.* 106:236–241.

Szabo, T.I. and Smith, M.V. 1972. The influence of light intensity and temperature on the activity of alfalfa leaf cutter bee. *Megachile rotundata* under field conditions. *J. Apic. Res.* 11:157–165.

Tahmasebi, G., Ebadi, R., Tajabadi, N., Akhoundi, M. and Faraji, S. 2002. The effects of geographical and climatic conditions on the morphological variation and separation of Iranian small honeybee (*Apis florea* F.) populations. *J. Sci. Technol. Agric. Nat. Res.* 6:169–176.

Tamm, S. and. Gass. C.L. 1986. Energy intake rates and nectar concentration preferences by hummingbirds. *Oecologia* 70:20–23.

Thakar, S.S. and Tonapi, K.V. 1962. Nesting behaviour of Indian honeybees. II. Nesting habits and comb cell differentiation in *Apis florea* Fab. *Indian Bee J.* 24:27–31.

Thakur, A.K., Sharma, O.P., Garg, R. and Dogra, G.S. 1982. Comparative studies on foraging behaviour of *Apis mellifera* and *A. cerana indica* on mustard. *Indian Bee J.* 44(4):91–92.

Tirgari, S.H. 1971. On the biology and manipulation of *Apis (Micrapis) florea* F. in Iran. *XXIII Int. Beekeep. Congr.* 330–332.

Tirgari, S., Polous, H., Fouroughi, H., Fakhrai, F., Nabavi, ·M.A. 1969. On the occurrence and biology of little bee (*Hicrapis flo rea* F.) and the first account on their migration (Hymenoptera, Apidae). II. *Natn. Congr. Ent. Pest Control Iran*: 4pp, Evine, Teheran (In Persian).

Underwood, B.A. 1991. Thermoregulation and energetic decision-making by the honeybees *Apis cerana*, *Apis dorsata* and *Apis laboriosa*. *J Exp. Biol.* 157:19–34.
Van Riper, W. 1958. Hummingbird feeding preferences. *The Auk* 75:100–101.
Verma, S.K. 1983. Studies on their foraging behaviour of *Apis cerana Indica* Fabr. in Jeolikot (Nainital,India). *Indian Bee J.* 45(1):5–7.
Verma, S.K. and Phogat, K.P.S. 1994. Impact of pollination by honey bees *(Apis cerana)* on yield of radish under valley condition of Himalayan hills. *Indian Bee J.* 45:183–186.
Visscher, P.K. and Seeley, T.D. 1982. Foraging strategy of honeybee colonies in a temperate deciduous forest. *Ecology* 63:1790–1801.
Vogel, S. 1974. Ölblumen und ölsammelnde Bienen. Tropische und Subtropische Pflanzenwelt (Mainz). *Akad. Wiss. Lit. Franz Steiner Verlag* 7:1–267.
von Frisch, K. 1967. *The Dance Language and Orientation of Bees.* Harvard University Press, Cambridge, 566 pp.
von Frisch, K. 1971. *Bees: Their Vision, Chemical Senses, and Language.* Cornell University Press, Ithaca, NY, p. 152.
Waller, C.D. 1970. Attracting honey bees to alfalfa with citral, geraniol and anise. *J. Apic. Res.* 9:9–12.
Waller, G.D. 1972. Evaluating responses of honeybees to sugar solutions using an artificial flower feeder. *Ann. Entomol Soc. Am.* 65:857–862.
Waser, N.M., Chittka, L., Price, M.V., Williams, N.M. and Ollerton, J. 1996. Generalization in pollination systems and why it matters. *Ecology* 77:1043–1060.
Wcislo, W.T. and Cane, J.H. 1996. Resource utilization by solitary bees (Hymenoptera: Apoidea), and exploitation by their natural enemies. *Ann. Rev. Entomol* 41:257–286.
Wenning, C.J. 2000. The consequences of swarming. *Am. Bee J.* 140:205–209.
Whitcombe, R.P. 1981. Temperature regulation and foraging activity of two species of honey bee (*Apis mellifera* and *A. florea*) in Oman. Anonymous. Joint Honey Production Project Apicultural Research Programme, Durham University Khabura Development Project, pp. 1–45.
Whitcombe, R.P. 1984. The biology of apis SPP. in Oman with special reference to Apis _ florea fab, Durham theses, Durham University. Available at Durham E-Theses: http://etheses.dur.ac.uk/7211/
Williams, I.H. 1979. Bee-keeping in the Sultanate of Oman. *Am. Bee J.* 510–511, 532–533, 535.
Willmer, P.G. 1983. Thermal constraints on activity pattern of nectar feeding insects. *Ecol. Entomol.* 8:455–469.
Wongsiri, S., Chanchao, C., Deowanish, S., Aemprapa, S., Chaiyawong, T., Petersen, S. and Leepitakrat, S. 2000. Honey bee diversity and beekeeping in Thailand. *Bee World* 81:20–29.
Wongsiri, S., Deowanish, S. and Thapa, R. 2002. Some behaviour diversity of *Apis dorsata, Apis florea* and *Apis andreniformis*. In: *Proc. 14th Int. Congr.*, IUSSI, Sapporo, Japan, p. 211.
Wongsiri, S. and Tangkanasing. P. 2000. *Biology of Honeybee (Thai version).* Chualongkorn, Bangkok. 213 pp.
Wongsiri, S., Rinderer, T.E. and Sylvester, H.A. 1991. *Biodiversity of Honeybees in Thailand.* Prachachon, Bangkok, pp. 50–63.
Woyke, J. 1981. Effect of sex allele homo-heterozygosity on honeybee colony population and on their production. 2. Unfavourable development conditions and restricted queens. *J. Apic. Res.* 20(3):148–155.
Woyke, J., Wilde, J. and Wilde, M. 2001a. Swarming, migration and absconding of *Apis dorsata* colonies. *Proceedings VII Internat. Conf. on Tropical Bees, and V Asian Apicultural Assoc. Conf.*, Chiang Mai, 19–25 March, 2000, pp. 183–188.
Woyke, J., Wilde, J. and Wilde, M. 2001b. A scientific note on *Apis laboriosa* winter nesting and brood rearing in the Himalayas. *Apidologie* 32:601–602.
Woyke, J., Wilde, J. and Wilde, M. 2003b. Periodic mass flights of *Apis Laboriosa* in Nepal. *Apidologie* 34:121–127.
Wratt, E.C. 1968. The pollinating activities of bumble bees and honey bees in relation to temperature, competing forage plants, and competition from other foragers. *J. Apic. Res.* 7:61–66.
Wyke, G.R. 1952. The preferences of honeybees for solutions of various sugars which occur in nectar. *J. Exp. Biol.* 29:511–528.

6

Diseases and Enemies of Dwarf Honeybees Apis florea *and* Apis andreniformis

Dharam P. Abrol

CONTENTS

Introduction

Honeybees, like all other organisms, suffer from many diseases and pathogens. And many of these diseases have been documented. Aristotle (384–322 BC) observed some disorders in honeybees around 2300 years ago that he mentioned in his book *Historia Animalium.* The Roman writer, Virgil, also described honeybee diseases some 300 years ago. But none of these observations could identify the diseases with certainty. It was believed that the presence or absence of any disease was due to the presence or absence of pathogens. Like other honeybee species, *Apis florea* (Fabricius, 1787) is also subject to attack by various diseases, enemies, parasites and predators. The dwarf honeybee *A. florea* is found in warm climates in parts of Oman, Iran and Pakistan, through the Indian subcontinent and Sri Lanka to as far east as Indonesia, but it is primarily confined to South East Asia. In spite of its small size, it competes well with other *Apis* species (Koeniger 1976) and is able to survive in very hot and dry climates with ambient temperatures reaching 50°C or more. *Apis andreniformis* F. Smith (1858) is the second honeybee species recognized recently and separated from *Apis florea*. Both *A. andreniformis* and *A. florea* have been reported to live cospecifically. Both species are distributed throughout tropical and subtropical Asia, including Southeast China, India, Burma, Laos, Vietnam, Malaysia, Indonesia (Java and Borneo), and the Philippines (Palawan). This bee is generally more defensive than *A. florea* and is known to attack when there are disturbances 3–4 m from the hive. However, it is less aggressive and does not show 'trembling' or 'shimmering' behaviour as a defensive response.

Combs of the dwarf honeybee are well covered with layers of workers clinging to each other. About three-quarters of the colony's worker population are deployed in forming this protective curtain of bees. When disturbed, this curtain shows a 'shimmering' movement; the individual bees shake their abdomens from side to side in a synchronous manner; at the same time, a hissing sound is released. If the colony is further disturbed, the worker bees raise their abdomens and take off from the curtain to attack the intruder, but the area of defensive attack is not as far as that of other species of bees. To cope up with ant attacks, worker bees coat both ends of the nest support with sticky strips of propolis or 'plant resin', from 2.5 to 4 cm wide. The use of this method is true in the case of *A. florea* only. Dwarf honeybees have successfully exploited a diversity of habitats in which their direct competition with other insects or predators is minimized. This is why this species is able to survive among solid contenders. Morphologically, dwarf honeybees are considered smaller compared to other bee species, but they have proven to have unexpected potential in maintaining ecological balance.

Dwarf honeybees are not ferocious like the other wild honeybees (*Apis dorsata*), but they make a defensive attack when there is some movement around their nest. *A. florea* is generally found in plains or lowland areas and in tropical areas of South East Asia. The bees are absent north of the Himalayas. To date, two species from the genus *Euvarroa* are believed to be associated with five honeybee species in Asia: the open-air nesters *A. andreniformis*, *A. florea*, *A. dorsata* as well as the cavity nesters *A. cerana* and *A. mellifera*. Delfinado and Baker (1974) made the first record of *Euvarroa sinhai* from *A. florea* in India. Similarly, Lekprayoon and Tangkanasing (1991) recorded *E. wongsirii* from *A. andreniformis* in Thailand.

Mite Pests of *Apis florea*

A number of mite species are associated with *Apis florea* and *A. andreniformis*, which cause considerable damage (Table 6.1). Mesostigmated mites are free-living predators whereas a few such as *Varroa* have adopted a parasitic mode of life (Hunter and Rosario 1988). Some are found in honeybee colonies as feeding on other saprophagous mites, insects, fungi and pollen. *Neocypholaelaps indica* is a common phoretic mite of *A. cerana, A. florea* and *A. dorsata* colonies in Asia (Delfinado-Baker et al. 1989; Needham et al. 2001). This mite is found on trees flowers where it feeds on pollen; during foraging of bees on flowers, it is carried to the hive where it feeds on stored pollen (Delfinado-Baker et al. 1989; Haq et al. 2001). Another species belonging to genus *Afrocypholaelaps* (Delfinado-Baker et al. 1989) is found with *A. cerana* colonies and not reported from *A. florea* colonies.

Euvarroa Mites

To date, two species from the genus *Euvarroa* are believed to be associated with five honeybee species in Asia: the open-air nesters *A. andreniformis*, *A. florea*, *A. dorsata* as well as the cavity nesters *A. cerana* and *A. mellifera*. Delfinado and Baker (1974) made the first reco of *Euvarroa sinhai* from *A. florea* in India. Similarly, Lekprayoon and Tangkanasing (1991) recorded *E. wongsirii* from *A. andreniformis* in Thailand.

TABLE 6.1
Mite Species Associated with *Apis florea*

A. florea	Distribution	References
Euvarroa sinhai	The Middle East, Pakistan, India, Nepal, Burma, Indochina, Malay Peninsula	Delfinado and Baker (1974)
Tropilaelaps clareae	India	Kapil and Aggarwal (1987b), Makkar and Chhuneja (2011), Singh et al. (2010), Abrol and Kakroo (1998)
Acarapis woodi	India	Atwal (1971), Chhillar et al. (2007)
A. andreniformis		
Euvarroa wonsirii	Burma, Indochina, Malaysia, Indonesia except islands beyond east of the Wallace line, Palawan Island	Lekprayoon and Tangkanasing (1991)

These two *Eurvarroa* species can be easily distinguished by the shape of their body and number of marginal setae. *E. sinhai* is pear-shaped with 39–40 marginal setae whereas *E. wongsirii* is triangular or wider posteriorly with 47–54 long setae (Delfinado and Baker 1974; Lekprayoon and Tangkanasing 1991). Lekprayoon and Tangkanasing (1991) reported *E. wongsirii* infesting *andreniformis* in Thailand and Malaysia. This species has not been found infesting *Apis florea* colonies so far. Similarly, *E. sinhai* has been reported infesting *A. florea* in India, Sri Lanka, Iran and Thailand (Delfinado and Baker 1974; Koeniger et al. 1983; Mossadegh 1991) and *A. andreniformis* in Thailand (Delfinado-Baker et al. 1989). *Euvarroa sinhai* has also been reported from hive debris of *Apis mellifera* colonies in India and Thailand (Kapil and Aggarwal 1987a; Lekprayoon and Tangkanasing 1991). Koeniger et al. (2002) reported *E. wongsirii* from *Apis dorsata* colonies in Borneo.

Kitprasert (1995) reported that dwarf honeybees, because of their hygienic behaviour, are able to reduce *Euvarroa* mite infestations in their colonies. Limited food supply of the single- comb open-air nesters, as well as reduced brood cell diameter of dwarf honeybees, may also contribute to suppressed mite populations (Rinderer et al. 1996). Kitprasert (1995) reported that swarming reduces *Euvarroa* populations by disrupting production of bee broods. Furthermore, drone production is seasonal and not a regular source of food for *Euvarroa* mites, resulting in reduction of their population. Otis (1996) reported the occurrence of *Euvarroa sinhai* throughout the distribution range of *A. florea* ranging from the Middle East in Asia to the Malay Peninsula. Another species, *E. wongsirii*, was reported by Lekprayoon and Tangkanasing (1991) as infesting *A. andreniformis* in Thailand. Later, this species was reported infesting *A. andreniformis* throughout its entire range of distribution from South China to Palawan Island in the Philippines (Otis 1996). Both species of mites overlap each other in their distribution. They more or less resemble each other with the difference that *Varroa* are oval and *Euvarroa* are triangular in shape. Similarly, *E. wongsirii* can be distinguished from *E. sinhai* by the broader triangular body (Lekprayoon and Tangkanasing 1991, 1993; Morin and Otis 1993).

Euvarroa and *Varroa* have similar biology. They can reproduce in drone broods only and then disperse via workers and drone bees (Akratanakul and Burgett 1976; Mossadegh and Birjandi 1986; Aggarwal 1988; Kapil and Aggarwal 1989; Morin and Otis 1993). As with *Varroa*, *Euvarroa* causes little damage to its endemic host colonies. This may be because of the intensive grooming behaviour of the worker bees and the seasonal presence of drone brood cells that reduces the population of the mite to a minimal. However, evidence suggests that *E. sinhai* can reproduce in worker brood cells of *A. mellifera* (Mossadegh 1990a). Also, Koeniger et al. (1993) reported that *E. sinhai* can survive in adult workers of *A. mellifera* and *A. cerana* in Thailand.

The mite *Euvarroa sinhai* was first described by Delfinado and Baker (1974) from drone broods of florea in India. Later, this species was identified infesting dwarf honeybees in Thailand (Akratanakul and Burgett 1976). Koeniger et al. (1983) found the adult female mites on adult worker bees in Sri Lanka but not on adult drones (Koeniger et al. 1983). Mossadegh and Komeili (1986) found the adult female mites on both adult worker and drone bees in addition to drone broods. This mite has been reported in hive debris from *A. mellifera* colonies in India (Kapil and Aggarwal 1987a) and Iran (Mossadegh 1997b).

Recent molecular and biogeographical studies of *A. florea* and *A. andreniformis* suggest that populations of the dwarf honeybees in Asia may not be as homogenous as previously thought. This finding raises questions about the genetic homogeneity of *Euvarroa* associated with these bees. Morin and Otis (1993) have reported some morphological differences in *E. sinhai* collected from India/Sri Lanka compared with specimens collected in Thailand. Evidently, it is expected to have genetic and morphological variations among different populations across different locations. This line of study will further our knowledge of *Euvarroa* diversity in Asia.

Other Mites (Acari) Associated with *A. florea*

Apart from *E. sinhai*, which is a parasitic mite of *A. florea* colonies in many areas including Iran, few other mites have been reported from this honeybee species. *Neocypholaelaps indica*, a pollen feeder and phoretic on all four species of *Apis*, was first recorded on *A. florea* colonies in Sri Lanka (Koeniger et al. 1983). Two other mites, *Suidasia pontifica* and *Tropilaelaps claerae*, were also reported on *A. florea* (Eikwort 1988). Two acarid mites, *Tyrophagus putrescentiae* (Schrank) and *Caloglyphus oudemansi* (Zak.), were identified from 43 colonies at two different areas (Ahwaz and Dezful) in southwest Iran. In some combs, mites were often found in great numbers of specimens per comb in all developmental stages. In 32 out of 43 infested colonies, hypopi were found that belonged to both species (Mossadegh 1998a). *T. putrescentiae* is principally a fungivorous mite, although it will feed on other organic remains (Eikwort 1988; Mossadegh and Kocheili 2003). In this study, both species of *T. putrescentiae* and *C. oudemansi* were found mainly on comb parts that dead broods and bees were infected with *Aspergillus* spp. In Iran, up to now *E. sinhai* has been restricted to *A. florea* with its geographical distribution, and it has been reported before (Mossadegh and Komeili 1982; Mossadegh 1991, 1993). This was the first report of two other mite species on this honeybee in Iran. *A. florea* is also a new host record for *T. putrescentiae* and *C. oudemansi* (Mossadegh 1998a). Association of viruses with mites is another area of interest that shall lead into new insights in our understanding of interactions of parasites with hosts.

Investigation of Microbial Pathogens in *A. florea*

Much is known of the bacteria and viruses associated with *A. mellifera*, but nothing is known about their association with *A. florea*. A published report of a virus disease in *A. florea* concerns the identification of black queen cell virus (BQCV) in an Iranian sample of diseased drone larvae (Mossadegh 1990b). BQCV is associated with *A. mellifera* queen cells that develop dark brown to black cell walls. This virus was also identified for the first time in January 1986 by Dr. B.V. Ball from samples of dead honeybee drone larvae of *A. florea* sent from southwest Iran to Rothamsted Research in the UK (Mossadegh 1990b). Infected larvae fail to pupate, and no adult emerges from the cells. The body colour of the larvae changes from pearly white to pale brown. The dead body of the larva becomes soft and liquefied (Mossadegh 1990b).

The presence of BQCV in *A. florea* drones was unexpected. This was the first report of the natural occurrence of the virus in broods other than the developing queen larvae and the first virus that has been identified in this bee species (Mossadegh 1990b). Further effect of the black queen cell virus on honeybee drone broods of *A. florea* was also investigated (Mossadegh 1992). From March 1988 to September 1991, the distribution of the BQCV and percentage of infected drone cells in the *A. florea* colonies was investigated in Khuzestan, southwest Iran (Mossadegh 1992). A total of 32 colonies with drone brood section combs were studied. Of the total colonies studied, 43.75% were infected with the virus. The number of infected drone cells was from 1 to 20 with a mean of 10.36 ± 4.43/colony. This was 1.12%–7.24% total of the whole drone cells of the colonies (Mossadegh 1992).

Other Microbial Pathogen in *A. florea*

The appearance of samples removed from the cells ranged from apparently normal to grossly abnormal. The appearance of some samples was consistent with bacterial infections, and of others, with those of viruses (Mossadegh and Hunter 1998; Hunter-Fujita et al. 1999; Hunter-Fujita et al. 2001). Bacteria morphologically indistinguishable from the causative organism of European foul brood, *Mellissococcuse pluton*, were detected in more than 30% of sample homogenates negatively stained with nigrosine. Viable organisms of *M. pluton* were isolated from at least one homogenate. Some types of organism detected in the homogenates and cultured from them have not been recorded for *A. mellifera* (Mossadegh and Hunter 1998; Hunter-Fujita et al. 1999; Hunter-Fujita et al. 2001). No morphologically intact virus particles were visible by electron microscopy. One sample did, however, react with antiserum against deformed wing virus. This is the first known report of this virus in *A. florea* (Mossadegh and Hunter 1998; Hunter-Fujita et al. 1999; Hunter-Fujita et al. 2001).

Virus Infections in *A. florea*

Iflaviridae

DWV infections of *A. cerana* have been reported in China, South Korea, Japan and Vietnam, (Kojima et al. 2011; Ai et al. 2012; Li et al. 2012; Forsgren et al. 2015), and in *A. florea* and *A. dorsata* in China (Zhang et al. 2012). The virus was detected in *V. destructor* and *T. mercedesae* mites, and higher levels of DWV were detected in *A. mellifera* infested by the mites in Thailand (Chantawannakul et al. 2006; Khongphinitbun Jong et al. 2015).

Dicistroviridae

BQCV was first isolated from queen larvae and pupae of *A. mellifera* (Bailey and Woods 1977). This virus has been found in *A mellifera* colonies in Thailand, South Korea, China, Japan and Vietnam (Sanpa and Chantawannakul 2009; Kojima et al. 2011; Ai et al. 2012; Reddy et al. 2013; Yang et al. 2013; Forsgren et al. 2015); *A. cerana* colonies in South Korea, China, Vietnam, Thailand and Japan (Choe et al. 2012; Li et al. 2012; Yang et al. 2013; Forsgren et al. 2015; Mookhploy et al. 2015); and *A. florea* and *A. dorsata* colonies from China and Thailand (Zhang et al. 2012; Mookhploy et al. 2015). *A. florea* BQCV has been reported by several investigators (Abrol and Bhat 1990; Verma et al. 1990; Oldroyd and Wongsiri 2007). In Thailand, the disease is found in colonies experiencing stress due to lack of food, excessive humidity, low worker population, poor-laying queens etc.

Based on the capsid coding region, the phylogenetic analysis revealed that BQCV isolates from northern Thailand, China, South Korea and Japan displayed a close relationship within Asia and split separately from South Africa and European regions, regardless of the host bee species from which the samples originated (Mookhploy et al. 2015). Similar findings were reported with the BQCV strains in Korea and Japan where the viral isolates from the same country or continent showed high levels of similarity (Kojima et al. 2011; Noh et al. 2013). The role of BQCV in honeybee mortality is currently poorly understood, particularly in association with other parasites such as the microsporidian *Nosema apis* (Bailey et al. 1983).

Fungal Infections in *A. florea*

Suwannapong et al. (2010) first recorded *Nosema ceranae* infecting *A. florea* workers in Thailand. They found that when healthy bees were fed a sucrose solution containing nosema spores, their survival rates were significantly lower as compared to control bees. Protein content of control bee hypopharyngeal glands 14 days post-inoculation was significantly higher compared with all treatments. The infection ratio of treated bees increased with time after inoculation, indicating that *N. ceranae* has a significant negative effect on honeybee hypopharyngeal gland protein production and shortening their life span.

Suwannapong et al. (2018) reported that Fumagillin, which has been successfully used for treatment of nosema cerane in *A. mellifera*, has been banned in Europe and elsewhere due to development of antibiotic resistance. As an alternative, they explored the ability of natural propolis extract in the case of experimentally infected colonies of *A. florea* and found that propolis-treated colonies had significantly lower infection loads in comparison to untreated bees. The propolis was found very effective in maintaining higher haemolymph trehalose levels and hypopharyngeal gland protein content in infected bees at par with healthy bees. However, this treatment needs further investigations to develop it as a possible alternative to Fumagillin to treat *N. ceranae* infections.

Nosema spp.

N. apis and *N. ceranae* are known to parasitize *A. mellifera* and *A. cerana* (Fries 1993; Klee et al. 2007), but *N. ceranae* has been found in *A. mellifera, A. cerana, A. florea, A. dorsata* and *A. koschevnikovi* (Chaimanee et al. 2010; Botías et al. 2012). Studies comparing the phylogenetic relationships among *N. ceranae* isolated from four different honeybee species in Thailand using polar tube proteins, important for microsporidian host invasion, showed three distinct clades (Chaimanee et al. 2011). *N. ceranae* isolated from *A. mellifera* grouped into the same clade as *N. ceranae* isolated from *A. cerana*, while *N. ceranae* isolated from *A. florea* and *A. dorsata* formed distinct clades. *A. mellifera* and *A. cerana* are cavity-nesting honeybees and are therefore more related to each other than to open-air, single comb-nesting species such as *A. dorsata* and *A. florea* (Alexander 1991).

Nosema ceranae was first observed in Thailand by Suwannapong et al. (2007) in *A. cerana* and can now be found throughout the country infecting *A. florea, A. dorsata* and *A. andreniformis* (Suwannapong et al. 2010, 2011). *Nosema* spores can be found in pollen removed from pollen storage areas in honeybee combs of *A. cerana, A. florea* and *A. dorsata*. The presence of *Nosema* spores in corbicular pollen may be due to self-contamination during the process of pollen collection. As the forager bee moistens the surface of its body with its protruding tongue and brushes the collected pollen, the spores inside its body become mixed with the pollen via an unknown mechanism. Spores can either come directly from the intestines after regurgitation or be present in the saliva. *Nosema* spores also are found in the honey of four species, *A. cerana, A. florea, A. dorsata* and *A. mellifera*. Experimental cross-infection of *A. cerana* and *A. florea* with *N. ceranae* isolated from both *A. cerana* and *A. florea* workers has been reported to divide rapidly in the honeybee midgut (Suwannapong et al. 2010).

Suwannapong et al. (2010) recorded the infection of *Apis florea* by *Nosema ceranae*, a newly identified pathogen of honeybee in Thailand that was initially isolated from *A. florea* workers. Each *Nosema* free-bee was fed 2 μL of 50% (w/v) sucrose solution containing 0, 10,000 20,000 or 40,000 *Nosema* spores/bee. The survival rates of treated bees were significantly lower compared to control bees. The infection ratio of bees treated with 40,000 spores/bee increased with time after inoculation. These results suggest that *N. ceranae* has a significant negative effect on honeybee hypopharyngeal gland protein production and contributes to their shortened life span.

Ascosphaera apis

The fungus *Ascosphaera apis*, which causes chalkbrood disease in honeybees, is now found throughout the world. In Asia, incidence of chalkbrood has been reported in China, Israel, Japan, Korea, the Philippines, Russia, Thailand and Turkey (Oldroyd and Wongsiri 2006; Aronstein and Murray 2010). Ingestion of the fungus by honeybee larvae allows *A. apis* to germinate within the gut and eventually penetrate the body cavity, resulting in significant mechanical and enzymatic damage; dead, dried larvae, named chalkbrood mummies, can be observed within uncapped brood cells (Heath 1982; Aronstein and Murray 2010). The fungus can produce lytic enzymes (protease and beta-N-acetylglucosaminidase) that may be an important factor that assists fungal invasion in the bee larvae (Theantana and Chantawannakul

2008). In *A. mellifera*, chalkbrood disease can result in lowered colony productivity but rarely results in colony death (Heath 1982; Aronstein and Murray 2010). It is also not considered to be a serious disease of honeybees in Asia, even though it has been reported more widespread than stonebrood disease, which is caused by the fungi from genus *Aspergillus* (Akratanakul 1990; Ra et al. 2012). It has not been reported infesting colonies of dwarf honeybee so far.

Stonebrood and Some Other Fungi Associated with *A. florea* in Iran

This study was prompted by an initial survey (October 1986) on the geographical distribution of the parasitic mite *Euvarroa sinhai* on *A. florea* in Iran (Mossadegh 1991).

During that survey, numerous diseased brood samples with symptoms similar to stonebrood of *A. mellifera* were observed. This was the first extended report of the occurrence of stonebrood-causing fungi and other fungi on *A. florea* in Iran and, we assume, in the world (Alizadeh and Mossadegh 1994). The literature reviewed did not mention these species of *Aspergillus* as being present in *A. florea* colonies, and it did not mention the other three fungal species reported here. Maassen (1906) reported that stonebrood disease was a major problem in *A. mellifera* in Germany (48). Since then, the disease has been reported in *A. mellifera* from many parts of the world (Gilliam and Vandenberg 1990; Mossadegh 1990a). Despite these numerous reports, stonebrood is still considered of minor importance to *A. mellifera* (Gilliam and Vandenberg 1990; Bailey and Ball 1991). Our findings have focused attention on the vulnerability of the dwarf honeybee to stonebrood fungi. Since this group of fungi is relatively unspecialized, with a wide saprophytic capability on combs and other substrates, certain conditions could lead to a buildup of fungal populations that would favour increased mortality of larvae and pupae, and also the probability of aflatoxin production, which may contaminate stored honey and wax (Gilliam and Vandenberg 1990; Alizadeh and Mossadegh 1994).

The pathogenicity of these fungi to *A. florea* adults has not yet been studied. It is, however, possible that the fungi can multiply in adult bees, as they do in *A. mellifera.* Alizadeh and Mossadegh (1994) reported that both larvae and pupae of *A. florea*, especially those of drones, were susceptible to attack by these fungi; many naturally infected larvae and pupae of drones had died in their capped cells. Caps of cells containing infested brood were extraordinarily thickened, perhaps for the containment of the disease in individual cells. Infested worker broods were rarely observed to be capped (Mossadegh et al. 1997). This is perhaps due to the nature of worker brood rearing, which continues from February to November (Mossadegh 1990a). It is possible that adult bees uncap and remove diseased broods prior to egg laying by the queen in these cells.

These results indicate that stonebrood is a widespread and potentially destructive disease of the dwarf honeybee in southwest Iran. Of the fungi studied, *A. flavus* appeared most often associated with drone pupae. *A. niger* and *A. fumigates* were also major causes of the disease. It must be emphasized that *A. niger* has not been reported to cause stonebrood in *A. mellifera* (6). There were many diseased samples in which mixed infections by these species – differing in appearance, size and morphology – were noticed. The other three fungi (*Paecilomyces* sp., *Penicillium* sp. and *Phoma* sp.) found might weaken colonies and make them more susceptible to stonebrood and other viral or bacterial diseases. The pathogenesis of stonebrood on the dwarf honeybee was not investigated in the present study. In other bees, such as *A. mellifera*, mortality involving stonebrood is believed to be caused by toxic products of the fungus (Trung 2010) or by both physical and chemical actions (Burnside 1930).

It is assumed that similar processes may occur in the dwarf honeybee. We clearly observed deep penetration of all interior tissues of affected larvae and pupae by the fungal hyphae, but no chemical analyses of diseased tissues were carried out. Therefore, additional study of these aspects is needed (Alizadeh and Mossadegh 1994; Mossadegh et al. 1997b).

Tropilaelaps spp.

Tropilaelaps spp. mites have been observed with colonies of *Apis cerana* and *Apis florea* (Delfinado-Baker 1982; Wongsiri et al. 1989) and *T. mercedesae* individuals with adult bees of *A. florea* (Kapil) and Aggarwal 1987b.

Bacterial Infections in *A. florea*

Paenibacillus larvae

Paenibacillus larvae, the causative agent of a disease called American Foulbrood (AFB), has a nearly ubiquitous distribution, including in Asia (Genersch 2010). The disease was found in *A. cerana* in India (Singh 1961) and *A. mellifera* in Taiwan (Yen and Chyn 1971). Bee larvae become infected when *P. larvae* spores are ingested; disease results in larvae that are brownish, semifluid and glue-like, before drying to a hard scale on the lower cell wall (Genersch 2010). Scales are highly infective and contain millions of spores that may be infective for several decades (Hasemann 1961). When artificially fed spores of *P. larvae*, *A. cerana* showed more resistance to infection than *A. mellifera* (Ho and Chen 2001). The occurrence of *Paenibacillus larvae* in *Apis florea* has not been reported so far.

Melissococcus plutonius

Melissococcus plutonius is a ubiquitously distributed bacterium of honeybees that has been detected throughout Asia, including in *A. mellifera* (Akratanakul 1990), *A. cerana* (Diwan et al. 1971; Bailey 1974; Zhou et al. 2000; Rana et al. 2012), *A. laboriosa* (Allen et al. 1990) and *A. florea* (Saraithong et al. 2015). *M. plutonius* primarily affects honeybee larvae, causing a disease known as European foulbrood (EFB) that is characterized by dead, twisted broods within their cells. The resistance of Asian honeybees towards some bacterial pathogens may be due to their unique bacterial community structures inhibiting pathogen growth or infection. Studies have shown that gut bacteria in *A. mellifera, A. cerana, A. florea* and *A. dorsata* (Makkar and Chhuneja 2011). Saraithong et al. (2015) analysed the different castes of *A. florea* for the presence of bacteria. The bacteria identified in the foregut of worker bees are *Bacillus cereus* and also in the queen foregut and the drone foregut. All three castes' foreguts have the same microbe *Bacillus cereus* as a predominant organism. Likewise in the midgut of worker bees, the bacteria identified was *Bacillus* sp.; in the queen midgut the bacteria found was *Bacillus licheniformis*, and in the drone midgut *Acinetobacter iwoffi* was found. Diverse microflora were found in the midgut of all the castes, *Bacillus licheniformis* in the midgut of the queen, *Bacillus* sp. in the workers' midgut and *Acinetobacter iwoffi* in the drones' midgut and in the hindgut of workers; in the queen's midgut we found *Bacillus cereus*, and in the drone hindgut *Klebsiella varricola* was found. In *A. florea*, the *Bacillus* sp. was the predominant organism found, and worker and queen castes have the same *Bacillus* genus organism.

The bacteria identified differed from gut to gut within the species and also differed from caste to caste. The bacteria that were found in drones differed from bacteria found in workers and queen. The bacteria isolated from different guts of *A. florea* were put to construct the phylogenetic tree by the neighbour joining method, and in this similarity probabilities were shown. Bacterial species with maximum query coverage are identified, and those sequences are submitted to GenBank for the accession numbers to determine the diversity in gut microflora by constructing a phylogenetic tree in the future. Suma (2015) found two species of gut bacteria, *Acinetobacter iwoffi* and *Klebsiella varricola*, only in drones of *Apis florea* as compared with other species.

Saraithong et al. (2015) studied the colonization and composition of bacterial flora in dwarf Asian honeybee (*Apis florea*) larvae from three locations in Chiang Mai province, Thailand. *A. florea* larvae harbour bacteria belonging to two phyla (*Firmicutes* and *Proteobacteria*), five classes (*Alphaproteobacteria*, *Betaproteobacteria*, *Gammaproteobacteria*, *Bacilli*, and *Clostridia*), six genera (*Clostridium*, *Gilliamella*, *Melissococcus*, *Lactobacillus*, *Saccharibacter* and *Snodgrassella*) and an unknown genus from uncultured bacterial species. Parichehreh et al. (2017) showed that the species *E. faecium, E. faecalis* and *E. hirae*, which are associated with *A. florea*, occur in different parts of Iran. They successfully isolated the species of *Enterococcus* associated with *A. florea* similar to *Lactobacillus* isolations from *A. florea*.

Apis species that engage in symbiotic association with lactic acid bacteria (LAB) have diverse functions on their hosts. The results suggested that *Enterococcus faecium*, the most common bacteria in the *Apis* species, was associated with *A. florea* populations of Roudan, Bandar Abbas and Iranshahr, and

Enterococcus faecalis was found in the gastrointestines of bees of Gachsaran. *E. hirae* was isolated from the Qeshm population. The bees in Behbahan, Dehloran, Jahrom, Bushehr, Kangan, Ahvaz, Jiroft, Kahnouj and Fasa had no *Enterococcus* bacteria.

According to phylogenetic analysis, five different phenotypes belonging to three different species, including *Enterococcus faecalis*, *Enterococcus faecium* and *Enterococcus hirae*, formed *Enterococcus* flora found in the gastrointestinal tract of dwarf honeybee (*A. florea*). *Apis florea* is one of the most important pollinators in the southern provinces of Khuzestan, Bushehr and Sistan, and Balouchestan (Parichehreh et al. 2014). One hundred isolates were gram-stained by Parichehreh et al. (2017) and tested for catalase reaction in the gastrointestinal tract of Asian dwarf honeybee, *Apis florea*, in Iran. By using bacterial universal primers, the 16S rDNA gene of bacterial colonies was amplified. 16S rDNA genes from 30 bacteria were sequenced. Phylogenetic analysis showed that *Enterococcus* flora in the gastrointestinal tract of *A. florea* contained five phenotypes that were classified in the species *E. faecium*, *E. faecalis* and *E. hirae*. Based on the specific association between bacteria and *A. florea*, we divided the Asian dwarf honeybee populations into four categories.

These results showed that *Enterococcus* has a modest fauna in *Apis florea*. Overall, it can be concluded that vegetation, climate conditions and latitude have a considerable effect on *Enterococcus* bacteria diversity in dwarf honeybees' gastrointestinal tracts. Based on this fact that each region has diverse flora, we could hypothesize that *Enterococcus* diversity might be due to differences in nutrient content of nectar and pollen, or microbes found on the flowers of each region. A study by Parichehreh et al. (2017) led to the isolation of *E. faecium, E. faecalis* and *E. hirae* from the stomach of honey of *A. florea* from different areas of Iran. Further research is required to identify new isolates of native LABs with commercial properties. These results can be documented in a bacteria bank for the future.

Insect Pests and Predators

Moths Attacking Honeybee Colonies

Two species of moths known as the greater wax moth *Galleria mellonella* (L.) and the Indian meal moth *Plodia interpunctella* (Hub.) are reported to attack nests of *Apis florea*. The former consumes wax and often does considerable damage to combs of traditional and modern beehives, while the latter feeds on cereals, dried fruits, meal, dead insects, nuts and many other stored products (Mossadegh, and Komeili 1982; Mossadegh 1997a). *Galleria* occurs throughout the world, almost wherever honeybees are kept. It infests colonies of *A. cerana, A. dorsata* and *A. mellifera* (Morse and Nowogrodzki 1990) as well as three Asian *Apis* species in India (Southwick 1985). The greater wax moth *G. mellonella* is a pest of bees all over areas where beekeeping with *A. mellifera* is practised. It has been widely spread by migratory beekeeping over the regions where *A. mellifera* does not exist. When the bees of *A. florea* colonies desert their comb or become weak, *Galleria* larvae can quickly destroy the comb to a mass of webbing and debris (Mossadegh 1998b).

Conopid Fly

Physocephala paralliventris Kröber (Conopidae) is a parasitic Conopid fly of three *Apis* species. It grasps bees in flight deposits a tiny larva on its integument which penetrates bees cuticle consuming internal contents of the bees (Tingek et al. 2004). Oldroyd and Nanork (2009) observed fly larvae in abdomens of *Apis florea* workers in Thailand, which could be potentially devastating. Hunting of *Apis florea* colonies is not as devastating as in the case of *Apis dorsata* and *A. laboriosa*, as the level of harvesting is far lees compared to the growth rate of colonies (Oldroyd and Wongsiri 2006).

Wasp Attack

The drones of *Apis florea* were preyed upon by *Vespa affnis* in their congregation areas (Koeniger et al. 1994). The attack of wasps was not restricted to individuals in flight, but occasionally whole colonies of honeybees were wiped out. Seeley et al. (1982) reported *Vespa tropica* destroyed an *Apis florea* colony in three hours.

Attack of Ants

The weaver ant, *Oecophylla smaragdina*, is a more serious enemy of dwarf honeybees (Seeley et al. 1982; Duangphakdee et al. 2005, 2009). It shares the tree environment with bees and grabs the foragers that are leaving or retuning. Once grabed by a single *O. smaragdina* worker, the bees have little chance of freeing themselves, but if an ant falls in the nest, then workers of *A. florea* in 10–15 numbers can mob and crush it (Seeley et al. 1982). The bees also use air blasts and buzzing flights to deter ants from attack (Seeley et al. 1982). Seeley et al. (1982) found that the worker bees line up side by side in several rows, forming a permanent barrier that is effective against any walking insects. Also, the workers of dwarf honeybees cover the branches with sticky plant resins to immobilize the intruders (Seeley et al. 1982). This resinous substance of *A. florea* and *A. andreniformis* has a strong repellent effect on *O. smaragdina* ants (Duangphakdee et al. 2005, 2009). This system is a constant feature of *Apis florea* (Duangphakdee et al. 2005, 2009). If this barrier is disturbed, the bees push with their heads and mandibles to remove the object (SenSarma et al. 2000).

Bee Eater Birds

Mosadegh reported three species of bee-eater birds belonging to genus *Merops* preying on bees in flight as well as near bee hives. They include the golden bee-eater, *Merops apiaster* L.; the Persian bee-eater or blue chick bee-eater, *M. superciliosus persicus*; and the little green bee-eater, *M. orientalis baluchicus* Neuman. All three species are migrants in Iran. They can be seen from early March to November depending on location and species. The golden bee-eater and Persian bee-eater occupy almost the same habitat and ecological niche; however, the latter prefer rather semi-arid plains with sparse vegetation. The third species seeks warmer, desert-like habitats.

Bee eater birds feed almost on flying insects (Fry 1983, 1991). When these birds locate an apiary, they can be seen every day around and above the hives. They invade the hives at certain hours of the day, particularly in the morning and late afternoon. On cold and cloudy days, they feed on bees and stay much longer around the beehives. Flocks of up to 400 birds flying around the apiaries were seen catching foraging bees arriving or leaving the hives. Sometimes they sit on the trees, walls, telephone wires or even on the hives or ground as a group, and they watch for flying bees. They are a nuisance of queen rearing in the apiaries. When the birds leave the apiary, the bees start their activities again with huge crowding around the hives and water sources (Fry 1983, 1991). Fry (1984) reported that the gut contents of bee-eater birds such as *Merops orientalis* and *M. superciliosus* contained 30% of *Apis florea* as a part of their diet.

Retreat Strategies

Apis florea mostly depend upon absconding and retreat strategies to defend against predators (Seeley et al. 1982), but when constantly disturbed, they may attack in small groups that may be joined by a large number of workers (Seeley et al. 1982). Once the stinging begins, it continues with the release of 2-decen-1-yl-acetate, the 'enemy-marking pheromone' (Koeniger et al. 1979).

The stings are relatively painless to humans and may not repel decisive attacks by tree shrews, monkeys or humans (Seeley et al. 1982). Wongsiri et al. (1996) reported that *A. andreniformis* is more prone to attacking and possesses a separate alarm pheromone.

Hissing

Hissing has been reported to occur in *A. florea* and *A. cerana* (Seeley et al. 1982; SenSarma et al. 2002). After hissing nest activities of the bees cease, foraging flights commence (SenSarma et al. 2002) – making the presence less conspicuous. Pirk et al. (2002) described testudo behaviour in *Apis florea*. Oldroyd and Wongsiri (2006) reported that attacking appears to be more easily elicited in *A. andreniformis* than in *A. florea* whose colonies are more bent on attacking predators.

Defence Strategies

Defence strategies are more important in open nesting Asian bees. Abdomen shaking was observed in *A. florea* and *A. andreniformis* (Butler 1954; Koeniger and Fuchs 1973; SenSarma et al. 2002; Oldroyd and Wongsiri 2006), while the giant honeybees, *A. dorsata* and *A. laboriosa*, showed the related 'shimmering' display of 'body-twisting'; the energetic worker piping of returning foragers obviously carries a strong message of imminent danger. *A. florea* workers detect and report enemies over distances of more than 10 m from the nest, thus truly functioning as far-reaching sensory organs for the colony (Seeley 1994). Collecting and communicating information on the equally essential environmental conditions and threats to the colony has, however, been widely overlooked and may reveal fascinating details in future research.

Hygienic Behaviour in the Dwarf Honeybee *Apis florea*

Woyke et al. (2012) found the first evidence of the hygienic behaviour of *A. florea* honeybee colonies in Thailand. They removed 89% of freeze-killed and 95% of pin-killed brood in 24 and 48 hours, respectively. Thus, the *A. florea* colonies could be recognized as hygienic honeybees. Woyke et al. (2011) suggested that *A. florea* removes dead broods. *Euvarroa* infestations are generally low in honeybee colonies, with decreased infestation likely due to host hygienic behaviour of dwarf honeybee workers removing dead broods together with their infesting mites (Kitprasert 1995). Limited food supply of the single comb open-air nesters, as well as reduced brood cell diameter of dwarf honeybees, may also contribute to suppressed mite populations (Rinderer et al. 1996). Drone reproduction is also seasonal, and swarming further reduces *Euvarroa* populations within colonies by disrupting bee brood and hence mite reproduction (Kitprasert 1995).

Conclusion

Apis florea honeybees have an important role in the pollination of plant species. Removing these bees from the ecosystem causes great damage to agricultural production. These honeybees have a high ability to tolerate high temperatures and a higher ability to adapt to the ecosystem conditions in comparison to *Apis mellifera* honeybees. Unfortunately, no specific studies have been conducted regarding the biological, phylogenetic and biochemical status of this species, and these studies are necessary to complete information about this species. *Apis florea* needs conservation, as this species is the most important pollinator for agriculture production and sustainability of ecosystems.

REFERENCES

Aarestrup, F.M., Seyfarth, A.M., Emborg, H.D., Pedersen, K., Hendriksen, R.S. Bager, F. 2001. Effect of abolishment of the use of antimicrobial agents for growth promotion on occurrence of antimicrobial resistance in fecal enterococci from food animals in Denmark. *Antimicrob. Agents Chemother.* 45(7), 2054–2059.

Abrol, D.P., Bhat, A.A. 1990. Studies on Thai sac brood virus affecting indigenous honey bee *Apis cerana* indica Fab. Colonies-prospects and future strategies. *J. Anim. Morphol. Physiol.* 37, 101–108.

Abrol, D.P., Kakroo, S.K. 1998. Pests, pathogen and predators of honeybees *Apis cerana* in Jammu. *Indian Bee J.* 60, 147–150.

Abrol, D.P., Putatunda, B.N. 1995. Discovery of an ectoparasitic mite Tropilaelaps koenigerum Delifinado-Baker & Baker on *Apis dorsata*, *A. meliffera* L. and *Apis cerana* F. in Jammu and Kashmir, India. *Current Sci.* 68, 90.

Aggarwal, K. 1988. Incidence of *Tropilaelaps clareae* on three Apis species in Hisar (India). In: Needham, G.R., Page, R.E.Jr, Delfinado-Baker, M., Bowmann, C.E. (eds) *Africanized Honeybees and Bee Mites*, Ellis Horwood, Chichester, pp. 396–403.

Ahmadi, A.A. 1988. Parasites of *Apis florea* in Iran. *Bee World* 69: 159–161.

Ai, H., Yan, X., Han, R. 2012. Occurrence and prevalence of seven bee viruses in *Apis mellifera* and *Apis cerana* apiaries in China. *J. Invertebr. Pathol.* 109, 160–164.

Akratanakul, P. 1990. Beekeeping in Asia. FAO (Food and Agriculture Organisation of the United Nations), Agricultural Services. Bulletin 68/4. Rome, Italy.

Akratanakul, P., Burgett, M. 1976. *Euvarroa sinhai* Delfindao & Baker (Acarina: Mesostigmata) a parasitic mite of *Apis florea*. *J. Apic. Res.* 15, 11–13.

Alexander, B. 1991. A cladistics analysis of the genus Apis. In: Smith, D.R. (ed.) *Diversity in the Genus Apis.* Westview Press, Boulder, pp. 1–28.

Alippi, A.M. 1995. Detection of Bacillus larvae spores in Argentinian honeys by using a semiselective medium. *Microbiologia* 11, 343–350.

Alizadeh, A., Mossadegh, M.S. 1994. Stonebrood and some other fungi associated with Apis florea in Iran. *J. Apic. Res.* 33(4), 213–218.

Allen, M.R., Ball, B.V. 1996. The incidence and world distribution of honey bee viruses. *Bee World* 77, 141–162.

Allen, M.F., Ball, B.V., Underwood, B.A. 1990. An isolate of Melissococcus pluton from *Apis laboriosa. J. Invertebr. Pathol.* 55, 439–440.

Anderson, D.L. 1991. Kashmir bee virus – A relatively harmless virus of honey bee colonies. *Am. Bee J.* 131, 767–770.

Anderson, D.L., Morgan, M.J. 2007. Genetic and morphological variation of bee-parasitic Tropilaelaps mites (Acari: Laelapidae): New and re-defined species. *Exp. Appl. Acarol.* 43, 1–24.

Arai, R., Miyoshi-Akiyama, T., Okumura, K., Morinaga, Y., Wu, M., Sugimura, Y., Yoshiyama, M., Okura, M., Kirakae, T., Takamatsu, D. 2014. Development of duplex PCR assay for detection and differentiation of typical and atypical Melissococcus plutonius strains. *J. Vet. Med. Sci.* 76, 491–498.

Arai, R., Tominaga, K., Wu, M., Okura, M., Ito, K. et al. 2012. Diversity of Melissococcus plutonius from honey bee larvae in Japan and experimental reproduction of European foulbrood with cultured atypical isolates. *PLoS ONE* 7, e33708.

Aronstein, K.A., Murray, K.D. 2010. Chalkbrood disease in honey bees. *J. Invertebr. Pathol.* 103, 520–525.

Atwal, A.S. 1971. Acarine disease problem of the Indian honey bee, *Apis indica* F. *Am. Bee J.* 111, 186–187.

Bailey, L. 1974. An unusal type of Streptococcus pluton from eastern hive bee. *J. Invertebr. Pathol.* 23, 246–247.

Bailey, L., Ball, B.V. 1991. *Honey Bee Pathology.* Academic Press Ltd., London. p. 193.

Bailey, L., Ball, B.V., Perry, J.N. 1983.Association of viruses with two protozoal pathogens of the honey bee. *An. Appl. Biol.* 103, 13–20.

Bailey, L., Collins, M.D. 1982a. Reclassification of 'Streptococcus pluton' (White) in a new genus Melissococcus, as Melissococcus pluton nom. rev.; comb. nov. *J. Appl. Bacteriol.* 53, 215–217.

Bailey, L., Collins, M.D. 1982b. Taxonomic studies on Streptococcus pluton. *J. Appl. Bacteriol.* 53, 209–213.

Bailey, L., Woods, R.D. 1977. Two more small RNA viruses from honey bees and further observations on sacbrood and acute bee-paralysis viruses. *J. Gen. Virol.* 37(1), 175–182.

Botías, C., Anderson, D.L., Meana, A., Garrido-Bail6n, E., Martín-Hernández, R., Higes, M. 2012. Furthur evidence of an oriental origin for *Nosema ceranae* (Microsporidia: Nosematidae). *J. Invertebr. Pathol.* 110(1), 108–113.

Budge, G.E., Barrett, B., Jones, B., Pietravalle, S., Marris, G., Chantawannakul, P., Thwaites, R., Hall, J., Cuthbertson, A.G., Brown, M.A. 2010. The occurrence of *Melissococcus plutonius* in healthy colonies of *Apis mellifera* and the efficacy of European foulbrood control measures. *J. Invertebr. Pathol.* 105, 164–170.

Burnside, C.E. 1930. Fungous diseases of the honeybee. Technical Bulletin United State Department of Agriculture 149, 1–43.

Butler, C.G. 1954. *The World of the Honeybee.* Collins, London.

Cervancia, C.R. 1993. Philippines beekeeping status of research and development. In: *Proc. Beenet Asia: Workshop on priorities in R&D on beekeeping in tropical Asia.* Kuala Lumpur, Malaysia. pp. 49–63.

Chaimanee, V., Chen, Y., Pettis, J.S., Scott Cornman, R., Chantawannakul, P. 2011. Phylogenetic analysis of *Nosema ceranae* isolated from European and Asian honeybees in Northern Thailand. *J. Invertebr. Pathol.* 107, 229–233.

Chaimanee, V., Warrit, N., Chantawannakul, P. 2010. Infections of *Nosema ceranae* in four different honeybee species. *J. Invertebr Pathol.* 105, 207–210.

Chantawannakul, P., Puchanichanthranon, T. 2005. Inhibitory effects of some medicinal plant extracts on the growth of *Ascosphaera apis. Acta Hort.* 678, 183–186.

Chantawannakul, P., Ward, L., Boonham, N., Brown, M. 2006. A scientific note on the detection of honeybee viruses using real-time PCR (TaqMan) in *Varroa mites* collected from a Thai honeybee (*Apis mellifera*) apiary. *J. Invertebr. Pathol.* 91, 69–73.

Chhillar, B.S., Gulati, R., Bhatnagar, P. 2007. Mite pests of honeybees In: *Agricultural Acarology.* Daya Publishing House, Delhi, India, pp. 117–40.

Chinh, P.H. 1998. Thai sac brood virus disease control in Vietnam. In: Matsuka, M., Verma, L.R., Wongsiri, S., Shrestha, K.K., Partap, U. (eds.) Asian Bees and Beekeeping. Proceedings of the 4th Asian Apiculture Association International Conference Science Publishers, Enfield, New Hampshire, Kathmandu, Nepal, pp. 57–59.

Choe, S.E., Nguyen, L.T.K., Noh, J.H., Koh, H.B., Jean, Y.H., Kweon, C.H., Kang, S.W. 2012. Prevalence and distribution of six bee viruses in Korean *Apis cerana* populations. *J. Invertebr. Pathol.* 109(3): 330–333.

Cintas, L.M., Herranz, C., Hernández, P.E., Casaus, M.P., Nes, L.F. 2001. Review: Bacteriocins of lactic acid bacteria. *Food Sci. Tech. Int.* 7, 281–305.

Delfinado, M.D., Baker, E.W. 1974. Varroidae, a new family of Mites on honey bees (Mesostigmata: Acarina). *J. Wash. Acad. Sci.* 64, 4–10.

Delfinado-Baker, M. 1982. New records for Tropilaelaps clareae from colonies of *Apis cerana indica. Am. Bee J.* 122, 382.

Delfinado-Baker, M., Baker, E.W., Phoon, A.C.G. 1989. Mites (Acari) associated with bees (Apidae) in Asia, with description of a new species. *Am. Bee J.* 129, 609–610.

Disayathanoowat, T., Yoshiyama, M., Kimura, K., Chantawannakul, P. 2012. Isolation and characterization of bacteria from the midgut of the Asian honey bee (*Apis cerana indica*). *J. Apic. Res.* 51(4), 312–319.

Disayathanoowat, T., Young, J.P.W., Helgason, T., Chantawannakul, P. 2011. T-RFLP analysis of bacterial communities in the midguts of *Apis mellifera* and *Apis cerana* honey bees in Thailand. *FEMS Microbiol. Ecol.* 79(2), 273–281.

Diwan, V.V., Kshirsagar, K.K., Ramama Rao, A.V., Raghunath, D., Bhambure, C.S., Godbole, S.H. 1971. Occurrence of a new bacterial disease of Indian honey bee *Apis indica.* F. *Curr. Sci.* 40, 196–197.

Duangphakdee, O., Koeniger, N., Deowanish, S., Hepburn, H.R., Wongsiri, S. 2009. Ant repellent resins of honeybees and stingless bees. *Insectes Soc.* 56, 333–339.

Duangphakdee, O., Koeniger, N., Koeniger, G., Wongsiri, S., Deowanish, S. 2005. Reinforcing a barrier – A specific social defense of the dwarf honeybee (*Apis florea*) released by the weaver ant (*Oecophylla smaragdina*). *Apidologie* 36, 505–511.

Eickwort, G.C. 1988. The origin of mites associated with honey bees. In: Needham, G.R., Page, R.E. Jr, Delfinado-Baker, M., Bowman, C.E. (eds) *Africanized honey bees and bee mites.* Ellis Horwood, Chichester, UK, pp. 327–384.

Forsgren, E., Wei, S., Guiling, D. et al. 2015. Preliminary observations on possible pathogen spillover from *Apis mellifera* to *Apis cerana. Apidologie* 46, 265–276.

Fries, I. 1993. Nosema apis – A parasite in the honey bee colony. *Bee World* 74(1), 5–19.

Fries, I., Feng, F., Da Silva, A., Slemenda, S.B., Pieniazek, N.J. 1996. *Nosema ceranae* n. sp. (Microspora, Nosematidae), morphological and molecular characterization of a microsporidian parasite of the Asian honey bee *Apis cerana* (Hymenoptera, Apidae). *Eur. J.Protistol.* 32(3), 356–65.

Fry, C.H. 1983. Honeybee predation by bee-eaters, with economic consideration. *Bee World* 64(2), 65–78.

Fry, C.H. 1984. *The Bee-Eaters.* T&AD Poyser, Town Head House, Calton.

Fry, C.H. 1991. *The Bee Eaters.* Mackays of Chatham PLC, Chatham, Kent, UK, 304pp.

Genersch, E. 2010. American Foulbrood in honeybees and its causative agent, *Paenibacillus larvae. J. Invertebr. Pathol.* 103, S10–S19.

Genersch, E., Ashiralieva, A., Fries, I. 2005. Strain- and genotype-specific differences in virulence of *Paenibacillus larvae* subsp larvae, a bacterial pathogen causing American foulbrood disease in honeybees. *Appl. Environ. Microbiol.* 71, 7551–7555.

Genersch, E., Forsgren, E., Pentikainen, J., Ashiralieva, A., Rauch, S., Kilwinski, J., Fries, I. 2006a. Reclassification of *Paenibacillus larvae* subsp pulvifaciens and *Paenibacillus larvae* subsp larvae as *Paenibacillus larvae* without subspecies differentiation. *Int. J. Syst. Evol. Microbiol.* 56, 501–511.

Genersch, E., Yue, C., Fries, I., de Miranda, J.R. 2006b. Detection of deformed wing virus, a honeybee viral pathogen, in bumble bees (*Bombus terrestris* and *Bombus pascuorum*) with wing deformities. *J. Invertebr. Pathol.* 91, 61–63.

Gilliam, M., Vandenberg, J.D. 1990. Fungi. In: Morse, R.A., Nowogro- dzki, R. (eds.) *Honey bee pests, predators, and diseases.* Cornvell University Press, Ithaca, NY (2nd ed.), pp. 64–90.

Hansen, H., Brødsgaard, C. 1999. American foulbrood: A review of its biology, diagnosis and control. *Bee World* 80, 5–23.

Haq, M.A., Halliday, R.B., Walter, D.E., Proctor, H.C., Norton, R.A. 2001. Life cycle and behavior of the coconut mite *Neocypholaelaps stridulans.* Acarology. In: Proceedings of 10th international congress, CSIRO, Melbourne, pp. 361–365.

Hasemann, L. 1961. How long can spores of American foulbrood live? *Am. Bee J.* 101, 298–299.

Heath, L.A.F. 1982. Development of chalk brood in a honey bee colony; Chalkbrood pathogens: A review. *Bee World* 63(3), 119–135.

Hepburn, R., Radloff, S.E. 2011. *Honeybees of Asia.* Springer, Berlin/Heidelberg. p. 914.

Ho, K., Chen, Y. 2001. Susceptibility of the Asian honey bee (*Apis cerana*) to American foulbrood (*Paenibacillus larvae*). In: Proceeding of the 37th International Apicultural Congress. Durban, South Africa, pp. 1–8.

Hunter, P.E., Rosario, R.M.T. 1988. Associations of Mesostigmata with other arthropods. *Annu. Rev. Entomol.* 33, 393–417.

Hunter Fujita, F.R., Mossadegh, M.S., Ball, B.V. 1999. Investigation of the microbial pathogens in *Apis florae.* In: Proceedings of XXXVI International Apicultural Congress (Apimondia), Vancouver, 12–17 September, Canada, p. 230.

Hunter-Fujita, R.F., Mossadegh, M.S., Taylor, S.C., Lawson, P.A., Ball, B.V. 2001. Microbiological investigation of the dwarf honey bee *Apis florea.* 37th International Apicultural Congress, 28 October–1 November. Durban, South Africa.

Kapil, R.P., Aggarwal, K. 1987a. First record of *Euvarroa sinhai* from *Apis mellifera* colonies. *Bee World* 68(4), 189.

Kapil, R.P., Aggarwal, K. 1987b. Some observations on the concurrent parasitization of *Apis florea* by *Tropilaelaps clareae* and *Euvarroa sinhai. Exp. Appl. Acarol.* 3, 267–269.

Kapil, R.P., Aggarwal, K. 1989. Observations on reproduction and seasonal population trends of *Euvarroa sinhai* (Mesostigmata: Varroidae) in India. In: Channabasavanna, G.P., Viraktamath, C.A. (eds) *Progress in Acarology.* Oxford and IBH, New Delhi, pp. 277–281.

Khongphinitbun Jong, Kitiphong, de Guzman, Lilia I., Burgett, Michael D., Rinderer, Thomas E., Chantawannakul, Panuwan. 2012. Behavioral responses underpinning resistance and susceptibility of honeybees to Tropilaelaps mercedesae. Apidologie INRA, DIB and Springer-Verlag, France, doi:10.1007/s13592-012-0129-x.

Kitprasert, C. 1995. Parasitism by the brood mite, *Euvarroa sinhai* Delfinado and Baker (Acari: Varroidae) on the dwarf honey bee, *Apis florea* F. (Hymenoptera: Apidae) in Thailand. Ph.D. Dissertation. Oregon State University. pp. 1–96.

Klee, J., Besana, A.M., Genersch, E., Gisder, S., Nanetti, A. et al. 2007. Widespread dispersal of the microsporadian *Nosema ceranae* and emergent pathogen of western honey bee, *Apis mellifera. J. Invertebr. Pathol.* 96, 1–10.

Koeniger, G., Koeniger, N., Anderson, D.L., Lekprayoon, C., Tingek, S. 2002. Mites from debris and sealed brood cells of *Apis dorsata* colonies in Sabah (Borneo) Malaysia, including a new haplotype of *Varroa jacobsoni. Apidologie* 33, 15–24.

Koeniger, N. 1976. Neue Aspekete der Phylogenie innerhalb der Gattung Apis. *Apidologie* 7, 357–366.

Koeniger, N., Fuchs, S. 1973. Sound production as colony defence in *Apis cerana.* In: Proceedings of 7th international congress, IUSSI, London, pp. 199–204.

Koeniger, N., Koeniger, G., Delfinado-Baker, M. 1983. Observation on mites of the Asian honeybee species (*Apis Cerana, Apis dorsata, Apis florea*). *Apidologie* 14, 197–204.

Koeniger, N., Koeniger, G., Mardan, M. 1994. Mimicking a honeybee queen? *Vespa affinis* indosiniensis Perez 1910 hunts drones of *Apis cerana* F. *Apidologie* 27, 353–359.

Koeniger, G., Koeniger, N., Mardan, M., Wongsiri, S. 1993. Variance in weight of sexuals and workers within and between four Apis species (*Apis florea, Apis dorsata, Apis cerana and Apis mellifera*). In: *Proceedings of the 1st International Conference on Asian Honeybees Bee Mites,* Bangkok, Wicwas, Cheshire, pp. 104–109.

Koeniger, N., Musaffar, N. 1988. Lifespan of the parasitic honeybee mite, *Tropilaelaps clareae*, on *Apis cerana*, *A. dorsata* and *A. mellifera*. *J. Apic. Res.* 27, 207–212.

Koeniger, N., Weiss, J., Maschwitz, U. 1979. Alarm pheromones of the sting in the genus Apis. *J. Insect Physiol.* 25, 467–476.

Kojima, Y., Toki, T., Morimoto, T., Yoshiyama, M., Kimura, K., Kadowaki, T. 2011. Infestation of Japanese native honey bees by tracheal mite and virus from non-native European honey bees in Japan. *Microbiol. Ecol.* 62, 895–906.

Lekprayoon, C., Tangkanasing, P. 1991. Euvarroa wongsirii, a new species of bee mite from Thailand. *Int. J. Acarol.* 17, 255–258.

Lekprayoon, C., Tangkanasing, P. 1993. Comparative morphology of *Euvarroa sinhai* and *Euvarroa wongsirii*: parasites of *Apis florea* and *Apis andreniformis*. In: Conner, L.J., Rinderer, T.E., Sylvester, A.H., Wongsiri, S. (eds.) *Asian apiculture*. Wicwas, Cheshire, pp. 427–433.

Li, J., Qin, H., Wu, J., Sadd, B.M., Wang, X., Evans, J.D., Peng, W. 2012. The prevalence of parasites and pathogens in Asian honey bees *Apis cerana* in China. *PLOS one* 7(11), e47955, 1–12.

Maassen, A. 1906. Die Aspergillusmykose der Bienen, Mitteilugen aus der Kaiserlichen. *Biologischen Anstat fur Land-Und Forstwirtschaft* 2, 30–31.

Maassen, A. 1913. Weitere Mitteilungen uber der seuchenhaften Brutkrankheiten der Bienen [Further communication on the epidemic brood disease of bees]. Mitteilungen aus der Kaiserlichen Biologischen Anstalt fur Land- und Forstwirtschaft 14, 48–58.

Makkar, G.S., Chhuneja, P.K. 2011. In: Dhawan, A.K., Sharma, D.R., Kumar, R. (eds.) *Extended Abstracts of 3rd Congress on Insect Science on Pest Management for Food Security and Environment Health, PAU, India, 18–20 April, 2011.* Indian Society for Advancement of Insect Science, Department of Entomology, PAU, Ludhiana, India, p. 272.

Mookhploy, W., Kimura, K., Disayathanoowat, T., Yoshiyama, M., Hondo, K. Chantawannakul, P. 2015. Capsid gene divergence of Black queen cell virus isolates in Thailand and Japan honey bee species. *J. Econ. Entomol.* 1–5, doi:10.1093/jee/tov102.

Morin, C.E., Otis, G.W. 1993. Observations on the morphology and biology of *Euvarroa wongsirii* (Mesostigmata: Varroidae), a parasite of *Apis andreniformis* (Hymenoptera: Apidae). *Int. J. Acarol.* 19, 167–172.

Morse, R.A., Nowogrodzki, N. 1990. Honey bee pests, predators, and diseases (2nd ed.). Comstack Publishing association, Cornell University, 474 pp.

Mossadegh, M.S. 1990a. Nesting behavior of *Apis florea* F. (Hymenoptera: Apidae) in Khuzestan, Iran. 11th International Congress, IUSSI, India, pp. 669–670.

Mossadegh, M.S. 1990b. Some honey bee virus diseases in Iran. Sci. J. Agric. 13(13), 64–72.

Mossadegh, M.S. 1991. Geographical distribution, levels of infestation and population density of the mite *Euvarroa sinhai* Delfinado and Baker (Acarina: Mesostigmata) in *Apis florea* F. Colonies in Iran. *Apidologie* 22: 127–134.

Mossadegh, M.S. 1992. Effect of Perizin (Comafos) a systemic acaricide used against *Varroa jacobsoni* Oud. (Acari: Varroidae) on the life bee-eaters (Coraciiforms: Meropidae). XIX International Congress of Entomology, June 25–July 4, Bejing, China, p. 567.

Mossadegh, M.S. 1993. Fresh news on *Euvarroa*. *Bee World* 74(1), 44–45.

Mossadegh, M.S. 1997a. The Indian-meal moth *Plodia interpunctella* (Hub.). *J. Sci. Agric.* 6, 52–68.

Mossadegh, M.S. 1997b. Some mites of the honey bee *Apis mellifera* L. hives in Iran. *Sci. J. Agric.* 19(1&2), 7–16.

Mossadegh, M.S. 1998a. Mites (Acari) associated with *Apis florea* F. in Iran. In: *Proceeding of the 13th Iranian Plant Protection Congress*, 23–27 August, Karaj, vol. 1, p. 242.

Mossadegh, M.S. 1998b. *Galleria mellonella* (L.) and *Plodia interpunctella* (Hub.), (Lep.: Pyralidae), pests of *Apis florea* F. colonies in Iran. In: *Proceeding of the 13th Iranian Plant Protection Congress*, 23–27 August, Karaj, Vol. 1, p. 241.

Mossadegh, M.S. 1992. Effect of the black queen cell virus on honeybee drone of *Apis florea* F. (Hym.: Apidae). Fifth International conference on Apiculture in Tropical Climate, Trinidad and Tobago, 7–12 September.

Mossadegh, M.S., Alizadeh, A., Eslamizadeh, R. 1997. A pathogenicity study of *Aspergillus* spp. in *Apis florea* F. and fungal agents causing brood and some other fungi associated with *Apis mellifera meda* Skorikov in Iran. *J Agric. Sci.* 3(11 & 12), 79–90.

Mossadegh, M.S., Birjandi, K. 1986. A., *Euvarroasinhai* Delfinado & Baker (Acarina: Mesostigmata): A parasitic mite on *Apis florea* F. in Iran. *Am. Bee J.* 126, 684–685.

Mossadegh, M.S., Hunter, F.R. 1998. An investigation of bacteria and viruses associated with disease of *Apis florae* F. in Iran. Plant Protection Department, College of Agriculture, Shahid Chamran University of Ahwaz, Iran, (71pp. in English and 30pp. in Persian).

Mossadegh, M.S., Kocheili, F. 2003. A semi description checklist of identified species of Arthropods (Agricultural, medical) and other pests from Khuzestan, Iran. Shahid Chamran University of Ahwaz, No. 266, 476 p (In Persian).

Mossadegh, M.S., Komeili, B.A. 1982. Wax moth. Ministry of Agriculture, Extention Service. 6: 1–9 (In Persian).

Nakamura, J., Matsuka, M. 2007. Geographic variation in the Japanese islands of *Apis cerana japonica* and in *A. cerana* populations bordering its geographic range. *Apidologie* 38, 335–340.

Naug, D. 2009. Nutritional stress due to habitat loss may explain recent honeybee colony collapses. *Biol. Conserv.* 142, 2369–2372.

Needham, G.R., Gerson, U., Sammataro, D. 2001. Mite biology. In: Webster, T.C., Delaplane, K.S. (eds.) *Mites of the honey bee.* Dadant, Hamilton, IL, pp. 1–16.

Noh, J.H., Reddy, K.E., Choe, S.E., Yoo, M.S., Doan, H.T.T. et al. 2013. Phylogenetic analysis of black queen cell virus genotypes in South Korea. *Virus Genes* 46(2), 362–368.

Oldroyd, B., Wongsiri, S. 2006. *Asian honey bees: Biology, Conservation, and Human Interactions.* Harvard University Press, London Oldroyd and Wongsiri, 2007.

Oldroyd, B.P., Nanork, P. 2009. Conservation of Asian honey bees. *Apidologie* 40, 296–312.

Oldroyd, B.P., Wongsiri, S. 2006. *Asian Honey Bees: Biology, Conservation and Human Interactions.* Harvard University Press, Cambridge, MA, p. 340.

Otis, G.W. 1996. Distribution of recently recognized species of honey bees (Hymenoptera: Apidae; Apis). *J. Kans. Entomol. Soc.* 69, 311–333.

Parichehreh, S., Farshineh, A.M., Fallahzadeh, M. 2014. Study and comparison of morphological characteristics of dwarf honey bees, *Apis florea* F. (Hymenoptera, Apidae) in Iran. *J. Entomol. Res.* 5(4), 315–330.

Parichehreh, S., Tahmasbi, G., Sarafrazi, A., Imani, S., Tajabadi, N. 2017. Identification of *Enterococcus* bacteria in gastrointestinal tract of dwarf honey bee, *Apis florea* Fabricius, 1973 (Hymenoptera: Apidae). *J Entomol. Soc. Iran* 37(2), 263–272.

Pirk, C.W.W., Hepburn, H.R., Radloff, S.E., Erlandson, J. 2002. Defense posture in the dwarf honeybee, *Apis florea. Apidologie* 33(3), 289–294.

Promnuan, Y., Kudo, T., Chantawannakul, P. 2009. Actinomycetes isolated from beehives in Thailand. *World J. Microb. Biot.* 25(9), 1685–1689.

Promnuan, Y., Kudo, T., Ohkuma, M., Chantawannakul, P. 2011. Actinomadura apis sp. nov., isolated from a honey bee (*Apis mellifera*) hive, and the reclassification of *Actinomadura cremea* subsp. rifamycini Gauze et al. 1987 as Actinomadura rifamycini (Gauze et al. 1987) sp. nov., comb. nov. *Int. J. Syst. Evol. Microbiol.* 61(9), 2271–2277.

Ra, D.K., Jeong, C., Lee, J.H., Lee, Y.M., Kim, K.H., Han, T.H., Lee, S.M. 2012. Prevalence of honeybee diseases in Incheon area in 2011. *Korean J. Vet. Serv.* 35, 111–117.

Rana, B.S., Rao, K.M., Chakravarty, S.K., Katna, S. 2012. Characterization of *Melisococcus plutonius* causing European foulbrood disease in *Apis cerana* F. *J. Apic. Res.* 51, 306–311.

Reddy, K.E., Noh, J.H., Kim, Y.-H., Yoo, M.S., Doan, H.T.T. et al. 2013. Analysis of the nonstructural and structural polyprotein regions, and complete genome sequences of Israel acute paralysis viruses identified from honeybees (*Apis mellifera*) in Korea. *Virology* 444(1–2), 211–217.

Rice, R. 2001. Nosema diseases in honeybees. Genetic variation and control RIRDC 1/46. Rural Industries Research and Development Corporation, Kingston, 36 pp.

Rinderer, T.E., Wongsiri, S., Kuang, B., Liu, J., Oldroyd, B., Sylvester, H.A., de Guzman, L.I. 1996. Comparative nest architecture of the dwarf honey bees. *J. Apic. Res.* 35, 19–26.

Sanpa, S., Chantawannakul, P. 2009. Survey of six bee viruses using RT-PCR in Northern Thailand. *J. Invertebr. Pathol.* 100, 116–119.

Saraithong, P., Li, Y., Saenphet, K., Chen, Z., Chantawannakul, P. 2015. Bacterial community structure in *Apis florae* larvae analyzed by denaturing gradient gel electrophoresis and 16S rRNA gene sequencing. *Insect Sci.* 22(5), 606–618.

Seeley, T.D. 1994. Honey bee foragers as sensory units of their colonies. *Behav. Ecol. Sociobiol.* 34, 51–62.

Seeley, T.D., Seeley, R.H., Akratanakul, P. 1982. Colony defense strategies of the honeybees in Thailand. *Ecol. Monogr.* 52, 43–63.
Sen Sarma, M., Fuchs, S., Tautz J. 2000. Debris removal by head pushing in Apis florea F. honeybees, *Naturwissenschaften* 87, 241–243.
SenSarma, M., Fuchs, S., Tautz, J. 2002.Worker piping triggers hissing for coordinated colony defence in the dwarf honeybee *Apis florea. Zoology* 105, 215–223.
Shimanuki, H. 1990. Bacteria. In Morse, R.A., Nowogrodzki, R. (eds.) *Honey Bee Pests, Predators and Diseases* (2nd ed.) . Cornell University Press, Ithaca, NY, pp. 27–47.
Simone, M., Evans, J.D., Spivak, M. 2009. Resin collection and social immunity in honey bees. *Evolution* 63, 3016–3022.
Simone-Finstrom, M.D., Spivak, M. 2012. Increased resin collection after parasite challenge: A case of self-medication in honey bees? *PLoS One* 7(3), e34601. doi:10.1371/journal.pone.0034601 1146.
Singh, R., Levitt, A.L., Rajotte, E.G., Holmes, E.C., Ostiguy, N. et al. 2010. RNA viruses in Hymenopteran Pollinators: Evidence of Inter-Taxa Virus Transmission via pollen and potential impact on Non-*Apis* Hymenopteran Species. *PLoS One* 5(12), e14357.
Singh, S. 1961. Appearance of American foulbrood disease in Indian honey bee (*Apis indica* Fabr.). *Indian Bee J.* 23(7/9), 46–50.
Southwick, E.E. 1985. Allometric relations, metabolism and heat conductane in clusters of honeybees at cool temperatures. *J. Comp. Physiol. B* 156, 143–149.
Stanley, J., Sah, K., Jain, S.K., Bhatt, J.C., Sushil, S.N. 2015. Evaluation of pesticide toxicity at their field recommended doses to honeybees, *Apis cerana* and *A. mellifera* through laboratory, semi-field and field studies. *Chemosphere* 119, 668–674.
Suma, K.P. 2015. Culturable microbial diversity associated with gut of apis spp. (Hymenoptera: Apidae) M.Sc Thesis Department of Apiculture University of Agricultural Sciences GKVK, Bengaluru, India, p. 122.
Suwannapong, G., Maksong, S., Phainchajoen, M., Benbow, M., Mayack, C. 2018. Survival and health improvement of Nosema infected *Apis florea* (Hymenoptera: Apidae) bees after treatment with propolis extract. *J. Asia-Pac. Entomol.* 21(2), 437–444.
Suwannapong, G., Maksong, S., Seanbualuang, P., Benbow, M.E. 2010. Experimental infection of red dwarf honeybee, *Apis florea*, with *Nosema ceranae. J. Asia-Pac. Entomol.* 13(4), 361–364.
Suwannapong, G., Seanbualuang, P. and Wongsiri, S.,. 2007. A histochemical study of the hyphopharyngeal glands of the dwarf honeybee *Apis andreniformis and Apis florea. J. Apicult. Res.* 46(2), 260–264.
Suwannapong, G., Yemor, T., Boonpakdee, C., Benbow, M.E. 2011. Nosema ceranae, a new parasite in Thai honeybees. *J. Invertebr. Pathol.* 106(2), 236–241.
Takamatsu, D., Morinishi, K., Arai, R., Sakamoto, A., Okura, M., Osaki, M. 2014. Typing of Melissococcus plutonius isolated from European and Japanese honey bees suggests spread of sequence types across borders and between different Apis species. *Vet. Microbiol.* 171, 221–226.
Theantana, T., Chantawannakul, P. 2008. Protease and β-N acetylglucosaminidase of honey bee chalkbrood pathogen *Ascosphaera apis. J. Apic. Res.*, 47(1), 68–76.
Tingek, S., Koeniger, G., Koeniger, N., Gries, M. 2004. Physocephala paralliventris Kröber (Conopidae), a parasitic fly of three Apis species in northern Borneo. In: *Bees for new Asia, Proceedings of the 7th Asian Apicultural Association Conference*, Los Baños, Philippines, pp. 364–365.
Trung, L.Q. 2010. Destructive forest fire making *Apis laboriosa* Giant bees in Vietnam more endangered. Organistion Bee Reseach and Development Centre of Vietnam, 1 p.
Tutkun, E., Maden, S., Inci, A., Yilmarz, B. 1993. General situation of chalkbrood disease in honeybees in Turkey. *Turk. Entomol. Derg.* 17(2), 65–68.
Verma, L.R., Rana, B.S., Verma, S. 1990. Observations on *Apis cerana* colonies surviving from Thai sacbrood virus infestation. *Apidologie* 21, 169–174.
Waite, R. 2003. Exotic pest legislation. *Br. Beekeep. Assoc.* 143, 7.
Wongsiri, S., Lekprayoon, C., Thapa, R., Rinderer, T.E., Sylvester, H.A., Oldroyd, B.P., Booncham, U. 1996. Comparative biology of *Apis andreniformis* and *Apis florea* in Thailand. *Bee World* 77, 23–35.
Wongsiri, S., Tangkanasing, P., Sylvester, H.A. 1989. The resistance behavior of *Apis cerana* against *Tropilaelaps clareae.* In: *Proceedings of the First Asia-Pacific Conference of Entomology*, Chiang Mai, Thailand, pp. 25–34.
Woyke, J., Wilde, J., Mananya, P. 2012. First evidence of hygienic behaviour in the dwarf honey bee *Apis florea. J. Apic. Res.* 51(4), 359–361.

Woyke, J., Wilde, J., Siviram, N., Nagaraja, N. 2011. Does open-air nesting honey bee *Apis florea* differ from the other bees of genus Apis in brood hygienic behaviour? In: *Global Conference on Entomology*, Chiang Mai, Thailand, 5–9 March, p. 164.

Yan, D., Tang, K.F.J., Lightner, D.V. 2009. Development of a real-time PCR assay for detection of monodon baculovirus (MBV) in penaeid shrimp. *J. Invertebr. Pathol.* 102, 97–100.

Yang, B., Peng, G., Li, T., Kadowaki, T. 2013. Molecular and phylogenetic characterization of honey bee viruses, *Nosema microsporidia*, protozoan parasites, and parasitic mites in China. *Ecol. Evol.* 3(2), 298–311.

Yang, G.H. 2005. Harm of introducing the western honeybee *Apis mellifera* L. to the Chinese honeybee *Apis cerana* F. and its ecological impact. *Acta Entomol. Sin.* 48, 401–406.

Yen, D.F., Chyn, L.C. 1971. Studies on a bacterial disease of honeybee in Taiwan. *Plant Protection Bulletin* 13, 12–17.

Zander, E. 1909. Tierische Parasiten als Krankenheitserreger bei der Biene. *Münchener Bienenzeitung* 31, 196–204.

Zhang, X., He, S.Y., Evans, J.D., Pettis, J.S., Yin, G.F., Chen, Y.P. 2012. New evidence that deformed wing virus and black queen cell virus are multi-host pathogens. *J. Invertebr. Pathol.* 109(1), 156–159.

Zhou, T., Feng, F., Dong, B. 2000. Study on the pathogen of European foulbrood in the Chinese honey bee (*Apis cerana* F.) *Acta Entomol. Sinica* 43, 104–108.

7

Nesting Behaviour, Structure and Architecture of Apis florea

B. V. Shwetha, N. S. Bhat and T. Neethu

CONTENTS

Introduction

Apis florea is commonly known as red dwarf bee/little honeybee and is one of the smallest species among the 11 species of honeybees belonging to the genus *Apis*. It spreads from the Middle East to Peninsular Malaysia where a warm climate prevails (Oldroyd and Wongsiri, 2006). It is primarily distributed across countries such as Saudi Arabia, Iraq, Iran, Afghanistan, Pakistan, India, Sri Lanka, Bangladesh, Myanmar, Thailand, South China and Malaysia. *A. florea* lives sympatric with *A. dorsata* and *A. cerana* in major parts of their distribution (Ruttner, 1988). The ability of *A. florea* to survive in very hot and dry climatic conditions is one of its most outstanding traits (Ruttner, 1988). In the Persian Gulf, it tolerates a temperature of 50°C and higher during the summer where no other honeybee species occurs in this region (Whitcombe, 1984; Ruttner, 1985). Seasonal migration and absconding are the characteristics of *A. florea* (Duangphakdee et al., 2009), and it moves from plains to hills and hills to plains.

A. florea is the most important pollinator of many field, plantation and fruit crops. In addition, the species has its importance in producing honey and wax. The honey from *A. florea* is reputed for its flavour and medicinal properties and is collected by honey hunters without damaging the brood area and attachment point of comb by cutting the side portions of the crown (Dutton and Free, 1979; Whitcombe, 1984). This approach facilitates the repeated harvest from the same colony in a few days and is quite close to domestication, even though *A. florea* is not a domesticated species.

Nesting Sites

A. florea is usually found to nest in open conditions. It is highly adaptable and locates a suitable nesting site under extremely varied conditions (Mogga et al., 1989; Moritz et al., 2010). Though nest

FIGURE 7.1 *A. florea* nest in a bush.

aggregations of *A. florea* are not found in India, aggregations are recorded in Thailand (Rinderer et al., 2002). The nesting sites of *A. florea* are usually arboreal and rarely terrestrial. In the former case, the honeybees build their nests on branches or twigs of trees, bushes, shrubs, hedges, wooded areas, house chimneys, empty caves and piles of dried sticks, areas with intensive agricultural activity as well as in the savanna or thickets of thorny creepers, which provide protection from rain and heat (Figure 7.1) (Franssen, 1932; Free, 1981; Matsuura, 1983; Booncham et al., 1995; Wongsiri et al., 1997; Nagaraja and Rajagopal, 1999). The *florea* nests being built low on the branches of dense, shrub vegetation are probably difficult for many of the bees' predators to find. Less frequently, the bees may build their nests in cavities such as caves, hollow trees, boxes, huts, on rocks, building walls or roofs, inside houses in partial light, on discarded metal scrap and even in dumped cycle rims where the crown is absent (Home and Smith, 1872; Douglas, 1886; Ghosh, 1919; Ayyar, 1940; Thakar and Tonapi, 1962; Dutton and Simpson, 1977; Free and Williams, 1979; Kshirsagar et al., 1980; Soman and Chawda, 1996). *A. florea* always builds combs in locations that offer a substantial amount of shade such that the sky is visible from the top of the nests, which favours the performance of communication dances on a horizontal platform – not on a vertical surface (Whitcombe, 1984). The site-favoured nesting will often have one surface of the comb exposed to direct sunlight for several hours a day, abundant twigs of varied thickness, proximity to food and water, and protection from predators and adverse weather conditions (Seeley and Buhrman, 1999; Franks et al., 2003; Oldroyd et al., 2008). The totally concealed or completely shaded sites would disorient dances unless the dancing bees had a clear view of the sky from the honeycomb (Whitcombe, 1984).

The habitats for *A. florea* are different for plain and hilly areas. In plain areas, 50% of the dwarf honeybee colonies preferably nest in bushes, 30% on small tree branches and 20% in wooden heaps on the ground. But in hilly areas, 60%–80% of honeybee colonies preferred heaped wood bundles on the ground, and the remaining 20%–40% of bees preferred to nest in bushes (Kumar, R. and Kumar, N. R., 2014). The majority of the colonies populate shrubs (48%) and trees (44%). They were seldom found in dark places such as wooden crates and chimney walls (Malathi, 2005). The maximum colonisation of 90.6% (Soman and Chawda, 1996) and 85.5% (Khan et al., 2002) by *A. florea* occurred on arboreal supports.

The colonies nesting in a cave or other cavity usually build their comb such that one side faces out of the entrance. This may be the most suitable position for the landing and dances of returning foragers, and it may also be the most effective method for thermoregulation. Such colonies will have only a vertical dance floor. *Florea* dancers certainly cannot use the plane of vibration of polarized light as a reference for dances on a vertical surface, so the dancers would be dependent on the ability to transform forage information based on the sun's position to a vertical surface. Such vertical dances may not be as accurate or satisfactory (Whitcombe, 1984).

Host Plants

In a study conducted around Bangalore (Basavarajappa, 2010), of the 14 different plants species belonging to 13 genera, 12 families were chosen by *A. florea* for nesting at different places. Of these, more often the colonies were found on *Pongamia pinnata, Mangifera indica, Cebex petandra, Tamarindus indica* and *Millettia ovalifolia*. Many of the trees such as *Acacia tortilis, Acacia nilotica, Prosopiscineraria* and *Ziziphusspina-christi, Citrus* sp. have large spines and thorns on their branches and twigs, which did not seem to deter *A. florea* from building nests on them. These thorns, rather than being a problem, may even strengthen the bonding of the comb to the tree and provide some protection from larger predators. Most of the nests of *A. florea* were located at a minimum distance of 30 m apart in bushes of Duranta, Cariosa, Lantana and in the piles of firewood.

In another study (Basavarajappa, 2010), *A. florea* colonies most often sited on *Pongamia pinnata* at short intervals, and the nesting frequency (number of nests observed at short intervals) was 16% followed by *Mangifera indica* (13.3%) and *Bouganvillea* sp. (10.7%). Because of repeated instances of *A. florea* colonies occurring on *P. pinnata, M. indica* and *Bouganvillea* sp., they are referred to as 'preferred nesting plants'. Though *Cebex petandra, Tamarindus indica, Hibiscus rosa-sinensis, Nerrium* sp. and *Millettia ovalifolia* were more abundant, they were less frequently used for nesting by *A. florea* and are designated as 'common nesting plants'. The other plants such as *Artocarpus heterophylus, Tecoma stans, Bambusa bambos, Michelia champaca* and *Polyalthia longifolia* rarely hosted the *A. florea* colonies. The colonies were not observed often on these plants, and hence the frequency was less than 2%; they are classified as 'uncommon nesting plants'.

Further, 24 plants belonging to 19 plant families were selected for nesting. Fabaceae members supported the highest number (14.8%) of colonies followed by the Moraceae family (74%). *Saraca asoca, Bougainvillea spectabilis, Muntiagia calabura* and *Terminalia catappa* were more preferred for nesting at Manasagangotri and Mysore, Karnataka, as they had necessary attributes such as height to have suitable flight range, foliage thickness and thickly interspersed slender branches covered with foliage that may provide protective curtain over the comb; this avoids direct visibility from predators and enemies and in turn protects the colony from them (Narayanaswamy and Basavarajappa, 2013).

Capprisse piaris was found to be the most preferred nesting shrub in Hissar (Khan et al., 2002), *Prosopis juliflora* (32.0%) in the Kutch area of Gujarat (Soman and Chawda, 1996). Higher numbers of colonies (12.0%) were found on *Bougainvilla* sp and *Lantana camera* and 8.0% each on *Croton variegatum, Jasminum sambac* and lemon (Malathi, 2005). Among the tree species, 8.0 per cent each of the colonies were found in *Mangifera indica* and *Artocarpus heterophyllus* and 4.0% each on *Cocas nucifera, Psidium gajava, Punica granatum, Arucariacucci, Pongamia pinnata, Syzigium cumini* and *Callistemon chinensis* in Hebbal and GKVK, Bangalore (Malathi, 2005).

Height of the Nest from Ground Level

The height of the *A. florea* nests from the ground level varies. This difference in height largely seems to be due to variation in the prevailing vegetation in different geographical areas, which provide natural nesting places. The nests are constructed at 3 to 5 m height from the ground level (Free, 1981; Soman and Chawda, 1996), between 2 and 7 m from the ground level with a mean height of 3.28 m in Oman (Whitcombe, 1984). The nest height varied from 0.3 to 14.0 m ground level in Thailand (Mossadegh, 1990) and averaged about 4.0 m above ground level (Rinderer et al.,

2002), 2 to 10 m from the ground (average 3 m) (Kshirsagar et al., 1983) and 2.0 to 3.0 m in Hissar (Khan et al., 2002).

The height of the nests from the ground level varied from 1 to 5 m. But, 86.7% nests were found below 4.0 m height; of these, 40.0% of colonies were found at the height between 2.0 and 3.0 m and 6.7% of colonies were recorded below 1.0 m and between 4.0 and 5.0 m from ground level in Bengaluru (Malathi, 2005).

A. florea selected different elevations during nest establishment in Manasagangotri and Mysore, Karnataka, that ranged from ground level to up to 15.24 m but preferred lower elevations (i.e., ground level to up to 4.57 m height) compared to higher elevations (i.e., 4.60 m onwards) for nesting, and at latter heights the nesting preference gradually reduced (Narayanaswamy and Basavarajappa, 2013).

Size of the Supportive Branch

The diameter of the nest support branches used by *A. florea* varies with the type of the host plant. In trees, the combs of colonies did encircle supports as large as 30 mm in mean diameter (40 mm wide by 20 mm high, in this case, of a date palm frond midrib). In most cases, the colonies were observed on thin- to medium-sized (1–4 in dia.) tree limbs/branches at different elevations (Narayanaswamy and Basavarajappa, 2013). The support diameter for nests on palm fronds is rather higher than for nests on other trees. Colonies have to nest on such a broad support that can hold the fully loaded comb with honey, pollen and brood; they also require a bare minimum of shade so they are less exposed to greater movement in the wind.

Orientation of Nests

A. florea is known to nest in all directions but prefers to nest in northwest, northeast, north-south and southeast directions (Whitcombe, 1984; Khan et al., 2002; Malathi, 2005). In India, northwest direction orientation of combs is the most preferred direction for comb construction by *A. florea* (Khan et al., 2002; Malathi, 2005). *A. florea* colonies orientated at four directions, namely north-south (NS), east-west (EW), northeast-southwest (NE-SW) and northwest-southeast (NW-SE) in Mysore. But nests with northeast-southwest orientations were more frequent compared to other orientations (Narayanaswamy and Basavarajappa, 2013).

Comb Construction

Honeybee nests are the result of the interactions and interplay among numerous individuals performing various building and construction operations. The end product, the comb, is used for storing resources, rearing the next generation of brood, and chemical and physical communication, which ensures the social cohesion of the honeybee colony (Hepburn et al., 2014). Each species has a unique nest architecture, and there can be radical differences between species that are nonetheless phylogenetically similar. Indeed, nest architecture alone is often used to distinguish the two dwarf honeybees, *A. florea* and *A. andreniformis* (Rinderer et al., 1996; Oldroyd and Wongsiri, 2006).

A. florea builds a single, semicircular shaped, exposed comb that is vertically attached to one or two thin branches in trees or bushes throughout South East Asia, which has been described quite thoroughly many times (Hepburn and Hepburn, 2011). It constructs a single comb suspending from a thin branch of small trees or bushes located horizontally with the honeycomb area enclosing the supporting twig (Akratanakul, 1977), but double-comb nests have rarely been reported in *A. florea* (Douglas, 1886; Akratanakul, 1977; Free, 1981; Whitcombe, 1984). Occasionally, *A. florea* nests are also built on a vertical surface (Figure 7.2). In these unusual nests, the crown is attached directly to a vertical surface. The bees form a 'curtain', 1–3 or more workers thick, that envelopes the comb in strong colonies for brood thermoregulation to protect the colony, its brood and honey from predators (Seeley et al., 1982).

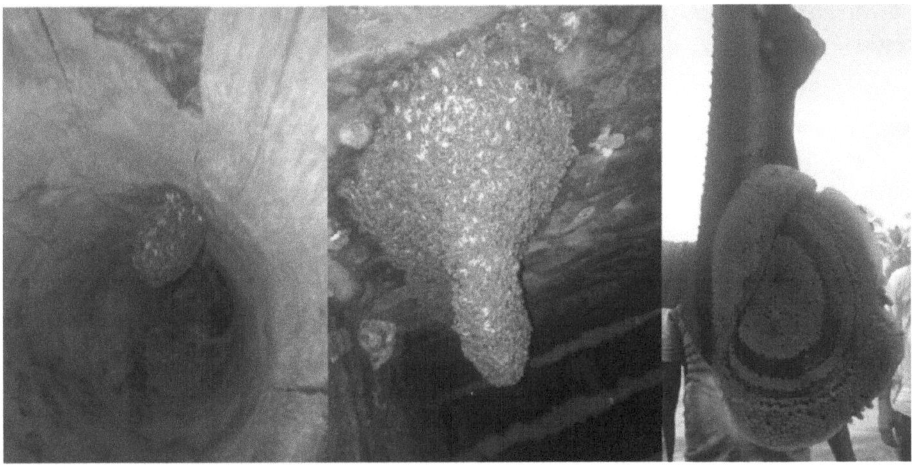

FIGURE 7.2 *A. florea* nests built on a vertical surface.

The comb is composed entirely of beeswax. The progressive building of the comb by a recently settled *A. florea* colony is vividly described (Pirk et al., 2004; Karihaloo et al., 2013). The comb-building commenced almost immediately after landing and settling on the twig at 2 h after settlement; a small row of about seven nascent cells were subtending the twig at 2 h. The second lower rows of cells were added after 6 h. After 9 h, four rows of nascent cells were added. These nascent cells are not polygonal but virtually circular. The row at the base of the twig consists of truncated hexagons and two rows of crude hexagons. The third row is exactly the same as in the first row, circular burrows in the wax. Twelve hours after settling, the comb acquires a more crystalline structure as regular hexagons begin to form, which is most likely a result of the wax flowing into proper shape with equal pressure from adjacent cells to shape a hexagon in the centre of the developing comb (Bergman and Ishay, 2007). As a rule, the first row of cells appears anomalously different from the hexagons of comb cells with which we are familiar. It generally consists of regular pentagons and circles; the site of support forms one side from which two vertical walls are suspended and then two oblique ones. Ordinarily, the growth of the comb then progresses at a faster downward than lateral rate, so the combs tend to be ellipsoid in the early stages of construction.

The comb is two-sided with a midrib in the pollen storage area and the brood area having a regular array of hexagonal cells constructed back to back and joined by triple-rhomboid bases that maximizes the volume for minimum surface area (Thompson, 1917; Tóth, 1964; Taylor, 1976). The hexagonal shape of cells is common among cell-building social insects, and there is a sound architectural reason for this style. Round, octagonal or pentagonal cell arrangements leave empty spaces between cells, and triangles or squares have a greater circumference than hexagons. Thus, the greatest number of cells per area can be arranged in a comb using the hexagonal shape. By staggering each side of the comb slightly from the opposing side, the bees can construct cells back to back, further maximizing the number of cells per area. Unlike most other social insects, honeybees build their cells horizontally rather than hanging vertically, although they are angled up at about 13° from the base to the opening to prevent honey from running out (von Frisch, 1967).

The comb is protected by a curtain of interlinked workers. *A. florea* workers deposit a sticky material, the 'insect lime/resin', round the supporting branch on either side of the comb. This insect lime is a kind of propolis but more sticky than that used by *A. mellifera* and will protect the colony from ants and other walking insect pests and enemies.

The dimensional growth curves of the nests in *A. florea* show daily changes in comb length, width and area from inception of the nests to their maturity and completion. The comb expands at about the same rates in changes of lengths and widths, providing a final structure that is close to circular except for the addition of drone cells in groups, which protrudes downwardly from the bottom of the combs.

The development of the crown region seems to consist of both structurally and biologically independent processes (Preecha et al., 2015).

Comb Dimensions

The observations were recorded on 15 colonies of *A. florea* in Hebbal and GKVK, Bangalore, regarding the comb dimensions. The mean comb breadth and length were 23.60 cm (15–30 cm) and 31.60 cm (18–45 cm), respectively. The mean length above attachment was 4.90 cm (1.5–6.5 cm), and the thickness at the honey portion was 7.95 cm (3.2–16.5 cm) (Malathi, 2005).

The colony size varies considerably during different months at various elevations under urban conditions. In general, the length and width of *A. florea* colonies were big in size during March, April, May and June **at 7.65–9.15 m elevation, and this was followed by 1.56–3.05 m elevation**. However, the minimum comb length was 16.3 ± 5.1 cm and width was 15.6 ± 4.3 cm, respectively, during April and May. The maximum comb length and width were 20.5 ± 0.5 and 19.0 ± 4.3 cm, respectively, during January and February at **1.56–3.05 m** elevations. Further, at ground level to up to 1.56 m height, *A. florea* gained good colony size compared to 3.08 to 4.57; 4.60 to 6.10; 6.13 to 7.62; 10.10 to 10.67 and 13.75 to 15.24 m elevations (Narayanaswamy and Basavarajappa, 2013).

Figure 7.2 shows the per cent occurrence of normal colonies at different elevations. It is evident from the figure that *A. florea* preferred lower elevations to the higher (ground level to up to 4.57 m height). Prevalence of 33% of colonies was at a height below 1.53 m. About 23.3% were found at a height of 1.55 to 3.04 m, and only 16.5% of colonies were at 3.04 to 4.57 m height.

Different Zones in the Comb

The worker bee builds different kinds of cells for different purposes and must apply different rules so that cells of the right kind are built at the right time in the right place (Oldroyd and Pratt, 2015). The region above the supportive structure is commonly called a 'crown' and is used for storing the honey and is also a dancing floor, which is a part of communication in bees (Dyer, 1985; Rinderer et al., 1992; Oldroyd and Wongsiri, 2006). The crown is integral to the structural integrity of the nest and supports the protective curtain of worker bees that shields the nest from the elements (Dyer and Seeley, 1991). The crown region does not have a midrib in *A. florea*. The construction of the crown mainly depends on the availability of the space in the supportive structure. *A. florea* builds nests on either horizontal or vertical surfaces of the supportive structure based on the suitability of the site with a slight difference in the comb structure. The nests constructed on a horizontal surface will have a crown region around the supportive structure where the cells are uneven in size and shape. The nests constructed on a vertical surface will not have a crown around the supportive structure, but the cells below the supportive structure are elongated and bulged for storing honey. In the combs where the crown is absent, the vertical surface of the comb is used for dancing (Free and Williams, 1979). This means that the rules that normally result in the construction of the crown of the nest can be modified if the founding swarm somehow finds itself on a vertical surface (Oldroyd et al., 2015).

All cells in the crown are joined directly to the branch without a midrib. The cells in the honey storage area seem to be oriented inward towards the supporting branch when viewed from an outer surface (Figure 7.3). The honey cell is the largest of the three types of cells. The crown is generally round and tapered to the supporting branch at the ends; few cells are distorted from the standard hexagon. A cross section of the crown of an *A. florea* nest reveals three levels of internal organization. First, cells from the side are very long and extend to the supporting branch. Above this area, cells coming from the opposite sides have their base at the sides of the cells coming from the other side. Cells coming from the top of the crown have the same pattern; however, the use of an adjacent side wall as a base is more extreme, with some cells open to the top surface having their base well away from the base of the supporting cell (Rinderer et al., 1996). The crown building, using hexagonal cells, poses serious geometrical problems

FIGURE 7.3 Crown portion of *A. florea* comb.

because it is not possible to encircle a regular cylinder, like a twig, with hexagons. Close examination of such combs revealed a combination of various other polygons (Phiancharoen et al., 2011); real solutions to the problem of comb geometry are yet to be determined. Honey is heavy, so the honey cells are usually at the highest part of the comb, attached to the support lacking a midrib. The honey cell is the largest, and its depth can vary greatly due to the storage volume required. When more honey is collected, the depth of the honey cells is increased by the bees for more storage and stronger attachment.

Pollen is stored in the cells below the crown region where it is easily accessible to nurse bees. The pollen, workers and drone cells are horizontal with a slight upward slant and are hexagonal in shape with rhomboid bases. The worker and pollen cells are arranged exactly opposite to the cells of the other side.

A very conspicuous characteristic of *A. florea* is the striking difference in size between worker and drone bees. This inequality is much smaller in other species. The worker and drone brood cells are similar, being hexagonal and horizontal, but the differences in body sizes are reflected in the diameter and depth of brood cells. Malathi (2005) worked out cell dimensions is great details from the colonies observed in and around Bangalore. The mean numbers of cells/cm^2 of worker, drone and honey cells were 13.10, 6.25 and 10.45, respectively. The number of worker drone and honey cells per linear 3 cm length were 10, 5 and 8, respectively. Further, the mean inner and outer diameters of the cells were 2.90 and 3.08 mm in worker cells, 4.18 and 4.37 mm in drone cells and 3.18 and 3.43 mm in honey cells, respectively (Malathi, 2005). Moreover, the capping of the drone cell is distinguishable. During seasons of abundant forage and when a colony is strong, drones are reared in larger cells in groups below the worker brood (Figure 7.4). The drone cells are much larger (4.2–4.8 mm) in size than the worker cells and generally on the lower part of the comb contiguous to worker brood cells (Whitcombe, 1984).

FIGURE 7.4 *A. florea* comb with drone cells in the lower portion.

The placement of a drone comb on the outer edges may be due to the drone brood being more expendable than the worker brood. The drone cells in cluster may help the queen in laying of fertilized and unfertilized eggs separate comb areas, thus allowing easier sex determination (Seeley and Morse, 1976). In the case of drone cells, their centres will not meet the other side **ref.**

The queen cells are found at the bottom or periphery of the comb (Whitcombe, 1984). During emergency situations, the queen cells are constructed in the centre of the comb by modifying worker cells (Kshirsagar et al., 1983). Queen cells are vertical in position and cylindrical in shape, tapering from the base towards the open end (Malathi, 2005). The mean inner and outer diameters of the queen cell were 5.23 and 5.46 mm, respectively. Similarly, outer and inner lengths were 15.40 and 14.93 mm, respectively.

Superseding a failing queen or swarm, queen cells were usually attached to the lower edge of the drone comb. Akratanakul (1977) found 3–13 queen cells per colony among 22 colonies examined, but up to 15–20 have been recorded by Ghatge (1949). Colonies that have lost their queen may produce up to 19 emergency queen cells by extending worker brood cells (Sakagami and Yoshikawa, 1973; Akratanakul, 1977; Free and Williams, 1979). During emergency queen rearing, the queen cells are constructed on the face of the comb by modifying the hexagonal-shaped worker cell to a cylindrical-shaped queen cell temporarily.

Brood capping in all three castes are porous. Worker cell cappings are flat whereas drone and queen cell cappings are convex. More than 60% of the wax used to cap brood cells was salvaged from previously used brood cell cappings following the emergence of young adult bees (Meyer and Ulrich, 1952). These minor constructions, based on recycling cappings wax (Lineburg, 1923a, 1923b), are generally performed by young bees, 3–9 days old, most of whom have not yet reached the peak of wax secretion (Rösch, 1927). After the emergence of the adult drones, wax would often be deposited around the periphery of the cells, restricting the entrance and making them circular rather than hexagonal in appearance.

As the colony expands, a new worker comb is added at the margin of the comb and the drone comb at the bottom. The drone cells must be removed by cell-constructing workers before new worker cells can be constructed (Oldroyd and Pratt, 2015).

Cell diameter may diminish slightly over a period of time because of accumulated cocoons, castoff larval and pupal skins, and treatment given to cells to prepare for the next brood cycle. Such an old comb appears brittle and is considerably darker than a fresh comb (Gary, 1975), and workers reared in old cells may be smaller than those reared in new cells larger in diameter (Abdellatif, 1965).

Comb Wax Salvage

The secretion of wax and construction of combs represent a large metabolic investment in honeybees so that desertion of the nest, for whatever reason, constitutes an energetically hefty expenditure (Hepburn et al., 1984; Pirk et al., 2011). The behaviour of cannibalising wax from a deserted nest and reusing it in the construction of a new one (Pirk et al., 2011) is only known for three species of honeybees – *A. andreniformis* (Duangphakdee, pers. obs.; Wongvilas, pers. obs.), *A. florea* (Akratanakul, 1977; Dutton and Free, 1979; Wongsiri et al., 1997; Hepburn et al., 2009, 2010) and *A. mellifera capensis* (Hepburn and Radloff, 1998). Because *A. florae* colonies also tend to nest near one another (Rinderer et al., 2002; Wattanachaiyingcharoen et al., 2008), this demographic characteristic invites competition for accessible, freestanding, empty combs, the wax of which is a valuable and metabolically expensive resource (Hepburn et al., 1984; Pirk et al., 2011). *A. florea* only salvages the crown of the comb, which is where honey and pollen are stored, most likely for efficiency reasons.

REFERENCES

Abdellatif, M. A., "Comb cell size and its effect on the body weight of the worker bee, *Apis mellifera* L." *Am. Bee J.*, 105 (1965): 86–87.

Akratanakul, P. "*The natural history of the dwarf Honeybee (Apis florea F.) in Thailand.*" Ph. D. Thesis. (Cornell University, Ithaca, New York, 1977).

Ayyar, T. V. R., *Handbook of economic entomology for South India.* (Madras Superintendent, Government Press, Chennai, 1940).

Basavarajappa, S. "Nesting plants of dwarf honeybee, *Apis florea* F. under tropical conditions of Karnataka, India." *Animal Biology* 60 (2010): 437–447.

Bergman, D. J. and Ishay, J. S, "Do bees and hornets use acoustic resonance in order to monitor and coordinate comb construction?" *Bull. Math. Biol.* 69 (2007): 1777–1790.

Booncham, U., Wongsiri, S., Thirakupt, K., "Niche differences between *A. florea* and *A. andreniformis* in dry-evergreen forest Thailand." (*Proceedings of International conference tropical bees environment*, Pedu Lake, Kedah, 1995: 33).

Douglas, J. C. "The hive bees indigenous to India and the introduction of the Italian bee." *J. Asiatic Soc. Bengal II Nat. Sci.* 55 (1886): 83–96.

Duangphakdee, O., Radloff, S. E., Pirk, C. W. W. and Hepburn, R., "Sun angle time windows for absconding by the dwarf honeybee, *Apis florea*." *J. Insect Physiol.*, 55 (2009): 1009–1012.

Dutton, R. and Simpson, J., "Producing honey with *Apis florea* in Oman." *Bee World.* 58(2) (1977): 71–76.

Dutton, R. W. and Free, J. B. "The present status of beekeeping in Oman." *Bee World* 60(4) (1979): 176–185.

Dyer, F. C., "Mechanisms of dance orientation in the Asian honey bee, *Apis florea*." *J. Comp. Physiol. A Neuroethol. Sens. Neural Behav. Physiol.* 157 (1985): 183–198.

Dyer, F. C. and Seeley, T. D., "Nesting behavior and the evolution of worker tempo in four honey bee species." *Ecology*, 72 (1991): 156–170.

Franks, N. R., Dornhaus, A., Fitzsimmons, J.P., Stevens, M., "Speedversus accuracy in collective decision making." *Proc. R Soc. Lond. B Biol. Sci.*, 270 (2003): 2457–2463.

Franssen, C. J. H., "Aanteekeningen over *Micrapis florea* F. (Notes on *Micrapis florea*)." *Natuurk. Tijdschr. Ned. – Indie*, 92 (1932): 55–63. (Dutch and BRA Translation E1170).

Free, J. B., "Biology and behaviour of the honeybee *Apis florea*, and possibilities for beekeeping." *Bee World*, 62(2) (1981): 46–59.

Free, J. B. and Williams, I. H., "Communication by pheromones and other means in *Apis florea* colonies." *J. apic. Res.*, 18(1) (1979): 16–25.

Gary, N. E., "Activities and behaviour of honey bees." In *The Hive and the Honey Bee* ed. Dadant, C.P. (Dadant and Sons, Hamilton, IL, 1975), pp. 185–264.

Ghatge, A., "Some research work on *Apis florea*." *Indian Bee J.*, 11(1/2) (1949): 5–6.

Ghosh, C. C., "Beekeeping in India." *Pmc. Ill ejtt. Meet Calcutta* 2 (1919): 770–782.

Hepburn, H. R, Duangphakdee, O., Phiancharoen, M., Radloff, S. E. Comb wax salvage by the red dwarf honeybee, *Apis florea* F. *J. Insect. Behav.* 23 (2010): 159–164.

Hepburn, H. R. and Hepburn, C., "Bibliography of the Asian species of honeybees." In *Honeybees of Asia*, Hepburn, H. R. and Radloff, S. E., Eds., Springer, Berlin, 2011, pp. 473–657.

Hepburn, H. R., Hugo, J. J., Mitchell, D. et al., "On the energetic costs of wax production by the African honeybee, *Apis melliferaadansonii*." *S. Afr. J. Sci.* 80 (1984): 363–368.

Hepburn, H. R., Pirk, C. W. W. and Duangphakdee, O., *Honeybee nests- composition, structure, function.* (Springer Verlag, Berlin, Heidelberg, New York, 2014).

Hepburn, H. R. and Radloff, S. E., *Honeybees of Africa.* (Springer, Berlin, 1998).

Hepburn, H., Radloff, S. E., Duangphakdee, O. and Phiancharoen, M. "Interspecific utilization of wax in comb-building by honeybees." *Naturwissenschaften*, 96 (2009): 719–723.

Home, C. and Smith, F., "Notes on the habits of some Hymenopterous insects from the north-west provinces of India." *Tram, Zoot Soc. Lond*, 7 (1872): 161–185.

Karihaloo, B. L., Zhang, K. and Wang, J., "Honey bee combs: How the circular cells transform into rounded hexagons." *J. R. Soc. Interface* 10 (2013): 0299.

Khan, M. S., Kaushik, H. D. and Rohilla, H. R., "Nesting parameters and comb dimensions of the dwarf honeybee, *A. florea* F. at Hisar, India." *Indian Bee J.*, 64 (2002): 12–19.

Kshirsagar, K. K., Muvel, K. S., Mittal, M. C. and Phadke, R. P., "Some observations on *Apis florea* F." (*Proceedings of IInd International Conference on Apiculture in Tropical Climates*, Delhi, 1980).

Kshirsagar, K. K., Muvel, K. S., Mittal, M. C. and Phadke, R. P., "Unusual nesting choice by dwarf honeybee, *Apis florea* Fab." *Indian Bee J.*, 60(4) (1983): 217–218.

Kumar, R. and Kumar, N. R., "Influence of changing cultural practices on nesting behaviour of *Apis florea* Fabricius." *Univers. J. Environ. Res. Technol.*, 4(2) (2014): 63–71.

Lineburg, B., "Conservation of wax by the bees." *Am. Bee J.*, 63 (1923a): 615–616.

Lineburg, B., "What do bees do with brood cappings?" *Am. Bee J.*, 63 (1923b): 235–236.
Malathi, K., Bioecology of little bee, *Apis florea* Fab. (M.Sc thesis, University of Agricultural Sciences, Bangalore, 2005).
Matsuura, M., "Notes on the nesting habits of the three honeybee species in Sumatra." *Honeybee Sci.*, 4 (1983): 117–122 (in Japanese).
Meyer, W. and Ulrich, W., "Zur Analyse der Bauinstinkte unserer Honigbiene. Untersuchungen Liber die "Kleinbauarbeiten." *Naturwissenschaften*, 39 (1952): 264.
Mogga, J. B., Abdin, A. M. Z. E., Nagi, S. K. A. and Ali, A. M., "*Apis florea* in Sudan: Some biological observations." (*Proceedings of 4th International Conference on Apiculture in Tropical Climates*, Cairo, Egypt, 1989, pp. 422–424).
Moritz, R. F. A., Haddad, N., Bataieneh, A., Shalmon, B. and Hefetz, A., "Invasion of the dwarf honeybee *Apis florea* into the near East." *Biol. Invasions*, 12 (2010): 1093–1099.
Mossadegh, M. S., "Nesting behaviour of *Apis florea* F. (Hymenoptera: Apidae) in Khuzestan, Iran." (*Proceedings of the 11th International Conference IUSSI*, New Delhi, 1990, pp. 669–670).
Nagaraja, N. and Rajagopal, D. "Colony establishment, nesting, and foraging activity of little honey bee *Apis florea* F. (Hymenoptera: Apidae)." *J. Entomol. Res.*, 23(4) (1999): 331–338.
Narayanaswamy, R. and Basavarajappa, S., "Nesting attributes of dwarf bee, *Apis florea* F. under urban ecosystem of Manasagangotri campus, Mysore, India." *European J. Zoo. Res.*, 2 (6) (2013): 6–15.
Oldroyd, B. P., Gloag, R. S. and Even, N. et al., "Nest site selection in the open-nesting Honeybee *Apisflorea*." *Behav. Ecol. Sociobiol.*, 62 (2008): 1643–1653.
Oldroyd, B. P., Pratt, S. C., "Comb architecture of the eusocial bees arises from simple rules used during cell building." Adv. Insect Physiol., 49 (2015): 101–121. https://doi.org/10.1016/bs.aiip.2015.06.001
Oldroyd, B. P. and Wongsiri, S. *Asian honey bees: Biology, conservation and human interactions.* (Harvard University Press, Cambridge, MA, 2006).
Phiancharoen, M., Duangphakdee, O. and Hepburn, H. R., "The biology of nesting." In: Hepburn, H.R., Radloff, S. E. (Eds.) *Honeybees of Asia.* (Springer, Berlin, 2011, pp. 109–132).
Pirk, C. W. W., Crous, K. L., Duangphakdee, O. et al., "Economics of comb wax salvage by the red dwarf honeybee, *Apis florea*." *J. Comp. Physiol. B*, 181(3) (2011): 353–359.
Pirk, C. W. W., Hepburn, H. R., Radloff, S. E. and Tautz, J., "Honey bee combs: Construction through a liquid equilibrium process?" *Naturwissenschaften*, 91 (2004): 350–353.
Preecha, Rod-I. M, Hepburn, R., and Duangphakdee, O., Developmental Cycles of the Red Dwarf Honeybee, *Apis florea. International Conference on Agricultural, Ecological and Medical Sciences (AEMS-2015)*, Phuket, Thailand, April 7–8, 2015.
Rinderer, T. E., Oldroyd, B. P., Guzman Li, D. E. et al., "Spatial distribution of the dwarf honey bees in an agroecosystem in southeastern Thailand." *Apidologie* 33 (2002): 539–543.
Rinderer, T. E., Oldroyd, B. P., Sylvester, A. H., Wongsiri, S., de Guzman, L. I., "Evolution of bee dances." *Nature*, 360 (1992): 305.
Rinderer, T. E., Wongsiri, S., Kuang, B. Liu, J., Oldroyd, B. P., Sylvester, H. A., De Guzman, L., Comparative nest architecture of the dwarf honey bees. *J. Apic. Res.* 35 (1996): 19–26.
Rösch, G. A., Über die Bautätigkeit im Bienenvolk und das Alter der Baubienen. Weiterer Beitrag zur Frage nach der Arbeitsteilung im Bienenstaat. *Zeit. Vergl. Physiol.* 6 (1927): 265–298.
Ruttner, F., *Biogeography and taxonomy of honeybees.* (Springer-Verlag, Berlin, 1988).
Ruttner, F., "Graded geographic variability in honey bees and environment." *Pszczelnicze Zeszyty Naukowe*, 29 (1985): 81–92.
Sakagami, S. F. and Yoshikawa, K., "Additional observation on the nest of the dwarf honeybee, *Apis florea* (Hymenoptera, Apidae)." *Kontyu* 41(2) (1973): 217–219.
Seeley, T. D. and Buhrman, S. C., "Group decision making in swarms of honey bees." *Behav. Ecol. Sociobiol.*, 45 (1999): 19–31.
Seeley, T. D. and Morse, R. A., "The nest of the honey bee (*Apis mellifera* L)." *Insectes Soc.*, 23 (1976): 495–512.
Seeley, T. D., Seeley, R. H. and Akratanakul, P., "Strategies of the honeybee in Thailand." *Ecol. Monogr.*, 52(1) (1982): 43–63.

Soman, A. G. and Chawda, S. S., “A contribution to the biology and behaviour of the dwarf bee. *Apis florea* F. and its economic importance in Kutch, Gujrat, India.” *Indian Bee J.*, 58(2) (1996): 81–88.
Taylor, J. E., “The structure of singluarities in soap bubble like and soap-film-like minimal surfaces.” *Ann. Math.*, 103 (1976): 489–539.
Thakar, C. V. and Tonapi, K. V., “Nesting behaviour of Indian honey bees-II. Nesting habits and comb cell differentiation in *Apis florea* Fab.” *Indian Bee. J.*, 24 (1962): 27–31.
Thompson, D. W., *On Growth and Form.* (Cambridge University Press, London, 1917).
Tóth, F., “What the bees know and what they do not know.” *Bull. Am. Math. Soc.* 70 (1964): 468–481.
von Frisch, K., *The dance language and orientation of bees.* (Harvard University Press, Cambridge, MA, 1967).
Wattanachaiyingcharoen, W., Wongsiri, S. and Oldroyd, B. P., “Aggregations of unrelated *Apis florea* colonies.” *Apidologie*, 39 (2008): 531–536.
Whitcombe, R. P., “The biology of Apis spp. in Oman with special reference to *Apis florea* Fab.” Thesis, (University of Durham, Durham, 1984).
Wongsiri, S., Lekprayoon, C., Thapa, R. et al., “Comparative biology of *Apis andreniformis* and *Apis florea* in Thailand.” *Bee World*, 78 (1997): 23–35.

8

Utilization of Apis florea *in Crop Pollination*

B. V. Shwetha, N. S. Bhat and T. Neethu

CONTENTS

Introduction

Honeybees are marvellous insects known to humankind since prehistoric times. Perhaps the bees are the only insects that are more beneficial to humans, as they produce honey, wax and venom and are also involved in increasing crop production through pollination. Pollination is an ecosystem service in that pollinators, in particular bees, contribute significantly to the pollination of a large array of crops (Morandin and Winston, 2005; Greenleaf and Kremen, 2006; Winfree et al., 2007). The pollinators are primarily flower visitors in search of food. Pollination is a side effect of their foraging activities. The individual insects have to make decisions on when to forage, where to forage and what to collect (Biesmeijer et al., 2006). Bees visit more than 90% of the globally most important crop types (Klein et al., 2007; Potts et al., 2016).

Bees are often not recognized for their contribution to modern agriculture. According to the USDA, a honeybee colony is worth 100 times more to the community than to the beekeeper, meaning the value they deliver extends well beyond their actual price (McGregor, 1976). Experiments have confirmed the above facts that efficient use of effective pollinators can significantly increase the yield level (Melnichenko, 1977). Bee pollination not only increases the yield level, but it also increases hybrid vigour, creates variation and maintains the gene flow in the ecosystem thereby conserving the crop varietal diversity. One out of three bites of food that graces our tables does so thanks to pollinators. In fact, when comparing the broad spectrum of flavours and nutrients we consume, considerably more of both come from pollinated fruits and vegetables. The wonderful diversity of our diets owes a considerable debt to the humble pollinator. Beyond their contribution to food production, bees provide a wide range of benefits to society (Matias et al., 2017). They support farmer and beekeeper livelihoods, represent social and cultural values and contribute to maintaining ecosystem stability (Potts et al., 2016; Potts, 2016).

Agriculture has become increasingly pollinator dependent (Aizen et al., 2009), and recent findings of declines in both wild and managed bees have raised concerns about the potential impact on pollination services (Allen-Wardell et al., 1998; Aizen & Harder, 2009; Gallai et al., 2009; Potts et al., 2010a, 2010b; Cameron et al., 2011). Worldwide, wild bees along with managed honeybees are the main and most economically important group of pollinators (Kremen et al., 2007). Honeybees are the most effective pollinators compared to other biotic pollinating agents because of their morphological adaptation, combined with their abundance, high flower reliability, long foraging distance and flower constancy (Roubik, 1995; Garibaldi et al., 2013; Couvillon et al., 2014; Rader et al., 2016). Bees actively hunt for and are able to identify the flowers of exact species they prefer to visit even in diverse settings (Denise et al., 2017). Before the intensification of agriculture over the last 50 years, native bees handled broad crop pollination requirements. In total, four honeybee species (*A. mellifera*, *A. cerana*, *A. dorsata*, *A. florea*) and two closely related bumblebee species (*Bombus impatiens* and *B. terrestris*) were identified as key social bee crop pollinators that contribute substantially to crop pollination at the global scale. A decrease of this pollination service could potentially reduce yields by 40% (Klein et al., 2007).

Initially the focus has been primarily on managed honeybees (*A. mellifera, A. cerana*), as they were regarded as the economically more important pollinators (Southwick and Southwick, 1992; Morse and Calderone, 2000; Allsopp et al., 2008 as quoted in Jaffé et al., 2009; Garibaldi et al., 2013). For a large number of crop species, pollination is provided by *A. mellifera* (Klein et al., 2007) For the world's top 15 crop types pollinated by bees, the European honeybee *A. mellifera* was the most reported pollinator (164 of 203 studies) and visited all crop types (Klein et al., 2018). Feral and wild bees, however, are important because they are reservoirs of local adaptations, which are said to ultimately determine the survival of honeybees in the wild (Matheson et al., 1996 as quoted in Jaffé et al., 2009). In the past century, conversion of natural areas to agriculture and urban development, along with natural habitat fragmentation, has caused more than 50 native bee pollinators to be placed on a Red List, or identified as being at risk of extinction in the near or long term. Thus, increased attention has been placed on wild bees as alternative pollinators. Wild bees currently play a significant role in crop pollination and are estimated to provide $150 billion in pollination services globally (Gallai et al., 2009). There is growing evidence that wild, unmanaged bees can provide effective pollination services where sufficient habitat exists to support their populations (Winfree et al., 2007; Garibaldi et al., 2013). They can also contribute to the long-term stability of crop pollination, thereby reducing the risk of pollination deficits from variable supply or activity of honeybees (Brittain et al., 2013b; Rader et al., 2013). As a result, wild pollinators should be integrated into crop pollination management plans as a supplement or alternative to managed bees (Kremen, 2005).

A. florea, commonly called the dwarf honeybee, provides a valuable supplement to commercial honeybee colonies by pollinating certain crops and forest plants more efficiently than their domesticated counterparts. *A. florea* has shown to significantly affect the yields of many crops, though its ability to provide full pollination services is dependent on its population sizes and the pollination requirements of the crop. *A. florea* is a better pollinator compared with *A. mellifera* and *A. cerana*, and it can offer insurance against changes in the availability of *A. mellifera* and *A. cerana* for many crops such as *Cucurbita maxima* L. (Chaudhury, 1974), *Cucurbita pepo* L. (Grewal and Sidhu, 1978), *Citrus reticulate* (Rakesh Kumar and Lenin, 1998), *Guizotia abyssinica* Cass., *Raphanus sativus* L., *Simaruba glauca* DC, *Syzigium jambos* L. (Shwetha, 2007) and *Zizyphus mauritiana* Lamark (Dhara Jyothi and Tandon, 1993).

Tending of *A. florea* Nest

In dry areas with a hot summer where other honeybees species are unlikely to survive and perform, *A. florea* can be used successfully because of its ability to survive in very hot and dry climates (Whitcombe, 1984; Ruttner et al., 1985). These bees arrive at no cost to farmers. *A. mellifera* and *A. cerana* colonies have to be purchased, maintained in a hive box, fed, moved, treated against diseases, easily inspected and manipulated where some amount of expenditure is incurred whereas feral *A. florea* colonies can be shifted to the field as and when the pollination service is required, and no hive box is needed for maintenance. The colony shifts during dusk. A cut is given on either side of the supportive structure, and the entire comb along with the supportive structure is transferred into a net bag that will enclose all the bees and at the same time provide aeration to the bees. The nest should be hung on small trees to maintain the vertical position of the combs at the desired location or field where pollination is intended. The colony should be placed in shade, or an artificial shade structure should be provided. The bees will take a day or two to adapt to the new ambient conditions and to resume normal activities and foraging. The workers will start foraging to the nearby best available food source, resulting in more numbers of foraging trips to collect nectar or pollen and/or both nectar and pollen depending on the quality of the resource available. As a result, there will be swift growth of the colony, and the symptoms of swarming will be exhibited. As the intention of placing the colony in the field is pollination, swarming is not a desirable behaviour. If the colony issues a swarm, then a part of the population will be lost, and there will be only a few bees available for pollination. Therefore, the swarming process has to be delayed by punching the drone cells and queen cells. By doing so, the entire colony can be retained in the same place for many days.

Foraging Preference

Individual bees integrate several sources of information, including personal and social, while making decisions about foraging (Biesmeijer and Seeley, 2005). Honeybee foragers use information gained in their own experience – such as memory of time and place, sugar concentration and amount of nectar previously collected – to decide whether to continue or resume foraging on a particular resource (Wainselboim et al., 2002; Grüter and Farina, 2009a; Van Nest and Moore, 2012; Al Toufailia et al., 2013). Bee species also differ in preference for flower height, time of flower visitation and within flower behaviour, which seems to be related to body size.

Floral Size Preference

Foragers exhibit higher rates of visitation to larger flowers (Galen and Newport, 1987; Campbell, 1989; Galen, 1989; Cresswell and Galen, 1991; Eckhart, 1991; Ohara and Higashi, 1994; Conner and Rush, 1996; Shykoff et al., 1997), and flower size is frequently correlated with the size of the floral reward (nectar and/or pollen) (Teuber and Barnes, 1979; Stanton and Preston, 1988; Cresswell and Galen, 1991; Duffield et al., 1993). Furthermore, foragers tend to select and visit more flowers on plants or patches with large floral display sizes in order to reduce travel times between flowers (Schmitt, 1983; Geber, 1985; Schmid-Hempel and Speiser, 1988; Klinkhamer et al., 1989; Goulson et al., 1998). Variations in the size of morphological characteristics of pollinators can affect their ability to extract rewards from flowers of different sizes and may affect pollen-collecting habits (Brian, 1952; Dafni and Kevan, 1997). Large foragers with broad heads or shorter tongues cannot reach as far into corolla tubes as smaller ones with narrow heads. Larger foragers are therefore limited to more open flowers or flowers with shorter, wider corolla tubes (Suzuki, 1994).

Flowers that are difficult to work on by *A. cerana* can be easily pollinated by *A. florea*, as *A. florea* has the necessary skill to probe variously shaped flowers; this probing behaviour is very important to harvest floral rewards. Based on their proboscis length, bees have adaptations to feed nectar of specific plants. *A. florea*, having much shorter tongue length (3.31 ± 0.234 mm, Mean ± S.D., $N = 50$) (Abrol, 2016), prefers smaller-sized flowers with smaller corolla tube (Kalpana and Ramanujam, 1990) that are not harvested by other bee species having much longer tongue lengths (6.45 ± 0.34 mm, Mean ± S.D., $N = 50$; Abrol, 2016).

Nectar Preference

When the sugar concentration of nectar is below a certain level (which has been estimated at 20%), the amount of energy needed to evaporate its water content to produce honey may make its collection uneconomical, and it is obviously to the bee's advantage to collect nectar with the greatest amount of sugar as quickly as possible. Therefore, the most important factors influencing the attractiveness of nectar are its abundance and sugar concentration. There may be great differences in the average sugar concentration of nectar of different species (Percival, 1965). Different varieties of the same species may also differ greatly in nectar concentration. But although different species and varieties may have nectar with different average sugar concentration, even within a single flower, and especially shallow open ones, the sugar concentration is subject to considerable fluctuations as a result of exposure to wind and rain, and changes in temperature and relative humidity. Therefore, the attractiveness of a species may vary at different times of the day and at different stages of following. Although a flower species may exhibit a characteristic daily secretory rhythm, bees may become conditioned to the time of daily periodic nectar production of the particular species they are visiting and spend the remainder of their day in the hive entrance.

Irrespective of temperature, nectar secretion is greater on a sunny than a dull day, reflecting a fact that the nectar sugar is a product of photosynthesis, which in turn is influenced by sunlight (Shuel, 1955). Soil moisture, atmosphere pressure, size of nectary and position of the flower on the plant may also influence the amount of nectar secreted (Free, 1970).

A. florea is known to harvest the nectar irrespective of the quantity and sugar concentration. It utilizes even trace quantities of nectar and also both diluted and concentrated nectar. Most of the flowers foraged for nectar had a nectar concentration ranging from 0.4 to 11 μL per flower (Raju et al., 2003). Of the two wild honeybee species, *A. florea* prefers flowers dominated with glucose and fructose sugars whereas *A. dorsata* prefers flowers dominated by sucrose content. This indicates *A. dorsata*, which is larger in size and has a much higher body weight (123.00 ± 5.21, Mean ± S.D., $N = 50$), prefers flowers providing high energy. *A. florea*, having a much lower body weight (22.6 ± 3.21, Mean ± S.D., $N = 50$), prefers to forage low- energy rewarding flowers (Abrol, 2016). However, foraging behaviour is also influenced by non-size-related variables such as plant distribution and density and plant attributes.

Nectar foragers sometimes prefer one food source over another as well as the specific position of one flower over another. Mayer and Lunden (1988) found more nectar foragers on the top of the flowers of *Manchurian crabapple* than red delicious apple. Fohouo et al. (2008) found the highest number of forager workers on *Syzygium guineense* var. *guineense* and the lowest number on *Psorospermum febrifugum*. Also, Weaver (1965) detected differences in honeybee foraging behaviour on hairy vetch (*Vicia vitlosa* Roth) flowers; some bees used the flower base while others used the flower mouth. Honeybees have a preference for apple tree branches located in the middle of trees rather than for those branches located higher up or lower (Mattu et al., 2012).

A. florea prefers lower energy rewarding flowers of *Brassica chinensis*, *B. campestris* and *B. rapa* over higher energy rewarding flowers of *Eruca sativa*, *Cajanus cajan* and *Raphanus sativus*. The flowers of the former three crops were simple as compared to more tunnelled flowers of the latter three crops, which enables the bees to have higher foraging and energy harvest rates on former crops as compared to the later ones (Sihag, 1997).

A. florea and *A. dorsata* were found to be secondary nectar robbers of *Hibiscus rosa-sinensis*. The holes left by sun – bird, *N. asiatica* and the night-foraging bats provide these bees an opportunity to forage on the leftover nectar in the flowers (Sihag, 1997). *A. florea* acted as a nectar thief on the cultivated cruciferous and leguminous crops with only a few bees resorting to front foraging. *A. florea* was observed to steal nectar from cucurbit flowers where it crawls along the corolla to reach the nectaries without coming in contact with the anthers or stigma (Sihag, 1993).

The bees also make use of various sources of social information, such as information about resource location, quality transmitted via the waggle dance (von Frisch, 1967; Grüter and Farina, 2009b) and information about resource quality and type from nectar samples unloaded in the hive (Grüter and Farina, 2009a).

Pollen Preference

Pollen is a critical resource for bees, providing protein and lipids that fuel development and survival. The selection of pollen is not influenced by colour, moisture or protein content, and there is no evidence that bees select pollen for its nutritive value (Free, 1967). Pollen contains phytosteroles that attract bees (Louveaux, 1959). Hexane or ethyl ether extract of pollen is very attractive to the foraging bees and initiates the behavioural response of packing the corbiculae (Taber, 1963).

The pollen need of a colony is closely correlated with the relative amounts of larval brood and stored pollen present in the hive (Dreller and Tarpy, 2000). If the number of cells containing larvae is high and the number containing pollen is low, then the colony's need for pollen is high. Pollen foragers can adjust trip time and pollen load size to adaptively tune the pollen collection rate of their colony (Fewell and Winston 1992; Janmaat et al., 2000). Pollen foragers spent significantly less time in the hive between collecting trips when pollen need was high. When pollen need was low, the pollen foragers stayed in the hive for a longer period after unloading the pollen and spent considerable time grooming and walking around slowly (Weidenmüller and Tautz, 2002).

The initiation of pollen-foraging activity was around 0800 h in November (0.75 bees/5 min) and 0900 h in December and January (1.50 and 1.00 bees/5 min, respectively). The pollen foraging activity was maximum between 1100 and 1400 h in November (17.75–23.00 bees/5 min), during 1400 h in December (44.00 bees/5 min) and between 1000 and 1300 h in the January (18.50–37.00 bees/5 min). The pollen-foraging activity ceased at 1700 h in November and January, which ceased one hour later (1800 h) during December (Malathi, 2005). Reyes-Carrillo et al. (2007) found high pollen collection in the early morning, while low amounts of pollen were collected in the afternoon.

In irrigated fields, the number of the pollen foragers is relatively more than nectar foragers in *A. florea* (Shuel, 1970).

The number of pollen grains sticking to the body of *A. florea* averaged 4700 in carrot (Sharma and Singh, 1999) and 11,437 in *Brassica campestris* (Sharma et al., 2000). There were 4710.9 pollen grains on phalsa Gill et al. (2001) and 5625 on onion flowers (Sharma et al., 2001).

The amount of pollen load carried was 4 mg for *A. florea*; the amount of pollen depleted was 520 pollen grains/insect visit; the amount of pollen deposited was 350 pollen grains with a pollen deposition efficiency of 67% in pumpkin, *Cucurbita maxima* Duch (Pandian et al., 2012).

Foraging Behaviour

The first foraging flight of *A. florea* worker bees was observed at an age range of 25–29 days (mean age of 27 days) during winter and of 18–20 days (mean age of 19 days) during summer. Comparison of the foraging activity during the two seasons of summer and winter revealed that the foraging activity was higher during summer than in winter (Malathi, 2005). *A. florea* is less active during the pre-winter months but is most active in the post-winter period (Ahmad and Srivastava, 2002). Pernal and Currie (2010) reported a higher foraging rate mean during the afternoon period (36.02 foragers/min) than during the morning period (17.66 foragers/min).

The foraging activity is impacted by colony strength and brood-rearing activity (Amdam et al., 2009; Abou-Shaara et al., 2013), the degree of pollen need (Weidenmüller and Tautz, 2002), the time of the year and the presence of other bee species. Worker bees collect these resources from a large variety of flower species and can fly several kilometres between the hive and good nectar or pollen sources. Like the vast majority of bees, *A. florea* and *A. cerana*, as well as *A. mellifera ligustica*, are restricted to foraging during the day (Hema, 2009). *A. florea* commenced its foraging activity on *Ocium basilicum* L. at about 0730 h while its cessation time was 1730 h. Average abundance of bees on sweet basal bloom was 12 (Standard Error 2.03) bees/m^2/min. Maximum abundance (24 ± 3.2 bees/m^2/min.) was recorded at about 0945 h (Dalio, 2018).

Survival of *A. florea* in a very hot and dry climate is one of its most outstanding traits. On the Persian Gulf, it tolerates a summer temperature of 50°C and more (Whitcombe, 1984; Ruttner et al., 1985). The abundance of *A. florea* has a positive relationship with temperature. *A. florea* is active at 46.5°C and did not forage at temperatures below 18°C (Whitcombe, 1981). The lower threshold for *A. florea* flight was 18°C, and peak activities were at 30°C Whitcombe (1984). Peak activity of *A. florea* on *Brassica campestris* var. toria was at 21°C–25°C (Kapil and Brar, 1971). In Bengaluru, India, the foraging activity of *A. cerana* declined significantly above 30°C (Reddy et al., 2015), showing that the species is adapted to a lower range of temperature. In contrast, temperature had a positive influence on the activity of *A. florea*, which is consistent with both its tropical distribution and the tolerance of very high temperatures (Cui and Corlett, 2016). The peak of *A. florea* activity was later in the day than that of *A. mellifera*, thus reducing possible competition for floral resources (Whitcombe, 1981). *A. florea* and *A. dorsata* foraged up to 1700 and 1630 h, respectively. However, on the day of the solar eclipse, both species stopped at 1540 h when the eclipse was at its peak and did not resume even when the eclipse was over at 1650 h (Jadhav, 1982).

The peak abundance of *A. florae* and floral resources (i.e., number of plant species in flowering and abundance of floral units) was recorded during early March to late May followed by a gradual decline until December. Monthly abundance of both bee species was positively related to the floral resources and negatively related to relative humidity, but it was not significantly related to temperature (Sajjad et al., 2017).

Foraging Time

The foraging activity of honeybees starts early in the morning and ceases in the evening, but the commencement and cessation times vary from region to region. *A. florea* with its smaller size and body weight is physiologically and morphologically better adapted to extract maximum reward from the flowers (Sihag, 1993). *A. mellifera* and *A. cerana* are able to forage earlier during the cooler hours of the day, while *A. florea* forages later in the day, resulting in intra-flower pollen transfer to the stigma. *A. florea*, with its small energetic needs, commenced activity between 1000 and 1100 h and dominated throughout the day. Pernal and Currie (2010) reported a mean higher foraging rate during the afternoon hours (36.02 foragers/min) than during the morning hours (17.66 foragers/min). *A. dorsata* and *A. florea* started their activity in the morning (0700–0800 h) and remained active till evening (1800 h) according to Chaudhary (2004). *A. florea* was more active at about 1000 h and was less active from 1600 h (Palanichamy et al. 1995). *A. dorsata* and *A. cerana indica* were relatively more active in the morning hours, while *A. florea* was active in the afternoon (Panda et al., 1993). *A. mellifera* and *A. florea* were active between 1100 to 1600 h on *Brasica juncea* (Naresh et al., 1998). Murrel and Nash (1981) observed maximum number of foragers of *A. florea* at 1230 h on toria.

During winter, the foraging activity of *A. florea* began at 0900 h (4.50–5.75 bees/5 min), which increased gradually, reaching the peak (122.75–128.50 bees/5 min) at 1100 h. *A. florea* then showed a declining trend and reached the lowest level (1.25–4.00 bees/5 min) at 1600 h, and activity ceased at 1700 h (Malathi, 2005).

Peak activity for *A. florea* is in the afternoon between 1100 and 1200 h (Panda et al., 1993). Foragers have the ability to remember the time of the day at which the higher food resources are available (Silva et al., 2013), and such ability may correlate with foraging activity peaks. According to Jadhav (1981), *A. florea* started foraging activity a little earlier (0855 h) than *A. dorsata* (0930 h) and continued at late hours in the evening (1700 h) on onion. Mohan Rao and Suryanarayana (1989) observed that *A. cerana* and *A. florea* initiated foraging at 0950–1130 h and ceased foraging at 1710–1815 h.

Foraging Distance

The foraging distance of honeybees strongly influences the sexual reproduction of most flowering plants and can determine the genetic structure of plant populations (Campbell, 1985; Waser et al., 1996). The flight distance is one of the biggest costs in terms of time and energy that a worker bee must weigh against the gain of the food (Seeley, 1994; Couvillon et al., 2014). The bees will not collect food from far away unnecessarily; therefore, forage distance can be used as a proxy for availability. In other words, the farther a bee flies to forage, lesser is the availability of forage closer by (Seeley, 1994, 1995).

Foraging ranges should be positively correlated with body size (Gathmann and Tscharntke, 2002; Beil, 2007; Greenleaf et al., 2007). Foraging distance has been shown to increase with body size for various taxa (Greenleaf et al., 2007). The larger-bodied species, *A. dorsata*, has the largest foraging range (95% of dances were for distances estimated to be less than 3.8 km) compared with the small-bodied *A. florea* (1.3 km for the 95% point) (Dyer and Seeley, 1991).

The maximum distance that *A. florea* workers foraged was 101.8 m, and the minimum distance was 13.4 m (Shwetha, 2007). *A. florea* bees were considerably higher, up to 150 m from the nesting site, but thereafter the numbers declined gradually, none being observed at more than 500 m (Abrol, 1988).

The mean foraging distances vary significantly for both nectar and pollen (Couvillon et al., 2015). The bees fly farther for nectar than for pollen, and the mean distance foragers travelled for pollen was 1074 m and 1408 m for nectar (Couvillon et al., 2015). Foraging distance varies with the month for the two forage types. The bees forage for both nectar and pollen at greater distances during summer (defined as pre-autumnal bloom, August 2009; July 2010; July 2011) than in early spring (March 2010; March 2011) and autumn (defined as during autumnal bloom, September–October 2009; September–October 2010). Under desert conditions, water foragers can fly up to 2 km from their colonies to collect water (Visscher et al., 1996).

Foraging distances will also vary with environmental conditions, such as the density and distribution of floral resources and the general physical resistance of the different habitats to flight (e.g., Ricketts, 2001) and the quantity and quality of available floral resources (Cresswell et al., 2000). If the high-reward resource is at a greater distance from the hive (Bacon et al., 1965; Beekman and Ratnieks, 2000), and there is resource scarcity (Steffan-Dwenter and Kuhn, 2003), the bees are observed to increase their foraging distance. Counts of marked *A. florae* bees were considerably higher, up to 150 m from the nesting site, but thereafter the numbers declined gradually, none being observed at more than 500 m. The foraging distance for colonies in the same region is impacted by race, colony strength, food resource, month and time of the day (Abou-Shaara, 2014). *A. florea* is known to forage within 1 km radius from the nest, but pollination efficiency is the best at 183–275 m from the nest (Free, 1993; Corlett, 2004; Abrol, 2012). *A. florea* bees cover a range of 400 m during their foraging activity (Lindauer, 1957). All the pollen types recovered in the present study were from plants situated within 400 m from the combs of these bees (Ramanujam and Kalpana, 1992). Visits by *A. florea* increased the fruit set even where substantial numbers of *A. cerana* were present, suggesting that the pollination contribution of *A. florea* is unique and additive to that of *A. cerana.*

Foraging Efficiency/Pollination Efficiency

The plant offers the floral reward to the pollinator and, in return, tries to take the benefit of pollination of its flowers by the pollinator. The real problem arises when the pollinator takes the benefit of floral reward but does not return equal benefit to the plant in terms of its reproductive success. Under such a situation, the major concern of the pollination biologists has been the measurement of degree of interdependence of the two components of this system (Sihag, 1997; Johnson and Steiner, 2000; Ne'eman, et al., 2010). This degree of interdependence can be measured by a parameter of the pollinator that is popularly known as pollination efficiency (PE). PE is a method used for determining the optimum number of pollinators needed to maximize pollination in a plant population (Sihag, 1997) and is an important parameter for the risk assessment in the pollinator conservation and sustainable agriculture (Ne'eman et al., 2010).

The seed set resulting from a single visit (Spears, 1983) and the fruit set resulting from multiple visits (Lamont and Wiens, 2003) of the pollinators were considered suitable methods of measuring the pollination efficiency. However, all the visits of a pollinator to a flower type are not always alike; there are qualitative as well as quantitative temporal variations in these visits (Sihag and Rathi, 1994; Chaudhary et al., 2009; Wadhwa and Sihag, 2012, 2015).

The number of bee visits to each flower significantly influences the quantitative parameters of the produce. A minimum of two visits by *A. cerana* and *A. florea* per flower decreases per cent flower drop and increases fruit set, fruit weight, fruit volume, number of sound seeds per fruit and 100-seed weight. A maximum of eight bee visits of *A. cerana* and *A. florea* per flower resulted in 100.00% fruit set; no flower drop; higher fruit weight, fruit volume, number of sound seeds per fruit and 100-seed weight in summer squash (Manjula, 2007). The number of seeds per capitulum and per cent oil content was highest in *A. florea* pollinated plots, which clearly indicated that *A. florea* is the best pollinator in chamomile crop. Overall, there is a general positive relationship between total visitation rates and yield but not with species richness. If the remaining species that thrive in intensively cultivated agricultural areas, including the managed honeybee, are effective pollinators, yield losses can be partly decoupled from losses of species (Bartomeus and Winfree, 2013).

Quality is also an important factor in crop production, especially from an economic standpoint. The quality of the produce/fruit can be negatively correlated with quantity when the fruit load is too high (Ferguson and Watkins, 1992), but it is not so otherwise, especially in crops with indeterminate flowering (Bommarco et al., 2012). Indeed, adequate pollination often leads to produce with enhanced quality in entomophilous crops such as orchard fruit production (Garratt et al., 2013), as well as in field crops (Bommarco et al., 2012) and small fruits and vegetables (Chagnon et al., 1993; Roselino et al., 2009; Andersson et al., 2012).

Quantitative effectiveness of a pollinator depends on visitation frequency as well as the abundance of pollen transferred. Effectiveness relates the overall visits of a pollinator to set seeds (Motten et al., 1981). Efficiency, on the other hand, is generally assessed as the amount of seed set resulting from each visit: a high-efficiency pollinator can generate a high seed set with fewer individual visits (Rader et al., 2009).

Pollen load carried by *A. cerana* and *A. florea* is positively correlated in relation to their body weight. *A. cerana* carried a maximum quantity of pollen (5.12 mg/bee), compared with *A. florea* (5.01 mg/bee). The greater the body weight, the higher is the pollen load carried. But, the pollen load-carrying efficiency is negatively correlated in relation to the bees' body weight and was high in *A. florea*, it being a smaller-sized bee that has high pollen-carrying ability (22.42%) compared with *A. cerana* (14.02%) in Chamomile (Shilpa, 2006).

Abundance of *A. florea* in Different Crops

On fennel, *Foeniculum vulgare*, *A. florea* was the most abundant (Shilpa, 2006) and constituted 81% of the visitors to fennel in India (Narayanan et al., 1960), and its peak activity was found between 1100 and 1400 h, which was synchronized with peak hour anthesis of flower (Basawana, 1982). Batra (1967) reported that *A. florea* visits more in number to flowers of *Cucurbita maxima* L. and constituted 70.7% (Grewal and Sidhu, 1978), 33.1% (Kapil and Chaudhary, 1974) and 3.0% (Girish, 1981) of the total bee population on *Cucurbita pepo*. The number of *A. florea* was maximum (Ahmad and Aslam, 2002), constituting 59.63% (Anonymous, 2000), and remained active for a longer period (Priti and Sihag, 1997) during the flowering period on carrot umbels. *A. florea* was the principal pollinating insect on coriander (Shekar and Suryanaryana, 1981), and its peak activity was found between 1100 and 1400 h, which was synchronized with peak hour anthesis of flower (Basawana, 1982). Apoidea were the predominant flower visitors (96%) of sesamum; among these, *A. florea* constituted 19.7% (Rakesh Kumar and Lenin, 2000). Of the total pollinators, 61.6% were Hymenoptera and *A. florea* constituted 5.68% in sesamum (Panda et al., 1989). Panda et al. (1991) observed *A. florea* foraged for 44.09 sec/flower on the flower head of sunflower. Honeybees were the main pollinators and represented 91.3%,of which *A. florea* constituted 51.5% (Mohan Rao and Suryanarayana, 1990), 38.18% (Panda et al. 1995), 27.35% (Guruprasad, 2001) and 6.73% (Kulkarni and Dhanorkar, 1998) on niger. Among the total insect pollinators visiting sesamum, the abundance of honeybee species constituted 77.67% of which *A. florea* constituted 29.00% (Vijaykumar, 2011). The foraging activity of *A. florea* started from 0700 h and terminated at 1800 h in fennel (*F. vulgare* Mill.), lavender (*Lavandula stoechas*), African marigold (*Tagetes minuta*) and White caroline snow creeper (*Porana volubilis*) – with a peak foraging activity at 1100 h and a mean number of 12.77/plant/5 min on fennel, with two peaks at 1200 h and 1000 h on lavender with a mean number of 4.72/plant/5 min and 4.62/plant/5 min, respectively. The peak activity in African marigold was at 1200 h with 5.74/plant/5 min, and there were two peaks at 1400 and 1100 h with 12.46 and 10.64/5 inflorescence/5 min, respectively, in White caroline snow creeper (*P. volubilis*).

Impact of Pollination by *A. florea*

Utilization of pollinators, especially honeybees, is considered one of the cheapest and eco-friendly approaches in maximizing the yield of cross-pollinated crops (Free, 1970). Many experiments have consistently confirmed that the yield levels can be increased to the extent of 50%–60% in fruits and plantation crops, 45%–60% in sunflower, sesamum and niger and 100%–150% in cucurbitaceous crops through good management of pollinators (Meinichenko and Khalifman, 1960).

The effect of honeybee pollination in coriander, *Coriandrum sativum*, on quantitative parameters such as number of seeds per umbel, number of seeds per plant, seed size and seed weight per umbel was maximum in *A. florea* pollinated plot (25.80, 232.40, 5.03 mm^3, and 0.31 g respectively) compared with open plot, *A. cerana, T. iridipennis*, and control plot. The qualitative parameters such as germination percentage, shoot length, root length and seedling vigour index were maximum in *A. florea* pollinated plot (80.00%, 10.66 cm, 7.76 cm and 1468) compared with open plot, *A. cerana, T. iridipennis,* and control plot at Bengaluru, India (Table 8.1) (Roopashree, 2011).

The efficiency of different honeybee species, viz., *A. cerana, A. dorsata, A. florea* and *T. iridipennis,* on different parameters such as per cent capitulum drop, number of seeds per capitulum and oil content in chamomile, *Matricaria chamomilla*, revealed that *A. florea* is a better pollinator compared with other honeybee species in Bengaluru, India (Table 8.2) (Shilpa, 2006).

The quantitative parameters such as flower drop, fruit set, fruit weight, number of sound seeds per fruit and germination (%) showed significant variation among different modes of pollination in summer squash. The effects of different modes of pollination on quantitative parameters of summer squash are given in Table 8.3 (Manjula, 2007).

The number of seeds per umbel, oil content and germination percent were highest in *A. florea* pollinated plots, which clearly indicated that among different honeybee species, *A. florea* is the best pollinator in honey plants in improvement of crops.

TABLE 8.1

Effect of Different Pollination Modes on Quantitative and Qualitative Parameters of Coriander, *Coriandrum sativum*

Pollination Treatments	Germination (%)	Shoot Length (cm)	Root Length (cm)	Seedling Vigour Index
Open pollination	79.00^a	9.96^b	7.26^b	1360^a
A. florea pollination	80.00^a	10.60^a	7.76^a	1468^a
A. cerana pollination	75.00^b	9.63^b	7.00^c	1247^b
T. iridipennis pollination	74.00^b	8.99^c	6.80^d	1168^b
Control	69.66^c	8.58^c	6.16^e	1026^c
Mean	75.33	9.55	7.00	1254.35
SEM ±	0.57	0.17	0.06	0.03
CD at 5%	1.85	0.55	0.21	0.15

TABLE 8.2

Effect of Different Pollination Modes on Yield Parameters of Chamomile *Matricaria chamomilla* L. Crop

Pollination Type	Per Cent Capitulum Drop	No. of Seeds per Capitulum	% Oil Content in Capitulum
Open pollination	10.54(3.39)^e	71.83(8.53)^a	0.17(1.08)^a
A. florea	12.48(3.67)^d	67.53(8.27)^b	0.15(1.07)^{ab}
A. cerana	15.30(4.03)^c	64.72(8.10)^{bc}	0.11(1.05)^c
T. iridipennis	20.35(4.62)^b	58.64(7.72)^d	0.09(1.04)^{cd}
Control	23.34(4.93)^a	52.72(7.32)^e	0.05(1.02)^e
Mean	16.40	63.08	0.11
SEM ±	0.015	0.054	0.000
CD at 5%	0.050	0.201	0.001

Notes: Figures in the parentheses are $\sqrt{X+1}$ transformed values.
Means followed by same letter in a column do not differ significantly by DMRT at 5% level.

TABLE 8.3

Impact of Different Modes of Pollination on the Quantitative Parameters of Summer Squash, *Cucurbita pepo*

Pollination Type	Flower Drop (%)	Fruit Set (%)	Fruit Weight (g)	No. of Sound Seeds/Fruit	Germination (%)
Open pollination	6.00[ab]	94.00[c]	1360.20[e]	412.00[cd]	86.40[de]
A. cerana pollination	12.00[b]	88.00[bc]	833.60[cd]	382.00[bc]	73.60[c]
A. florea pollination	14.00[bc]	86.00[bc]	602.20[c]	357.40[bc]	70.00[bc]
T. iridipennis pollination	16.00[bc]	84.00[b]	272.60[b]	314.60[b]	64.40[b]
Hand pollination at 0900 h	2.00[a]	98.00[cd]	1040.60[de]	436.60[e]	84.40[d]
Hand pollination at 1000 h	4.00[ab]	96.00[d]	1014.00[d]	420.00[cd]	80.00[cd]
Control	100.00[c]	0.00[a]	0.00[a]	0.00[a]	0.00[a]
Mean	22.00	78.00	731.88	331.80	65.54
SEM ±	3.41	3.41	16.06	5.38	1.03
CD at 5%	9.95	9.95	46.90	15.71	3.00

Note: Mean values followed by the same alphabet in the columns do not differ significantly by DMRT at 5% level.

Conclusion

Wild and managed bees are under increasing pressure. We highlight their importance for pollination of wild and cultivated plants as well as for other bee mediated ecosystem services. The much reduced provision of food and natural nesting resources, increasing fragmentation of sites rich in floral resources and habitat features have led to a strong decline in abundance and diversity of wild bees. The loss of bee pollinators will likely change cropping patterns towards non-pollinator-dependent crop types with consequences for the producer and consumer welfare (Bauer and Wing, 2016). Conservation, restoration and protection of bee habitats across many different landscapes are needed to increase bee abundances and ensure pollination for both agricultural and non-agricultural native plant communities (Megan and Frank, 2018). However, increasing or maintaining high pollinator diversity can enhance yield quantity and stability by improving the pollination efficiency of honeybees (Greenleaf and Kremen, 2006) and reduce the risk of pollination failure due to climate change (Bartomeus et al., 2013; Rader et al., 2013) or environmental disturbances such as extreme weather events (Brittain et al., 2013a).

REFERENCES

Abou-Shaara, H. F., "The foraging behaviour of honey bees, *Apis mellifera*: A review." *Veterinarni medicina* 59(1) (2014): 1–10.

Abou-Shaara, H. F., Al-Ghamdi, A. A. and Mohamed, A. A., "Honey bee colonies performance enhance by newly modified beehives." *Journal of Apicultural Science* 57 (2013): 45–57.

Abrol, D. P., "Foraging range of subtropical bees, *Megachile flavipes, Megachile nana* Hymenoptera: Megachilidae) and *Apis florea* (Hymenoptera: Apidae)." *Journal of the Indian Institute of Science* 68 (1988): 43–47.

Abrol, D. P., "Foraging strategies in honeybees, *Apis dorsata* F. and *Apis florea* F. in relation to availability of energy rewards." *Journal of Apiculture* 31(1) (2016): 9–18.

Abrol, D. P., *Pollination Biology: Biodiversity Conservation and Agricultural Production.* Springer, Dordrecht, the Netherlands, 2012.

Ahmad, M. and Aslam, M., "Pollinators visiting carrot (*Daucus carota* L.) seed crop." *Journal of Research (Science)* 13(1) (2002): 31–35.

Ahmad, R. and Srivastava, D. P., "Relative visitation of *Hymenopteran* bees to pigeonpea." *Indian Journal of Pulses Research* 15(2) (2002): 203.

Aizen, M. A., Garibaldi, L. A., Cunningham, S. A. and Klein, A. M., "How much does agriculture depend on pollinators? Lessons from long-term trends in crop production." *Annals of Botany* 103 (2009): 1579–1588.

Aizen, M. A. and Harder, L. D., "The global stock of domesticated honey bees is growing slower than agricultural demand for pollination." *Current Biology* 19 (2009): 915–918.

Al Toufailia, H., Grüter, C. and Ratnieks, L. W., "Persistence to unrewarding feeding locations by honeybee foragers (*Apis mellifera*): The effect of experience, resource profitability, and season." *Ethology* 119 (2013): 1096–1106. doi:10.1111/eth.12170.

Allen-Wardell, G., Bernhardt, P. and Bitner, R. et al., "The potential consequences of pollinator declines on the conservation of biodiversity and stability of food crop yields." *Conservation Biology* 12 (1998): 8–17.

Allsopp, M. H., de Lange, W. J. and Veldtman, R., "Valuing insect pollination services with cost of replacement." *PLoS ONE* 3 (2008): 3128.

Amdam, G. V., Rueppell, O., Fondrk, M. K., Page, R. E. and Nelson, C. M., "The nurse's load: Early-life exposure to brood-rearing affects behavior and lifespan in honey bees (*Apis mellifera*)." *Experimental Gerontology* 44 (2009): 447–452.

Andersson, G. K. S., Rundlöf, M. and Smith, H. G., "Organic farming improves pollination success in strawberries." *PLoS ONE* 7(2) (2012): 31599. doi:10.1371/journal.pone.0031599.

Bacon, O. G., Burton, V. E., McSwain, J. W. et al., *Pollinating Alfalfa with Leaf-cutting Bees (AXT 160).* University of California Agricultural Extension Service, Berkeley, CA, 1965.

Bartomeus, I., Park, M. G., Gibbs, J. et al., "Biodiversity ensures plant–pollinator phenological synchrony against climate change." *Ecology Letters* 16(11) (2013): 1331–1338. doi:10.1111/ele.12170.

Bartomeus, I. and Winfree, R., "Pollinator declines: Reconciling scales and implications for ecosystem services." *F1000Research* 146 (2013): 2–4. doi:10.12688/f1000research.2-146.v1.

Basawana, K. S., "Role of insect pollinators on seed production in coriander and fennel." *South Indian Horticulture* 56 (1982): 117–118.

Batra, S. W. T., "Crop pollination and the flower relationship of the wild bees of Ludhiana, India (Hym: Apoidae)." *Journal of the Kansas Entomological Society* 40 (1967): 164–177.

Bauer, D. M. and Wing, I. S., "The macroeconomic cost of catastrophic pollinator declines." *Ecological Economics* 126 (2016): 1–13.

Beekman, M. and Ratnieks, F. L. W., "Long-range foraging by the honeybee, *Apis mellifera* L." *Functional Ecology* 14 (2000): 490–496.

Beil, M., "Blütenbesuchende Wildbienengemeinschaften (Hymenoptera, Apoidea) in Sand-Ökosystemen." Dissertation, FB Biologie, TU Darmstadt, 2007.

Biesmeijer, J. C., Roberts, S. P. M., Reemer M. et al., "Parallel declines in pollinators and insect-pollinated plants in Britain and the Netherlands." *Science* 313 (5785) (2006): 351–354.

Biesmeijer, J. C. and Seeley, T. D. "The use of waggle dance information by honey bees throughout their foraging careers." *Behavioral Ecology and Sociobiology* 59 (2005): 133–142.

Bommarco, R., Marini, L. and Vaissière, B. E., "Insect pollination enhances seed yield, quality, and market value in oilseed rape." *Oecologia* 169(4) (2012): 1025–1032. doi:10.1007/s00442-012-2271-6.

Brian, A. D., "Division of labour and foraging in *Bombus agrorum* Fabricius." *The Journal of Animal Ecology* 21 (1952): 223–240.

Brink, D. and de Wet, J. M. J., "Interpopulation variation in nectar production in *Aconitum columbianum* (Ranunculaceae)." *Oecologia* 47 (1980): 160–163.

Brittain, C., Kremen, C. and Klein, A. M., "Biodiversity buffers pollination from changes in environmental conditions." *Global Change Biology* 19(2) (2013a): 540–547.

Brittain, C., Williams, N., Kremen, C. and Klein, A. M., "Synergistic effects of non-Apis bees and honey bees for pollination services." *Proceedings of the Royal Society B: Biological Sciences* 280 (2013b): 2012–2767.

Cameron, S. A., Lozier, J. D., Strange, J. P. et al., "Patterns of widespread decline in North American bumble bees." *Proceedings of the National Academy of Sciences* 108 (2011): 662–667.

Campbell, D. R., "Pollen and gene dispersal the influences of competition for pollination." *Evolution* 39 (1985): 418–431.

Campbell, D. R., "Inflorescence size, test of the male function hypothesis." *American Journal of Botany* 76 (1989): 730–738.
Chagnon, M., Gingras, J. and Deoliveira, D., "Complementary aspects of strawberry pollination by honey and indigenous bees (Hymenoptera)." *Journal of Economic Entomology* 86 (1993): 416–420.
Chaudhary, N., "Factors influencing inter and intra specific preferences of honey bee for three concurrently flowering vegetable/spice crops." PhD Thesis, CCSHAU, Hisar, India, 2004: 96+xxviii.
Chaudhary, N., Sihag, R. C. and Pandey, M. C., "Relative abundance, diversity and dominance concentration of two honeybee species on three concurrently flowering crops onion (*Allium cepa* L.), carrot (*Daucus carota* L.) and fennel (*Foeniculum vulgare* L.)." *Pestology* 33 (7) (2009): 9–13.
Conner, J. K. and Rush, S., "Effects of flower size and number on pollinator visitation to wild radish, *Raphanus raphanistrum*." *Oecologia* 105 (1996): 509–516.
Corlett, R. T., "Flower visitors and pollination in the Oriental (Indomalayan) Region." *Biological Review* 79 (2004): 497–532.
Couvillon, M. J., Schürch, R. and Ratnieks F. L. W., "Waggle dance distances as integrative indicators of seasonal foraging challenges." *PLoS ONE* 9(4) 2014: 93495.
Couvillon, M. J., Walter, C. M., Blows, E. M. et al., "Busy bees: Variation in insect flower-visiting rates across multiple plant species." *Psyche: A Journal of Entomology* 2015: 1–7.
Cresswell, J. E. and Galen, C., "Frequency-dependent selection and adaptive surfaces for floral character combinations: The pollination of *Polemonium viscosum*." *The American Naturalist* 138 (1991): 1342–1353.
Cresswell, J. E., Osborne, J. L. and Goulson, D., "An economic model of the limits to foraging range in central place foragers with numerical solutions for bumblebees." *Ecological Entomology* 25 (2000): 249–255.
Cui, Q. and Corlett, R. T., "Seasonal and diurnal patterns of activity in honeybees (*Apis* spp.) on the northern edge of the Asian tropics; their implications for the climate-change resilience of pollination." *Tropical Conservation Science* 9(3) (2016): 1–9. doi:10.1177/1940082916667142.
Dafni, A. and Kevan, P. G., "Flower size and shape: Implications in pollination." *Israel Journal of Plant Sciences* 45 (1997): 201–212.
Dalio, J. S., "Foraging activity of dwarf honey bee (*Apis Florea*) on bloom of *Ocimum basilicum* L." *Journal of Research in Agriculture and Animal Science* 5(1) (2018): 11–14.
Dhara Jyothi, B. and Tandon, P. L., "Insect pollinator activity on ber (*Zizyphus mauritiana* Lamk.)." *Pollination in Tropics: Proceedings of the International Symposium on Pollination in Tropics*, Bangalore, India, August 8–13, 1993: 210–211.
Dreller, C. and Tarpy, D. R., "Perception of the pollen need by foragers in a honeybee colony." *Animal Behaviour* 59 (2000): 91–96.
Duffield, G. E., Gibson, R. C., Gilhooly, P. M. et al., "Choice of flowers by foraging honeybees (*Apis mellifera*) – Possible morphological cues." *Ecological Entomology* 18 (1993): 191–197.
Dyer, F. C. and Seeley, T. D., "Dance dialects and foraging range in three Asian honey bee species." *Behavioral Ecology and Sociobiology* 28 (1991): 227–233.
Eckhart, V. M., "The effects of floral display on pollinator visitation vary among populations of *Phacelia linearis* (Hydrophyllaceae)." *Evolutionary Ecology* 5 (1991): 370–384.
Ferguson, L. B. and Watkins, C. B., "Crop load affects mineral concentrations and incidence of bitter pit in 'Cox's Orange Pippin' apple fruit." *Journal of the American Society of Horticultural Sciences* 117 (1992): 373–376.
Fewell, J. and Winston, M. L., "Colony state and regulation of pollen foraging in the honey bee, *Apis mellifera* L." *Behavioral Ecology and Sociobiology* 30(6) (1992): 387–393. doi:10.1007/BF00176173.
Fohouo, F. T., Djonwangwe, D. and Bruckner, D., "Foraging behavior of the African honey bee (*Apis mellifera adansonii*) on *Annona senegalensis, Croton macrostachyus, Psorospermum febrifugum* and *Syzygium guineense* var. *guineense* flowers at Ngaoundere (Cameroon)." *Pakistan Journal of Biological Sciences* 11 (2008): 719–725.
Free, J. B., "Factors determining the collection of pollen by Honeybee foragers." *Animal Behaviour* 15 (1967): 134–144.
Free, J. B., *Insect Pollination of Crops.* 2nd ed., Academic Press, London, UK, 1970: 544.
Free, J. B., *Insect Pollination of Crops.* 2nd ed., Academic Press, London, UK, 1993.
Galen, C., "Measuring pollinator-mediated selection on morphometric floral traits – Bumblebees and the alpine sky pilot *Polemonium viscosum*." *Evolution* 43 (1989): 882–890.

Galen, C. and Newport, M. E. A., "Bumblebee behaviour and selection on flower size in the sky pilot, *Polemonium viscosum*." *Oecologia* 74 (1987): 20–23.

Gallai, N., Salles, J. M., Settele, J. et al., "Economic valuation of the vulnerability of world agriculture confronted with pollinator decline." *Ecological Economics* 68(3) (2009): 810–821.

Garibaldi, L. A., Steffan-Dewenter, I., Winfree, R. et al., "Wild pollinators enhance fruit set of crops regardless of honey bee abundance." *Science* 339 (2013): 1608–1611.

Garratt, M. P. D., Breeze, T., Jenner, N. et al., "Avoiding a bad apple: Insect pollination enhances fruit quality and economic value." *Agriculture, Ecosystem and Environment* 184 (2013): 4–40.

Gathmann, A. and Tscharntke, T., "Foraging ranges of solitary bees." *Journal of Animal Ecology* 71 (2002): 757–764.

Geber, M. A., "The relationship of plant size to self-pollination in *Mertensia ciliate*." *Ecology* 66 (1985): 762–772.

Gill, M. S. N. and Malik, C. P., "Pollen biotechnology storage and viability." In *Pollen Physiology and Biotechnology*, C. P. Malik (Ed.). Today and Tommorrow's Printer and Publisher, New Delhi, India, 1992.

Gill, S. S., Kaushik, H. D. and Sharma, S. K., Effect of modes of pollination on fruit set and insect pollinators of phalsa (*Grewia subinaequalis* D.C.). *Research on Crops* 2(2) (2001): 193–196.

Girish, P. P., "Role of bees and pollination of summer squash (*Cucurbita pepo* Linne) with special reference to *Apis cerana* F." MSc Thesis, University of Agril. Sciences, Bangalore, India, 1981.

Goulson, D., Stout, J. C., Hawson, S. A. et al., "Floral display size in comfrey, *Symphytum officinale* L. (Boraginaceae): Relationships with visitation by three Bumblebee species and subsequent seed set." *Oecologia* 113 (1998): 502–508.

Greenleaf, S. S. and Kremen, C., "Wild bee species increase tomato production but respond differently to surrounding land use in Northern California." *Biological Conservation* 133 (2006a): 81–87.

Greenleaf, S. S. and Kremen, C., "Wild bees enhance honey bee's pollination of hybrid sunflower." *Proceedings of the National Academy of Sciences* 103 (2006b): 13890–13895.

Greenleaf, S. S., Williams, N. M., Winfree, R. and Kremen, C., "Bee foraging ranges and their relationship to body size." *Oecologia* 153 (2007): 589–596.

Grewal, G. S. and Sidhu, A. S., "Insect pollinators of some cucurbits in Punjab." *Indian Journal of Agricultural Sciences* 48 (1978): 79–83.

Grüter, C. and Farina, W. M., "Past experiences affect interaction patterns among foragers and hive-mates in honey bees." *Ethology* 115 (2009a): 790–797.

Grüter, C. and Farina, W. M., "The honey bee waggle dance: can we follow the steps?" *Trends in Ecology & Evolution* 24 (2009b): 242–247.

Guruprasad, G. S., "Maximization of productivity of Niger through enhancement of pollination." MSc Thesis, University of Agricultural Sciences, Dharwad, India, 2001.

Jadhav, L. D., "Role of insects in the pollination of onion (*Allium cepa* L.) in the Maharashtra State, India." *Indian Bee Journal* 43 (1981): 61–63.

Jadhav, L. D., "Effect of solar eclipse on the foraging activity of honeybees." *Biovigyanam* 8(2) (1982): 157–161.

Jaffé, R., Dietemann, V., Allsopp, M. H. et al., "Estimating the density of honeybee colonies across their natural range to fill the gap in pollinator decline censuses." *Conservation Biology* 24 (2009): 583–593.

Janmaat, A. F., Winston, M. L. and Ydenberg, R. C., "Condition dependent response to changes in pollen stores by honey bee (*Apis mellifera*) colonies with different parasitic loads." *Behavioral Ecology and Sociobiology* 47 (2000): 171–179.

Johnson, S. D. and Steiner, K. E. "Generalization versus specialization in plant pollination systems." *Trends in Ecology & Evolution* 15(4) (2000): 140–143.

Kapil, R. P. and Brar, H. S., "Foraging behaviour of *Apis florea* F. on *Brassica campestris* var toria." *Proceedings of XXIIth International Congress of Apiculture*, Moscow, Russia, 1971: 335–339.

Kapil, R. P. and Chaudhary, J. P., "Biology and utilization of insect pollinators for crop production (A7-Ent-118)." First Annual Report, Haryana Agricultural University, Hisar, India, 1973–1974: 30.

Klein, A. M., Vaissiere, B. E., Cane, J. H. et al., "Importance of pollinators in changing landscapes for world crops." *Proceedings of the Royal Society B: Biological Sciences* 274 (2007): 303–313.

Klein, A.-M., Boreux, V., Fornoff, F. et al., "Relevance of wild and managed bees for human well-being." *Current Opinion in Insect Science* 26 (2018): 82–88. doi:10.1016/j.cois.2018.02.011.

Klinkhamer, P. G. L., de Jong, T. J. and De Bruyn, G. J., "Plant size and pollinator visitation in *Cynoglossum officinale*." *Oikos* 54 (1989): 201–204.
Kremen, C., "Managing ecosystem services: What do we need to know about their ecology?" *Ecology Letters* 8(5) (2005): 468–479.
Kremen, C., Williams, N. M., Aizen, M. A. et al., "Pollination and other ecosystem services produced by mobile organisms: A conceptual framework for the effects of land-use change." *Ecology Letters* 10(4) (2007): 299–314.
Kulkarni, S. N. and Dhanorkar, B. K., "Effect of *Apis cerana indica* on niger seed production in Marathwada region (abstract)." *FAO Workshop on Sustainable Bee Keeping Development*, Dharwad, India, 1–5 August 1998.
Kumar, P., Baskaran, S., Sundaravadivelan, C. et al., "Insect visitors of pumpkin, *Cucurbita maxima* Duch., in relation to temperature and relative humidity." *Journal of Agricultural Technology* 8(2) (2012): 501–513.
Kumar, R. and Lenin J. K., "Studies on foraging behaviour of honey bees and pollination in orange, citrus reticulate Blanco." *Indian Bee Journal* 60(2) (1998): 106–107.
Kumar, R. and Lenin, J. K., "Insect pollinator and effect of cross pollination on yield attributes of sesame (*Sesamum indicum* L.)." *Indian Bee Journal* 62 (2000): 67–69.
Lamont, B. B. and Wiens, D., "Are seed set and speciation rates always low among species that resprout after fire, and why?" *Evolutionary Ecology* 17(3) (2003): 277–292.
Lindauer, M., "Communication among the honeybees and stingless bees of India." *Bee World* 38 (1957): 3–39.
Louveaux, J., "Recherches sur la récolte du pollen par les abeilles (*Apis mellifera* L.) (Fin)." *Annales des Abeilles* 2(1) (1959): 13–111.
Malathi, K., "Bioecology of little bee, *Apis florea* Fab." MSc thesis, University of Agricultural Sciences, Bangalore, India, 2005.
Manjula, S. C., "Pollination potentiality of honeybee species in summer squash (*Cucurbita pepo* L.)." MSc thesis, University of Agricultural Sciences, Bangalore, India, 2007.
Matheson, A., Buchmann, S. L., O'Toole, C. et al., *The Conservation of Bees*. Academic Press, London, UK, 1996.
Matias, D. M. S., Leventon, J., Rau, A.-L. et al., "A review of ecosystem service benefits from wild bees across social contexts." *Ambio* 46 (2017): 456–467.
Mattu, V. K., Raj, H. and Thakur, M. L., "Foraging behavior of honeybees on apple crop and its variation with altitude in Shimla hills of western Himalaya." *International Journal of Science and Nature* 3 (2012): 296–301.
Mayer, D. F. and Lunden, J. D., "Foraging behavior of honey bees on Manchurian crabapple and red delicious apple." *Journal of the Entomological Society of British Columbia* 83 (1988): 67–71.
McGregor, S. E., *Insect Pollination of Cultivated Crop Plants by S.E.* USDA Originally published, 1976.
Megan, E. L. and Drummond, F., "A review of native wild bee nutritional health." *International Journal of Ecology* 2018 (2018): Article ID 9607246, 10 pages.
Melnichenko, A. N., *Pollination of Agricultural Crops*. American Publication Co. Pvt. Ltd., New Delhi, India, 3(1977): 406.
Mohan Rao, G. and Suryanarayana, M. C., "Studies on pollination of watermelon (*Citrulus lanatus* (Thumb.) Mansf.)." *Indian Bee Journal* 50(1) (1988): 5–8.
Mohan Rao, G. and Suryanarayan, M. C., "Effect of honey bee pollination on seed yield in onion (*Allium cepa* L)." *Indian Bee Journal* 51(1) (1989): 9–11.
Mohan Rao, G. and Suryanarayana, M. C., "Studies on the foraging behaviour of honey bees and its effect on the seed yield in niger." *Indian bee Journal* 52(1–4) (1990): 31–33.
Morandin, L. A. and Winston, M. L., "Wild bee abundance and seed production in conventional, organic, and genetically modified canola." *Ecological Applications* 15 (2005): 871–881.
Morse, R. A. and Calderone, N. W., "The value of honey bees as pollinators of U.S. crops in 2000." *Bee Culture* 128 (2000): 15.
Motten, A. F., Campbell, D. R., Alexander, D. E. and Miller, H. L., "Pollination effectiveness of specialist and generalist visitors to a North Carolina population of *Claytonia virginica*." *Ecology* 62 (1981): 1278–1287.
Murrel, D. C. and Nash, W. T., "Nectar secretion by toria (*Brassica campestris* L.V. toria) and foraging behaviour of three *Apis* species on toria in Bangladesh." *Journal of Apicultural Research* 20(1) (1981): 34–38.

Narayanan, E. S., Sharma, P. L. and Phadke, K. G., "Studies on requirements of various crops for insect pollinators. Insect pollinators of Saunf (*Foeniculum vulgare*) with particular reference to the honeybees at Pusa (Bihar)." *Indian Bee Journal* 22 (1960): 7–11.

Naresh, M., Gurdip, S. and Grewal, G. S., "Comparative abundance and foraging behaviour of insect pollinators of Raya, *Brassica juncea* L. and role of *Apis mellifera* L. in crop pollination." *Journal of Insect Science* 11(1) (1998): 34–37.

Ne'eman, G., Jürgens, A., Newstrom-Lloyd, L. et al., "A framework for comparing pollinator performance: Effectiveness and efficiency." *Biological Reviews* 85(3) (2010): 435–451.

Ohara, M. and Higashi, S., "Effects of inflorescence size on visits from pollinators and seed set of *Corydalis ambigua* (Papaveraceae)." *Oecologia* 98 (1994): 25–30.

Palanichamy, P. S., Baskaran, S. and Mohandoss, A., "Insect pollination of the moringa plant *Moringa concanensis Nimmo* Linn." *Environment and Ecology* 13(1) (1995): 47–51.

Panda, P., Nanda, U. K., Mohpatra, H. and Padhi, J., "Insect pollination in some oilseed crops in Orissa, India." *Indian Bee Journal* 51(3) (1989): 97–98.

Panda, P., Panda, B. and Sontakke, B. K., "Foraging behaviour of honey bees on sunflower (*Helianthus annus* L.) sown on different dates." *Indian Bee Journal* 53(1–4) (1991): 44–49.

Panda, P., Rakh, L. K., Padhi, J. and Panigrahi, D., "Relative abundance and foraging behaviour of common bee species on niger in Phulbani District, Orissa, India." *Indian Bee Journal* 57 (1995): 10–14.

Panda, P., Sontakke, B. K. and Panda, B., "Foraging behaviour of honey bee species on different varieties of niger, in Orissa." *Journal of Insect Science* 6(1) (1993): 104–106.

Percival, M., "Pollen collection by *Apis mellifera*." *The New Phytologist* 46 (1947): 142–173.

Percival, M., *Floral Biology*. Pergamon Press, London, UK, 1965.

Pernal, S. F. and Currie, R. W., "The influence of pollen quality on foraging behavior in honeybees (*Apis mellifera* L.)." *Behavioral Ecology and Sociobiology* 51 (2010): 53–68.

Potts, S. G., Biesmeijer, J. C., Kremen, C. et al., "Global pollinator declines: Trends, impacts and drivers." *Trends in Ecology & Evolution* 25(6) (2010a): 345–353.

Potts, S. G., Imperatriz-Fonseca, V. L. and Ngo, H. T., *Assessment Report of the Intergovernmental Science-Policy Platform on Biodiversity and Ecosystem Services on Pollinators, Pollination and Food Production*. Secretariat of the Intergovernmental Science-Policy Platform on Biodiversity and Ecosystem Services, Bonn, Germany, 2016a.

Potts, S. G., Imperatriz-Fonseca, V. L., Ngo, H. T. et al., "Safeguarding pollinators and their values to human well-being." *Nature* 540 (2016b): 220–229.

Potts, S. G., Roberts, S. P. M., Dean, R. et al., "Declines of managed honey bees and beekeepers in Europe." *Journal of Apicultural Research* 49 (2010b): 15–22.

Prakash, K. B., "Pollination potentiality of Indian honeybee viz., *Apis cerana* on the production of cucumber (*Cucumis sativus* (Linn.) S.W. Cucurbitaceae)." MSc thesis, University of Agricultural Sciences, Bangalore, India, 2002.

Priti and Sihag, R. C., "Diversity visitation, frequency, foraging behaviour and pollinating efficiency of different insect pollinator visiting carrot, *Daucus carota* L. Var HC-1 blossoms." *Indian Bee Journal* 59 (1997): 1–4.

Rader, R., Bartomeus, I., Garibaldi, L. et al., "Non bee insects are important contributors to global crop pollination." *Proceedings of the National Academy of Sciences of the United States of America* 113 (2016): 146–151. doi:10.1073/pnas.1517092112.

Rader, R., Howlett, B. G., Cunningham, S. A. et al., "Alternative pollinator taxa are equally efficient but not as effective as the honeybee in a mass flowering crop." *Journal of Applied Ecology* 46 (2009): 1080–1087.

Rader, R., Reilly, J., Bartomeus, I. and Winfree, R., "Native bees buffer the negative impact of climate warming on honey bee pollination of watermelon crops." *Global Change Biology* 19(10) (2013): 3103–3110.

Raju, A. J. S., Rao, S. P. and Kumari, P. K., "The role of *Apis florea* in the pollination of some plant species in Andhra Pradesh, India." *Korean Journal of Apiculture* 18(1) (2003): 37–42.

Ramanujam, C. G. K. and Kalpana, T. P., "*Tamarindus indica* L: An important forage plant for *Apis florea* F. in south central India." *Apidologie* 23(5) (1992): 403–413.

Rasmussen, C., Nieh, J. and Biesmeijer, J. C., "Foraging biology of neglected bee pollinators." *Psyche: A Journal of Entomology* 2010 (2010): Article ID 134028, 2 pages. doi:10.1155/2010/134028.

Reddy, P. V. R., Rashmi, T. and Verghese, A., "Foraging activity of Indian honey bee, *Apis cerana* in relation to ambient climate variables under tropical conditions." *Journal of Environmental Biology* 3 (2015): 577–581.

Reyes-Carrillo, J. L., Eischen, F. A., CanoRios, P. et al., "Pollen collection and honey bee forage distribution in Cantaloupe." *Acta Zoologica Mexicana* 23 (2007): 29–36.

Ricketts, T. H., "The matrix matters: Effective isolation in fragmented landscapes." *The American Naturalist* 158 (2001): 87–99.

Roopashree, N., "Role of honey bees in pollination of Coriander (*Coriandrum sativum* L.)." MSc thesis, University of Agricultural Sciences, Bangalore, India, 2011.

Roselino, A. C., Santos, S. B., Hrncir, M. and Bego, L. R., "Differences between the quality of strawberries (Fragaria x ananassa) pollinated by the stingless bees *Scaptotrigona affidepilis* and *Nannotrigona testaceicornis. Genetics and Molecular Research* 8 (2009): 539–545.

Roubik, D. W., *Pollination of Cultivated Plants in the Tropics.* FAO Agricultural Services Bulletin No. 118. FAO, Rome, Italy, 1995.

Rueppell, O., Fondrk, M. K., Page, R. E. and Nelson, C. M., "The nurse's load: Early-life exposure to brood-rearing affects behavior and lifespan in honey bees (*Apis mellifera*)." *Experimental Gerontology* 44 (2009): 447–452.

Ruttner, F., Pourasghar, D. and Kauhausen, D., "Die Honigbiener des Iran 1. *Apis florea* Fabr." *Apidologie* 16 (1985): 119–138.

Sajjad, A., Ali, M. and Saeed, S., "Yearlong association of *Apis dorsata* and *Apis florea* with flowering plants: Planted forest vs. agricultural landscape." *Sociobiology* 64(1) (2017): 18–25.

Schmid-Hempel, P. and Speiser, B., "Effects of inflorescence size on pollination in *Epilobium angustifolium*." *Oikos* 53 (1988): 98–104.

Schmitt, J., "Flowering plant density and pollinator visitation in Senecio." *Oecologia* 60 (1983): 97–102.

Seeley, T. D., "Honeybee foragers as sensory units of their colonies." *Behavioral Ecology and Sociobiology* 34 (1994): 51–62. doi:10.1007/BF00175458.

Seeley, T. D., *The Wisdom of the Hive: The Social Physiology of Honey Bee Colonies.* Harvard University Press, Cambridge, MA, 1995.

Sharma, S. K. and Singh, J. R., "Pollination efficiency of *Apis dorsata* F. and *A. florea* F. on carrot (*Daucus carota* L.)." *Indian Bee Journal* 61(1) (1999): 1–4.

Sharma, S. K., Singh, J. R. and Kaushik, H. D., "Efficiency of two wild honey bees (*Apis* spp.) as pollinators of onion (*Allium cepa* L.)." *Haryana Journal of Horticultural Sciences* 30(1&2) (2001): 134–135.

Sharma, S. K., Singh, J. R. and Ombir, "Comparative efficiency of three *Apis* spp. on *Brassica campestris* var. '*sarson*' BSH 1." *Crop Research* 19(1) (2000): 125–127.

Shilpa, P., "Insect pollinators of selected aromatic crops with special reference to chamomile, *Matricaria chamomilla* L." MSc (Agri) thesis, University of Agricultural Sciences, GKVK, Bangalore, India, 2006.

Shuel, R. W., "Nectar secretion in relation to nitrogen supply, nutritional status and growth of plant." *Canadian Journal of Agricultural Science* 35 (1955): 124–138.

Shuel, R. W., "Current research on nectar." *Bee World* 51 (1970): 63–70.

Shwetha, B. V., "Studies on some behavioural aspects of Little bee, *Apis florea* Fab. (Hymenoptera: Apidae)." MSc thesis, University of Agricultural Sciences, Bangalore, India, 2007.

Shykoff, J. A., Bucheli, E. and Kaltz, O., "Anther smut disease in *Dianthus silvester* (Caryophyllaceae): Natural selection on floral traits." *Evolution* 51 (1997): 383–392.

Sihag, R. C., "Behaviour and ecology of sub-tropical carpenter bee, *Xylocopa fenestrate* F. 6. Crop hosts, foraging dynamics and pollination potential." *Journal of Apicultural Research* 32(2) (1993): 94–101.

Sihag, R. C., "Determining the efficiency of pollinators." In *Pollination Biology: Basic and Applied Principles*, R. C. Sihag (Ed.). Rajendra Scientific Publishers, Hisar, India, 1997a: 171–186.

Sihag, R. C., "Determining the required number of pollinators for crop pollination." In *Pollination Biology: Basic and Applied Principles*, R. C. Sihag (Ed.). Rajendra Scientific Publishers, Hisar, India, 1997b: 187–187.

Sihag, R. C., "Pollination pesticide interaction." In *Pollination Biology: Basic and Applied Principles*, R. C. Sihag (Ed.). Rajendra Scientific Publishers, Hisar, India, 1997c: 132–142.

Sihag, R. C., "Some unresolved issues of measuring the efficiency of pollinators: Experimentally testing and assessing the predictive power of different methods." *International Journal of Ecology* 2018 (2018): Article ID 3904973, 13 pages. doi:10.1155/2018/3904973.

Sihag, R. C. and Rathi, A., "Diversity, abundance, foraging behaviour and pollinating efficiency of different bees visiting pigeon pea (*Cajanus cajan* (L) Millsp.) blossoms." *Indian Bee Journal* 56 (1994): 187–201.

Silva, D. P., Moisan-De-Serres, J., Souza, D. C. et al., "Efficiency in pollen foraging by honey bees: Time, motion, and pollen depletion on flowers of *Sysirinchium palmifolium* (Asparagales: Iridaceae)." *Journal of Pollination Ecology* 11 (2013): 27–32.

Somanathan, H., Warran, E. J., Borges, R. M. et al., "Resolution and sensitivity of the eyes of the Asian honeybees *Apis florea, Apis cerana* and *Apis dorsata*." *The Journal of Experimental Biology* 212 (2009): 2448–2453.

Southwick, E. E. and Southwick Jr., L., "Estimating the economic value of honey bees (Hymenoptera: Apidae) as agricultural pollinators in the United States." *Journal of Economic Entomology* 85 (1992): 621–633. doi:10.1093/jee/85.3.621.

Spears Jr., E. E., "A direct measure of pollinator effectiveness." *Oecologia* 57(1–2) (1983): 196–199.

Stanton, M. L. and Preston R., "A qualitative model for evaluating the effects of flower attractiveness on male and female fitness in plants." *American Journal of Botany* 75 (1988): 540–544.

Steffan-Dewenter, I. and Kuhn, A., "Honeybee foraging in differentially structured landscapes." *Proceedings of the Royal Society B* 270 (2003): 569–575.

Suzuki, K., "Pollinator restriction in the narrow tube flower type of *Mertensia ciliata* (James) G. Don (Boraginaceae)." *Plant Species Biology* 9 (1994): 69–73.

Taber, St., *Warum die Bienen Pollen sammeln. Ref. 19. Internat Bienenzüchter-Kongr Prag (Vollst Texte II: 518).* Minist Landwirtsch, Prag, Czech Republic, 1963.

Teuber, L. R. and Barnes D. K., "Breeding alfalfa for increased nectar production." *Proceedings of the IV International Symposium on Pollination*, Maryland, 1979: 109–116.

Van Nest, B. N. and Moore, D., "Energetically optimal foraging strategy is emergent property of time-keeping behavior in honey bees." *Behavioral Ecology* 23 (2012): 649–658. doi:10.1093/beheco/ars010.

Vijaykumar, P. V., "Pollinators diversity with special reference to role of honey bee in quantitative and qualitative improvement of sesamum." MSc thesis, University of Agricultural Sciences, Bangalore, India, 2011.

Visscher, P., Crailsheim, K. and Sherman, G., "How do honeybees (*Apis mellifera*) fuel their water foraging flights?" *Journal of Insect Physiology* 42 (1996): 1089–1094.

von Frisch, K., *The Dance Language and Orientation of Bees*. Harvard University Press, Cambridge, MA, 1967, 566pp.

Wadhwa, N. and Sihag, R. C., "Psychophilous mode of pollination predominates in sarpagandha (*Rouvolfia serpentina*)," *Journal of Entomology* 9(4) (2012): 187–207.

Wadhwa, N. and Sihag, R. C., "Melittophilous mode of pollination predominates in European plum (*Prunus domestica* L.) in the semi-arid environment of Northwest India." *Asian Journal of Agricultural Research* 9(5) (2015): 189–207.

Wainselboim, A. J., Roces, F. and Farina, W. M., "Honey bees assess changes in nectar flow within a single foraging bout." *Animal Behaviour* 63 (2002): 1–6. doi:10.1006/anbe.2001.1879.

Waser, N. M., Chittka, L., Price, M. V. et al., "Generalization in pollination systems and why it matters." *Ecology* 77 (1996): 1043–1060.

Weaver, N., "The foraging behavior of honeybees on hairy vetch." *Insectes Sociaux* 12 (1965): 321–326.

Weidenmüller, A. and Tautz, J., "In-hive behavior of pollen foragers (*Apis mellifera*) in honey bee colonies under conditions of high and low pollen need." *Ethology* 108 (2002): 205–221.

Whitcombe, R. P., "Temperature regulation and foraging activity of two species of honeybee (*Apis mellifera* and *Apis florea*) in Oman." Preliminary Report, Joint Honey Production Project, *Apicultural Research Program* 20(2) (1981): 45.

Whitcombe, R. P., "The biology of *Apis* spp. in Oman with special reference to *Apis florea* Fab." PhD thesis, University of Durham, Durham, UK, 12 (1984): 621.

Winfree, R., Williams, N. M., Dushoff, J. and Kremen C., "Native bees provide insurance against ongoing honey bee losses." *Ecology Letters* 10(11) (2007): 1105–1113.

9

Safety of Dwarf Honeybee, Apis florea *in Relation with Agricultural Pest Management*

Chuleui Jung, Muhammad Noor-ul-Ane and Saeed Mohamadzade Namin

CONTENTS

Introduction

Bees are vital to global biodiversity and food security through their pollination of plants including crops. Honeybees, wild and domesticated, are important pollinators for agricultural crop production. While collecting nectar and pollens for the colony over the large area, foragers encounter toxic materials in agricultural or even in natural ecosystems. Bees collect nectar and pollen often from pesticide-contaminated flowers, and they collect propolis from tree buds or the flowerless trees such as pine trees with pesticide drift. Even with many factors such as habitat degradation and fragmentation, land-use changes, invasive pests and diseases, malnutrition from scarcity of foraging resources, climate changes and chemical pollution, most honeybee poisoning is often associated with pesticides (Johansen, 1977; Atkins, 1992; Ellis et al., 1998). Since the 1980s, introduction of parasitic honeybee mites, *Varroa destructor* (Choi et al., 1986) and *Tropilaelaps mercedesae* (Jung and Lee, 2018), and disease prevalence (e.g., sacbrood virus in Asian honeybees) have imposed even higher risk to bee health (Jung and Lee, 2018). Recently, new classes of systemic pesticides, neonicotinoids, are in high public issue due to honeybee toxicity, as shown in Figure 9.1 (Ghosh and Jung, 2017). Not only insecticides but other types of pesticides are also harmful to honeybees. Honeybees possess markedly lower numbers of genes encoding detoxification enzymes, including cytochrome P450 monooxygenases (P450s), glutathione-S-transferases and carboxylesterases (Claudianos et al., 2006). Here we provide some information on the interaction of pesticides and honeybee health, especially *Apis florea*.

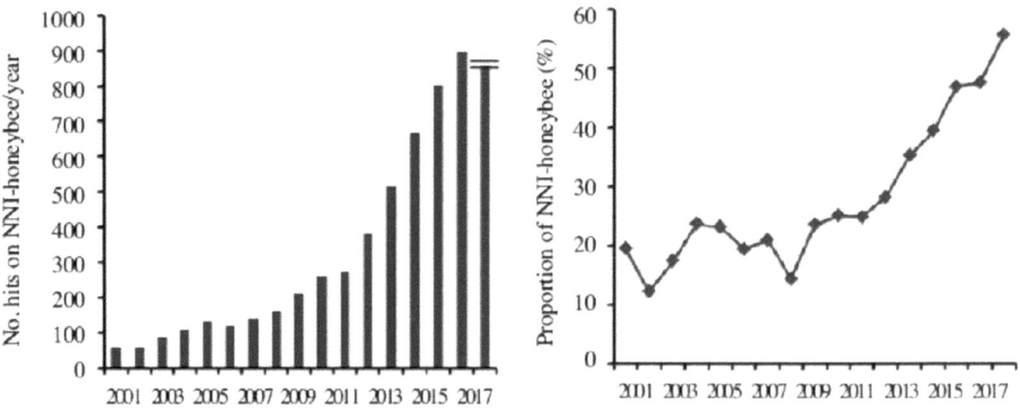

FIGURE 9.1 Trend of global issues on neonicotinoid insecticides and honeybee toxicity. Results of Google scholar search on neonicotinoid and honeybee (left) and proportion of neonicotinoid and honeybee among insecticides and honeybee. (From Ghosh, S. and Jung, C. *J. Apicult.* 32, 333–344, 2017.)

Apis florea is a dwarf honeybee mostly distributed in the forest area in South Asia. It is similar in appearance to *A. andreniformes*, but the male genitalia, mating time and some behaviours are different (Oldroyd and Wonsiri, 2006). *Apis florea's* distribution area covers diverse eco-climatic conditions and was recorded from Cambodia, China, India, Iran, Myanmar, Nepal, Oman, Pakistan, Saudi Arabia, Sri Lanka, Sudan, Thailand and Vietnam (Hepburn et al., 2005). The honeybee is not quite domesticated, but it is an important pollinator in some agricultural crops (Abrol, 2010), which pose potential risk due to agricultural chemical pesticides.

The phenomena of bee poisoning can be diverse. Honeybees seem poor at detecting the presence of toxic compounds in food, possibly in association with reduced gustatory receptors in the bee genome (Robertson and Wanner, 2006). Instead, learning capacity could reduce the negative physiological effects after exposure to a toxic compound (Ayestaran et al., 2010). The toxic effect can vary depending on the genetics, physiology, life stage, caste, age, season, temperature, feeding history and experiences (Johnson et al., 2010). Intercolonial differences or regional differences to pesticide susceptibility could be due to genetic differences (Lee et al., 2016). Repeated uses of pyrethroid acaricides against *Varroa* mites may be responsible for the elevated tolerance to pyrethroids used in agriculture (Elzen et al., 2003). Well-fed and young bees have higher antioxidant capacity of vitellogenin (Seehuus et al., 2006).

The interaction between toxic compounds and pathogens could have synergistic impact. Exposure to pesticides often makes bees more susceptible to pathogen infection. Chemicals applied for *Varroa* mite control can also alter honeybee gene expression, resulting in higher susceptibility to *Nosema* disease infection (Boncristiani et al., 2012; Wu et al., 2017).

Here we provide the general features of pesticides with mode of actions. We then address how these chemicals could be exposed to the nontarget organisms of honeybees and summarize reported toxic effects. And lastly, the life strategies on detoxification and avoidance to chemicals are followed, and how to protect and conserve *A. florea* against pesticides in agriculture and natural system is discussed.

Agricultural Pesticides Relative to Mode of Action

Herbicides and Fungicides

Herbicides and fungicides are not designed to kill insects; thus, they are generally low in toxicity to honeybees in a direct manner (e.g., acute toxicity). But bees could suffer from herbicides through an indirect route via the loss of flowering plants that produce nectar and pollen (Johansen, 1977). Recently, caged bee experiments demonstrated that the herbicide, paraquat, could affect honeybee life span by giving oxidative stress (Cousin et al., 2013), and glyphosate could impact honeybee gut microbiomes, leaving easy infection and death (Motta et al., 2018). Fungicides are widely detected in honeybee colonies (Mussen et al., 2004), and their impacts are higher on larvae than adults, since fungicides are not restricted in use during blooming. High levels of the fungicide chlorothalonil often found from bee-collected pollen and wax cause mortality to larvae by oxidative stress (Zhu et al., 2014).

Insecticides

Acetylcholinesterase inhibitors: Carbamates (CA) and organophosphates (OP) are insecticides that target the nervous system by inhibiting acetylcholinesterase (AChE) activity. AChE is the enzyme deactivating acetylchloline, a neurotransmitter in the synapses of an insect's nervous system. These insecticides were developed in the 1950s and widely used until the 1990s. These are broad-spectrum insecticides, which implies high risk to honeybees, too. Dimethoate (OP) and bendiocarb (CA) are the two mostly associated with bee poisoning in the UK (Barnett et al., 2007), and carbaryl (CA) is used in Korea during apple blooming (Kim and Jung, 2013a, 2013b; Kang and Jung, 2017). Also, these insecticides are absorbed and accumulated in the beeswax and pose a greater risk (Chauzat and Faucon, 2007).

GABA-gated Cl^- channel agonists: Fipronil in the phenylpyrazole chemical family binds to the GABA-gated Cl^- channel in insect neurons and leads to hyperpolarization of the neuron and an inability to transmit action potentials; fipronil is a broad-spectrum insecticide and highly toxic to honeybees (oral LD_{50} = 4 ng/bee, Casida and Durkin, 2013). This causes hyperexcitation of contaminated insects' nerves and muscles, but GluCl channels do not exist in mammals. Honeybee massive poisoning occurred when the chemical was sprayed to control the rice water weevil in Korea during 1999 (Kim et al., 2009; Jung and Lee, 2018).

Nicotinic acetylcholine receptor agonists: Developed in the 1980s, neonicotinoids (NNIs) are synthetic analogs of nicotine that activate the nicotinic acetylcholine receptor (nAChR), and promote the action potentials in postsynaptic nerve cells (Nauen et al., 2001). Two groups, nitro ($-NO_2$) and cyan (–CN), are different in honeybee toxicity (Ghosh and Jung, 2017). The nitro-NNIs – imidacloprid, clothianidin, thiamethoxam and dinotefuran – are highly toxic, while the cyano-NNIs – thiacloprid and acetamiprid – are much less toxic with acute LD_{50S} from 0.004 to 0.075 μg/bee versus contact LD_{50S} from 7.1 to 14.6 μg/bee. The difference is likely related to rapid cytochrome P450 detoxification (Iwasa et al., 2004). These systemic insecticides are blamed for their sublethal effects on honeybee learning and memories related to bee decline. Even these lower concentrations do not evoke the lethal acute responses (LD_{50} = 4.5 ng) or chronic responses (LC_{50} = 1,760 μg/L); the field-relevant dose of imidacloprid (0.02–0.30 ng per nectar load or 0.7–10 μg/L) may impair the performance and ability of foraging honeybees (Guez, 2013).

Other neurotransmitter receptor agonists: Amide acaricides are common in spider mite control, acting as an agonist of the octopamine receptor. Amitraz is also used in *Varroa* mite control (Gracia et al., 2017). A new class of anthranilic and phthalic diamide insecticides activate the ryanodine receptors in muscles by stimulating the release of Ca^{2+} from sarcoplasmic reticulum (Dinter et al., 2009). The chlorantraniliprole has a low acute toxicity to bees (oral LD_{50} >104 μg/bee; Dinter et al., 2009) possibly from lower affinity.

Insect growth regulators (IGRs): The insect growth regulator (IGR) insecticides are more toxic to immature than adult bees because they are related to juvenile hormone (JH) or the ecdysteroids or chitin synthesis (Tasei, 2001). Diflubenzuron, a chitin synthesis inhibitor, affects larval survival at concentrations as low as 1 ppm (Tasei, 2001). Fenoxycarb, a JH mimic, can be toxic to larvae but also to adults by an increase in JH titer affecting the adults prematurely (Heylen et al., 2011). Whole colonies treated with a field rate of fenoxycarb immediately experienced high mortality of immature bees and reduced brood abundance a year following treatment (Thompson et al., 2005) (Table 9.1).

TABLE 9.1

Developmental History of Major Insecticide Chemical Groups

Development	Chemical Group	Target	Mode of Action	Example	Target Spectrum
1950s	Organochlorine	Axon	Sodium (Na^+) channel	DDT	Broad
1950s	Carbamate	Synapse	Acetylcholinesterase (AChE)	Bendiocarb, Cabaryl	Broad
1950s	Organophosphate	Synapse	Acetylcholinesterase (AChE)	Dimethoate	Broad
1970s	Pyrethroids	Axon	Sodium (Na^+) channel	Fluvalinate	Selective
1980s	Neonicotinoids	Postsynapse	Nicotinic Acetylcholine Receptor Agonists	Imidacloprid	Broad
1990s	Insect growth regulators	Development	Hormone analogs or developmental disorder	Pyriproxyfen	Selective

Exposure Pathways of Pesticides to Honeybees

Direct body exposure: Honeybees can be exposed to spray directly when they are flying or while sitting on flowers during pesticides spraying through contact to the body surface or intake through an oral pathway (Zhu et al., 2015). This direct exposure could kill or harm honeybees immediately, or it can be a source of carrying pesticides to the colony.

Indirect body exposure: The honeybees can also be exposed to pesticides while sitting or walking on pesticide-treated surfaces such as foliar and soil (Tasei, 1996). Air movement may drift pesticides to adjacent untreated crops, weeds and nearby honeybee hives (May et al., 2015). The abraded dust falls on flowers and weeds, then the honeybees can get exposure to abraded dust either orally or in contact during foraging (Figure 9.2).

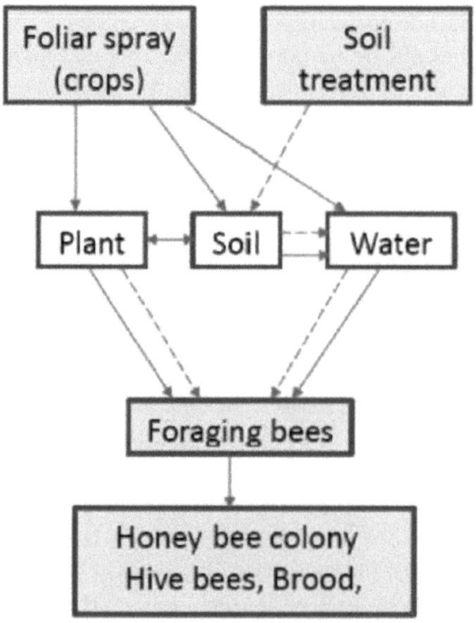

FIGURE 9.2 Possible routes of pesticide exposure to honeybee including *A. florea*.

Indirect oral exposure: Pollen and nectar can be contaminated by pesticides through foliar spray, systemic pesticides and dust deposition on flowers. Ponds and streams that are near agricultural areas are usually contaminated with agrochemicals by means of spray drift or water runoff. The systemic pesticides are also present in this guttation water (Tapparo et al., 2011). *Apis melifera* and *A. florea* share almost all routes of exposure except hive pest management like *Varroa*. Acaricides are not applied to *A. florea*, as these are wild bees; therefore, *A. florae* does not share this route of exposure to in-hive chemicals.

Risk Assessment Levels

The impact of pesticides on bees can be evaluated through hierarchical risk assessment system in a stepwise approach. In the case of honeybees, a tier system is widely adopted based on the system provided from OECD (1998a, 1998b) and USEPA (1996a, 1996b). The detection of the pesticides' toxicity levels against honeybees and other pollinators is the prerequisite before registration of a new pesticide. The tier system is composed of three steps (Figure 9.3). In Tier I, acute oral and contact toxicity test determines the LD_{50} (median lethal dose) and compare this to the field rate (ai g/ha) to estimate the hazard quotient. In Tier II, a toxicity of residues on foliage test is conducted to estimate if the RT_{25} (residual time to 25% mortality) is greater than 21 days, and in Tier III, field and semi-field pollinator tests are performed with brood test included to check the colony level effects. In Korea, the regulation of pesticide toxicity detection was introduced in 2000 only by an acute toxicity test. Recently, a residual toxicity test was included as an additional test in the regulation process in 2007 (RDA, 2009).

$$HQ = \frac{\text{Application rate } (\text{g a.i./ha})}{\text{Contact or Oral } LD_{50} \ (\mu\text{g a.i./bee})}$$

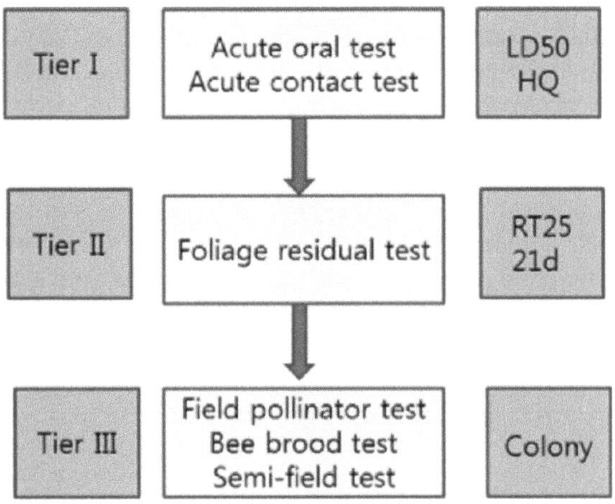

FIGURE 9.3 Conceptual test system of honeybee ecotoxicology in the Tier system adopted from OECD, EPA and RDA of Korea.

Pesticide Effects on Honeybees

Lethal Effects

The risk of pesticides to bees on its own is also a complex phenomenon. Basic toxicity information about new pesticides to bees is a prerequisite for pesticides registration most countries. Toxicological effects including mortality after exposure of pesticides in short period is termed acute toxicity. Pesticides are categorized as following relative to the acute LD_{50} per honeybee: extremely toxic (<2 µg), moderately toxic (2–10.99 µg), toxic (11–100 µg) and nontoxic (>100 µg) (Antonina and Marian, 2013).

Acute oral toxicity was tested among four species of honeybees – *A. mellifera, A. cerana, A. florea* and *A. dorsata* – using three neonicotinoid and one carbamate insecticides (Lee et al., 2016). LC_{50} values (ppm) of *A. florea* were 0.031, 2.947, 0.036, and 2.50 against thiamethoxam, imidacloprid, clothiandin and carbaryl, respectively. Researchers reported *A. florea* was more tolerant to tested insecticides than *A. melifera* (Lee et al., 2016). Similar results were reported that *A. florea* is more tolerant to imidacloprid than *A. mellifera* (Husain et al., 2014). Most residues of hexachlorocyclohexane (HCH) (67%) and endosulfan (58%) pesticides were present in honey of *A. florea* as compared to other organochlorines and organophosphates insecticides. Acephate, cholorpyriphos, dimethoate and carbaryl residues among other organophosphates were only present in *A. florea* (Khan et al., 2004).

Sublethal Effects

Measurement of the acute toxicity of pesticides largely relies on determination of an acute median lethal dose or concentration. However, lethal effect can be a partial measure of the deleterious effects. Pesticide-induced physiological and behavioural negative impacts are termed as sublethal effects.

Immune depression: When exposed to pesticides, a sublethal dose of pesticide exposure can reduce the immune response of honeybees, making them more susceptible to disease infection or to environmental stressors (Di Prisco et al., 2013; Brandt et al., 2016).

Enzymes as bioindicators: Acetylcholinesterase (AChE), Na^+/K^+ ATPase and cytochrome oxidase are associated when bees are exposed to deltamethrin, cypermethrin and imidacloprid, respectively (Bendahou et al., 1999; Decourtye et al., 2005; Badiou et al., 2008). In the case of *A. florea*, there was no change in Catalase (CAT), but reduced amounts of GST (glutathione transferases) were observed (Hayat et al., 2018).

Foraging failure: Schneider et al. (2012) reported that foraging flight and activity were reduced when bees were exposed to sublethal doses of clothianidin (≥0.5 ng/bee) and imidacloprid (≥1.5 ng/bee). The forager bees exposed to different concentrations of imidacloprid reduced their rate of returning (Bortolotti et al., 2003). Foraging behaviour change may also depend on the time period of the pesticide to cause toxicological effect (Guez et al., 2005). Neonicotinoids caused reduction in visit rate to food sources and increased time intervals between visits to food sources (Belzunces et al., 2012).

Orientation and navigation interference: Honeybees can use different cues (e.g., orientation, homing flights and olfaction etc.) to navigate their food sources. They increased time for homing flights and the number of indirect flights after exposure to glyphosate and thiamethoxam (Henry et al., 2012; Balbuena et al., 2015). Tosi et al. (2017) also reported negative impact of thiamethoxam on duration, velocity and distance of flight after chronic exposure to bees. Hesselbach and Scheiner (2018) claimed blocking of acetylcholine receptor (AchR) by flupyradifurone was the actual reason for reduction of sugar identification and olfaction in bees.

Detoxification

Detoxification is a survival strategy to escape the adverse effects of toxic chemicals. The major enzyme groups playing important roles are the cytochrome P450 monooxygenases (P450s), glutathione transferases (GSTs) and carboxylesterases (CCEs) (Papadopoulos et al., 2004; Isman, 2006; Mao et al., 2011;

TABLE 9.2

Number of Genes Encoding Detoxification Enzymes in Honeybees and Other Insects

Enzyme Group	*A. florea*	*A. mellifera*	*B. impatiens*	*B. terrestris*	*Megalochile rotudata*	*N. vitripennis*	*D. melanogaster*	*A. gambiae*
P450	–	46	44	44	52	92	85	106
CCE	25	24	24	24	22	46	35	51
GST	12	12	13	13	9	18	38	31
Total	–	82	81	81	83	156	157	188
Order	Hymeoptera	Hymeoptera	Hymeoptera	Hymeoptera	Hymeoptera	Hymeoptera	Diptera	Diptera

Xu et al., 2013). Recent studies on honeybee and *Bombus terrestris* indicated that the sensitivity of these bee species to neonicotinoids is determined by cytochrome P450s (Manjon et al., 2018). The genome of honeybee contains about 11,000 protein coding genes (Holt et al., 2002; Claudianos et al., 2006). Studies on honeybees (*A. florea* and *A. mellifera*) and bumblebees (*Bombus impatiens* and *B. terrestris*) genome (Table 9.2) indicated that honeybees and bumblebees having a greater reduction in the number of genes encoding for xenobiotic detoxification enzyme families in comparison to several insect pests of agricultural and medical importance and even parasitoids genome (Claudianos et al., 2006; Berenbaum and Johnson, 2015; Sadd et al., 2015). In dwarf honeybee, the genome contains 12 GSTs and 25 CCEs genes that show high similarity with *A. mellifera*, even on the level of clades and their classes. In *A. florea* and *A. mellifera* genome, relative to bumblebees, hormone and semiochemical processing associated CCE genes are reduced while dietary/detoxification associated CCE genes are increased (Sadd et al., 2015). Recent experiments on several neonicotinoid insecticides (thiamethoxam, imidacloprid and clothianidin) and one carbamate (carbaryl) showed that LC_{50} value for *A. florea* and *A. dorsata* was higher than in *A. mellifera* and *A. cerana* (Lee et al., 2016). The whole genome sequence of *A. florea* and *A. dorsata* has not been published yet, and perhaps they have more P450 detoxification enzymes compared with managed honeybees.

The nervous system is the mode of action of most insecticides. Therefore, it is not surprising that insecticides at levels that do not lead to mortality can influence insect behaviour (Haynes, 1988). Behavioural escape is the ability of an insect to move away from an insecticide-treated area, often without lethal consequence (Grieco et al., 2007). An insect leaves an insecticide-treated area only after making physical contact with the chemical (Chareonviriyaphap et al., 1997). There is information available of behavioural escape in honeybee, *A. mellifera*, to commonly used pesticides including fungicides, acaricides and insecticides (Kang and Jung, 2017). Recently, *A. florea* was reported to have higher learning and memory capacity than *A. mellifera*, which might be functioning for better behavioural escape from the hazard (Kasp and Shafir, 2013). Understanding the behavioural advantages of honeybees to reduce exposure to pesticides can assist in making better pollinator management programs.

Management for Protecting *A. florea* – Challenge

In conventional agricultural practices, pesticides are continuously used for crop protection while the wild and domesticated pollinators are important for crop production (Ghosh and Jung, 2016). Most insecticides adversely affect this mutual cooperation (Frazier et al., 2008). *A. florea* is a wild honeybee, and the protection from pesticides requires sophisticated use from farmers in addition to the habitat conservation. Several suggestions can help protect and conserve *A. florea* against pesticide use in agriculture:

Low-toxic pesticides: Pesticides are extremely variable in toxicity, durability, ability to bioaccumulate in the food chain and the range of plants and animals that are affected in the application. Extensive toxicity databases for honeybees are available, e.g., Agritox, ECOTOX; research on *A. florea* (Sanchez-Bayo and Goka, 2014; Lee et al., 2016) can be consulted, and choosing selective or low-toxic insecticides can be helpful (Abrol, 2011).

Less toxic formulation: Different formulations of the same insecticide have different effects on honeybees. Proper selection of pesticide formulation minimizes honeybee losses. Dusts are more hazardous than liquid formulations, and emulsifiable concentrates are less hazardous than wettable powders. Microencapsulated insecticides are much more toxic to honeybees than any other formulation (Sanford, 2003; Abrol, 2011). These capsules are carried to the colony by honeybees and can remain poisonous for a long time.

Reducing pesticide drift: Pesticide drift can damage foraging worker bees (Barnett et al., 2007; Mineau et al., 2008). Spraying pesticides when pollinators are not active – e.g., in the evening – reduces drift by avoiding spray on windy weather, and modifying the droplet size and distribution would reduce exposure risk (Craig et al., 1998; Beekman and Ratnieks, 2000; Sanchez-Bayo and Goka, 2016). Using low-pressure spraying, maintaining high water volumes during application and spraying close to the crop surface are useful to reduce pesticide drift (Abrol, 2011).

Using repellent: Many chemicals can repel insects. The first use of repellents to protect bees was the addition of carbolic acid to insecticide sprays (Callbreth, 1900). After that, many synthetic compounds including alcohols, aldehydes, aliphatic ketones, alicyclic ketones, aryl alkyl ketones, carboxylic acids, anhydrides and esters have been used as repellents for honeybees (Kasana et al., 1997). Several natural substances including extracts of tobacco, rue, garlic, parsley and century (Nicodemo and Nogueiracouto, 2004) have demonstrated the repellency to bees. In addition, 2-Heptanone, the major substance of alarm pheromone secreted by the mandibular glands of *A. mellifera* and *Trigona gribodoi* (Keeping et al., 1982), has shown a repellent property to *A. florea* (Naik et al., 2002). Application of the essential oil from propolis as a repellent against *A. florea* was also demonstrated (Naik et al., 2013). Recently, linalool and α–terpineol extracted from essential oil of *Swertia densifolia* showed repellent properties on *A. florea* behaviour (Naik et al., 2015). Further research can lead to finding a good repellent to prevent honeybee poisoning from pesticide spray.

Conclusion

Apis florea mainly inhabits the forested wild habitats but often forages on crops blooming such as mustard, onion, longan, jatropha fruits and many others. Increasing risks to pesticide poisoning threaten valuable pollinators not only in agricultural landscapes but also in urban areas. Many insecticides cause direct and indirect impacts to foragers of *A. florea*, resulting in impairment of development of broods and ovipositional failure or lowered performance of queens in the nest. Also, lower doses of pesticides could inflict the behavioural or physiological negative effects. Since honeybees possess a lower number of genes responsible for detoxification, careful use of pesticides is recommended for the conservation and protection of the valuable ecosystem service provider *A. florea*. Pesticide drift is the main cause of mortality in honeybees, but evening spraying with selective pesticides for honeybees and spraying with low-pressure targeting only to the crop surface can reduce risk. In addition, combined use of pesticide with chemicals that repel honeybees could lower the poisoning risk. Moreover, facilitated communication between farmers and beekeepers or conservation advocators for the protection and conservation of the pollinators is more demanding.

REFERENCES

Abrol, D. P. "Foraging behaviors of *Apis florea* F. and important pollinator of *Allium cepa* L." *Journal of Apicultural Research* 49 (2010): 318–325.

Abrol, D. P. *Pollination Biology: Biodiversity Conservation and Agricultural Production*. Springer, Dordrecht, the Netherlands, 2011.

Antonina, J., and B. Marian. "The impact of pesticides on honey bees and hence on humans." *Journal of Animal Science and Biotechnology* 46 no. 2 (2013): 272–277.

Atkins, E. L. "Injury to honey bees by poisoning." In *The Hive and the Honey Bee*, 1153–1208. Rev. Dadant and Sons, Hamilton, IL, 1992.

Ayestaran, A., M. Giurfa, and M. G. de Brito Sanchez. "Toxic but drank: Gustatory aversive compounds induce post-ingestional malaise in harnessed honeybees." *PLoS One* 5 no. 10 (2010): e15000.

Badiou, A., M. Meled, and L. P. Belzunces. "Honeybee *Apis mellifera* acetylcholinesterase-A biomarker to detect deltamethrin exposure." *Ecotoxicology and Environmental Safety* 69 no. 2 (2008): 246–253.

Balbuena, M. S., L. Tison, M. L. Hahn, U. Greggers, R. Menzel, and W. M. Farina. "Effects of sublethal doses of glyphosate on honeybee navigation." *Journal of Experimental Biology* 218 no. 17 (2015): 2799–2805.

Barnett, E.A., A. J. Charlton, and M. R. Fletcher. "Incidents of bee poisoning with pesticides in the United Kingdom, 1994–2003." *Pest Management Science* 63 no. 11 (2007): 1051–1057.

Beekman, M., and F. L. W. Ratnieks. "Long-range foraging by the honey-bee, *Apis mellifera* L." *Functional Ecology* 14 no. 4 (2000): 490–496.

Belzunces, L., S. Tchamitchian, and J. L. Brunet. "Neural effects of insecticides in the honey bee." *Apidologie* 43 no. 3 (2012): 348–370.

Bendahou, N., M. Bounias, and C. Fleche, "Toxicity of cypermethrin and fenitrothion on the hemolymph carbohydrates, head acetylcholinesterase, and thoracic muscle Na^+, K^+ -ATPase of emerging honeybees (*Apis mellifera* L.)." *Ecotoxicology and Environmental Safety* 44 no. 2 (1999): 139–146.

Berenbaum, M. R., and R. M. Johnson. "Xenobiotic detoxification pathways in honey bees." *Current Opinion in Insect Science* 10 (2015): 51–58.

Boncristiani, H., R. Underwood, R. Schwarz, J. D. Evans, J. Pettis, and D. vanEngelsdorp. "Direct effect of acaricides on pathogen loads and gene expression levels in honey bees *Apis mellifera*." *Journal of Insect Physiology* 58 no. 5 (2012): 613–620.

Bortolotti, L., R. Montanari, J. Marcelino, P. Medrzycki, S. Maini, and C. Porrini. "Effect of sub-lethal imidacloprid doses on the homing rate and foraging activity of honey bees." *Bulletin of Insectology* 56 no. 1 (2003): 63–67.

Brandt, A., A. Gorenflo, R. Siede, M. Meixner, and R. Büchler, "The neonicotinoids thiacloprid, imidacloprid, and clothianidin affect the immunocompetence of honey bees (*Apis mellifera* L.)." *Journal of Insect Physiology* 86 (2016): 40–47.

Callbreth, J. S. "The use of carbolic acid during spraying time." *Gleaning in Bee Culture* 28 (1900): 84.

Casida, J. E., and K. A. Durkin. "Neuroactive insecticides: Targets, selectivity, resistance, and secondary effects." *Annual Review Entomology* 58 (2013): 99–117.

Chareonviriyaphap, T., D. R. Roberts, R. G. Andre, H. J. Harlan, S. Manguin, and M. J. Bangs. "Pesticide avoidance behavior in *Anopheles albimanus*, a malaria vector in the Americas." *Journal of the American Mosquito Control Association* 13 (1997): 171–183.

Chauzat, M.-P., and J.-P. Faucon. "Pesticide residues in beeswax samples collected from honey bee colonies (*Apis mellifera* L.) in France." *Pest Management Science* 63 no. 11 (2007): 1100–1106.

Choi, S. Y., K. S. Woo, and Y. S. Kim. "Questionary survey on the varroa mites in Korean beekeeping." *Korean Journal of Apiculture* 1 (1986): 62–75.

Claudianos, C., H. Ranson, R. M. Johnson, S. Biswas, M. A. Schuler, M. R. Berenbaum et al. "A deficit of detoxification enzymes: Pesticide sensitivity and environmental response in the honeybee." *Insect Molecular Biology* 15 (2006): 615–636.

Cousin, M., E. Silva-Zacarin, A. Kretzschmar, M. El Maataoui, J.-L. Brunet, and L. P. Belzunces. "Size changes in honey bee larvae oenocytes induced by exposure to paraquat at very low concentrations." *PLoS One* 8 no. 5 (2013): e65693.

Craig, I., N. Woods, and G. A. Dorr. "Simple guide to predicting aircraft spray drift." *Crop Protection* 17 no. 6 (1998): 475–482.

Decourtye, A., J. Devillers, E. Genecque, K. Le Menach, H. Budzinski, and S. Cluzeau. "Comparative sublethal toxicity of none pesticides on olfactory learning performances of the honeybee *Apis mellifera*." *Archives of Environmental Contamination and Toxicology* 48 no. 2 (2005): 242–250.

Di Prisco, G., V. Cavaliere, D. Annoscia, P. Varricchio, E. Caprio, and F. Nazzi et al. "Neonicotinoid clothianidin adversely affects insect immunity and promotes replication of a viral pathogen in honey bees." *Proceedings of the National Academy of Sciences of the United States of America* 110 no. 46 (2013). 18466–18471.

Dinter, A., K. E. Brugger, N. M. Frost, and M. D. Woodward. "Chlorantraniliprole (Rynaxypyr): A novel DuPontTM insecticide with low toxicity and low risk for honey bees (*Apis mellifera*) and bumble bees (*Bombus terrestris*) providing excellent tools for uses in integrated pest management." *Julius-K ¨ uhn-Archiv* 423 (2009): 84–96.

Ellis, M. D., F. P. Baxendale, and D. L. Keith. *Protecting bees when using insecticides*. University of Nebraska Cooperative Extension: NebGuide G98-1347, 1998.

Elzen, P. J., G. W. Elzen, and W. Rubink. "Comparative susceptibility of European and Africanized honey bee (Hymenoptera: Apidae) ecotypes to several insecticide classes." *Southwestern Entomologist* 28 no. 4 (2003): 255–260.

Frazier, M., C. Mullin, J. Frazier, and S. Ashcraft. "What have pesticides got to do with it?" *American Bee Journal* 148 (2008): 521–524.

Ghosh, S., and C. Jung. "Global honeybee colony trend is positively related to crop yields of medium pollination dependence." *Journal of Apiculture* 31 no. 1 (2016): 85–95.

Ghosh, S., and C. Jung. "A short review on neonicotinoids; Use in crop protection and issues on honeybee and hive." *Journal of Apiculture* 32 no. 4 (2017): 333–344.

Gracia, M. J., C. Moreno, M. Ferrer, A. Sanz, M. A. Peribáñez, and R. Estrada. "Field efficacy of acaricides against *Varroa destructor*." *PLoS One* 12 (2017): e0171633.

Grieco, J. P., N. L. Achee, T. Chareonviriyaphap, W. Suwonkerd, K. Chauhan, M. R. Sardelis, and D. R. Robert. "A new classification system for the actions of IRS chemicals traditionally used for malaria control." *PLoS One* 2 no. 8 (2007): e716.

Guez, D. "A common pesticide decreases foraging success and survival in honey bees: Questioning the ecological relevance." *Frontiers in Physiology* 4 (2013): 37.

Guez, D., S. W. Zhang, and M. V. Srinivasan, "Methyl parathion modifies foraging behaviour in honeybees (*Apis mellifera*)." *Ecotoxicology* 14 no. 4 (2005): 431–437.

Hayat, K., A. Muhammad, A. A. Muhammad, A. Sajjad, F. S. Muhammad and M. K. Qaiser et al. "Insecticide exposure affects DNA and antioxidant enzymes activity in honey bee species *Apis florea* and *A. dorsata*: Evidence from Punjab, Pakistan." *Science of the Total Environment* 635 (2018): 1292–1301.

Haynes, K. F. "Sublethal effects of insecticides on the behavioral responses of insects." *Annual Review of Entomology* 33 (1988): 149–168.

Henry, M., M. Béguin, F. Requier, O. Rollin, J. F. Odoux, P. Aupinel, et al. "A common pesticide decreases foraging success and survival in honey bees." *Science* 336 no. 6079 (2012): 348–350.

Hepburn, H. R., S. E. Radloff, G. W. Otis, S. Fuchs, L. R. Verma, T. Ken et al. "*Apis florea*: morphometrics, classification and biogeography." *Apidologie* 36 no. 3 (2005): 359–376.

Hesselbach, H., and R. Scheiner. Effects of the novel pesticide flupyradifurone (Sivanto) on honeybee taste and cognition. *Scientific Reports* 8 no. 4954 (2018): 1–8.

Heylen, K., B. Gobin, L. Arckens, R. Huybrechts, and J. Billen. "The effects of four crop protection products on the morphology and ultrastructure of the hypopharyngeal gland of the European honeybee." *Apis mellifera. Apidologie* 42 no. 1 (2011): 103–116.

Holt, R. A., G. M. Subramanian, A. Halpern, G. G. Sutton, R. Charlab, D. R. Nusskern et al. "The genome sequence of the malaria mosquito *Anopheles gambiae*." *Science* 298 (2002): 129–149.

Husain, D., M. Qasim, M. Saleem, M. Akhter, and K. A. Khan. "Bioassay of insecticides against three honey bee species in laboratory conditions." *Cercetări Agronomice în Moldova* 47 no. 2 (2014): 69–79.

Isman, M. B. "Botanical insecticides, deterrents, and repellents in modern agriculture and an increasingly regulated world." *Annual Review of Entomology* 51 (2006): 45–66.

Iwasa, T., N. Motoyama, J. Ambrose, and R. M. Roe. "Mechanism for the differential toxicity of neonicotinoid insecticides in the honey bee, *Apis mellifera*." *Crop Protection* 23 no. 5 (2004): 371–378.

Johansen, C. A. "Pesticides and pollinators." *Annual Review of Entomology* 22 (1977): 177–192.

Johnson, R. M., M. D. Ellis, C. A. Mullin, and M. Frazier. "Pesticides and honey bee toxicity-USA." *Apidologie* 41 no. 3 (2010): 312–331.

Jung, C., and M. Lee. "Beekeeping in Korea: Past, present, and future challenges." In *Asian Beekeeping in the 21st Century*, Chantawannakul, P., G. Williams, and P. Neumann (Eds.), 175–197. Springer, Singapore, 2018.

Kang, M., and C. Jung. "Avoidance behavior of honey bee, *Apis mellifera* from commonly used fungicides, acaricides and insecticides in apple orchards." *Journal of Apiculture* 32 no. 4 (2017): 295–302.

Kasana, V., M. Malik, and O. Malik. "Status of honeybee repellents: Research and application." In *Perspectives in Indian Apiculture*, Mishra, R. C. (Ed.), 363–386. Agro botanicaa Publishers, Bikaner, India, 1997.

Kasp, R., and S. Shafir. "Associative olfactory learning of the red dwarf honey bee *Apis florea*." *Apidologie* 44 (2013): 100–109.

Keeping, M. G., R. M. Crewe, and B. I. Field. "Mandibular gland secretion of the stingless bee, *Trigona gribodoi* Magretti: Isolation, identification and compositional changes with age." *Journal of Apicultural Research* 21 no. 2 (1982): 65–73.

Khan, M. S., B. Kumari, H. R. Rohilla, H. D. Kaushik, and R. K. Arora. "Analysis of insecticide residues in honeys from apiary (*Apis mellifera*) and wild honey bee (*Apis dorsata* and *Apis florea*) colonies in India." *Journal of Apicultural Research* 43 no. 3 (2004): 79–82.

Kim, B., Y. Yang, Y. Park, M. Joeng, A. You, K. Park et al. "Risk assessment of fipronil on honey bee (*Apis mellifera*)." *The Korean Journal of Pesticide Science* 13 no. 1 (2009): 39–44.

Kim, D. W., and C. Jung. "Comparative analysis of the perception on the honeybee damage by the pesticides between beekeepers and apple growers." *Korean Journal of Apiculture* 28 (2013a): 33–38.

Kim, D. W., and C. Jung. "Oral acute toxicity of chemical compounds used for flower/fruit thinner of apple against buff-tailed bumblebee, *Bombus terrestris* (Hymenoptera: Apidae)." *Korean Journal of Apiculture* 28 (2013b): 25–32.

Lee, C., S. Jeong, C. Jung, and M. Burgett. "Acute oral toxicity of neonicotinoid insecticides to four species of honey bee, *Apisflorea, A. cerana, A. mellifera,* and *A. dorsata.*" *Journal of Apiculture* 31 no. 1 (2016): 51–58.

Manjon, C., B. J. Troczka, M. Zaworra, K. Beadle, E. Randall, G. Hertlein et al. "Unravelling the molecular determinants of bee sensitivity to neonicotinoid insecticides." *Current Biology* 28 (2018): 1137–1143.

Mao, W., M. A. Schuler, and M. R. Berenbaum. "CYP9Q-mediated detoxification of acaricides in the honey bee (*Apis mellifera*)." *Proceedings of the National Academy of Sciences of the United States of America* 108 (2011): 12657–12662.

May, E., J. Wilson, and J. Isaacs. *Minimizing Pesticide Risk to Bees in Fruit Crops.* Extension Bulletin Michigan State University-E3245, East Lansing, MI, 2015, pp. 1–16.

Mineau, P., K. M. Harding, M. Whiteside, M. R. Fletcher, D. Garthwaite, and L. D. Knopper. "Using reports of bee mortality in the field to calibrate laboratory-derived pesticide risk indices." *Environmental Entomology* 37 no. 2 (2008): 546–554.

Motta, E. V. S., K. Raymann, and N. A. Moran. "Glyphosate perturbs the gut microbiota of honey bees." *Proceedings of the National Academy of Sciences of the United States of America* 115 no. 41 (2018): 10305–10310.

Mussen, E. C., J. E. Lopez, and C. Y. S. Peng. "Effects of selected fungicides on growth and development of larval honey bees, *Apis mellifera* L. (Hymenoptera: Apidae)." *Environmental Entomology* 33 no. 5 (2004): 1151–1154

Naik, D. G., P. Banhatti, S. S. Chawda, and D. Thomas. "2-Heptanone as a repellent for *Apis florea.*" *Journal of Apicultural Research* 40 (2002): 59–61.

Naik, D. G., H. S. Vaidya, and T. P. Namjoshi. "Essential oil of Indian propolis: Chemical composition and repellency against the honeybee *Apis florea.*" *Chemistry & Biodiversity* 10 (2013): 649–657.

Naik, D. G., H. Vaidya-Kannur, P. V. Deshpande, C. N. Dandge, and G. V. P. Reddy. "Potential use of an essential oil from the flower of *Swertia densifolia* as a repellent for *Apis florea* (Hymenoptera: Apidae)." *Annals of the Entomological Society of America* 108 (2015): 18–25.

Nauen, R., U. Ebbinghaus-Kintscher, and R. Schmuck. "Toxicity and nicotinic acetylcholine receptor interaction of imidacloprid and its metabolites in *Apis mellifera* (Hymenoptera: Apidae)." *Pest Management Science* 57 no. 7 (2001): 577–586.

Nicodemo, D., and R. H. Nogueiracouto. "Use of repellents for honeybees (*Apis mellifera* L.) in vitro in the yellow passion-fruit (*Passiflora edulis* Deg.) crop and in confined beef cattle feeders." *Journal of Venomous Animals and Toxins Including Tropical Diseases* 10 (2004): 77–85.

OECD. "Guidelines for the testing of chemical number 213 honey bees, acute oral toxicity test." OECD, Environmental Health and Safety Division, Paris (1998a).

OECD. "Guidelines for the testing of chemical number 214 honey bees, acute contact toxicity test." OECD, Environmental Health and Safety Division, Paris (1998b).

Oldroyd, B. P., and D. Wonsiri. *Asian Honey Bees: Biology, Conservation and Human Interaction.* Harvard University Press, Cambridge, MA, 2006, p. 340.

Papadopoulos, A. I., I. Polemitou, P. Laifi, A. Yiangou, and C. Tananaki. "Glutathione S-transferase in the insect *Apis mellifera macedonica.*" *Comparative Biochemistry and Physiology PartC: Toxicology & Pharmacology* 139 (2004): 93–97.

RDA. Rural Development Administration. "The acts for management of agriculture chemicals." www.rda.go.kr (2009).

Robertson, H. M., and K. W. Wanner. "The chemoreceptor superfamily in the honey bee, *Apis mellifera*: Expansion of the odorant, but not gustatory, receptor family." *Genome Research* 16 no. 11 (2006): 1395–1403.

Sadd, B. M., S. M. Barribeau, G. Bloch, D. C. de Graaf, P. Dearden, C. G. Elsik, et al. "The genomes of two key bumblebee species with primitive eusocial organization." *Genome Biology* 16 (2015): 76.

Sanchez-Bayo, F., and K. Goka. "Impacts of pesticides on honey bees." In *Beekeeping and Bee Conservation – Advances in Research*, 77–97. InTech, Rijeka, 2016.

Sanchez-Bayo, F., and K. Goka. "Pesticide residues and bees–a risk assessment." *PLoS One* 9 no. 4 (2014): e94482.

Sanford, M. T. "Protecting honeybees from pesticides." University of Florida Cooperative Extension Service Institute of food and Agriculture Sciences, 2003: CIR534.

Schneider, C. W., J. Tautz, B. Grünewald, and S. Fuchs. "RFID tracking of sublethal effects of two neonicotinoid insecticides on the foraging behavior of *Apis mellifera*." *PLoS One* 7 no. 1 (2012): e30023.

Seehuus, S.-C., K. Norberg, U. Gimsa, T. Krekling, and G. V. Amdam. "Reproductive protein protects functionally sterile honey bee workers from oxidative stress." *Proceedings of the National Academy of Sciences of the United States of America* 103 no. 4 (2006): 962–967.

Tapparo, A., C. Giorio, M. Marzaro, D. Marton, L. Solda, and V. Girolami. "Rapid analysis of neonicotinoid insecticides in guttation drops of corn seedlings obtained from coated seeds." *Journal of Environmental Monitoring* 13 no. 6 (2011): 1564–1568.

Tasei, J.-N. "Effects of insect growth regulators on honey bees and non-*Apis* bees: A review." *Apidologie* 32 no. 6 (2001): 527–545.

Tasei, J. N. "Impact des pesticides sur les abeilles et les autres pollinisateurs." Lusignan: INRA, Fichier informatique. Rapport. http://www.beekeeping.com/articles/fr/tasei.htm (1996).

Thompson, H. M., S. Wilkins, A. H. Battersby, R. J. Waite, and D. Wilkinson. "The effects of four insect growth-regulating (IGR) insecticides on honeybee (*Apis mellifera* L.) colony development, queen rearing and drone sperm production." *Ecotoxicology* 14 no. 7 (2005): 757–769.

Tosi, S., B. G. Burgio and J. C. Nieh. "A common neonicotinoid pesticide, thiamethoxam, impairs honey bee flight ability." *Scientific Reports* 7 no. 1 (2017): 1201.

USEPA. "Ecological effects test guidelines OPPTS 850.3020. Honey bee acute contact toxicity." EPA 712-C-96-147 (1996a).

USEPA. "Ecological effects test guidelines OPPTS 850.3030. Honey bee toxicity of residues on foliage." EPA 712-C-96-148 (1996b).

Wu, M. C., Y. W. Chang, K. H. Lu, and E. C. Yang. "Gene expression changes in honey bees induced by sublethal imidacloprid exposure during the larval stage." *Insect Biochemistry and Molecular Biology* 88 (2017): 12–20.

Xu, J., J. P. Strange, D. L. Welker, and R. R. James. "Detoxification and stress response genes expressed in a western North American bumble bee, *Bombus huntii* (Hymenoptera: Apidae)." *BMC Genomics* 14 (2013): 874.

Zhu, W., D. R. Schmehl, C. A. Mullin, and J. L. Frazier. "Four common pesticides, their mixtures and a formulation solvent in the hive environment have high oral toxicity to honey bee larvae." *PLoS One* 9 no. 1 (2014): e77547.

Zhu, Y. C., J. Adamczyk, T. Rinderer, J. Yao, R. Danka, R. Luttrell, and J. Gore. "Spray toxicity and risk potential of 42 commonly used formulations of row crop pesticides to adult honey bees (Hymenoptera: Apidae)." *Journal of Economic Entomology* 108 no. 6 (2015): 2640–2647.

10

*Nutritional Aspects of the Dwarf Honeybee (*Apis florea *F.) for Human Consumption*

Sampat Ghosh, Chuleui Jung, Bajaree Chuttong and Michael Burgett

CONTENTS

Introduction

The use of insects as food (entomophagy) has been recognized as playing an increasing role in human nutritional needs and food security (FAO, 2012; Van Huis, 2013). The honeybee species have traditionally bccn incorporated into the diet in numerous cultures, most especially in East and South East Asia. Chen et al. (1998) specifically address the use of honeybees as human food in Thailand.

Apis florea, one of two described dwarf honeybee species, is confined to tropical and subtropical regions of Asia including West Asia (Oman, Iran, Pakistan, Afghanistan), the Indian subcontinent and Sri Lanka, the Far East (Indonesia, Malaysia, the Philippines) and South East Asia (Free, 1981; Wongsiri et al., 1997; Hepburn et al., 2005). Several scientific reports indicate that *A. florea* is an efficient pollinator especially in orchard and field crop monocultures (Free, 1981; Ruttner, 1988; Crane, 1991). Although *A. florea* is a proficient pollinator, it does not receive much attention because of its feral existence and hence cannot be used as a supplemental pollinator in a management sense. This behavioural limitation is largely based on the bees' practice of open-air nesting, which precludes them from being maintained in manmade hives.

A colony of *A. florea* consists of a single comb whose upper part expands to form the crest that surrounds the horizontal branch substrate from which the comb is suspended (Figure 10.1). Indigenous peoples throughout the bees' biogeographical range harvest dwarf honeybees from the wild. It is estimated that annually approximately 10,000 *A. florea* colonies are harvested by honey hunters in Thailand (Wongsiri et al., 1997). The exploitation of colonies is not just for obtaining honey but for the consumption

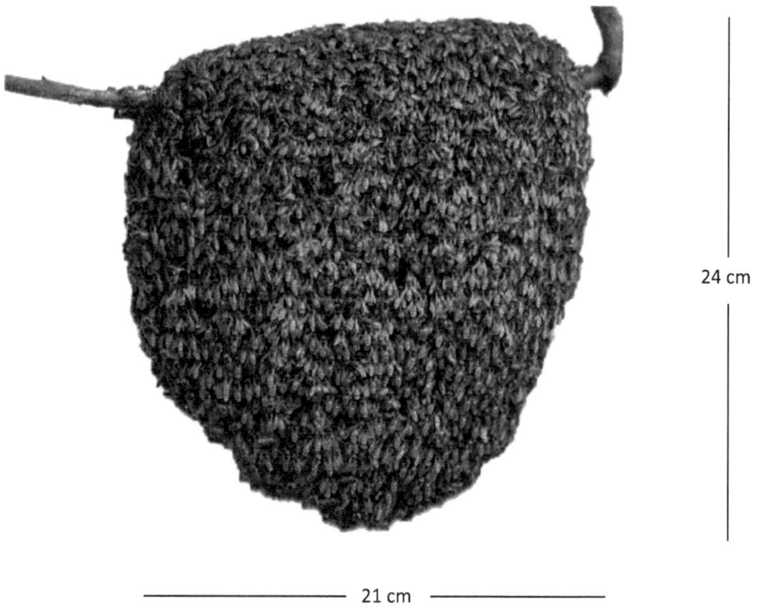

FIGURE 10.1 Mature *Apis florea* colony.

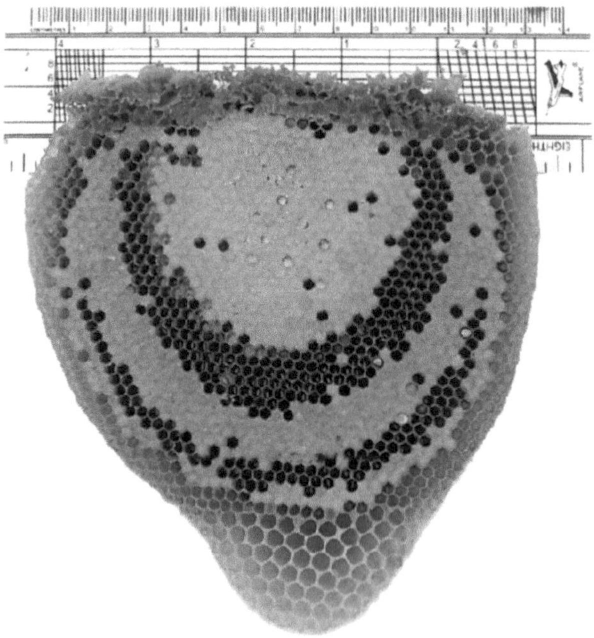

FIGURE 10.2 *Apis florea* brood comb; sealed cells are pupae.

of broods, primarily larvae and pupae. Figure 10.2 shows a typical *A. florea* brood comb. Wongsiri et al. (1997) reports on the consumption of different species of honeybee broods and adult bees including *A. florea*. The most comprehensive analysis of the nutritional value of honeybee broods is that of Ghosh et al. (2016), who elaborated on the nutritional profile of *A. mellifera*, the western honeybee. Nutrition information concerning the additional honeybee species, including *A. florea* broods, is nonexistent.

Materials and Methods

Sample Collection and Preparation

Feral *Apis florea* colonies were collected from the Chiang Mai province in northern Thailand in 2017 and 2018. The brood combs and adult bees were taken to the Chiang Mai University Postharvest Technology Research Center for sample processing. Bee pupae were individually removed from the combs *via* fine tweezers and then stored at –20°C. Wings of the adults were removed prior to freezing. Both adult worker bees and pupae were freeze-dried, ground to powder and prepared as dry matter (DM) for further analysis. All the solvents and chemicals used in the study were of analytical grade.

Amino Acid Analysis

Amino acid composition was determined by Sykam Amino Acid Analyzer S433 (Sykam GmbH, Germany) following the standard method of AOAC (1990) (Table 10.1). The ground samples were hydrolyzed in 6 N HCl for 24 h at 110°C under nitrogen atmosphere and then concentrated with a rota-evaporator. The concentrated samples were reconstituted with sample dilution buffer supplied by the manufacturer (0.12N, pH 2.20). The hydrolyzed samples were analyzed for amino acid composition. The operating condition of the amino acid analyzer was as the following:

Instrument:	Amino Acid Analyzer (Sykam GmbH, Germany)
Column:	LCA K07/Li (PEEK – column 4.6 × 150 mm)
Application:	Physiological
Detector:	Photometer (570 nm, 440 nm)
Detection principle:	Ninhydrin reaction
Inert gas:	N_2

The amino acid score was calculated based on FAO/WHO/UNU (2007).

TABLE 10.1

Amino Acid Composition of Pupae of *Apis florea*

Amino Acid	g/100 g DM	% of Total Amino Acids
Aspartic acid	3.675	9.79
Threonine	1.726	4.60
Serine	1.776	4.73
Glutamic acid	5.029	13.40
Glycine	2.227	5.93
Cysteine	0.577	1.54
Valine	2.119	5.65
Methionine	1.163	3.10
Isoleucine	1.66	4.42
Leucine	3.339	8.90
Tyrosine	1.616	4.31
Phenylalanine	1.715	4.57
Lysine	2.328	6.20
Histidine	1.022	2.72
Arginine	1.954	5.21
Proline	2.715	7.24
Alanine	2.884	7.69

Fatty Acid Analysis

Fatty acid composition was analyzed by gas chromatography flame ionization detection (GC-14B, Shimadzu, Tokyo, Japan), following the standard method of the Korean Food Standard Codex (2010). The samples were derivatized into fatty acid methyl esters (FAMEs) following the method described by Lepage and Roy (1986). Identification and quantification of FAMEs were accomplished by comparing the retention times of peaks with those of pure standards purchased from Sigma and analyzed under the same conditions. The results were expressed as a percentage of individual fatty acids in the lipid (i.e., sum of total fatty acids fraction) (Table 10.2).

Minerals Analyses

Minerals were analyzed following the standard method of the Korean Food Standard Codex (2010). The dried powder samples were digested with nitric and hydrochloric acid (1:3) at 200°C for 30 minutes. Each sample was filtered using Whatman filter paper (0.45 micron) and stored in washed glass vials before analysis. Minerals were analyzed by an inductively coupled plasma optical emission spectrophotometer (ICP-OES 720 series Agilent) (Table 10.3).

TABLE 10.2

Fatty Acid Composition of Pupae and Adult *Apis florea*

	***Apis florea* (Pupa)**		***Apis florea* (Adult)**	
Fatty Acid	**mg/100 g DM**	**%**	**mg/100 g DM**	**%**
Lauric acid (C12:0)	20	0.28	20	0.37
Tridecanoic acid (C13:0)	ND		10	0.19
Myristic acid (C14:0)	130	1.81	80	1.49
Palmitic acid (C16:0)	2540	35.33	1650	30.73
Stearic acid (C18:0)	630	8.76	520	9.68
Arachidic acid (C20:0)	30	0.42	40	0.74
SFA (Subtotal)	**3350**	**46.59**	**2320**	**43.20**
Palmitoleic acid (C16:1)	30	0.42	30	0.56
Oleic acid (C18:1n-9, Cis)	3420	47.57	2670	49.72
cis-11-Eicosenoic acid (C20:1)	310	4.31	290	5.40
MUFA (Subtotal)	**3760**	**52.29**	**2990**	**55.68**
Linoleic acid (C18:2n-6, Cis)	70	0.97	60	1.12
Linolenic acid (C18:3n-3)	10	0.14		ND
PUFA (Subtotal)	**80**	**1.11**	**60**	**1.12**

TABLE 10.3

Mineral Content of Pupa and Adult *Apis florea* (mg/100 g DM)

	Pupa	**Adult**
Calcium	67.1	80
Sodium	72.9	84.9
Potassium	1235	1519.3
Magnesium	117.2	144.2
Phosphorus	1105	1368.9
Iron	7.9	9.4
Zinc	7.7	8.7
Copper	2.9	2.2
Manganese	0.4	0.1

Results and Discussion

Amino Acid Profile

Protein receives the most attention when determining the quality of food. The quality of protein as related to human health and nutrition largely depends on its amino acid composition. A total of 17 amino acids including almost all essential amino acids were determined. Tryptophan was not estimated presumably due to acid hydrolysis. The total amino acids – i.e., protein content – was found to be 37.53 g per 100 g dry matter, which is somewhat less than that found for *A. mellifera ligustica* pupae (Ghosh et al., 2016). Glutamic acid was the most predominant among all estimated proteinergic amino acids followed by aspartic acid. Among the essential amino acids, leucine predominated. The trend agrees with several other published reports on the amino acid composition of edible insects (De Guevara et al., 1995; Finke, 2002; Chakravorty et al., 2014, 2016; Ghosh et al., 2016, 2017). Glutamine helps in the transportation of free ammonia, which is toxic to the human body. Glutamate is also a component of glutathione and a precursor of neurotransmitter γ-amino butyric acid (GABA). Asparagine is important for maintenance, proper functioning and chemical balance in many different tissues including the brain. Lysine, an entirely ketogenic amino acid, was found in high content. It is important particularly in regions where the cereal-based staple foods such as wheat, cassava and maize etc. are lysine limiting. Lysine synthesizes carnitine, which is required for the transportation of fatty acids into the mitochondria for β-oxidation. Threonine is known to be the second limiting amino acid in the maintenance requirement of the human body (Said and Hegsted, 1970; Hegsted, 1973; Fuller et al., 1989) and was found in significant amounts. Alanine was found to be the second most abundant essential amino acid in *A. florea* pupa, which is in agreement with many other insects (Finke, 2002). Alanine plays a crucial role in glucose metabolism and also is important in muscle health and fatigue.

People generally use foods of animal origin as well as legumes as sources of protein. The amount of individual essential amino acid content of *A. florea* pupa was found to be comparable with the conventional sources of protein (Figure 10.3).

Fatty Acid Profile

Considering the accumulation of fatty acids as fat content, *A. florea* pupae contained higher amounts of fat than adult bees. This value is a little higher than that found in the case of *A. mellifera* (Ghosh et al., 2016). Oleic acid dominated, and among the saturated fatty acids (SFA) palmitic acid was found to be the most

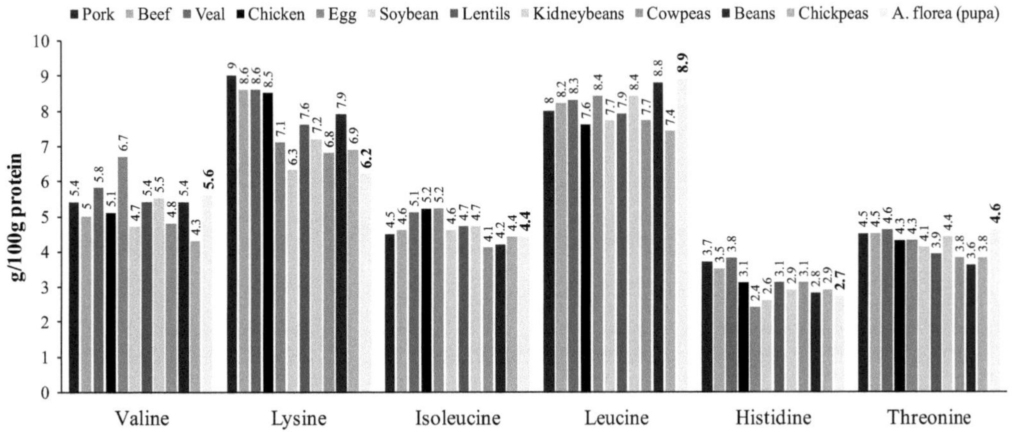

FIGURE 10.3 Comparative account of several essential amino acids with conventional sources of protein and *Apis florea* pupa. (Data other than *Apis florea* were obtained and calculated from www.ndb.nal.usda.gov.)

abundant followed by stearic acid. This trend was similar to that of *A. mellifera* (Ghosh et al., 2016). High levels of SFA are not desirable because of their linkage to atherosclerotic disorder. However, not all the saturated fatty acids elevate the cholesterol level. Lauric, myristic and palmitic acid have been shown to do that and among them, myristic acid elevates the cholesterol the most (Mensink, 1993). All fatty acid levels were found higher in the pupal stage than that found in adults except arachidic acid whereas tridecanoic acid was found in only adult honeybees. The proportion of MUFA was increased in the adult honeybee; in contrast, the proportion of SFA was decreased and PUFA proportion remained the same. Stearic acid, which has been shown to lower LDL cholesterol (Mensink, 2005), was present in moderate amounts. The proportion of MUFA was highest in adult bees compared with pupae. MUFA lowers LDL cholesterol but does not lower high-density lipoprotein (HDL) or cause increases of triglycerides, as a low-fat/carbohydrate diet does (Grundy, 1986; Mensink and Katan, 1987). The results were found to agree with other studies (Finke, 2005; Ghosh et al., 2016). The small amount of linoleic acid recovery might be due to the contamination of the feed.

Minerals Profile

All examined minerals were found to be higher in the adult bees opposed to the pupal stage except for copper and manganese. The calcium content of the dwarf honeybee was higher than that of conventional foods of animal origin except for eggs (Figure 10.4). By comparison, sodium was less in the dwarf bee than that in conventional meat sources and eggs (Figure 10.4), but essential sodium is easily accessible through the consumption of table salt (NaCl: sodium chloride). In contrast, *A. florea* contained higher amounts of potassium 1519.3 and 1235 mg/100 g for adult and pupal form, respectively. Other scientific

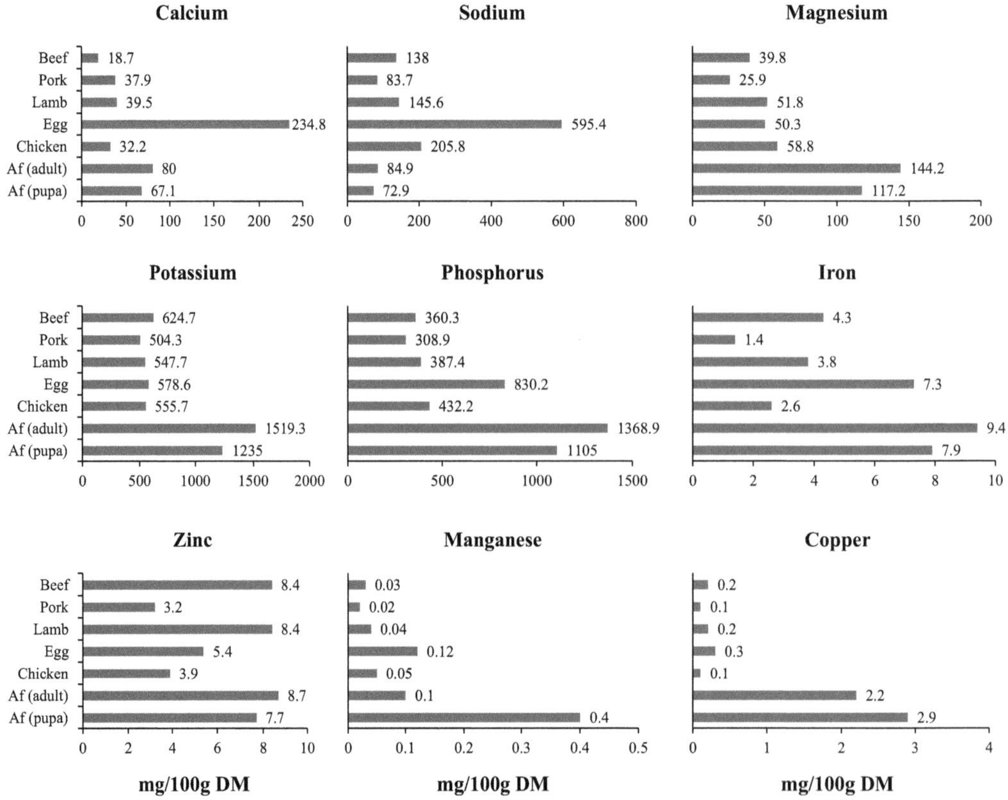

FIGURE 10.4 Comparative mineral content of *Apis florea* pupae and adults and conventional foods of animal origin. (Value other than *A. florea* were obtained and calculated from www.ndb.nal.usda.gov.)

reports indicated that the honeybee *A. mellifera* contained high levels of potassium, and drone pupae contained the highest concentrations of all the stages analyzed, higher than values known from most other edible insects (Bukkens, 1997; Finke, 2002; Chakravorty et al., 2011, 2014; Kim and Jung, 2013). However, similar values of potassium content of different developmental stages of *A. mellifera* were reported by Studier and Sevick (1992) [1275 mg/100 g for adult honeybee (*A. mellifera*) of unknown sex] and Ghosh et al. (2016) [1585.4 for worker *A. mellifera*]. The high K/Na ratio (16.9 and 17.9 for pupae and adults, respectively) makes the species potentially valuable as a low potassium source for humans, which is associated with a variety of physiological disorders and respiratory tract and kidneys, and hypertension (Cohn et al., 2000). Compared with other edible insects (Finke, 2002, 2004), phosphorus concentrations were relatively high irrespective of the developmental stage. The amount of magnesium found for pupa and adult *A. florea* was similar to that found for different developmental stages of *A. mellifera* (Levy and Cromroy, 1972; Studier and Sevick, 1992; Ghosh et al., 2016). Magnesium plays a crucial role as a cofactor for various enzymatic reactions in the body.

Zinc, an important micromineral and component of a large number (>300) of enzymes, is an essential participant in the metabolism of carbohydrates, proteins, fats and nucleic acids. Furthermore, it has a crucial role to play in polynucleotide transcription and thus the process of genetic expression. The zinc content of the dwarf honeybee ranged from 7.7 mg/100 g in the pupa to 8.7 mg/100 g in adults, both values higher than those reported for conventional foods such as pork, chicken and egg (Figure 10.3). Copper, present in small but significant amounts, plays an important role in cellular respiration, peptide amidation, neurotransmitter biosynthesis, pigment formation and connective tissue strength, but it is toxic in excess. Amounts of this metal found in honeybees are somewhat higher than those of conventional meat sources and eggs, but they are still well below levels one needs to be concerned about (Figure 10.4). Iron deficiency is a widespread concern especially in developing and underdeveloped countries. The women of childbearing ages, infants of the weaning stages, young children and sickly adults are the most vulnerable and often receive insufficient amounts of iron in their diets. In comparison to conventional foods of animal origin, the iron content of both the developmental stages – i.e., pupae and adults – were found higher (Figure 10.4). Assuming good bioavailability of the minerals, especially iron, the consumption of *A. florea* could be regarded as useful in mitigating the risk of iron-deficiency anemia.

Conclusion

Based on the findings, the dwarf honeybee could be an important nutritional source. The brood contained almost all the essential amino acids that are often lacking in staple cereal-based foods. Access to animal protein is also sometimes beyond the buying capacity of poor people, particularly the developing and underdeveloped regions of the world. Moreover, less amount of fat and higher content of minerals of nutritional importance make dwarf honeybees suitable as food. However, the development of the rearing technique in order to farm the dwarf honeybee is utterly lacking.

ACKNOWLEDGEMENT

SG and CJ were partly supported by the Basic Science Research Program through the National Research Foundation of Korea (NRF) funded by the Ministry of Education (NRF-2018R1A6A1A03024862). We thank Dr. Rewat Phongphisutthinant, Dr. Supakit Chaipoot, Science and Technology Research Institute, Chiang Mai University and Ms. Lakkhika Panyaraksa for sample preparation, and Insect Industry R & D Center, Andong National University for the technical analysis.

REFERENCES

AOAC. *Official Methods of Analysis*, 15th ed. Association of Official Analytical Chemists, Washington, DC, 1990.

Bukkens, S.G.F. "The nutritional value of edible insects." *Ecology of Food and Nutrition* 36 (1997): 287–319.

Chakravorty, J.; Ghosh, S.; Jung, C.; Meyer-Rochow, V.B. "Nutritional composition of *Chondacris rosea* and *Brachytrupes orientalis*: Two common insects used as food by tribes of Arunachal Pradesh, India." *Journal of Asia Pacific Entomology* 17 (2014): 407–415.

Chakravorty, J.; Ghosh, S.; Megu, K.; Jung, C.; Meyer-Rochow, V.B. "Nutritional and anti-nutritional composition of *Oecophylla smaragdina* (Hymenoptera: Formicidae) and *Odontotermes* sp. (Isoptera: Termitidae): Two preferred edible insects of Arunachal Pradesh, India." *Journal of Asia-Pacific Entomology* 19 (2016): 711–720.

Chakravorty, J.; Ghosh, S.; Meyer-Rochow, V.B. "Chemical composition of *Aspongopus nepalensis* Westwood 1837 (Hemiptera; Pentatomidae), a common insect food insect of tribal people in Arunachal Pradesh (India)." *International Journal of Vitamin and Nutrition Research* 81 (2011): 48–56.

Chen, P.P.; Wongsiri, S.; Jamyanya, T.; Rinderer, T.E.; Vongsamanode, S.; Matsuka, M.; Sylvester, H.A.; Oldroyd, B.P. "Honey bees and other edible insects used as human food in Thailand." *American Entomologist* 44 (1998): 24–29.

Cohn, J.N.; Kowey, P.R.; Whelton, P.K.; Prisant, M. "New guidelines for potassium replacement in clinical practice, A complementary review by the National Council on Potassium in Clinical Practice. *Archives of Internal Medicine* 160 (2000): 2429e2436.

Crane, E. "*Apis* species of tropical Asia as pollinators, and some rearing methods for them. In: Heemert, C van; Ruijter, A de (Eds.) *The Sixth International Symposium on Pollination*. Tilburg, Netherlands, 27–31 August, 1990." *Acta Horticulturea* 288 (1991): 29–48.

De Guevara, O.L.; Padilla, P.; Gracia, L.; Pino, J.M.; Ramos-Elorduy, J. "Amino acid determination in some Mexican insects". *Amino Acids* 9 (1995): 161–173.

FAO. Assessing the potential of insects as food and feed in assuring food security. Presented at Technical Consulting Meeting. 23–25 January, FAO, Rome, Italy, 2012.

FAO/WHO/UNU. Protein and amino acid requirements in human nutrition; Report of a joint WHO/FAO/UNU Expert Consultation, WHO Technical Report Series no. 935. WHO, 2007.

Finke, M.D. "Complete nutrient composition of selected invertebrates commonly fed to insectivores." *Zoo Biology* 21 (2002): 269–285.

Finke, M.D. "Nutrient composition of bee brood and its potential as human food." *Ecology of Food and Nutrition* 44 (2005): 257–270.

Finke, M.D. "Nutrients content of insects: Encyclopedia of Entomology." Available at: http://www.organicvaluerecovery.com/studies/studies_nutrient_content_of_insects.htm, 2004.

Free, John. B. "Biology and behaviour of the Honeybee *Apis florea* and possibilities for Beekeeping." *Bee World* 62 no. 2 (1981): 46–59.

Fuller, M.F.; McWilliam, R.; Wang, T.C.; Giles, L.R. "The optimum dietary amino acid pattern for growing pigs; requirements for maintenance and tissue protein accretion." *British Journal of Nutrition* 62 (1989): 255–267.

Ghosh, S.; Jung, C.; Meyer-Rochow, V.B. "Nutritional value and chemical composition of larvae, pupae, and adults of worker honeybee, *Apis melliferaligustica* as a sustainable food source." *Journal of Asia-Pacific Entomology* 19 (2016): 487–495.

Ghosh, S.; Lee, S.-M.; Jung, C.; Meyer-Rochow, V.B. "Nutritional composition of five commercial edible insects in South Korea." *Journal of Asia-Pacific Entomology* 20 (2017): 686–694.

Grundy, S.M. "Comparison of monounsaturated fatty acid and carbohydrates for lowering plasma cholesterol." *New England Journal of Medicine* 314 (1986): 745–748.

Hegsted, D.M. "The amino acid requirements of rat and human being". In: Porter, J.W.G.; Rolls, B.A. (Eds.) *Proteins in Human Nutrition*. Academic Press, London, UK, 1973.

Hepburn, H.R.; Radloff, S.E.; Otis, G.W.; Fuchs, S.; Verma, L.R.; Ken, T.; Chaiyawong, T.; Tahmasebi, G.; Ebadi, R.; Wongsiri, S. "*Apis florea*: morphometrics, classification and biogeography." *Apidologie* 36 no. 3 (2005): 359–376.

Kim, H.; Jung, C. "Nutritional characteristics of edible insects as potential food materials." *Korean Journal of Apiculture* 28 (2013): 1–8.

Korean Food Standard Codex. Ministry of Food and Drug safety. Republic of Korea, 2010.

Lepage, G.; Roy, C.C. "Direct transesterification of classes of lipids in a one-step reaction." *Journal of Lipid Research* 27 (1986): 114–120.

Levy, R.; Cromroy, H.L. "Concentration of some major and trace elements in forty-one species of adult and immature insects determined by atomic absorption spectroscopy." *Annals of Entomology Society of America* 66 (1972): 523–526.

Mensink, R.P. "Effects of stearic acid on plasma lipid and lipoproteins in human." *Lipids* 40 (2005): 1201–1205.

Mensink, R.P. "Effects of the individual saturated fatty acids on serum lipids and lipoprotein concentrations." *American Journal of Clinical Nutrition* 57 (1993): 711S–714S.

Mensink, R.P.; Katan, M.B. "Effect of monounsaturated fatty acids versus complex carbohydrate on high density lipoproteins in healthy men and women." *Lancet* 1 (1987): 122–125.

Ruttner, F. *Biogeography and Taxonomy of Honeybees.* Springer-Verlag, Berlin, Germany, 1988.

Said, A.K.; Hegsted, D.M. "Response of adult rats to low dietary levels of essential amino acids. *Journal of Nutrition* 100 (1970): 1363–1375.

Studier, E.H.; Sevick, S.H. "Live mass, water content, nitrogen and minerals levels of some insects from south-central lower Michigan." *Comparative Biochemistry and Physiology* 103A (1992): 579–595.

Van Huis, A. "Potential of insects as food and feed in assuring food security." *Annual Review of Entomology* 58 (2013): 563–583.

Wongsiri, S.; Lekprayoon, C.; Thapa, R.; Thirakupt, K.; Rinderer, T.E.; Sylvester, H.A.; Oldroyd, B.P.; Booncham, U. "Comparative biology of *Apis andreniformis* and *Apis florea* in Thailand." *Bee World* 78 no. 1 (1997): 23–35.

11

The Antibacterial Activity of Honey with Special Reference to Apis florea

Dharam P. Abrol

CONTENTS

Introduction

Antibacterial activity has been known since ancient times. Egyptian papyri and Sumerian clay tablets from 900 to 1250 BC used more than 30% of treatments based on honey as medicine (Stomfay-Stitz 1960). The antibacterial activities of honey were first reported in 1892. The ancient Greeks used honey before major athletic events to treat various types of fatigue and as a drink consisting of honey and water (Wilson and Crane 1975). Hippocrates (460–357 BC) found honey to cure sores and ulcers of the lips and cure wound infections. Honey as medicine has also been used throughout Europe, Arabia and China. It has been employed to cure gastric and intestinal and kidney problems, eye disorders, skin diseases and smallpox. Honey has been used in surgical dressings. It has also been utilized for coughs, pulmonary issues and gastric disorders. German women used a mixture of honey and crushed bees for beautification and strengthening effect. Chinese, Hindu, Arabic and African cultures still consider honey a valuable remedy for internal and external problems (Beck and Smedley 1997). The honey as antibacterial medicine is gaining recognition in western countries (Molan 2006). Honey has a number of laxative and beneficial effects on blood glucose levels, as well as anti-inflammatory and immune stimulating properties. The increasing resistance to drugs and side effects thereof throughout the world (Bogdanov 2009b; IHC 2002) have once again resulted in renewed interest for honey's use. The information compounds involved in the antibacterial activity of honey are still far from complete. The honey is mainly composed of sugars, water and other compounds that are less than 1% composed of proteins, minerals, enzymes, vitamins, 5-hydroxymethylfurfural (HMF), volatile compounds, flavonoids and phenolic acids (Bogdanov 2009a).

The antibacterial activity of honey depends upon the composition, botanical and geographical origin (IHC 2002). Other factors such as season of harvesting, environmental factors, processing and storage conditions also influence the antibacterial activity (Souza et al. 2006). The number of antioxidant compounds present in honey are ascorbic acid, flavonoids, glucose oxidase, amino acids, proteins phenolic acids, organic acids, carotenoid derivatives and maillard reaction products (Bogdanov 2009; Souza et al. 2006). The biological properties that make the honey an ideal source of medicine are antibacterial, bacteriostatic and antioxidant activity (Souza et al. 2006; Singh et al. 2012).

The antioxidant properties of honey destroy biologically destructive chemicals linked with many diseases. The phytochemicals such as organic acids, vitamins and enzymes serve as sources of dietary antioxidants. Antioxidant activity depends upon the floral source and type of honey. Gheldof et al. (2002) reported that darker honeys are generally more antioxoidant than the lighter honeys. Certain honeys exhibit broad-spectrum antimicrobial activity and are useful to treat antibiotic resistant bacteria. The use of honey against bacteria is increasing in modern medicine as a consequence of its antimicrobial activity. The level of hydrogen peroxide activity varies among honeys from within the same floral sources. This is due to geographical origin of the floral source and prevailing environmental conditions (Irish et al. 2011).

The characterization of honey is important to understand its therapeutic, antibacterial and antioxidant properties (Adebiyi et al. 2004). The physicochemical characteristics such as moisture content, electrical conductivity, reducing sugars, sucrose content, minerals, free acidity and HMF can be used to assess quality of honey (Baltrusaityte et al. 2007; Khan et al. 2006).

Apis florea

Apis florea has been reported from various countries for its role in crop pollination and production of quality honey, beeswax and honey for certain therapeutic applications (Cooper et al. 2002; Jaganathan et al. 2014; Popek 2002). It is most commonly distributed in arid and desert regions of Central India. The Kutch area of Gujarat is known for large quantities of *Apis florea* honey production (Soman and Chawda 1996). *Apis florea* is different from honeys of other bee species. It is superior in quality due to high amounts of a glucose oxidase enzyme that shows antibiotic activity (Johnston et al. 2003).

The use of honey as a therapeutic agent is being utilized as a treatment for injury in several countries such as Australia (Dustmann 1978), New Zealand (Marghitas et al. 2010) and the UK (Selçuk and Nevin 2003). The antimicrobial potential of honey has been reported against a number of gram-positive and gram-negative bacteria (Selçuk and Nevin 2003) such as *S. aureus, P. aeruginosa, E. coli, P. mirabilis, S. pyogenus, S. flexneri* and *S. typhi* (Kingsley 2001; Molan 1992; Nzeako et al. 2002). Honey has been reported as a medicine for dyspepsia and peptic ulcer (Kumar et al. 2010). It has been found to inhibit the development of wound-infecting bacterial strains, including *E. coil, S. aureus* and *Salmonella* (Cooper et al. 2002; Willix et al. 1992). The antibacterial potential of honey is due to osmolarity, acidity, presence of hydrogen peroxide and several other unknown substances (Akujobi et al. 2010).

Antioxidants

Antioxidants delay or inhibit the oxidation of lipids or other molecules by inhibiting the initiation or propagation of oxidative chain reactions (Velioglu et al. 1998) and this is due to their redox properties. They play an important role in absorbing and neutralizing free radicals (Javanmardi et al. 2003).

They inhibit the process of oxidation even at low concentrations and have diverse physiological roles in the body. Antioxidant materials act as radical scavengers and help in conversion of radicals to less reactive species (Mandal et al. 2009). They defend against both enzymatic and non-enzymatic reactions, protecting the body from oxidative damage. The addition of non-enzymatic antioxidants to foods prevents lipid oxidation.

Antioxidant Capacity of Honey

A number of methods have been used to study antioxidant activity in honey, which include determination of active oxygen species (viz. the superoxide anion, peroxyl and hydroxyl radicals) and their radical scavenging ability (Gheldof and Engeseth 2002), the 1,1-diphenyl-2-picrylhydrasyl (DPPH) antioxidant content, enzymatic or non-enzymatic measurements of lipid peroxidation inhibition (Chen et al. 2000), the ferric reducing/antioxidant power assay (Aljadi and Kamaruddin 2004) and the trolox equivalent antioxidant capacity (TEAC) assay (Huang et al. 2005). In general, darker honeys have been shown to be higher in antioxidant content than lighter honeys.

Ferric Reducing Antioxidant Power (FRAP)

The principle of this method is based on the reduction of a ferric 2,4,6- tripyridyl-s-triazine complex (Fe3+-TPTZ) to its ferrous coloured form (Fe2+-TPTZ) in the presence of antioxidants (Jose et al. 2009). The FRAP assay 'Ferric reducing ability of plasma' was described by Benzie and Strain (1996), and it has been used by several authors for the study of the antioxidant capacity of natural sources such as honey (Aljadi and Kamaruddin 2004; Prior et al. 2005). As the FRAP assay measures the reducing capacity based upon reduction of ferric ion, antioxidants that act by radical quenching (H transfer), particularly carotenoids (present in the honeybee), will not be determined. Another point to take into consideration is the concomitant production of Fe (II), which is a well-known prooxidant and may result in the generation of additional radicals in the reaction medium, as OH• from hydrogen peroxide. Finally, compounds that absorb at the wavelength of the determination may interfere, causing overestimation of the FRAP value (Jose et al. 2009). Any compound even with redox potential lower than that of the redox pair Fe(III)/Fe(II) can theoretically reduce Fe(III) to Fe(II), contributing to the FRAP value and inducing falsely high results. On the other hand, not all antioxidants reduce Fe (III) at a rate fast enough to allow its measurement within the observation time (typically 4 min). Indeed, many polyphenols react more slowly and require longer reaction times ($\geq$30 min) for total quantification and, depending on the analysis time, the order of their reactivity is changed. However, some polyphenolic compounds (caffeic acid, ferulic acid, quercetin and tannic acid) may have slower reactions, requiring more time (30 min) until completing the complex reduction. When used to determine the antioxidant potential of polyphenols in water and methanol (solvent used for their extraction form in honeybee), the change in absorbance continued after 4 min. Therefore, the FRAP values for these compounds cannot be accurately determined in 4 min, and for this reason, 10 min are strongly recommend as a useful reaction time (Jose et al. 2009).

Diphenyl-1-Picrylhydrazyl Radical Scavenging Activity (DPPH RSA) Assay

In the presence of an antioxidant, the purple colour of DPPH •fades and the change of absorbance can be followed spectrophotometrically at 515 nm (Schlesier et al. 2002). The DPPH analysis is a quick and simple test; it guarantees reliable results and needs only a UV-visible spectrophotometer to perform, which probably explains its widespread use in antioxidant screening. One more important disadvantage is many antioxidants that react quickly with the radical peroxide are almost or completely inert to DPPH. Despite the above-mentioned limitations, DPPH is stable, commercially available and does not have to be generated before assay like ABTS. For these reasons, it could be considered an easy and useful spectrophotometric method with regard to screening/measuring the antioxidant capacity in honey (Jose et al. 2009).

2,2′-Azino-Bis (3-ethylbenzothiazoline-6-sulfonic acid) Diamonium Salt (ABTS)

The fact that ABTS •+ is not found in biological systems and is not similar to radicals found in those systems is also a problem and thus represents a "nonphysiological" radical source. In spite of this, the ABTS assay is considered an easy and accurate method for use in the antioxidant capacity studies in honey because it allows determination of the radical scavenging ability present in the honey.

In Vitro Antioxidant Capacity of Honey

Rosa et al. (2011) examined 67 Spanish honeys from different floral origins, nectars and honeydews for potential antimicrobial and antioxidant activities. Antioxidant capacities were determined in terms of their antiradical capacity using the stable free radical DPPH. The bacterial species *Micrococcus luteus* and *Staphylococcus aureus* were used as the resistant microorganism in the antibacterial assays. The results showed that honeydew honeys presented higher antioxidant capacities than nectar honeys. The kind of honey, location and date of the honey harvest were shown to have a significant effect on the contents of phenolic antioxidants. In the Czech honey, ferulic acid (0.11 mg/100 g) and chrysin (0.06 mg/100 g) and other minority phenolics and flavonoids were identified and quantified as honey phenolic antioxidants contained (Alábová et al. 2003; Lachman et al. 2010).

In Vivo Antioxidant Capacity of Honey

More recently, a small number of studies have investigated the *in vivo* antioxidant capacity of honey (Busserroles et al. 2002; Gheldof et al. 2003; Schramm et al. 2003). Gheldof et al. (2003) found that the serum antioxidant capacity, measured using oxygen radical absorbance capacity (ORAC), was increased by 7% in individuals who had consumed a 500 mL solution of buckwheat honey (160 g/L) but not in those who consumed 500 ml of black tea or black tea plus 160 g/L of syrup of mixed sugars (45% fructose, 35% glucose, 20% water). Similarly, in another study, plasma antioxidant capacity increased by 12%–25% six hours after the ingestion of a single dose of Buckwheat honey (1.5 g/kg body weight), and the plasma total-phenolic content was increased by 4%–8% (Schramm et al. 2003). No significant effects on either parameter were reported in subjects who consumed an equivalent amount of corn syrup (Schramm et al. 2003). The plasma-reducing capacity was also shown to be improved by approximately 10% following ingestion of honey (Schramm et al. 2003). In a two-week rat feeding study in which animals were fed diets containing 65 g of carbohydrate (starch, honey or a mixture of fructose and glucose as in honey) per 100 g of diet, honey-fed rats had higher plasma α-tocopherol levels, lower plasma nitrogen oxide levels and a lower susceptibility of lipid peroxidation in the heart (Busserroles et al. 2002). Different formulations of honey have been found to significantly inhibit the growth of pathogenic microorganisms such as *Staphylococcus aureus*, *Escherichia coli*, *Candida albicans* and *Aspergillus niger*, which indicates that the honey has therapeutic properties (Gulfraz et al. 2010).

Effects of Honey on the Immune System

Human immune response and associated activities have been reported to improve with consumption of honey. Under experimental conditions, honey has been found to improve the multiplication of human peripheral blood cells (B- and T-lymphocytes) and the activation of neutrophils (Abuharfeil et al. 1999). In monocytic cell culture, honey has been found to stimulate the release of inflammatory cytokines, such as tumour necrosis factor-alpha, interleukin-1β and interleukin-6 (Tonks et al. 2003), which help in triggering a number of functions of the immune response to infection (Molan 2002). In the case of mice, a honey-supplemented diet stimulates the production of antibodies during the primary and secondary immune responses against thymus-dependent and thymus-independent antigens (Al-Waili and Haq 2004).

Antioxidant Properties of Honey

Honey is a natural product and carries no side effects upon consumption. The compounds found in honey such as vitamin C, phenol compounds, catalase, peroxides and glucose oxidase enzymes have antioxidant properties (Meda et al. 2005). Several polyphenols found in honey (Caffeic acid, Caffeic acid phenyl ester, Chrysin, Galangin, Quercetin, Acacetin, Kaempferol, Pinocembrin, Pinobanksin and Apigenin) are promising in the treatment of cancer (Bertoncelj et al. 2007).

Flavonoids and carotenoids present in honey act as an antidepressant during physical, intellectual and emotional stress (EU Council: European Union Council Directive 2001). Honey also helps to modulate the immune system, antimicrobial properties, lipid values and regulates glycaemic (Khalil et al. 2012). It has also been reported to protect against ochratoxin A-induced oxidative stress in liver and kidney in rats (Moniruzzaman et al. 2013). Antioxidants have preventive effects against various diseases such as cancer, cardiovascular diseases, inflammatory disorders, neurological degeneration, wound healing, infectious diseases and aging (Fahim et al. 2014), and they contain anti-inflammatory and antioxidant properties (Perna et al. 2013).

In addition to its health-related advantages, honey is also useful as a preservative due to its high osmolarity that can prevent microorganism growth. Its favourable smell and taste make it suitable to be used as a food's taste enhancer.

In preventing microbial growth, honey was reported to have a varying degree of antimicrobial activity. The level of its antimicrobial potency is directly related to its physicochemical characteristics and the microbes of interest. The physicochemical properties of honey are highly dependent on its origin (geographical area, climate, season and hive), foraging bees and floral sources. It, however, can be influenced by human activities during processing such as harvesting methods, storage, packaging, transportation and physical contact. Microbes such as Enterobacteriaceae, healthcare-acquired pathogens and food spoilage organisms were reported to be susceptible to some types of honey in different degrees and effects. Generally, bacteria such as *Staphylococcus aureus, Bacillus subtilis, Escherichia coli, Pseudomonas aeruginosa, Klebsiella pneumonia, Salmonella typhimurium, Enterococcus faecalis, Listeria monocytogenes*, and fungi such as *Penicillium expansum, Aspergillus niger* and *Alcaligenes faecalis* were killed by honey at certain concentrations (Kwakman et al. 2010; Mundo et al. 2004). Nevertheless, microbes such as spore-forming organisms and yeasts were frequently reported to be major honey contaminants due to their ability to survive high osmolarity and low pH of honey. *Clostridium botulinum* is one of the most significant contaminants that raises concern among medical practitioners and food providers because it could cause infant botulism due to its spores in honey given to children younger than 2 years of age (Midura 1996).

In the present millennium, antibiotic-resistant bacteria have been reported as a progressive critical challenge that needs an immediate robust solution. Rapid emergence of these super bugs together with a slow, new antibiotic discovery has initiated insecurity in the medical and health industry (Arias and Murray 2009). Limited resources, high cost and time are the reasons why this industry seems to fail to produce excellent antibiotics in time. Antibiotic multi-resistant bugs just add fuel to the flame. Startled by this phenomenon, scientists have begun to look for potential alternatives to combat this seemingly endless issue. Honey has been listed as one of the potential agents together with other herbs and natural products. This is due to the numerous studies reporting the antimicrobial activity of various types of honey across the globe, which may contribute to development of antimicrobial agents to cope with the drug resistance problem. Ironically, most of them claim to possess effective antibacterial potency against well-known clinically isolated multidrug-resistant bacteria strains.

Apart from antibacterial activity, honey is also popular for its antioxidant capacity. This is said to be reflected in its chemical components, which contain high amounts of phenolic compounds such as phenolic acids and flavonoids. These organic components are derived from the floral source where the bees collect nectar, which primarily comes from minerals of that particular geographical area. To date, many researchers are interested in the phenolic compounds due to their potential as a dietary supplement to overcome oxidative stress and many clinical ailments such as cardiovascular diseases (Vermerris and Nicholson 2009), neurodegenerative diseases (Uttara et al. 2009) and cancers (Sosa et al. 2013). Phenolic acids are ubiquitous in honey. These heat and light-stable compounds are suitable for honey floral authenticity. Recently, the honey phenolic component has been used as a reference to the standardization purpose by several researchers especially in antimicrobial quality of honey (Allen et al. 1991; Irish et al. 2011). Therefore, more extensive study of this particular subject should be conducted to enhance understanding of the phenolic acids' contribution to honey bioactivities.

In Malaysia, honey production is not a major contribution to its economic, health-related as well as food industries. However, due to its high availability, bioactivity potential, low cost and numerous selections, honey production in Malaysia has a bright potential to be commercialized to meet the market demand of the world, or at least to meet the local demand. Well known to the locals, there are various types of honey waiting to be investigated. Lack of scientific research that is confirmed by limited scientific literature (of our knowledge) makes these types invisible in the market. They include *tualang*, *gelam*, *kelulut*, coconut, pineapple, acacia, *hutan* (forest), *pucuk daun*, rambutan, durian, *getah* (rubber) and a few others. Only a few reports were found investigating different aspects of several batches of *tualang* and some *gelam* honey (Aljadi and Kamaruddin 2004; Aljadi 2003b; Ghashm et al. 2010; Moniruzzaman et al. 2013; Nasir et al. 2010; Nurul Syazana et al. 2011; Tan et al. 2009; Tumin et al. 2005; Zaid et al. 2010). Research on Malaysian dry summers and wet winters reveals that geographical difference may affect honey constituents and contribute to honey variations. As the world's second largest honey producer (as in 2012), Turkey should move together its honey production with its research progression (Eliot Masters and Noble 2014). However, to date, very limited data have been published by researchers investigating the analytical and bioactivity in respect to Turkish honey. The difference between Malaysian honey (tropical honey) and Turkish honey (Mediterranean honey) should provide a more comprehensive understanding about their bioactivities and physicochemical components, therefore justifying the present study.

Lack of scientific information limits the utilization of Malaysian and Turkish honey. To date, numerous studies have been done investigating the biological profile of honey around the world. This study should provide the same basic information in reference to 10 tested honeys from Malaysia and Turkey. The level of bioactivities analyzed such as antioxidant and antibacterial activities should assist in the application of these honeys. In addition to their bioactivity profile, the contaminant screening of unpasteurized honey will provide useful information regarding honey safety especially for those harvested from different origins such as Malaysia and Turkey.

Biological Activity of Honey

Most of the natural occurring products have specific biological activity. Similar to honey, it possesses numerous beneficial bioactivities that may enhance health. It includes antioxidant, antimicrobial (antibacterial, antifungal and antiviral), anti-inflammatory, wound healing promoter, energy booster and antiaging activities. Until now, the mechanisms of most of these activities remain unknown. Despite a lack in scientific evidence regarding to mechanism of actions, studies on honey bioactivities are gaining more attention from researchers around the globe. This may be due to several economic and environmental factors that persistently rose such as rapidly emerging multi-resistance bacteria strains, high cost and time-consuming antibiotics research as well as ethical issues regarding certain pharmaceutical research. Honey seems to be a perfect candidate to overcome these restraints since it has been reported to have the opposite effect against all those obstacles.

Antimicrobial Activity

Microbes such as bacteria, fungi and virus are among the major players in biosystems. They can be beneficial to humankind as well as a potential cause of harm. Overcolonization of microbes may lead to infections and worsen the pathological conditions of a patient, which may delay wound healing, cause bacteremia and sometimes cause death. Recent studies found that honey can be an excellent antimicrobial agent to overcome this infection and promote wound healing (Aljady et al. 2000; Kwakman et al. 2010; Subrahmanyam et al. 2001).

Antibacterial Potency

Antibacterial potency is a major biological activity of honey that is indisputably becoming the limelight in today's research field. Several studies have successfully proven the prominent antibacterial activity of honey against different opportunistic pathogens (Allen et al. 1991; Irish et al. 2011). Since the discovery of the first antibiotic, penicillin in 1928 by Alexander Fleming, the industry was blooming with commercialization of new antibiotics of different types and classes. Following this successful pharmaceutical era is a bacteria evolution and adaptation to survive the eradication attempt. The late 1930s was known as the starting point for the first specific bacterial resistance discovered against sulfonamide (Davies and Davies 2010). Two modes of bacterial adaptation are by accumulation of various genes on resistance plasmid which each one encoded to a single drug resistance mechanism or increase gene expression for the same purpose (Nikaido 2009). This is done through five possible mechanisms: (1) the production of antibiotics metabolizing enzymes, (ii) efflux pumps activity to eliminate antibiotics, (iii) modification of cellular target to prevent antibiotics binding, (iv) alternative pathways to bypass the antibiotics action and (v) elimination or downregulation of transmembrane porins of antibiotics transporter (Liu and Pop 2009).

This phenomenon leads to a serious clinical manifestation called superbugs, which literally means a high level of multiple drug-resistant bacteria species characterized by enhanced morbidity and mortility. There are strains that are becoming resistant to almost all antibiotics including disinfectants such as *Pseudomonas aeruginosa* and methicillin-resistant *Staphylococcus aureus* (MRSA). *P. aeruginosa* and *Acinetobacter baumanii* are referred to as pan-resistant gram-negative strains that are said to have no antibiotics that can be used effectively against them so far (Nikaido 2009). It was noted that super-resistant bacteria also have the capability to enhance virulence and transmissibility, which is considered their progressive virulence factor (Davies and Davies 2010).

Until now, multidrug resistance remains a challenging global public health problem that requires a continuous and robust solution. Due to this protracted impasse, especially the emergence of multidrug-resistant bacteria among the hospital-acquired strains, physicians were urged to seek alternative remedy since new antibiotics finding is taking too much time and cost as well as too laborious. This is where the utilization of herbs, ginseng, propolis, honey, aloe vera and many more natural products came into the picture. However, honey seems to be a perfect candidate based on several factors such as easy availability, affordable price (cheap in some places), no adverse effect reported so far and various health benefits. Studies on the antibacterial properties of honey revealed numerous factors that contribute to this biological action. They include H_2O_2 production, osmolarity (due to high sugar content), low pH, high acidity, low water activity, presence of phenolic compounds, enzymes, proline and antibacterial peptides (Aljadi 2003b; Bogdanov 1997; Bogdanov 2009b; Kwakman et al. 2010). All these factors depended on honey geographical origin, botanical source of nectar, bee species, season, foraging activity of bees as well as environmental influences. In addition, secondary factors may also affect the quality of honey including processing procedure, transport, harvesting method, packaging and time of processing. There are 26 osmolarity carbohydrates, which include glucose, fructose, sucrose, maltose, trehalose, etc. Hygroscopic water content is between 17% and 22%. Antibacterial factor Enzyme Presence of bee-origin glucose oxidase generate H_2O_2 Phytochemical phenolic compounds include phenolic acids (caffeic, ferulic, cinnamic, benzoic, syringic, etc.) and flavonoids (quercetin, pinocembrin, kaempferol, myricetin, catechin, rutin, etc.) Antibacterial peptide bee defensin-1 protein/proline is a lowering agent for melting temperature (*T*m) of DNA pH 3.5–4.5. High acidity H_2O_2 is generated from hydrolysis of glucose by glucose oxidase.

Antifungal Activity

Besides bacteria, honey has also been reported to be effective against fungi and yeast. Clinically significant fungi such as *Tinea capitis* and *Tinea versicolor*, which cause superficial skin infections especially in poverty-stricken and poor sanitation areas, were reported to be sensitive to honey (Ngatu et al. 2011). *Aspergillus Niger* is an opportunistic pathogen, which showed negative effects with the presence of honey; worsen by the addition of starch which increase the diastase number of that honey (Boukraâ et al.

2008; Mundo et al. 2004). Another *Aspergillus* species named *Aspergillus nidulans* also showed susceptibility against Saudi Arabian honey (Noori et al. 2013). Phenolic compounds, particularly flavonoids, were reported to have the ability to inhibit the dimorphic conversion of *Candida albicans* (Candiracci et al. 2012; Noori et al. 2013). Anticandidal activity was also demonstrated in Iranian honey against different *Candida* species including *C. parapsilosis*, *C. tropicalis*, *C. kefyr*, *C. globrata* and *C. dubliniensis* (Khosravi et al. 2008). Cassia (*Cassia javanica*), citrus (*Citrus reticulate*) and ziziphus (*Ziziphus spina-Christi*) honey were reported to have various degrees of antifungal activity against a wide range of dermatophytes (*trichophyton* spp. and *Microsporum* spp.) with ziziphus being the most effective ones (El-Gendy 2010). *Rhotorula* spp. is a pigmented yeast that was also reported to be susceptible to honey (Moussa et al. 2012).

The antioxidative property of honey has been studied extensively since the finding of phytochemicals content in honey and its role in disease pathogenicity. Associated with its phytochemical content, honey has been reported to have anticancer, anticarcinogenic, antimitogenic and anti-inflammatory activities (Khalil et al. 2010). Honey contains caffeic acid and other phenolic acids, which are known to be able to suppress colon tumours in rat. It also contains quercetin, which has antiproliferative effects against glioma and breast cancer cells. Acacetin is one type of flavonoid that can induce apoptosis and block the cell cycle progression at the G1 phase. The presence of proline has also contributed to the antiproliferative effect of honey. A proline-rich compound was reported to have specific inhibitory effects on proliferation of several cancer cell lines: human THP-1, liver cancer HepG2 and breast cancer MCF-7 cells (El-Gendy 2010).

Many researchers have reported the scavenging activity of honey against various pro-oxidants (Aljadi and Kamaruddin 2004; Baltrusaityte et al. 2007; Beretta et al. 2005; Bertoncelj et al. 2007; Blasa et al. 2006; Dobre 2010; Estevinho et al. 2008; Ferreira et al. 2009; Frankel et al. 1998; Islam et al. 2012; Küçük et al. 2007; Moniruzzaman et al. 2013). It happens either by electron transfer or hydrogen atom transfer to stabilize the reactive molecules, eventually breaking the free radical chain reaction to produce inert low-energy products. It is postulated that antioxidant capacity of honey depends on its botanical origin, season, and environmental and processing (secondary) factors (Baltrusaityte et al. 2007). Therefore, different types of honey contain different levels of antioxidant capacity. It is directly reflected by the amount of phytochemical as well as other plant-origin antioxidant component in a particular honey.

Microorganisms in Honey

Antibacterial activity of honey is attributed to several internal and external factors. Internal factors consist of physicochemical properties of honey and its freshness while external factors may include the medium, bacteria species and growth conditions. Despite strong inhibitory effects of physicochemical characteristics, there were microorganisms reported to be able to survive in honey. The concern of honey microorganism contamination was raised after a report on infant botulism (Arnon 1998). It is obvious that some microorganisms are able to survive the extreme osmotic pressure and low pH level of honey. The possible explanation was the microorganism has an adaptation mechanism to overcome the unfavourable conditions, in most cases, spore production ability.

The first infant botulism was reported in California, USA, which was firstly misdiagnosed as encephalitis in 1931 (Nevas 2006; Midura et al. 1979). The possible food items associated with infant botulism are honey and milk. The main reason for this clinical manifestation is the production of the toxin by *Clostridium botulinum* whenever in unfavourable conditions, especially in poorly developed immune systems of infants. This is the reason why detection of honey contaminants is becoming more important. Since then, a few research studies have been conducted to elucidate the possible honey contaminants. Until now, only simple cultivable microorganisms have been screened and isolated from unpasteurized honey. Sinacori et al. (2014) has successfully isolated 464 pure bacteria cultures and 117 filamentous fungi strains from 31 nectar and 7 honeydew honey. The molecular-based detection showed the dominancy of spore-forming *Bacillus* spp. including pH-tolerant lactic acid bacilli (LAB). Carvalho et al. (2010) have evaluated three different approaches of yeast identification from honey and found that partial sequence of the 26S rDNA method produced the best results. They were able to isolate *Rhodotorula mucilaginosa*, *Candida magnoliae* and *Zygosaccharomyces mellis* as the predominant species. Microbiological screening of honey has also been performed to detect the presence of *Paenibaccilus larvae*, an etiological agent of American foulbrood (AFB) disease in honeybees (Gilmore et al. 2010).

In spite of potential risk factors, screening and detection of honey microbiological contaminants may be conducted to evaluate the possible bioactivities. Lee et al. (2008) concluded that antimicrobial activity of honey may in part be attributed to the production of antimicrobial compounds by bacteria present, notably bacteriocin-like compounds and peptide antibiotics. A study devoted to isolation of *Lactobacillus acidophilus* from Malaysian honey revealed that their presence may contribute to antibacterial activity of those honeys (Mustafa Aween et al. 2012). It is worth noting that microbial contaminants in honey may directly relate to bee species, floral sources, geographical location and environmental factors, which play a significant role in microbial survival and spreading.

In general, there are two possible sources of honey microbial contaminations. The primary source includes the raw material such as nectar and pollen. Microbial contaminants may be originating from normal flora of honeybees. Environmental surroundings of the hive including air and dust also may be classified as a primary source of microbial contamination. Secondary source contamination involves the processing procedures of honey. It includes humans, equipment and containers as well as packaging materials (Snowdon and Cliver 1996). However, this mode of contamination usually involves soil- and environment-dwelling bacteria that seldom cause clinical manifestation on human beings. On the whole, there are many reasons for us to evaluate the possible contaminants in honey. It is becoming significantly important since honey has been proposed as an alternative remedy in various medical- and health-related disciplines.

Honey and Its Antioxidant Effect

The antioxidant content and activity of honey from many regions of the world have been described. Many of the phenolic acids and flavonoids found in these honeys have been identified, (Estevinho et al. 2008; Kaškonienė et al. 2009; Pyrzynska and Biesaga 2009; Truchado et al. 2009), but only a few studies involving the identification of polyphenols in honeys from Africa have been undertaken. These include honeys from Sudan (Makawi et al. 2009); major flavonoids found include quercetin, kaempferol, apigenin, hesperetin and isorhamnetin, and honeys from Tunisia (Martos et al. 1997); major flavonoids found include myricetin and quercetin. The studies of African honeys are mainly on antioxidant content and activity; these include honeys from Algeria (Ouchemoukh et al. 2007), Burkina Faso (Meda et al. 2005), Cameroon (Joseph et al. 2007) and Morocco (Malika et al. 2005). Most information available is related to the antioxidant effects on wound healing.

Some authors have evaluated the ability of honey or honey-derived polyphenols on membrane oxidation and DNA oxidation (Beretta et al. 2007; Viuda-Martos et al. 2008); however, almost no studies could be found that analyzed the effect of these antioxidants in cell culture. Therefore, the purpose of this study is to address the following gaps found in literature: determine the antioxidant activity of honey from the Southern Africa region and determine these antioxidant effects *in vitro*/cell culture. No studies have been undertaken to determine the anti-oxidant activity of honey from Southern Africa. Hayat et al. (2018) reported that insecticide exposure can affect honeybees in agroecosystems, posing behavioural stresses that can lead to population decline. They found that comet tail lengths of DNA in *A. florea* and *A. dorsata* were higher as compared to insecticide-free areas.

Moisture content in A. *florea* honey is much less compared with *A. mellifera*, which is due to different nesting habits and floral resources of nectar collection. *Apis florea* honey has higher acidity and ash content due to different foraging preferences of the two species. Many components of honey are unstable and thermolabile over time; however, *A. florea* honey tolerated temperature regimes applied regarding HMF production. HMF reached 16.30 ± 1.1 in *A. mellifera* and 7.41 ± 1.4 mg/100 g in *A. florea* honeys that were exposed for 30 min at 80°C. Honey from *A. florea* showed more heat tolerance to thermogeneration of HMF than honey from *A. mellifera*. It may be concluded that honey from *Apis florea* combs inhibit the growth of bacterial strains.

Vaghela and Reddy (2016) reported that honey of *Apis florea* collected from a wide range of arid zones of the Kachchh region in Gujarat showed significant in vitro antioxidant/radical scavenging activities. The honey contained a good amount of total protein, phenols and flavonoids and a fair amount of proline content, indicating that the honey collected from the Kachchh region of Gujarat has good source of antioxidants. Total phenolic content of honey was found least affected due to processing and also showed highest correlation with antioxidant activity. Pigment content (colour intensity) correlated well with phenol and flavonoid contents. Total proteins and flavonoids also showed significant correlation with antioxidant content whereas with proline content, the correlation of antioxidant properties was fairly significant. Therefore, the total phenolics were chiefly found to be influencing the antioxidant power of the Kachchh honey.

Conclusions

Honey from *Apis florea* is unique, as its foraging range is small and it collects nectar resources from nearby flowers. They bees forage on very small flowers on which large bees do not prefer to forage. Honey from *Apis florea* is multifloral, has excellent taste and possesses medicinal properties. It contains a good amount of total protein, phenols and flavonoids and a fair amount of proline content, indicating a good source of antioxidants.

REFERENCES

Abuharfeil, N., Al-Oran, R. and Abo-Shehada, M. 1999. The effect of bee honey on the proliferative activity of human B- and T-lymphocytes and the activity of phagocytes. *Food and Agricultural Immunology*, 11:169–177.

Adebiyi, F. M., Akpan, I., Obiajunwa, E. I. and Olaniyi, H. B. 2004. Chemical physical characterization of Nigeria honey. *Pakistan Journal of Nutrition*, 3:278–281.

Akujobi, C. O. and Njoku, H. O. 2010. Bioassay for the determination of Microbial Sensitivity to Nigerian Honey. *Global Journal of Pharmacology* 4(1):36–40.

Alábová, K., Vorlová, L., Borkovcová, I., Smutná, M. and Večerek, V. 2003. Hydroxymethylfurfural in Czech honeys. *Czech Journal of Animal Science*, 48.

Aljadi, A. M. and Kamaruddin, M. Y. 2004. Evaluation of the phenolic contents and antioxidant capacities of two Malaysian floral honeys. *Food Chemistry*, 85(4):513–518.

Aljadi, A. M. and Yusoff, K. M. 2003a. Biochemical Study on The Wound Healing Property of Honey. PhD Thesis and Dissertation, University of Malaya, Kuala Lumpur, Malaysia.

Aljadi, A. M. and Yusoff, K. M. 2003b. Isolation and Identification of Phenolic Acids in Malaysian Honey with Antibacterial Properties. *Turkish Journal of Medical Science*, 33:229–236.

Aljady, A. M., Kamaruddin, M. Y., Jamal, A. M. and MohdYassim, M. Y. 2000. Biochemical study on the efficacy of Malaysian honey in infected wounds. *Medical Journal of Islamic World Academy of Sciences*, 13(3):125–132.

Allen, K. L., Molan, P. C. and Reid, G. M. 1991. A survey of the antibacterial activity of some New Zealand honeys. *Journal of Pharmacy and Pharmacology*, 43:817–822.

Al-Waili, N. S. and Haq, A. 2004. Effect of honey on antibody production against thymus-dependent and thymus-independent antigens in primary and secondary immune responses. *Journal of Medicinal Food*, 7:491–494.

Arias, C. A. and Murray, B. E. 2009. Antibiotic-resistant bugs in the 21st century-a clinical super-challenge. *New England Journal of Medicine*, 360:439–443.

Arnon, S. S. 1998. Infant botulism. *Textbook of Pediatric Infectious Diseases*, 4th ed. Philadelphia, PA: WB Saunders Co, pp. 1570–1577.

Baltrusaityte, V., Venskutonis, P. R. and Ceksteryte, V. 2007. Radical scavenging activity of different floral origin honey and beebread phenolic extracts. *Food Chemistry*, 101:502–514.

Beck, B. F. and Smalley, D. 1997. *Honey and Your Health.* Silver Spring, MD: Health Resources Press.

Benzie, I. F. and Strain, J. J. 1996. The ferric reducing ability of plasma (FRAP) as a measure of "antioxidant power": The FRAP assay. *Analytical Biochemistry*, 239:70–76.

Beretta, G., Granata, P., Ferrero, M., Orioli, M. and Facino, R. M. 2005. Standardization of antioxidant properties of honey by a combination of spectrophotometric/fluorimetric assays and chemometrics. *Analytica Chimica Acta*, 533:185–191.

Beretta, G., Orioli, M. and Facino, R. M. 2007. Antioxidant and radical scavenging activity of honey in endothelial cell cultures (EA.hy926). *Planta Medica* 73(11):1182–1189.

Bertoncelj, J., Doberšek, U., Jamnik, M. and Golob, T. 2007. Evaluation of the phenolic content, antioxidant activity and colour of Slovenian honey. *Food Chemistry*, 105:822–828.

Bogdanov, S. 1997. Charakterisierung von Schweizer Sortenhonigen. *Agrarforschung*, 4:427–430.

Bogdanov, S. 2009a. Book of honey: Honey composition. *Bee Product Science*, 1–9. http://fantasticflavour.com/yahoo_site_admin/assets/docs/CompositionHoney.20105942.pdf. Accessed 13 September 2015.

Bogdanov, S. 2009b. Physical properties of honey. In: *Book of Honey*, Crane, E. (Ed.), Northern Bee Books, United Kingdom.

Boukraâ, L., Niar, A., Hama, B. and Benhanifia, M. 2008. Additive action of royal jelly and honey against *Staphylococcus aureus*. *Journal of Medicinal Food*, 11(1):190–192.

Busserolles, J., Gueux, E., Rock, E., Mazur, A. and Rayssiguier, Y. 2002 Substituting honey for refined carbohydrates protects rats from hypertriglyceridemic and prooxidative effects of fructose. *Journal of Nutrition*, 132(11):3379–3382.

Candiracci, M., Citterio, B. and Piatti, E. 2012. Antifungal activity of the honey flavonoid extract against Candida albicans. *Food Chemistry*, 131:493–499.

Carvalho, M., Estevinho, L. M. and Choupina, A. 2010. Yeast species associated with honey: Different identification methods. *Archivos de Zootecnia*, 59:103–113.

Chen, L., Mehta, A., Berenbaum, M., Zangerl, A. R. and Engeseth, N. J. 2000. Honeys from different floral sources as inhibitors of enzymatic browning in fruit and vegetable homogenates. *Journal of Agricultural and Food Chemistry*, 48:4997–5000.

Cooper, R., Molan, P. and Harding, K. 2002. The sensitivity to honey of Gram-positive cocci of clinical significance isolated from wounds. *Journal of Applied Microbiology*, 93:857–863.

Davies, J. and Davies, D. 2010. Origins and evolution of antibiotic resistance. *Microbiology & Molecular Biology Reviews*, 74:417–433.

Dobre, I., Gâdei, G., Patrascu, L., Elisei, A. M. and Segal, R. 2010. The antioxidant activity of selected Romanian honeys. *The Annals of the University Dunarea de Jos of Galati Fascicle VI–Food Technology*, 34(2):67–73.

Dustmann, J. H. 1978. Antibacterial effect of honey. *Apiacta*, 14(1):7–11.

El-Gendy, M. 2010. In vitro, evaluation of medicinal activity of Egyptian honey from different floral sources as anticancer and antimycotic infective agents. *Journal of Microbiology Biochemical Technology*, 5948:118–123.

Eliot Masters, M. W. and Noble, B. 2014. Value Chain assessment report: Honey and Beeswax. In: Dangay, D. (Ed.), *Trade Hub and African Partners Network*. Bethesda, MD: Abt Associates Inc in collaboration with J.E. Austin Associates.

Estevinho, L., Pereira, A. P., Moreira, L., Dias, L. G. and Pereira, E. 2008. Antioxidant and antimicrobial effects of phenolic compounds extracts of Northeast Portugal honey. *Food and Chemical Toxicology*, 46:3774–3779.

EU Council: Council Directive of the European Union. 2002. European Union Council directive 2001/110/EC of 20 December 2001 relating to honey. *Official Journal of the European Communities*, 10:47–52.

Fahim, H., Dasti, J. I., Ali, I., Ahmed, S. and Nadeem, M. 2014. Physico-chemical analysis and antimicrobial potential of *Apis dorsata*, *Apis mellifera* and *Ziziphus jujube* honey samples from Pakistan. *Asian Pacifii Journal of Tropical Biomedicine*, 4(8):633–641.

Ferreira, I. C. F. R., Aires, E., Barreira, J. C. M. and Estevinho, L. M. 2009. Antioxidant activity of Portuguese honey samples: Different contributions of the entire honey and phenolic extract. *Food Chemistry*, 144:1438–1443.

Frankel, S., Robinson, G. and Berenbaum, M. 1998. Antioxidant capacity and correlated characteristics of 14 unifloral honeys. *Journal of Apicultural Research*, 37:27–31.

Ghashm, A. A., Othman, N. H., Khattak, M. N., Ismail, N. M. and Saini, R. 2010. Antiproliferative effect of Tualang honey on oral squamous cell carcinoma and osteosarcoma cell lines. *BMC Complementary & Alternative Medicine*, 10:49.

Gheldof, N. and Engeseth, N. J. 2002. Antioxidant capacity of honeys from various floral sources based on the determination of oxygen radical absorbance capacity and inhibition of in vitro lipoprotein oxidation in human zout samples. *Journal of Agricultural and Food Chemistry*, 50(10):3050–3055.

Gheldof, N., Wang, X. H. and Engeseth, N. J. 2002. Identification and quantification of antioxidant components of honeys from various floral sources. *Journal of Agricultural and Food Chemistry*, 50(21):5870–5877.

Gheldof, N., Wang, X. H. and Engeseth, N. J. 2003. Buckwheat honey increases serum antioxidant capacity in humans. *Journal of Agricultural and Food Chemistry*, 51(5):1500–1505.

Gilmore, D., Link, L. and Fell, R. 2010. Analysis of bacterial pathogens in Virginia honeys. *Science of Bee Culture*, 2:11–14.

Gulfraz, M., Ifthkar, S., Raja, S., Asif, F., Mehmood, S., Anwar, Z. and Kaukoh, G. 2010. Quality assessment and antimicrobial activity of various honey types of Pakistan. *African Journal of Biotechnology*, 9:6902–6909.

Hayat, K., Afzal, M., Aqueel, M. A., Ali, S., Saeed, M. F., Khan, Q. M., Ashfaq, M. and Damalas, C. A. 2018. Insecticide exposure affects DNA and antioxidant enzymes activity in honey bee species Apis florea and A. dorsata: Evidence from Punjab, Pakistan. Science of the Total Environment, 635:1292–1301.

Huang, P., Gautschi, M., Walter, W., Rospert, S. and Craig, E. A. 2005. The Hsp70 Ssz1 modulates the function of the ribosome-associated J-protein Zuo1. *Nature Structural & Molecular Biology*, 12(6):497–504.

IHC. 2002. Harmonized methods of the international honey commission. Swiss Bee Research Centre. FAM, Switzerland: Liebefeld.

Irish, J., Blair, S. and Carter, D. A. 2011. The antibacterial activity of honey derived from Australian flora. *PLoS One*, 6:e18229.

Islam, A., Khalil, I., Islam, N., Moniruzzaman, M., Mottalib, A., Sulaiman, S. A. and Gan, S. H. 2012. Physicochemical and antioxidant properties of Bangladeshi honeys stored for more than one year. *BMC Complementary & Alternative Medicine*, 12:177.

Jaganathan, S. K., Mondhe, D., Wani, Z. A. and Supriyanto, E. 2014. Evaluation of selected honey and one of its phenolic constituent eugenol against L1210 lymphoid leukemia. *Scientific World Journal* Volume 2014, Article ID 912051, 3 pages. doi:10.1155/2014/912051.

Javanmardi, J., Stushnoff, C., Locke, E. and Vivanco, J. M. 2003. Antioxidant activity and total phenolic content of Iranian Ocimum accessions. *Journal of Food Chemistry* 83:574–550.

Johnston, M. D., Hanlon, G. W., Denyer, S. P. and Lambert, R. J. 2003. Membrane damage to bacteria caused by single and combined biocides. *Journal of Applied Microbiology* 94, 1015–1023.

José Miguel Alvarez-Suarez, Sara Tulipani Stefania Romandini and Vidal, A. 2009. Methodological aspects about determination of phenolic compounds and in vitro evaluation of antioxidant capacity in the honey: A review. Current Analytical Chemistry, 5(4):293–302.

Joseph, T., Awah-Ndukum, J., Fonteh-Florence, A., Delphine, N. D., Jonnas, P., Ze Antoine, M. 2007. Physicochemical and microbiological characteristics of honey from the Sudano-Guinean Zone of west Cameron. *African Journal of Biotechnology*, 6(7): 908–991.

Kaškonienė, V., Maruška, A., Kornyšova, O., Charczun, N., Ligor, M. and Buszewski, B. 2009. Quantitative and qualitative determination of phenolic compounds in honey. *Cheminė Technologija*, 52:74–80.

Khalil, I. M., Moniruzzaman, M., Boukraâ, L., Benhanifia, M., Islam, M. A., Islam, M. N., Sulaiman, S. A. and Gan, S. H. 2012. Physicochemical and antioxidant properties of Algerian honey. *Molecules*, 17(9):11199–11215.

Khalil, M., Sulaiman, S. and Boukraa, L. 2010. Antioxidant properties of honey and its role in preventing health disorder. *The Open Nutraceuticals Journal*, 3:6–16.

Khan, M. N., Qaiser, M., Raza, S. M. and Rehman, M. 2006. Physicochemical properties and pollen spectrum of imported and local samples of blossom honey from the Pakistani market. *International Journal of Food Science & Technology*, 41(7):775.

Khosravi, A. R., Shokri, H., Katiraee, F. and Ziglari, T. 2008. Fungicidal potential of different Iranian honeys against some pathogenic Candida species. *Journal of Apicultural Research* 47(4):260–264.

Kingsley, A. 2001 The use of honey in the treatment of infected wound. *British Journal of Nursing*, 10(22 Suppl):S13–6, S18, S20.

Küçük, M., Kolayh, S., Karaoğlu, Ş., Ulusoy, E., Baltact, C. and Candan, F. 2007 Biological activities and chemical composition of three honeys of different types from Anatolia. *Food Chemistry*, 100:526–534.

Kwakman, P. H., Te Velde, A. A., De Boer, L., Speijer, D., Vandenbroucke-Grauls, C. M. and Zaat, S. A. 2010. How honey kills bacteria. *The FASEB Journal*, 24:2576–2582.

Lachman, J., Orsak, M., Hejtmankova, A. and Kovarova, E. 2010. Evaluation of antioxidant activity and total phenolics of selected Czech honeys. *LW T-Food Science and Technology*, 43:52–58.

Lee, H., Churey, J. J. and Worobo, R. W. 2008. Antimicrobial activity of bacterial isolates from different floral sources of honey. *International Journal of Food Microbiology*, 126:240–244.

Liu, B. and Pop, M. 2009. ARDB—antibiotic resistance genes database. *Nucleic Acids Research*, 37:D443–D447.

Makawi, S. Z. A., Taha, M. I., Zakaria, B. A., Siddig, B., Mahmod, H., Elhussein, A. R. M. and Gad Kariem, E. A. 2009. Identification of 5-hydroxymethylfurfural HMF in some sugar containing food products by HPLC. *Pakistan Journal of Nutrition*, 8:1391–1396.

Malika, N., Faid, M. and Chkib, E. 2005. Microbiological and physico-chemical properties of Moroccan honey. *International Journal of Agricultural Biology*, 7(5):773–776.

Mandal, V., Sen, S. K. and Mandal, N. C. 2009. Effect of prebiotics on bacteriocin production and cholesterol lowering activity of *Pediococcus acidilactici* LAB 5. *World Journal of Microbiology and Biotechnology*, 25:1837–1841.

Marghitas, L. A., Dezmiren, D. S., Pocol, C. B., Ilea, M., Bobis, O. and Gergen, I. 2010. The development of a biochemical profile of acacia honey by identifying biochemical determinants of its quality. *Notulae Botanicae Horti Agrobotanici Cluj-Napoca*, 38(2):Special Issue, 9–13.

Martos, I., Cossentini, M., Ferreres, F. and Tomás-Barberán, F. A. 1997. Flavonoid composition of Tunisian honeys and propolis. *Journal of Agricultural and Food Chemistry*, 45:2824–2829.

Meda, A., Lamien, C. E., Romito, M., Millogo, J. and Nacoulma, O. G. 2005. Determination of the total phenolic, flavonoid and proline contents in Burkina Fasan honey, as well as their radical scavenging activity. *Food Chemistry*, 91:571–577.

Midura, T. F. 1996. Update: Infant botulism. *Clinical Microbiology Reviews*, 9:119–125.

Midura, T. F., Snowden, S., Wood, R. M. and Arnon, S. S. 1979. Isolation of Clostridium botulinum from honey. *Journal of Clinical Microbiology*, 9(2):282–283.

Molan, P. C. 1992. The antibacterial activity of honey: 1. The nature of antibacterial activity. *Bee World*, 73:5–28.

Molan, P. C. 2002. Re-introducing honey in the management of wounds and ulcers-theory and practice. *Ostomy Wound Management*, 48:28–40.

Molan, P. C. 2006. The evidence supporting the use of honey as a wound dressing. *The International Journal of Lower Extremity Wounds*, 5:40–54.

Moniruzzaman, M., Sulaiman, S. A., Khalil, M. I. and Gan, S. H. 2013. Evaluation of physicochemical and antioxidant properties of sourwood and other Malaysian honeys: A comparison with manuka honey. *Chemistry Central Journal*, 7(138):1–12.

Moussa, A., Noureddine, D., Saad, A. and Abdelmelek, M. 2012. Anti-Aspergillus niger of eucalyptus honey influenced by thermal treatment. *Open Access Scientific Reports*, 1:175. doi:10.4172/scientificreports.175.

Mundo, M. A., Padilla-Zakour, O. I. and Worobo, R. W. 2004. Growth inhibition of food borne pathogens and food spoilage organisms by select raw honeys. *International Journal of Food Microbiology*, 97:1–8.

Mustafa Aween, M., Hassan, Z., Muhialdin, B. J., Mohd Noor, H. and Eljamel, Y. A. 2012. Evaluation on antibacterial activity of *Lactobacillus acidophilus* strains isolated from honey. *American Journal of Applied Sciences*, 9:807.

Nasir, N. A., Halim, A. S., Singh, K. K., Dorai, A. A. and Haneef, M. N. 2010. Antibacterial properties of tualang honey and its effect in burn wound 117 management: A comparative study. *BMC Complementary & Alternative Medicine*, 10:31.

Nevas, M. 2006. Clostridium botulinum in honey production with respect to infant botulism. Academic dissertation, University of Helsinki, Finland.

Ngatu, N. R., Saruta, T., Hirota, R., Eitoku, M., Muzembo, B. A., Matsui, T., Nangana, L. S., Mbenza, M. A., Kumagai, N. and Suganuma, N. 2011. Antifungal efficacy of Brazilian green propolis extracts and honey on Tinea capitis and Tinea versicolor. *European Journal of Integrative Medicine* 3:281–287.

Nikaido, H. 2009. Multidrug resistance in bacteria. *Annual Review Biochemistry*, 78:119–146.

Noori, A. L., Al Ghamdi, A., Ansari, M. J., Al-Attal, Y., Al-Mubarak, A. and Salom, K. 2013. Differences in Composition of Honey Samples and Their Impact on the Antimicrobial Activities against Drug Multiresistant Bacteria and Pathogenic Fungi. *Archives of Medical Research*, 44:307–316.

Nurul Syazana, M. S., Halim, A. S., Gan, S. H. and Shamsuddin, S. 2011. Antiproliferative effect of methanolic extraction of tualang honey on human keloid fibroblasts. *BMC Complementary and Alternative Medicine*, 11:82.

Nzeako, B. C., Okafor, N. and Azikiwe, N. 2002. Prevalence of Aeromonas hydrophila in seasonal episodes of gastroenteritis in Nsukka Nigeria. *Kuwait Medical Journal*, 34(1):16–19.

Ouchemoukh, S., Louaileche, H. and Schweitzer, P. 2007. Physicochemical characteristics and pollen spectrum of some Algerian honeys. *Food Control*, 18:52–58.

Perna, A., Simonetti, A., Intaglietta, I. and Gambacorta, E. 2013. Antioxidant properties, polyphenol content and colorimetric characteristics of different floral origin honeys from different areas of Southern Italy. *Journal of Life Sciences*, 7(4):428–436.

Popek, S. 2002. The procedure to identify a honey type. *Food Chemistry*, 79:401–406.

Prior, R. L., Wu, X. and Schaich, K. 2005. Standardized methods for the determinations of antioxidant capacity and phenolics in foods and dietary supplements. *Journal of Agricultural and Food Chemistry*, 53:4290–4302. doi:10.1021/jf0502698.

Pyrzynska, K. and Biesaga, M. 2009. Analysis of phenolic compounds and flavonoids in honey. *Trends in Analytical Chemistry*, 28:893–903.

Rosa, A., Tuberoso, C. I. G., Atzeri, A., Melis, M. P., Bifulco, E. and Dessì, M. A. 2011. Antioxidant profile of strawberry tree honey and its marker homogentisic acid in several models of oxidative stress. *Food Chemistry*, 129(3):1045–1053.

Schlesier, K., Harwat, M., Böhm, V. and Bitsch, R. 2002. Assessment of antioxidant activity by using different in vitro methods. *Free Radical Research*, 36(2):177–187.

Schramm, D. D., Karim, M., Schrader, H. R., Holt, R. R., Cardetti, M. and Keen, C. L. 2003. Honey with high levels of antioxidants can provide protection to healthy human subjects. *Journal of Agricultural and Food Chemistry*, 51:1732–1735. doi:10.1021/jf025928k.

Selçuk, H. and Nevin, K. 2003. Investlgation of antimicrobial effect of honey and the role of osmolority Anadolu University *Journal of Science and Technology*, 4(1):85–88.

Sinacori, M., Francesca, N., Alfonzo, A. et al. 2014. Cultivable microorganisms associated with honeys of different geographical and botanical origin. *Food Microbiology*, 38:284–294.

Singh, M. P., Chourasia, H. R., Agarwal, M., Malhotra, A., Sharma, M., Sharma, D. and Khan, S. 2012. Honey as complementary medicine—A review. *International Journal of Pharma and Bio Sciences*, 3(2):12–31.

Snowdon, J. A. and Cliver, D. O. 1996. Microorganisms in honey. *International Journal of Food Microbiology*, 31:1–26.

Soman, A. G. and Chawda, S. S. 1996. A contribution to the biology and behaviour of dwarf bee, *A. florea* F. and its economic importance in Kutch, Gujarat, India. *Indian Bee Journal*, 58(2):81–88.

Sosa, V., Moliné, T., Somoza, R. et al. 2013. Oxidative stress and cancer: An overview. *Ageing Research Reviews*, 12:376–390.

Souza, B., Roubik, D., Barth, O. et al. 2006. Composition of stingless bee honey: Setting quality standards. *Interciencia*, 31(12):867–875.

Stomfay-Stitz, J. 1960. Honey: An ancient yet modern medicine. *Science Council* 23:110–125.

Subrahmanyam, M., Sahapure, A. G., Nagane, N. S., Bhagwat, V. R. and Ganu, J. V. 2001. Effects of topical application of honey on burn wound healing. *Annals of Burns and Fire Disasters*, 3:143–145.

Tan, H. T., Rahman, R. A., Gan, S. H., Halim, A. S., Hassan, S. A., Sulaiman, S. A. and Kirnpal-Kaur, B. 2009. The antibacterial properties of Malaysian tualang honey against wound and enteric microorganisms in comparison to manuka honey. *BMC Complementary and Alternative Medicine*, 9:34.

Tonks, A. J., Cooper, R. A., Jones, K. P., Blair, S., Parton, J. and Tonks, A. 2003. Honey stimulates inflammatory cytokine production from monocytes. *Cytokine*, 21:242–247.

Truchado, P., Gil-Izquierdo, A., Tomás-Barberán, F. and Allende, A. 2009. Inhibition by chestnut honey of *N*-Acyl-L-homoserine lactones and biofilm formation in *Erwinia carotovora*, *Yersinia enterocolitica*, and *Aeromonas hydrophila*. *Journal of Agricultural and Food Chemistry*, 57(23):11186–11193.

Tumin, N., Halim, N. A., Shahjahan, M., Noor Izani, N. J., Khan, A. H. and Mohsin, S. S. J. 2005. Antibacterial activity of local malaysian honey. *Malaysian Journal of Pharmaceutical Sciences*, 3(2):1–10.

Uttara, B., Singh, A. V., Zamboni, P. and Mahajan, R. T. 2009. Oxidative stress and neuro degenerative diseases: A review of upstream and downstream antioxidant therapeutic options. *Current Neuropharmacology*, 7(1):65–74.

Vaghela, J. and Reddy, A. 2016. Antioxidant potential of apis florea honey from dryland ecosystem in western India. *International Journal of Advanced Research*, 4(2):1392–1402.

Velioglu, Y. S., Mazza, G., Gao, L. and Oomah, B. D. 1998. Antioxidant activity and total phenolics in selected fruits, vegetables and grain products. *Journal of Agricultural and Food Chemistry*, 46:4113.

Vermerris, W. and Nicholson, R. 2009. *Phenolic Compound Biochemistry*. Dordrecht, the Netherlands: Springer.

Viuda-Martos, M., Ruiz-Navajas, Y., Fernández-López, J. and Pérez-Álvarez, J. A. 2008. Antibacterial activity of different essential oils obtained from spices widely used in Mediterranean diet. *International Journal of Food Science and Technology*, 43:526e531.

Willix, D. J., Molan, P. C. and Harfoot, C. G. 1992 A comparison of the sensitivity of wound-infecting species of bacteria to the antibacterial activity of manuka honey and other honey. *Journal of Applied Bacteriology*, 73:388–394.

Wilson, R. B. and Crane, E. 1975. Uses and products of honey in honey a comprehensive survey. In: *Honey a Comprehensive Survey*. Crane, E. (Eds.). Heinemann, London, UK, pp. 378–391.

Zaid, S. S., Sulaiman, S. A., Sirajudeen, K. N. S. and Othman, N. H. 2010. The effects of tualang honey on female reproductive organs, tibia bone and hormonal profile in ovariectomised rats-animal model for menopause. *BMC Complementary and Alternative Medicine*, 31(10):82.

12

Genetic Diversity of Apis florea *in India*

B. Fakrudin, K. Omkar, Jameel Khan and K. Nandini

CONTENTS

Introduction

India is one of the mega hotspots for flora and fauna diversity in the world. Many economically and ecologically important insect and noninsect species are sustained across the region. Honeybees have been providing several ecosystem services. They are the most dominant and best known pollinators as they have coevolved with the flowering plants. Both in natural and agricultural ecosystems, bees and their relationship with flowering plants occupy a keystone position (Dafni, 1997).

In India, among the three native species of honeybees viz. *A. florea*, *A. dorsata* and *A. cerana*, the *Apis florea* species (also known as dwarf bees) is mild in nature. Dwarf honeybees are the most ancestral in the genus *Apis* as evident from the morphological and molecular data (Arias and Sheppard, 2005). The individual members of *Apis florea* have a body length of 7–10 mm in worker bees, 13–15 mm in queen and 11–13 mm in male bees. Genetic diversity exits at the individual level of *Apis* spp. and at the population level, both of which contribute to fitness and survival (Figure 12.1). Striving in hot and arid climates, *Apis florea* maintains homeostasis of nest and caste development for colony survival and development (Suwannapong et al., 2012). Being native to Thailand, *Apis florea* is challenged by numerous pests and disease, such as hornets; maintaining enough diversity is considered to be the first key for survival (Akratanakul, 1976).

The diversity for hive-making location is high in this species. The hives of *A. florea* are bigger than those of *A. andreniformis*, and there seems to be existence of variability among the geographic populations.

The dwarf honeybee species, *Apis florea* (F.) (Fabricius 1787), shares similarity with *A. andreniformis* that is distinct in general behaviour, nest crown morphology and multiple body parameters including colour (Oldroyd and Wongsiri, 2006). Important parameters such as structural features of drone genitalia, chemical profile of bees wax, allozyme patterns, mtDNA sequence divergence, timing of mating

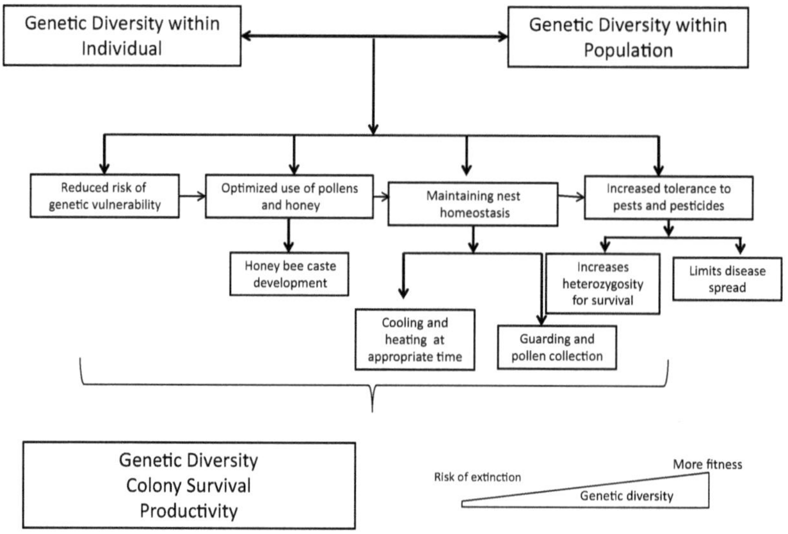

FIGURE 12.1 Genetic diversity as a strategy for survival at individual and population levels displayed by *Apis florea*.

flight and a number of other morphological traits have been found to be distinguishing features of *A. florea's* closely related species *A. andreniformis* (Wu and Kuang, 1986; Smith, 1991). These features have clearly placed these species as completely reproductively isolated species (Wongsiri et al., 1990). Factors such as sympatric behaviour over a wide area extending from northeast India to Indochina and classification based on workers do not show great morphological differentiation, and use of limited numbers of specimens have been the points of concern in classification of *A. florea* and other closely related *Apis* species (Radloff et al., 2011). This species prevails sympatrically with other closely related species of honeybees. *Apis florea* is distributed from the Persian Gulf region to the Himalayas and extends to the Asia pacific regions (Hepburn and Radloff, 2011). Further, the mitochondrial genome of *A. florea* collected from India and northern Thailand was genetically distinct, showing 5%–7% sequence divergence (Smith et al., 1991; Rinderer et al. 1995).

Polymorphism and Allelic Frequency

Indian populations of *Apis florea* sampled across different locations were subjected to genetic diversity assessment using DNA markers by Lazar et al. (2013). In all, individuals of about 35 populations were genotyped using microsatellite markers that revealed rich allelic diversity ranging from 53 to 114. A total of 504 alleles have been recorded from about 6 microsatellite loci screened in individuals of 35 populations. A microsatellite marker HBSSR48 recorded highest 114 alleles followed by HBSSR65 with 103 alleles with an average of 84. Further, the number of effective alleles (Ne) was maximum for the locus HBSSR05 (7.145) followed by HBSSR48 (6.855). The expected heterozygosity (He) was 0.743 with a mean value for loci ranging between 0.669 for HBSSR07 to 0.833 for HBSSR05 microsatellite loci. Further, the private alleles were observed across all the populations: a highest mean frequency of 1.108 for POP28 followed by 1.1078 for POP33 was reported. The grand mean Lazar et al. (2013) an of observed heterozygosity (Ho) and expected heterozygosity (He) across all the populations was 0.564 and 0.743, respectively. The populations of *Apis florea* were considered to be highly diverse with respect to allelic frequency (Figure 12.2).

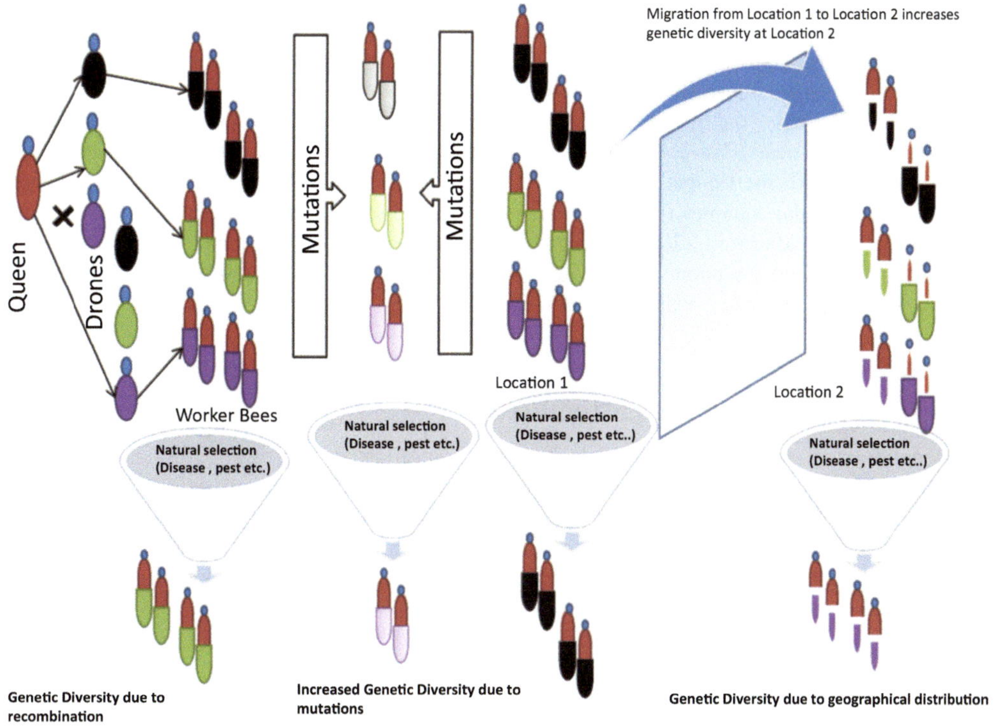

FIGURE 12.2 Graphical representation of different factors contributing to genetic diversity of *Apis florea*.

Analysis of Molecular Variation

Analysis of molecular variation in 35 populations of *Apis florea* sampled across India has been explored and reported by Lazar et al. (2013). The variation among population was found to be 16.52%, which is rather low, compared to variation within populations, which was at 83.48%. The mean FST value was recorded to be 0.16, implicating that a significant genetic structuring within populations has occurred over a period of time. The analysis of molecular variance calculated based on mitochondrial sequence diversity (part of the *COII* gene) across all the populations indicated that among the groups the variation stood at 35.34%, among populations within groups it was 22.15% and within population variation it was reported to be 42.15% with an overall mean Fst value of 0.565. These findings point at a fact that a significant genetic structuring among the sampled populations exists.

Further, the nucleotide sequence of the 5′ end of the mitochondrial *COII* gene from four *Apis* species – viz. *mellifera, cerana, dorsata* and *florea* and *Bombus lucorum* (Apidae) and *Xylocopa violacea* (Anthophoridae) as outgroups – indicated that the phylogenetic trees built using neighbour-joining and parsimony methods resulted in three branches corresponding to *A. dorsata, A. florea* and *A. cerana* (Garnery et al., 1991).

Principal Coordinate Analysis

Principal component analysis is an effective analysis method to study clustering and relatedness among the populations or individuals based on genetic distances. The PCA analysis has revealed that the populations sampled from Karnataka and Assam were diverse as observed through their separate and distant distribution in the quadrates. The populations from Tamil Nadu and Kerala were genetically more identical than the rest of the populations as indicated by their close grouping. The populations from Maharashtra – although distributed in a single quadrate – showed significant distribution in their populations, indicating genetic diversity within populations.

Similar studies carried out by Hepburn et al. (2005), using morphometric characters indicated samples from Iran were genetically distinct but distinctly related to majority of the samples from Oman, Pakistan, when compared with the sample populations collected from India, Myanamar, Nepal, Thailand and Vietnam. In particular, samples from southern India were reported to be genetically closer to samples from Sri Lanka. Haddad et al. (2009) reported similar morpho clusters in samples from southern India and Sri Lanka formed a separate cluster, and another cluster with samples from Iran, Sudan, Oman, Jordan and Pakistan formed a separate cluster.

Phylogenetic and Population Structure

The populations of most of the species show some level of genetic structuring due to variety of non-mutually exclusive agents. Understanding this structuring among the populations is very important for many fields of study including population genetics, population ecology and conservation biology. Among the parameters reflecting genetic structuring, Wright's F-statistics (1951) is one of the most widely used to evaluate the population structuring. The mean F_{IS} value of 0.245 across the populations ranged between 0.035 for HBSSR53 and 0.377 for HBSSR85. The F_{IT} mean value was 0.395, and the values ranged from 0.174 for HBSSR53 to 0.521 for HBSSR85 loci. The mean the F_{ST} value was found to be 0.203 (20.3%) among all the populations. For the loci, the values ranged between 0.144 for HBSSR53 and 0.266 for HBSSR07, indicating the presence of significant genetic differentiation among the populations with respect to the individual loci considered. The pairwise population F_{ST} values ranged from a minimum of 0.019 (between POP15 and POP16) to a maximum of 0.308 (between POP1 and POP2). The pairwise F_{ST} values indicated the presence of significant genetic differentiation among the populations.

A phylogenetic tree also known as an evolutionary tree is a branching diagram or tree indicating the inferred evolutionary relationship within or among the species based on their physical or genetic distances/relatedness. Similarly, accessing the population structure and assignment of individuals to these populations is an important aspect of population genetic studies. The genetic structure of the populations was investigated using 471 individuals sampled across the populations of India (Lazar et al., 2013). The K value was found to be 6, indicating the presence of six population clusters across the sampled populations. Phylogenetic analysis using microsatellite data formed six major clusters. The populations from Jammu and Assam formed a single group indicating that they are individual populations with shared genetic clustering. The populations from Andhra Pradesh and Tamil Nadu formed another independent single cluster with intermixing populations. The samples from Karnataka shared similarity with the populations of Andhra Pradesh and Tamil Nadu. The populations from Maharashtra and Kerala formed distinct clusters, indicating their distinct populations status. Similarly phylogenetic analysis was carried out using COII gene sequences, which indicated the presence of four major clusters among the sampled populations. Cluster one consisted of samples from Jammu and Kashmir, Assam, Maharashtra, Karnataka, Tamil Nadu and Andhra Pradesh. The majority of the samples in this cluster were intermixed except the samples from Assam, which exhibited clear grouping. The rest of the three clusters represented samples collected exclusively from the southern region of India. The phylogenetic tree indicated that there was an intermixing of the populations between northern and southern regions. Phylogenetic analysis using COII gene sequences in comparison to sequences from earlier reported haplotypes were carried out. The authors reported three major clusters among which two were exclusive to the sampled areas in India, and one cluster was similar to haplotypes reported from China (Cao et al., 2012), indicating gene flow among the sampled populations. The authors also carried out haplotype analysis and identified a total of 49 haplotypes distributed among the sampled regions.

In comparison to *Apis florea* populations of India, colonies from Myanmar, Thailand, Cambodia and Pakistan have shared varied levels of genetic similarity. A cluster analysis using the complete linkage procedure carried out on colony mean traits of diverse populations including Asian countries resulted in a dendrogram of two main clusters where samples of southern India and Sri Lanka were highly comparable (Hepburn et al., 2005; Haddad et al., 2009).

Using 32 Thailand and 5 Malaysian colonies of *A. andreniformis*, Rattanawannee et al. (2007) conducted a phylogeny and genetic variation study. They analyzed the cytochrome oxidase subunit b (*cyt b*) and found that there was no geographic structure in Thailand *A. andreniformis*. Higgs et al. (2010) demonstrated that the size polymorphism in this fragment of the *rrnL* gene reliably distinguishes *A. andreniformis* and *A. florea*. Most of the molecular phylogeny studies indicated that *A. florea* is the ancestral species of *Apis* genus. The climatic range of *A. florea* is much broader than that of *A. andreniformis* and includes both desert and semi-arid zones in the west that gradually change to subtropical, tropical dry and tropical wet climates (Hepburn and Radloff, 2011).

Conclusion

Apis florea populations are distributed throughout India, especially in southern and northeastern hilly region populations, which have rich diversity of these bees. Geographic populations of this species have varied levels of genetic relatedness and distinct genetic parameters. There is a need for concerted biological, phylogenetic, genomic and biochemical studies on this species. Conservation efforts towards protection of this species as a genetic treasure should be initiated in the whole of Asia including India to prevent extinction. The microsatellite analysis revealed significant observed and expected heterozygosity among the sampled populations, clearly indicating the presence of significant genetic diversity. The combined results of NJ tree, structure analysis and PCA indicated significant clustering across the populations of India. The *COII* sequence divergence has revealed that there is a significant gene flow between the populations from northern and southern regions in India. The analysis of the newly identified haplogroups from India with that of *A. florea* haplotypes reported from China identified many new haplotypes that are exclusive to the Indian region. However, there is a need for further studies with larger sampling sizes to access genetic diversity and eventually conserve these populations in India.

REFERENCES

Akratanakul, P. 1976. Honeybees in Thailand. *American Bee Journal*, 116:120–121.

Arias, M.C. and Sheppard, W.S. 2005. Phylogenetic relationships of honey bees (*Hymenoptera: Apinae; Apini*) inferred from nuclear and mitochondrial DNA sequence data. *Molecular Phylogenetics and Evolution*, 37:25–35.

Cao, L.-F., Zheng, H.-Q., Hu, C.-Y., He, S.-Y., Kuang, H.-O. and Hu, F.-L. 2012. Phylogenetic relationships of honey bees (Hymenoptera: Apinae) in-ferred from mitochondrial and nuclear DNA sequences. *Annals of the Entomological Society of America*, 105(2):298–304.

Dafni, A. 1997, The response of *Amphicoma* spp. (Coleoptera; Glaphyridae) beetles to red models differing in area, shape, and symmetry, Israel. *Journal of Plant Sciences*, 45:2–3, 247–254.

Garnery, L., Vautrin, D., Cornuet, J.M. and Solignac, M. 1991. Phylogenetic relationships in the genus *Apis* inferred from mitochondrial DNA sequence data. *Apidologie*, 22:87–92.

Haddad, N., Crutsinger, G.M., Kevin, G., Haarstad, J., Knops, J.M.H. and Tilman, D. 2009. Plant species loss decreases arthropod diversity and shifts trophic structure. *Ecology Letters*, 12(10):1029–1039.

Haddad, N., De Miranda, J.R., Bataena, A. 2008. Discovery of Apisfloreain Aqaba, Jordan. *Journal of Apiculture Research*, 47:173–174.

Hepburn, H.R. and Radloff, S.E. 2011. Biogeography. In Hepburn, H.R. and Radloff, S.E. (Eds). *Honey Bees of Asia*. Springer Press, Berlin, Germany, pp. 62–63.

Hepburn, H.R., Radloff, S.E., Otis, G.W., Fuchs, S., Verma, L.R., Tan, K., Chaiyawong, T., Tahmasebi, G., Wongsiri, S. 2005. Apisflorea: Morphometrics, classification and biogeography. *Apidologie*, 36:359–376.

Higgs, J.S., McHale, M. and Oldroyd, B.P. 2010. A scientific note on a rapid method for the molecular discrimination of *Apis andreniformis* and *A. florea*. *Apidologie*, 41:96–98.

Lazar, K.V., Fakrudin, B., Omkar Babu, K., Mohankumar, S., Parthiban, T.P., Chapalkar, S., Viraktamath, S. and Vastrad, A.S. 2013. Genetic diversity and phylogeography of Apis florea. In Viraktamath, S., Fakrudin, B., Vastrad, A.S. and Mohankumar, S. (Eds). *Monograph and Phylogeography of Honey Bees and Stingless Bees in India.* Department of Biotechnology, Ministry of Science and Technology Government of India, New Delhi, India, pp. 101–123.

Oldroyd, B.P. and Wongsiri, S. 2006. *Asian Honey Bees.* Harvard University Press, Cambridge, MA.

Radloff, S.E., Hepburn, H.R. and Engel, M.S. 2011. The Asian species of *Apis.* In Hepburn, H.R. and Radloff, S.E. (Eds). *Honey Bees of Asia.* Springer Press, Berlin, Germany, pp. 1–22.

Rattanawannee, A., Chanchao, C. and Wongsiri, S. 2007. Morphometric and genetic variation of small dwarf honey bees *Apis andreniformis Smith*, 1858 in Thailand. *Insect Science*, 14:451–460.

Rinderer, T.E., Oldroyd, B.P., Wongsiri, S., Sylvester, H.A., de Guzman, L.I., Stelzer, J.A. and Riggio, R.M. 1995. A morphological comparison of the dwarf honey bees of southeastern Thailand and Palawan, Philippines. *Apidologie*, 26:387–394.

Smith, D.R. 1991. Mitochondrial DNA and honey bee biogeography. In: Smith, D.R. (Ed.). *Diversity in the Genus Apis.* Westview Press, Boulder, Co., pp. 131–176.

Smith D.R., Palopoli, M.F., Taylor, B.R., Garnery, L., Cornuet, J.M., Solignac, M. and Brown, W.M. 1991. Geographical overlap of two mitochondrial genomes in Spanish honeybees (*Apis melliferaiberica*). *Journal of Heredity*, 82(2):96–100.

Suwannapong, G., Benbow, M.E. and Nieh, J.C. 2012. Biology of Thai honeybees: Natural history and threats. In Florio, R.M. (Ed). *Bees: Biology, Threats and Colonies.* Nova Science Publishers, Hauppauge, NY.

Wongsiri, S., Limbipichai, K., Tangkanasing, P., Mardan, M.B., Rinderer, T.E., Sylvester H.A. Koeniger, G. and Otis, G.W. 1990. Evidence of reproductive isolation confirms that *Apis andreniformis* (Smith, 1858) is a separate species from sympatric *Apis florea* (Fabricius, 1787). *Apidologie*, 21:47–52.

Wu, Y. and Kuang, B. 1986. A study of genus Micrapis (*Apidae*). *Zoological Research*, 7:99–102.

13

Morphometric Variation and Floral Resources of the Dwarf Honeybee, Apis florea *F. in Andhra Pradesh and Telangana, India*

A. J. Solomon Raju

CONTENTS

Introduction

Of the different honeybee species, *Apis mellifera* and *A. cerana* have been studied in detail throughout their distribution range for infraspecific classification. The biometric studies of *A. cerana* and *A. florea* were carried out by a few workers in Himachal Pradesh, Jammu and the Kashmir regions of India (Kshirsagar 1983; Singh et al. 1990; Verma et al. 1994). Based on morphometric studies, Ruttner et al. (1995) reported geography-related variability and climate-related variation in *A. florea.* Tahmasebi et al. (2002) defined two morphoclusters in *A. florea* from Iran. Hepburn et al. (2005) made a comparison of *A. florea* from India, Sri Lanka, Iran, India, Thailand and Vietnam and reported two morphoclusters. Ruttner (1988) and Tahmasebi et al. (2002) noted that *A. florea* bees of Iran are larger than those of South India and Sri Lanka and hence the morphocluster of bees of Iran is quite distinct from that of bees of South India and Sri Lanka. Haddad et al. (2009) carried out morphometric analysis of *A. florea* in Jordan and Sudan. This state of work on morphoclusters or biometric clusters of *A. florea* is not useful to understand the variation in this bee species throughout its distribution range.

Tan et al. (2007) showed that morphometric characters of honeybees are related to the nutrition levels and floral diversity of the habitats. Despite recognizing the fact that the flower species in any ecological region play an important role in maintaining and multiplying the colonies of honeybees, there have been hardly any studies on their floral calendars in different regions, especially in India. Floral calendars are needed for each ecological region, for what may apply in one place may not apply in another, even sometimes quite nearby (Roubik 1995). With this backdrop, the work on morphometrics and floral calendar of *A. florea* was carried out in Andhra Pradesh and Telangana of India.

Materials and Methods

Andhra Pradesh and Telangana are spread out over an area of 276,754 km^2 (latitude 12°41′ and 22°00′ N and longitude 77°00′ and 84°40′ E). The land area is broadly divided into seven distinct agroclimatic zones: Krishna-Godavari Zone, North Coastal Zone, Southern Zone, North Telangana Zone, South

Telangana Zone, Scarce Rainfall Zone, and High Altitudes and Tribal Areas Zone. Ten representative localities were selected in each agroclimatic zone based on the distance, topography and vegetation for the collection of samples of *Apis florea*. The distance between localities is 60–80 km. In each locality, 40 worker bees were collected of which 20 were used for morphometics, and 20 others were deposited in the department. The bees were collected by using insect nets, killed with petroleum ether, air-dried for 2–3 days and placed in butter paper bags containing para-dichloro-benzene crystals to avoid fungal contamination. Latitude, longitude and altitude were recorded for each locality using a Magellan eXplorist 210 Model Digital GP System.

Bees were dissected systematically into the head, thorax and abdomen using a Magnus Compound Microscope. The measurements were done using a Leica M165C Stereomicroscope with computer-aided measuring system and program. Twenty-one characters – head length and width, proboscis length, distance between two dorsal ocelli, dorsal ocello-ocular distance, antenna length, femur length, tibia length, metatarsus length and width, forewing length and width, cubital index, number of hamuli, 3rd tergite length, 3rd sternite length and width, wax mirror length and width, distance between two wax plates on 3rd sternite and lateral width of 4th tomentum – were measured. These characters were reduced to 16 characters by taking the ratios of head length and width, metatarsus length and width, forewing length and width, 3rd sternite length and width, and wax mirror length and width. All statistical calculations were performed using SPSS 16.0 for Windows and SYSTAT version 13.0 statistical package for Windows. The data obtained were subjected to multivariate morphometric analysis, hierarchial cluster analysis and principal component analysis. Field observations were made on *A. florea* regarding the plant species used by them as forage throughout the year in all localities/agroclimatic zones to prepare a floral calendar.

Results

Results of Morphometric Analysis

Twenty-one morphometric characters of *Apis florea* were measured for each locality in each agroclimatic zone of Andhra Pradesh and Telangana. The measured characters were then reduced to 16 characters, the average means and standard deviations of which were noted by agroclimatic zone in Table 13.1. The cluster analysis performed for the means of individual characters for each agroclimatic zone showed two clusters. The populations of Zones I and II followed by Zone VII represented cluster 1, and those of Zones III and IV represented cluster 2. Zone V is associated with both clusters 1 and 2, while Zone VI is totally isolated (Figure 13.1). The cluster analysis performed for the means of individual characters for 70 localities at once showed 15 clusters (Figure 13.2).

The means and standard deviations of 16 morphometric characters of *A. florea* from all localities were segregated into six latitudinal ranges: 13°12′–13°49′ N (three localities in Zone III and one in VI), 14°06′–14°54′ N (six localities in Zone III and two in Zone VI), 15°07′–15°52′ N (two localities in Zone I, one in Zone III and six in Zone VI), 16°01′–16°56′ N (five localities in Zone I, one in Zone IV, three in Zone V and one in Zone VI), 17°04′–17°58′ N (three localities in Zone I, five in Zone II, two in Zone IV, seven in Zone V and three in Zone VII) and 18°03′–19°40′ N (five localities in Zone II, seven in Zone IV and seven in Zone VII). The data, when subjected to hierarchial cluster analysis, showed one cluster in 13°12′–13°49′ N; two clusters each in 14°06′–14°54′ N, 15°07′–15°52′ N and 16°01′–16°56′ N; four clusters in 17°04′–17°58′ N and five clusters in 18°03′–19°40′ N.

The means and standard deviations of 16 morphometric characters of *A. florea* from all localities sampled were segregated into three altitudinal ranges: 0.61–296 m (10 localities each in Zones I and II, 6 each in Zones III and VI, 4 in Zone IV, 2 in Zone V and 8 in Zone VII), 316–595 m (2 localities in Zone III, 6 in Zone IV, 8 in Zone V and 4 in Zone VI) and 691–932 m (2 localities each in Zones III and VII). The data, when subjected to hierarchial cluster analysis for distinguishing morphometric clusters, showed 12 clusters in 0.61–296 m, 7 in 316–595 m and 1 in 691–932 m. The means and standard deviations of 16 morphometric characters for all localities are presented in Table 13.2. In principal component analysis, the communalities are estimates of the variance in each variable accounted for by

TABLE 13.1

Means and Standard Deviation of Morphometric Characters *Apis florea* Bees from All Seven Agroclimatic Zones

	Krishna–Godavari		North Coastal		Southern		North Telangana		South Telangana		Scarce Rainfall		High Altitudes & Tribal Areas	
Character	**Mean**	**SD**	**Mean**	**SD**	**Mean**	**SD**	**Mean**	**SD**	**Mean**	**SD**	**Mean**	**SD**	**Mean**	**SD**
L:W	0.61	0.03	0.62	0.05	0.65	0.01	0.67	0.03	0.68	0.04	0.65	0.01	0.65	0.02
PL	2.05	0.08	2.08	0.15	2.13	0.08	2.14	0.11	2.14	0.05	2.14	0.14	2.11	0.11
DBO	0.30	0.01	0.30	0.01	0.29	0.02	0.28	0.01	0.29	0.02	0.28	0.03	0.28	0.02
D00D	0.19	0.01	0.20	0.01	0.21	0.01	0.19	0.01	0.19	0.01	0.20	0.01	0.20	0.02
AL	2.19	0.05	2.20	0.04	2.23	0.04	2.28	0.09	2.23	0.05	2.23	0.04	2.21	0.10
FL	1.12	0.03	1.11	0.04	1.14	0.07	1.15	0.03	1.14	0.05	1.12	0.02	1.12	0.04
TL	1.53	0.07	1.50	0.09	1.58	0.11	1.56	0.04	1.56	0.05	1.54	0.04	1.53	0.06
ML:MW	2.21	0.15	2.24	0.16	2.11	0.09	2.23	0.17	2.17	0.11	2.11	0.07	2.17	0.09
FWL:FWW	2.75	0.08	2.80	0.13	2.46	0.04	2.46	0.04	2.46	0.05	2.43	0.04	2.64	0.15
CI a:b	3.08	0.23	3.03	0.20	3.07	0.30	3.17	0.22	3.15	0.23	3.18	0.26	3.14	0.26
HAM	11.11	0.49	11.13	0.40	11.26	0.21	11.23	0.54	11.49	0.30	10.84	0.34	11.10	0.25
3rd TL	1.05	0.03	1.02	0.03	1.02	0.03	1.01	0.02	1.02	0.02	0.99	0.02	1.01	0.03
3rd SL:SW	0.70	0.01	0.70	0.01	0.70	0.01	0.69	0.01	0.70	0.01	0.69	0.00	0.70	0.02
LWM:WWM	0.55	0.01	0.56	0.03	0.58	0.08	0.55	0.01	0.54	0.01	0.55	0.01	0.54	0.02
DBWP	0.13	0.02	0.13	0.03	0.09	0.01	0.08	0.01	0.09	0.02	0.09	0.01	0.11	0.04
4th TOMW	0.46	0.04	0.41	0.03	0.37	0.05	0.37	0.01	0.38	0.02	0.37	0.02	0.35	0.04

L:W, Ratio of Head Length and Width; PL, Proboscis Length; DBO, Distance between Two Dorsal Ocelli; D00D, Dorsal Ocello-Ocular Distance; AL, Antennal Length; FL, Femur Length; TL, Tibia Length; ML:MW, Ratio of Metatarsus Length and Width; FWL:FWW, Ratio of Forewing Length and Width; CI a:b, Ratio of Cubital Index(a:b); HAM, No. of Hamuli; 3rd TL, Tergite Length; 3rd SL:SW, Ratio of 3rd Sternite Length and Width; LWM:WWM, Ratio of Wax Mirror Length and Width; DBWP, Distance between Two Wax Plates on 3rd Sternite; 4th TOMW, Lateral Width of 4th Tomentum; SD, Standard Deviation.

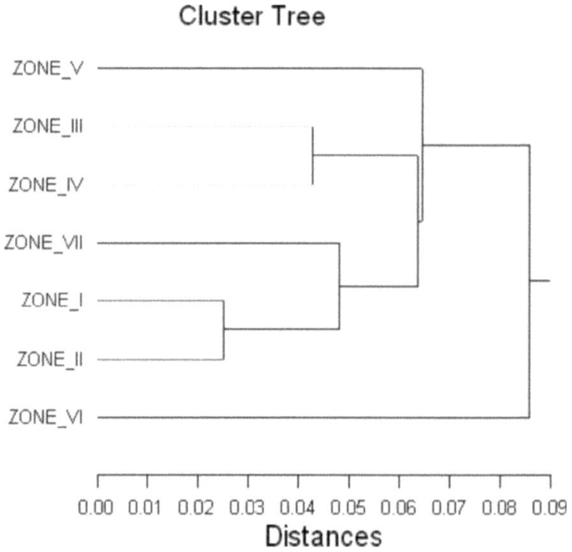

FIGURE 13.1 Characterization of *Apis florea* from all agroclimatic zones and derived hierarchical clustering dendrogram analysis using the single linkage method of squared Euclidean distance on means of 16 morphometric characters.

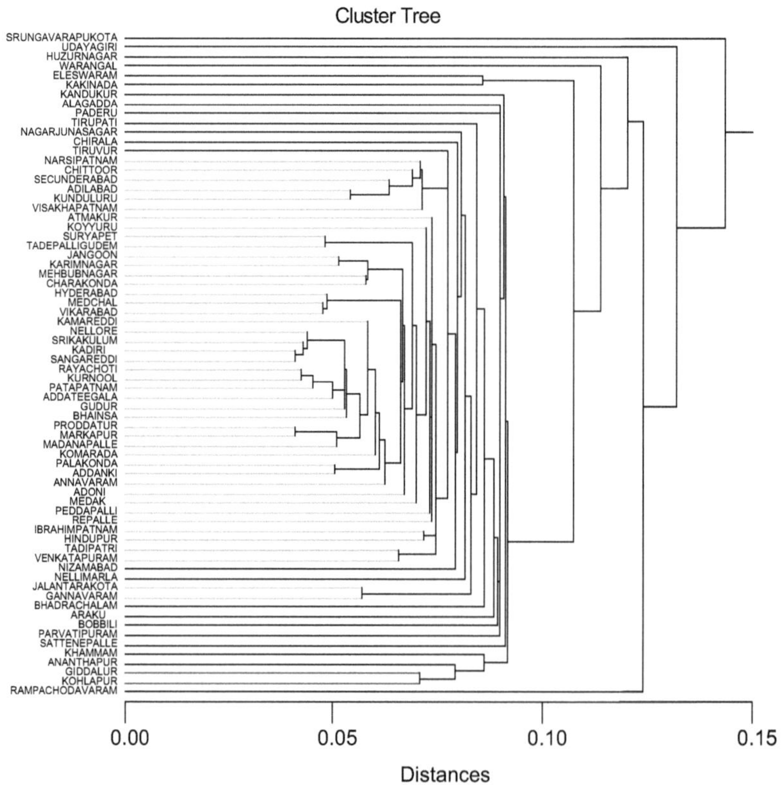

FIGURE 13.2 Characterization of *Apis florea* from 70 localities of seven agroclimatic zones derived from cluster analysis using the single linkage method of squared Euclidean distance on means of 16 morphometric characters.

TABLE 13.2

Descriptive Statistics, Means and Standard Deviations of 16 Morphometric Characters of *Apis florea* from Seventy Localities of Seven Agroclimatic Zones

Descriptive Statistics			
	Mean	**Std. Deviation**	**Analysis N**
L:W	0.64	0.03	70
PL	2.11	0.10	70
DBO	0.28	0.01	70
DOOD	0.19	0.01	70
AL	2.22	0.06	70
FL	1.13	0.04	70
TL	1.54	0.07	70
ML:MW	2.17	0.12	70
FWL:FWW	2.57	0.16	70
CI a:b	3.11	0.23	70
HAM	11.16	0.40	70
3rd TL	1.01	0.02	70
3rd SL:SW	0.69	0.01	70
LWM:WWM	0.55	0.03	70
DBWP	0.10	0.02	70
4th TOMW	0.38	0.04	70

TABLE 13.3

Apis florea Factor Analysis Communalities

Communalities		
	Initial	**Extraction**
L:W	1.000	0.669
PL	1.000	0.634
DBO	1.000	0.698
DOOD	1.000	0.872
AL	1.000	0.702
FL	1.000	0.824
TL	1.000	0.892
ML:MW	1.000	0.699
FWL:FWW	1.000	0.817
CI a:b	1.000	0.569
HAM	1.000	0.424
3rd TL	1.000	0.722
3rd SL:SW	1.000	0.506
LWM:WWM	1.000	0.610
DBWP	1.000	0.821
4th TOMW	1.000	0.742

Extraction Method: Principal Component Analysis.

the components and are always equal to 1.0 for correlation analysis. The communalities obtained for the data fed indicate high values (>0.610) and hence the extracted components represent the variables well (Table 13.3). The data fed for the Scree plot showed six components that are subsequently used for factor analysis (Figure 13.3).

Principal component analysis (PCA) was done using the means of 16 characters from all localities. Table 13.4 gives the data on variance accounted for by each component to the total variance in all of

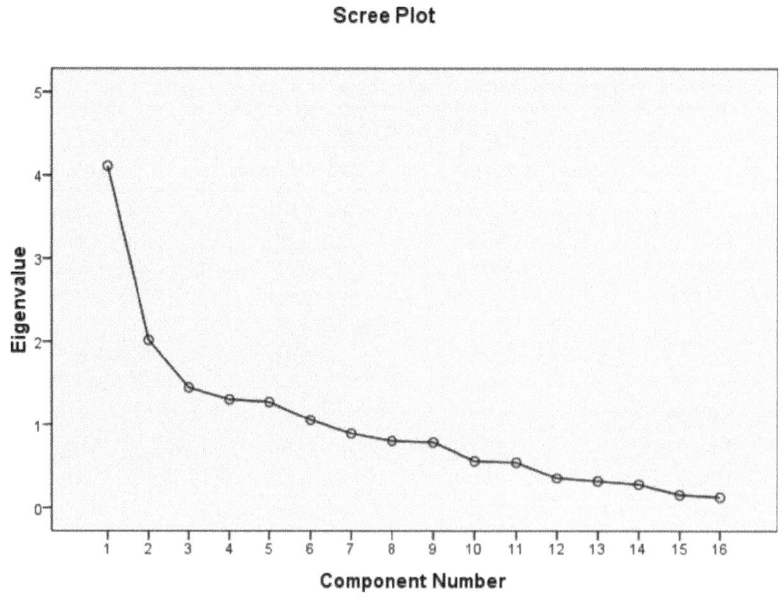

FIGURE 13.3 Multiple factor analysis: principal component analysis scree plot determines the optimal number of components from 16 morphometric characters of *Apis florea* from 70 localities in 7 agroclimatic zones of. The eigenvalue of each component in the initial solution is plotted. The six extracted components are on the steep slope, while those on the shallow slope contribute little to the solution. The last big drop occurs between the fifth and sixth components and hence the first six components are chosen for explanation.

TABLE 13.4

Factor Analysis of Total Variance for 16 Morphometric Characters of *Apis florea* from Seventy Localities in Seven Agroclimatic Zones

Total Variance Explained									
	Initial Eigenvalues			Extraction Sums of Squared Loadings			Rotation Sums of Squared Loadings		
Component	Total	% of Variance	Cumulative %	Total	% of Variance	Cumulative %	Total	% of Variance	Cumulative %
1	4.111	25.693	25.693	4.111	25.693	25.693	2.921	18.258	18.258
2	2.016	12.600	38.292	2.016	12.600	38.292	2.116	13.222	31.480
3	1.446	9.034	47.327	1.446	9.034	47.327	1.992	12.447	43.928
4	1.303	8.144	55.470	1.303	8.144	55.470	1.573	9.834	53.762
5	1.272	7.948	63.418	1.272	7.948	63.418	1.342	8.387	62.149
6	1.056	6.599	70.017	1.056	6.599	70.017	1.259	7.868	70.017
7	0.891	5.570	75.587						
8	0.800	4.998	80.585						
9	0.783	4.892	85.477						
10	0.559	3.495	88.972						
11	0.544	3.397	92.370						
12	0.354	2.214	94.584						
13	0.315	1.968	96.552						
14	0.276	1.725	98.277						
15	0.152	0.950	99.227						
16	0.124	0.773	100.000						

Extraction Method: Principal Component Analysis

TABLE 13.5

Apis florea Factor Analysis of the Rotated Component Matrix Indicating Six Components for Further Analysis for 16 Morphometric Characters of Seventy Localities in Seven Agroclimatic Zones

Rotated Component Matrix						
	Component					
	1	2	3	4	5	6
L:W	−0.746	0.216	−0.169	−0.157	−0.092	−0.067
PL	−0.577	−0.015	0.092	−0.457	0.232	0.173
DBO	0.725	0.047	−0.141	−0.005	0.077	0.380
DOOD	0.118	−0.009	0.070	−0.061	0.077	0.919
AL	−0.618	0.035	−0.396	0.256	0.309	−0.028
FL	−0.107	0.854	0.105	0.176	−0.178	0.100
TL	−0.070	0.895	0.040	−0.289	0.013	0.004
ML:MW	0.027	0.057	0.114	0.824	−0.043	−0.039
FWL:FWW	0.508	−0.288	0.508	0.440	0.137	0.072
CI a:b	0.168	−0.026	−0.716	−0.066	0.050	−0.146
HAM	−0.092	0.487	−0.175	0.234	0.255	−0.166
3rd TL	0.140	0.301	0.603	0.230	0.408	−0.170
3rd SL:SW	0.381	−0.033	0.592	−0.061	0.017	−0.072
LWM:WWM	−0.013	−0.020	0.047	−0.054	0.764	0.143
DBWP	0.530	−0.317	0.477	0.391	0.154	0.189
4th TOMW	0.595	−0.132	0.051	−0.109	0.527	−0.281

Extraction Method: Principal Component Analysis.
Rotation Method: Varimax with Kaiser Normalization.
a. Rotation converged in 10 iterations.

the variables. Table 13.5 gives the data of a rotated component matrix (component loading values) that enables researchers to identify the components for further analysis. In Table 13.6, the component score coefficient matrix data is presented, which enables the selection of the most important variables from 16 characters used for the study. The data was subjected to the Kaiser-Meyer-Olkin Measure of Sampling Adequacy and Bartlett's Test of Sphericity, and the result is shown in Table 13.7, which indicates the suitability of the data.

The factor analysis yielded six factors with eigenvalues greater than 1. The first factor with loading values between 0.50 and 0.74 accounted for 25.69% of variation, which was mainly associated with the ratio of head length and width, proboscis length, dorsal ocelli distance, antennal length, the ratio of forewing length and width, distance between two wax plates on the 3rd sternite and lateral width of the 4th tomentum. The second factor with loading values between 0.85 and 0.89 accounted for 12.60% of variation, which was mainly associated with femur length and tibial length. The third factor with loading values between 0.59 and 0.71 accounted for 9.03% of variation, which was mainly associated with cubital index, 3rd tergite length and ratio of the 3rd sternite length and width. The fourth factor with loading value 0.82 accounted for 8.14% of variation, which was mainly associated with the ratio of metatarsus length and width. The fifth factor with loading value 0.76 accounted for 7.94% of variation, which was mainly associated with the ratio of length and width of a wax mirror. The sixth factor with loading value 0.91 accounted for 6.59% of variation, which was mainly associated with the dorsal ocelli ocular distance. The proboscis length varied from 1.73 to 2.51 mm, hamuli number from 10.25 to 12.15 and cubital index from 2.5 to 3.68.

TABLE 13.6

Factor Analysis Component Score Coefficient Matrix for 16 Morphometric Characters of *Apis florea* from 70 Localities in 7 Agroclimatic Zones

Component Score Coefficient Matrix						
	Component					
	1	2	3	4	5	6
L:W	−0.262	0.016	0.049	−0.017	−0.012	−0.008
PL	−0.265	−0.084	0.193	−0.264	0.210	0.140
DBO	0.339	0.131	−0.245	−0.028	0.033	0.275
DOOD	−0.017	0.024	−0.028	0.014	0.067	0.739
AL	−0.259	−0.049	−0.200	0.303	0.312	0.070
FL	0.031	0.425	0.043	0.134	−0.132	0.109
TL	0.108	0.450	0.048	−0.203	0.014	−0.004
ML:MW	−0.082	0.032	−0.040	0.572	−0.037	0.031
FWL:FWW	0.049	−0.094	0.167	0.202	0.039	0.032
CI a:b	0.213	0.029	−0.468	0.018	0.079	−0.094
HAM	0.022	0.244	−0.151	0.192	0.218	−0.086
3rd TL	−0.040	0.155	0.296	0.060	0.260	−0.150
3rd SL:SW	0.079	0.013	0.320	−0.170	−0.059	−0.125
LWM:WWM	−0.070	−0.008	−0.027	−0.025	0.593	0.128
DBWP	0.058	−0.103	0.143	0.178	0.055	0.124
4th TOMW	0.246	0.006	−0.077	−0.169	0.359	−0.273

Extraction Method: Principal Component Analysis.
Rotation Method: Varimax with Kaiser Normalization.
Component Scores.

TABLE 13.7

Factor Analysis of Kaiser–Meyer–Olkin Measure of Sampling Adequacy (KMO) and Bartletts Test of Sphericity for 16 Morphometric Characters of *Apis florea* from 70 Localities in Seven Agroclimatic Zones

KMO and Bartlett's Test		
Kaiser–Meyer–Olkin Measure of Sampling Adequacy		0.669
Bartlett's Test of Sphericity	Approx. Chi-Square	388.703
	df	120
	Sig.	0.000

Floral Calendar of *Apis florea*

In Table 13.8, the numbers of flowering trees, shrubs, herbs and creepers utilized by bees as forage sources in each month are provided. The data showed that the trees dominated the other life forms during March–June and hence they were considered to be important in providing forage for *A. florea*. The total number of plants in flowering in each month ranged from 77 to 92. Most of the herbs occur as weeds and establish huge populations everywhere during the rainy season; individual plants of some herbs provide copious pollen and nectar daily while some other herbs collectively in a population provide adequate pollen and nectar for *A. florea*. Some mangrove flora also serve as pollen and nectar sources for *A. florea* in estuarine areas of Andhra Pradesh.

TABLE 13.8

Floral Calendar of *Apis florea* in Andhra Pradesh and Telangana

January	February	March	April
Cocos nucifera (T)	*Mangifera indica* (T)	*Alangium salvifolium* (T)	*Alangium salvifolium* (T)
Phoenix sylvestris (T)	*Cocos nucifera* (T)	*Anacardium occidentale* (T)	*Anacardium occidentale* (T)
Ageratum conyzoides (H)	*Phoenix sylvestris* (T)	*Mangifera indica* (T)	*Mangifera indica* (T)
Cosmos bipinnatus (H)	*Cosmos bipinnatus* (H)	*Cocos nucifera* (T)	*Wrightia tinctoria* (T)
Cosmos sulphureus (H)	*Cosmos sulphureus* (H)	*Phoenix sylvestris* (T)	*Cocos nucifera* (T)
Wedelia chinensis (H)	*Wedelia chinensis* (H)	*Cosmos bipinnatus* (H)	*Phoenix sylvestris* (T)
Tridax procumbens (H)	*Tridax procumbens* (H)	*Cosmos sulphureus* (H)	*Cosmos bipinnatus* (H)
Tecoma stans (T)	*Tecoma stans* (T)	*Wedelia chinensis* (H)	*Cosmos sulphureus* (H)
Ceiba pentandra (T)	*Ceiba pentandra* (T)	*Tridax procumbens* (H)	*Wedelia chinensis* (H)
Caesalpinia pulcherrima (S)	*Caesalpinia pulcherrima* (S)	*Careya arborea* (T)	*Tridax procumbens* (H)
Carica papaya (T)	*Carica papaya* (T)	*Tecoma stans* (T)	*Careya arborea* (T)
Evolvulus alsinoides (H)	*Terminalia catappa* (T)	*Kigelia pinnata* (T)	*Tecoma stans* (T)
Ipomoea cairica (C)	*Evolvulus alsinoides* (H)	*Spathodea campanulata* (T)	*Kigelia pinnata* (T)
Ipomoea pes-caprae (C)	*Ipomoea cairica* (C)	*Boswellia ovalifoliolata* (T)	*Spathodea campanulata* (T)
Croton bonplandianum (H)	*Ipomoea pes-caprae* (C)	*Garuga pinnata* (T)	*Boswellia ovalifoliolata* (T)
Jatropha curcas (S)	*Croton bonplandianum* (H)	*Caesalpinia pulcherrima* (S)	*Garuga pinnata* (T)
Jatropha pandurifolia (S)	*Jatropha curcas* (S)	*Bauhinia racemosa* (T)	*Caesalpinia pulcherrima* (S)
Gliricidia sepium (T)	*Jatropha pandurifolia* (S)	*Peltophorum pterocarpum* (T)	*Bauhinia racemosa* (T)
Rhynchosia albiflora (C)	*Gliricidia sepium* (T)	*Carica papaya* (T)	*Peltophorum pterocarpum* (T)
Rhynchosia beddomei (S)	*Rhynchosia albiflora* (C)	*Suaeda maritima* (H)	*Tamarindus indica* (T)
Rhynchosia cana (S)	*Rhynchosia cana* (S)	*Suaeda nudiflora* (H)	*Crataeva magna* (T)
Rhynchosia densiflora (S)	*Rhynchosia densiflora* (S)	*Terminalia bellirica* (T)	*Carica papaya* (T)
Rhynchosia heynei (S)	*Rhynchosia heynei* (S)	*Terminalia catappa* (T)	*Suaeda maritima* (H)
Rhynchosia suaveolens (S)	*Rhynchosia suaveolens* (S)	*Evolvulus alsinoides* (H)	*Suaeda nudiflora* (H)
Rhynchosia viscosa (C)	*Rhynchosia viscosa* (C)	*Ipomoea cairica* (C)	*Terminalia bellirica* (T)
Tephrosia purpurea (H)	*Tephrosia purpurea* (H)	*Shorea roxburghii* (T)	*Terminalia catappa* (T)
Anisomeles indica (S)	*Anisomeles indica* (S)	*Diospyros chloroxylon* (T)	*Terminalia pallida* (T)
Anisomeles malabarica (S)	*Anisomeles malabarica* (S)	*Croton bonplandianum* (H)	*Evolvulus alsinoides* (H)
Hyptis suaveolens (S)	*Hyptis suaveolens* (S)	*Jatropha curcas* (S)	*Ipomoea cairica* (C)
Leucas aspera (H)	*Leucas aspera* (H)	*Jatropha pandurifolia* (S)	Diospyros chloroxylon (T)
Plectranthus mollis (H)	*Plectranthus mollis* (H)	*Gliricidia sepium* (T)	*Croton bonplandianum* (H)
Thymus vulgaris (S)	*Thymus vulgaris* (S)	*Pongamia pinnata* (T)	*Jatropha curcas* (S)
Couroupita guianensis (T)	*Couroupita guianensis* (T)	*Pterocarpus santalinus* (T)	*Jatropha pandurifolia* (S)
Malvastrum coromandelicanum (S)	*Malvastrum coromandelicanum* (S)	*Tephrosia purpurea* (H)	*Pongamia pinnata* (T)
Triumfetta pentandra (H)	*Triumfetta pentandra* (H)	*Anisomeles indica* (S)	*Pterocarpus santalinus* (T)
Acacia mellifera (T)	*Acacia mellifera* (T)	*Plectranthus mollis* (H)	*Tephrosia purpurea* (H)
Acacia nilotica (T)	*Acacia nilotica* (T)	*Thymus vulgaris* (S)	*Thymus vulgaris* (S)
Albizia lebbeck (T)	*Acacia sinuata* (T)	*Couroupita guianensis* (T)	*Couroupita guianensis* (T)
Pithecellobium dulce (T)	*Albizia lebbeck* (T)	*Malvastrum coromandelicanum* (S)	*Malvastrum coromandelicanum* (S)
Mimosa pudica (C)	*Pithecellobium dulce* (T)	*Azadirachta indica* (T)	*Azadirachta indica* (T)
Psidium guajava (T)	*Mimosa pudica* (C)	*Walsura trifoliata* (T)	*Walsura trifoliata* (T)
Argemone mexicana (H)	*Moringa oleifera* (T)	*Acacia mellifera* (T)	*Acacia mellifera* (T)
Antigonon leptopus (C)	*Psidium guajava* (T)	*Acacia nilotica* (T)	*Acacia nilotica* (T)
Portulaca oleracea (H)	*Argemone mexicana* (H)	*Acacia sinuata* (C)	*Acacia sinuata* (C)
Punica granatum (T)	*Antigonon leptopus* (C)	*Albizia lebbeck* (T)	*Albizia lebbeck* (T)
Zizyphus mauritiana (T)	*Portulaca oleracea* (H)	*Pithecellobium dulce* (T)	*Mimosa pudica* (C)

(*Continued*)

TABLE 13.8 (*Continued*)
Floral Calendar of *Apis florea* in Andhra Pradesh and Telangana

January	February	March	April
Zizyphus oenoplia (S)	*Punica granatum* (T)	*Mimosa pudica* (C)	*Moringa oleifera* (T)
Bruguiera gymnorrhiza (T)	*Zizyphus mauritiana* (T)	*Moringa oleifera* (T)	*Psidium guajava* (T)
Bruguiera cylindrica (T)	*Zizyphus oenoplia* (S)	*Psidium guajava* (T)	*Syzygium alternifolium* (T)
Ceriops decandra (T)	*Bruguiera gymnorrhiza* (T)	*Syzygium alternifolium* (T)	*Syzygium cuminii* (T)
Ceriops tagal (T)	*Bruguiera cylindrica* (T)	*Syzygium cuminii* (T)	*Argemone mexicana* (H)
Rosa centifolia (S)	*Ceriops decandra* (T)	*Argemone mexicana* (H)	*Antigonon leptopus* (C)
Rosa sp. (S)	*Ceriops tagal* (T)	*Antigonon leptopus* (C)	*Portulaca oleracea* (H)
Hamelia patens (S)	*Rosa centifolia* (S)	*Portulaca oleracea* (H)	*Punica granatum* (T)
Mitragyna parviflora (T)	*Rosa* sp. (S)	*Punica granatum* (T)	*Zizyphus mauritiana* (T)
Tarenna asiatica (S)	*Hamelia patens* (S)	*Zizyphus mauritiana* (T)	*Zizyphus oenoplia* (S)
Santalum album (T)	*Mitragyna parviflora* (T)	*Zizyphus oenoplia* (S)	*Bruguiera gymnorrhiza* (T)
Sapindus emarginatus (T)	*Tarenna asiatica* (S)	*Bruguiera gymnorrhiza* (T)	*Bruguiera cylindrica* (T)
Solanum trilobatum (C)	*Santalum album* (T)	*Bruguiera cylindrica* (T)	*Ceriops decandra* (T)
Sterculia foetida (T)	*Madhuca longifolia* (T)	*Ceriops decandra* (T)	*Rosa centifolia* (S)
Sterculia urens (T)	*Mimusops elengi* (T)	*Rosa centifolia* (S)	*Rosa* sp. (S)
Clerodendrum inerme (S)	*Sterculia foetida* (T)	*Rosa* sp. (S)	*Hamelia patens* (S)
Lantana camara (S)	*Sterculia urens* (T)	*Hamelia patens* (S)	*Mitragyna parviflora* (T)
Duranta erecta (S)	*Clerodendrum inerme* (S)	*Mitragyna parviflora* (T)	*Morinda tomentosa* (T)
Vitex negundo (S)	*Lantana camara* (S)	*Tarenna asiatica* (S)	*Pavetta tomentosa* (S)
Achras zapota (T)	*Duranta erecta* (S)	*Wendlandia glabrata* (T)	*Tarenna asiatica* (S)
Dalbergia latifolia (T)	*Gmelina arborea* (T)	*Wendlandia tinctoria* (T)	*Wendlandia glabrata* (T)
Tecomaria capensis (T)	*Vitex negundo* (S)	*Madhuca longifolia* (T)	*Wendlandia tinctoria* (T)
Glycosmosis pentaphylla (T)	*Achras zapota* (T)	*Mimusops elengi* (T)	*Murraya koenigii* (T)
Tagetes erecta (S)	*Dalbergia latifolia* (T)	*Sterculia foetida* (T)	*Madhuca longifolia* (T)
Cestrum diurnum (S)	*Tecomaria capensis* (T)	*Clerodendrum inerme* (S)	*Mimusops elengi* (T)
Corallocarpus epigaeus (H)	*Prosopis juliflora* (T)	*Lantana camara* (S)	*Lantana camara* (S)
Muntingia calabura (T)	*Glycosmosis pentaphylla* (T)	*Duranta erecta* (S)	*Clerodendrum inerme* (S)
Dahlia sp. (H)	*Cestrum diurnum* (S)	*Gmelina arborea* (T)	*Gmelina arborea* (T)
Zinnia sp. (H)	*Corallocarpus epigaeus* (H)	*Gmelina asiatica* (S)	*Gmelina asiatica* (S)
Capsicum annuum (H)	*Muntingia calabura* (T)	*Vitex negundo* (S)	*Vitex negundo* (S)
Syndrella nodiflora (H)	*Dahlia* sp. (H)	*Achras zapota* (T)	*Achras zapota* (T)
	Zinnia sp. (H)	*Dalbergia latifolia* (T)	*Dalbergia latifolia* (T)
	Capsicum annuum (H)	*Tecomaria capensis* (T)	*Tecomaria capensis* (T)
	Syndrella nodiflora (H)	*Prosopis juliflora* (T)	*Calliandra haematocephala* (S)
		Glycosmosis pentaphylla (T)	*Corallocarpus epigaeus* (H)
		Cestrum diurnum (S)	*Muntingia calabura* (T)
		Corallocarpus epigaeus (H)	*Dahlia* sp. (H)
		Muntingia calabura (T)	*Zinnia* sp. (H)
		Dahlia sp. (H)	*Capsicum annuum* (H)
		Zinnia sp. (H)	
		Capsicum annuum (H)	
		Syndrella nodiflora (H)	

May	June	July	August
Sesuvium portulacastrum (H)	*Agave americana* (S)	*Justicia procumbens* (H)	*Asystasia gangetica* (H)
Alangium salvifolium (T)	*Sesuvium portulacastrum* (H)	*Agave americana* (S)	*Justicia procumbens* (H)
Anacardium occidentale (T)	*Buchanania lanzan* (T)	*Sesuvium portulacastrum* (H)	*Agave americana* (S)
Mangifera indica (T)	*Buchanania axillaris* (T)	*Buchanania lanzan* (T)	*Buchanania lanzan* (T)
Wrightia tinctoria (T)	*Semecarpus anacardium* (T)	*Buchanania axillaris* (T)	*Buchanania axillaris* (T)

(*Continued*)

TABLE 13.8 (*Continued*)

Floral Calendar of *Apis florea* in Andhra Pradesh and Telangana

May	June	July	August
Cocos nucifera (T)	*Wrightia tinctoria* (T)	*Semecarpus anacardium* (T)	*Semecarpus anacardium* (T)
Phoenix sylvestris (T)	*Cocos nucifera* (T)	*Cocos nucifera* (T)	*Cocos nucifera* (T)
Cosmos bipinnatus (H)	*Cosmos bipinnatus* (H)	*Cosmos bipinnatus* (H)	*Ageratum conyzoides* (H)
Cosmos sulphureus (H)	*Cosmos sulphureus* (H)	*Cosmos sulphureus* (H)	*Cosmos bipinnatus* (H)
Wedelia chinensis (H)	*Wedelia chinensis* (H)	*Wedelia chinensis* (H)	*Cosmos sulphureus* (H)
Tridax procumbens (H)	*Tridax procumbens* (H)	*Tridax procumbens* (H)	*Wedelia chinensis* (H)
Careya arborea (T)	*Avicennia officinalis* (T)	*Avicennia officinalis* (T)	*Tridax procumbens* (H)
Tecoma stans (T)	*Avicennia marina* (T)	*Avicennia marina* (T)	*Avicennia officinalis* (T)
Kigelia pinnata (T)	*Avicennia alba* (T)	*Avicennia alba* (T)	*Avicennia marina* (T)
Spathodea campanulata (T)	*Tecoma stans* (T)	*Tecoma stans* (T)	*Avicennia alba* (T)
Caesalpinia pulcherrima (S)	*Kigelia pinnata* (T)	*Spathodea campanulata* (T)	*Tecoma stans* (T)
Bauhinia racemosa (T)	*Spathodea campanulata* (T)	*Caesalpinia coriaria* (T)	*Millingtonia hortensis* (T)
Peltophorum pterocarpum (T)	*Caesalpinia crista* (S)	*Caesalpinia crista* (S)	*Caesalpinia coriaria* (T)
Tamarindus indica (T)	*Caesalpinia pulcherrima* (S)	*Caesalpinia pulcherrima* (S)	*Caesalpinia crista* (S)
Crataeva magna (T)	*Bauhinia racemosa* (T)	*Tamarindus indica* (T)	*Caesalpinia pulcherrima* (S)
Carica papaya (T)	*Peltophorum pterocarpum* (T)	*Cleome chelidonii* (H)	*Tamarindus indica* (T)
Suaeda maritima (H)	*Tamarindus indica* (T)	*Cleome viscosa* (H)	*Cleome viscosa* (H)
Suaeda nudiflora (H)	*Cleome chelidonii* (H)	*Crataeva magna* (T)	*Carica papaya* (T)
Terminalia bellirica (T)	*Cleome viscosa* (H)	*Carica papaya* (T)	*Lumnitzera racemosa* (T)
Terminalia catappa (T)	*Crataeva magna* (T)	*Lumnitzera racemosa* (T)	*Terminalia catappa* (T)
Terminalia pallida (T)	*Carica papaya* (T)	*Terminalia catappa* (T)	*Evolvulus alsinoides* (H)
Evolvulus alsinoides (H)	*Terminalia bellirica* (T)	*Terminalia tomentosa* (T)	*Ipomoea cairica* (C)
Ipomoea cairica (C)	*Terminalia catappa* (T)	*Evolvulus alsinoides* (H)	*Ipomoea pes-caprae* (C)
Croton bonplandianum (H)	*Terminalia pallida* (T)	*Ipomoea cairica* (C)	*Ipomoea tuba* (C)
Jatropha curcas (S)	*Terminalia tomentosa* (T)	*Croton bonplandianum* (H)	*Croton bonplandianum* (H)
Jatropha pandurifolia (S)	*Evolvulus alsinoides* (H)	*Excoecaria agallocha* (T)	*Excoecaria agallocha* (T)
Pongamia pinnata (T)	*Ipomoea cairica* (C)	*Jatropha curcas* (S)	*Jatropha curcas* (S)
Pterocarpus santalinus (T)	*Croton bonplandianum* (H)	*Jatropha gossypiifolia* (S)	*Jatropha gossypiifolia* (S)
Tephrosia purpurea (H)	*Excoecaria agallocha* (T)	*Jatropha pandurifolia* (S)	*Jatropha pandurifolia* (S)
Thymus vulgaris (S)	*Jatropha curcas* (S)	*Tephrosia purpurea* (H)	*Tephrosia purpurea* (H)
Couroupita guianensis (T)	*Jatropha pandurifolia* (S)	*Ocimum americanum* (H)	*Ocimum americanum* (H)
Malvastrum coromandelicanum (S)	*Pongamia pinnata* (T)	*Ocimum basilicum* (H)	*Ocimum basilicum* (H)
Walsura trifoliata (T)	*Tephrosia purpurea* (H)	*Thymus vulgaris* (S)	*Thymus vulgaris* (S)
Acacia mellifera (T)	*Thymus vulgaris* (S)	*Couroupita guianensis* (T)	*Couroupita guianensis* (T)
Acacia nilotica (T)	*Couroupita guianensis* (T)	*Hugonia mystax* (S)	*Hugonia mystax* (S)
Albizia lebbeck (T)	*Hugonia mystax* (S)	*Hibiscus tiliaceus* (T)	*Malvastrum coromandelicanum* (S)
Mimosa pudica (C)	*Hibiscus tiliaceus* (T)	*Malvastrum coromandelicanum* (S)	*Sida acuta* (S)
Moringa oleifera (T)	*Malvastrum coromandelicanum* (S)	*Sida acuta* (S)	*Sida cordifolia* (H)
Psidium guajava (T)	*Thespesia populneoides* (T)	*Sida cordifolia* (H)	*Triumfetta pentandra* (H)
Syzygium alternifolium (T)	*Acacia mellifera* (T)	*Thespesia populneoides* (T)	*Acacia caesia* (S)
Syzygium cuminii (T)	*Acacia nilotica* (T)	*Acacia caesia* (S)	*Acacia farnesiana* (S)
Argemone mexicana (H)	*Albizia lebbeck* (T)	*Acacia mellifera* (T)	*Acacia mellifera* (T)
Antigonon leptopus (C)	*Psidium guajava* (T)	*Acacia nilotica* (T)	*Acacia nilotica* (T)
Portulaca oleracea (H)	*Syzygium alternifolium* (T)	*Albizia lebbeck* (T)	*Albizia lebbeck* (T)

(*Continued*)

TABLE 13.8 (*Continued*)

Floral Calendar of *Apis florea* in Andhra Pradesh and Telangana

May	June	July	August
Punica granatum (H)	*Syzygium cuminii* (T)	*Aegiceras corniculatus* (T)	*Aegiceras corniculatus* (T)
Zizyphus mauritiana (T)	*Argemone mexicana* (H)	*Psidium guajava* (T)	*Psidium guajava* (T)
Zizyphus oenoplia (S)	*Antigonon leptopus* (C)	*Syzygium cuminii* (T)	*Syzygium cuminii* (T)
Bruguiera gymnorrhiza (T)	*Portulaca oleracea* (H)	*Argemone mexicana* (H)	*Argemone mexicana* (H)
Bruguiera cylindrica (T)	*Punica granatum* (T)	*Antigonon leptopus* (C)	*Pedalium murex* (H)
Ceriops decandra (T)	*Zizyphus mauritiana* (T)	*Portulaca oleracea* (H)	*Antigonon leptopus* (C)
Rosa centifolia (S)	*Zizyphus oenoplia* (S)	*Punica granatum* (T)	*Portulaca oleracea* (H)
Rosa sp. (S)	*Bruguiera gymnorrhiza* (T)	*Zizyphus mauritiana* (T)	*Punica granatum* (T)
Hamelia patens (S)	*Bruguiera cylindrica* (T)	*Zizyphus oenoplia* (S)	*Zizyphus mauritiana* (T)
Mitragyna parviflora (T)	*Ceriops decandra* (T)	*Bruguiera gymnorrhiza* (T)	*Zizyphus oenoplia* (S)
Morinda tomentosa (T)	*Rosa centifolia* (S)	*Bruguiera cylindrica* (T)	*Bruguiera gymnorrhiza* (T)
Pavetta tomentosa (S)	*Rosa* sp. (S)	*Ceriops decandra* (T)	*Bruguiera cylindrica* (T)
Tarenna asiatica (S)	*Hamelia patens* (S)	*Rosa centifolia* (S)	*Ceriops decandra* (T)
Wendlandia glabrata (T)	*Mitragyna parviflora* (T)	*Rosa* sp. (S)	*Rosa centifolia* (S)
Wendlandia tinctoria (T)	*Morinda tomentosa* (T)	*Hamelia patens* (S)	*Rosa* sp. (S)
Murraya koenigii (T)	*Pavetta tomentosa* (S)	*Mitragyna parviflora* (T)	*Borreria hispida* (H)
Santalum album (T)	*Tarenna asiatica* (S)	*Morinda tomentosa* (T)	*Hamelia patens* (S)
Madhuca longifolia (T)	*Murraya koenigii* (T)	*Pavetta tomentosa* (S)	*Mitragyna parviflora* (T)
Gmelina arborea (T)	*Santalum album* (T)	*Tarenna asiatica* (S)	*Morinda tomentosa* (T)
Gmelina asiatica (S)	*Duranta erecta* (S)	*Cardiospermum canescens* (C)	*Pavetta tomentosa* (S)
Lantana camara (S)	*Lantana camara* (S)	*Cardiospermum halicacabum* (C)	*Tarenna asiatica* (S)
Tectona grandis (T)	*Tectona grandis* (T)	*Solanum melongena* (H)	*Santalum album* (T)
Achras zapota (T)	*Vitex altissima* (T)	*Sonneratia apetala* (T)	*Allophylus serratus* (S)
Dalbergia latifolia (T)	*Tribulus terrestris* (H)	*Solanum trilobatum* (C)	*Cardiospermum canescens* (C)
Tecomaria capensis (T)	*Achras zapota* (T)	*Helicteres isora* (T)	*Cardiospermum halicacabum* (C)
Calliandra haematocephala (S)	*Dalbergia latifolia* (T)	*Duranta erecta* (S)	*Solanum melongena* (H)
Muntingia calabura (T)	*Tecomaria capensis* (T)	*Lantana camara* (S)	*Sonneratia apetala* (T)
Dahlia sp. (H)	*Muntingia calabura* (T)	*Tectona grandis* (T)	*Solanum trilobatum* (C)
Zinnia sp. (H)	*Dahlia* sp. (H)	*Vitex altissima* (T)	*Helicteres isora* (T)
Capsicum annuum (H)	*Zinnia* sp. (H)	*Tribulus terrestris* (H)	*Lantana camara* (S)
	Capsicum annuum (H)	*Achras zapota* (T)	*Duranta erecta* (S)
		Dalbergia latifolia (T)	*Vitex altissima* (T)
		Tecomaria capensis (T)	*Tribulus terrestris* (H)
		Muntingia calabura (T)	*Achras zapota* (T)
		Dahlia sp. (H)	*Dalbergia latifolia* (T)
		Zinnia sp. (H)	*Tecomaria capensis* (T)
		Capsicum annuum (H)	*Prosopis juliflora* (T)
			Glycosmosis pentaphylla (T)
			Muntingia calabura (T)
			Dahlia sp. (H)
			Zinnia sp. (H)
			Capsicum annuum (H)
			Achyranthes aspera (H)
			Syndrella nodiflora (H)

(*Continued*)

TABLE 13.8 (*Continued*)

Floral Calendar of *Apis florea* in Andhra Pradesh and Telangana

September	October	November	December
Asystasia gangetica (H)	*Asystasia gangetica* (H)	*Asystasia gangetica* (H)	*Asystasia gangetica* (H)
Justicia procumbens (H)	*Justicia procumbens* (H)	*Justicia procumbens* (H)	*Cocos nucifera* (T)
Agave americana (S)	*Agave americana* (S)	*Cocos nucifera* (T)	*Ageratum conyzoides* (H)
Buchanania lanzan (T)	*Cocos nucifera* (T)	*Ageratum conyzoides* (H)	*Chromolaena odorata* (S)
Buchanania axillaris (T)	*Ageratum conyzoides* (H)	*Chromolaena odorata* (S)	*Cosmos bipinnatus* (H)
Semecarpus anacardium (T)	*Chromolaena odorata* (S)	*Cosmos bipinnatus* (H)	*Cosmos sulphureus* (H)
Cocos nucifera (T)	*Cosmos bipinnatus* (H)	*Cosmos sulphureus* (H)	*Wedelia chinensis* (H)
Cosmos bipinnatus (H)	*Cosmos sulphureus* (H)	*Wedelia chinensis* (H)	*Tridax procumbens* (H)
Cosmos sulphureus (H)	*Wedelia chinensis* (H)	*Tridax procumbens* (H)	*Tecoma stans* (T)
Wedelia chinensis (H)	*Tridax procumbens* (H)	*Tecoma stans* (T)	*Millingtonia hortensis* (T)
Tridax procumbens (H)	*Tecoma stans* (T)	*Millingtonia hortensis* (T)	*Ceiba pentandra* (T)
Tecoma stans (T)	*Millingtonia hortensis* (T)	*Caesalpinia crista* (S)	*Caesalpinia pulcherrima* (S)
Millingtonia hortensis (T)	*Caesalpinia crista* (S)	*Caesalpinia pulcherrima* (S)	*Bauhinia purpurea* (T)
Caesalpinia coriaria (T)	*Caesalpinia pulcherrima* (S)	*Bauhinia purpurea* (T)	*Peltophorum pterocarpum* (T)
Caesalpinia crista (S)	*Bauhinia purpurea* (T)	*Peltophorum pterocarpum* (T)	*Carica papaya* (T)
Caesalpinia pulcherrima (S)	*Peltophorum pterocarpum* (T)	*Carica papaya* (T)	*Evolvulus alsinoides* (H)
Peltophorum pterocarpum (T)	*Cleome viscosa* (H)	*Evolvulus alsinoides* (H)	*Ipomoea cairica* (C)
Cleome viscosa (H)	*Carica papaya* (T)	*Ipomoea cairica* (C)	*Ipomoea pes-caprae* (C)
Carica papaya (T)	*Evolvulus alsinoides* (H)	*Ipomoea pes-caprae* (C)	*Croton bonplandianum* (H)
Evolvulus alsinoides (H)	*Ipomoea cairica* (C)	*Croton bonplandianum* (H)	*Jatropha curcas* (S)
Ipomoea cairica (C)	*Ipomoea pes-caprae* (C)	*Jatropha curcas* (S)	*Jatropha pandurifolia* (S)
Ipomoea pes-caprae (C)	*Croton bonplandianum* (H)	*Jatropha pandurifolia* (S)	*Rhynchosia albiflora* (C)
Ipomoea tuba (C)	*Jatropha curcas* (S)	*Derris trifoliata* (S)	*Rhynchosia beddomei* (S)
Croton bonplandianum (H)	*Jatropha pandurifolia* (S)	*Rhynchosia beddomei* (S)	*Rhynchosia densiflora* (S)
Excoecaria agallocha (T)	*Derris trifoliata* (S)	*Rhynchosia heynei* (S)	*Rhynchosia heynei* (S)
Jatropha curcas (S)	*Rhynchosia heynei* (S)	*Rhynchosia rothii* (S)	*Rhynchosia rothii* (S)
Jatropha pandurifolia (S)	*Rhynchosia rothii* (S)	*Rhynchosia viscosa* (C)	*Rhynchosia viscose* (C)
Rhynchosia rothii (S)	*Tephrosia purpurea* (H)	*Tephrosia purpurea* (H)	*Tephrosia purpurea* (H)
Tephrosia purpurea (H)	*Anisomeles malabarica* (S)	*Anisomeles indica* (S)	*Anisomeles indica* (S)
Ocimum americanum (H)	*Hyptis suaveolens* (S)	*Anisomeles malabarica* (S)	*Anisomeles malabarica* (S)
Ocimum basilicum (H)	*Leucas aspera* (H)	*Hyptis suaveolens* (S)	*Hyptis suaveolens* (S)
Thymus vulgaris (S)	*Ocimum americanum* (H)	*Leucas aspera* (H)	*Leucas aspera* (H)
Couroupita guianensis (T)	*Ocimum basilicum* (H)	*Ocimum americanum* (H)	*Plectranthus mollis* (H)
Malachra capitata (H)	*Thymus vulgaris* (S)	*Ocimum basilicum* (H)	*Thymus vulgaris* (S)
Malvastrum coromandelicanum (S)	*Couroupita guianensis* (T)	*Plectranthus mollis* (H)	*Couroupita guianensis* (T)
Pavonia zeylanica (S)	*Malachra capitata* (H)	*Malvastrum coromandelicanum* (S)	*Malvastrum coromandelicanum* (S)
Sida acuta (S)	*Malvastrum coromandelicanum* (S)	*Pavonia zeylanica* (S)	*Pavonia zeylanica* (S)
Sida cordifolia (H)	*Pavonia zeylanica* (S)	*Sida acuta* (S)	*Sida cordifolia* (H)
Triumfetta pentandra (H)	*Sida acuta* (S)	*Sida cordifolia* (H)	*Triumfetta pentandra* (H)
Acacia caesia (S)	*Sida cordifolia* (H)	*Triumfetta pentandra* (H)	*Acacia mellifera* (T)
Acacia farnesiana (T)	*Triumfetta pentandra* (H)	*Acacia mellifera* (T)	*Acacia nilotica* (T)
Acacia mellifera (T)	*Acacia farnesiana* (S)	*Acacia nilotica* (T)	*Albizia lebbeck* (T)
Acacia nilotica (T)	*Acacia mellifera* (T)	*Albizia lebbeck* (T)	*Pithecellobium dulce* (T)
Albizia lebbeck (T)	*Acacia nilotica* (T)	*Pithecellobium dulce* (T)	*Mimosa pudica* (C)
Moringa oleifera (T)	*Albizia lebbeck* (T)	*Mimosa pudica* (C)	*Psidium guajava* (T)
Psidium guajava (T)	*Mimosa pudica* (C)	*Moringa oleifera* (T)	*Argemone mexicana* (H)

(*Continued*)

TABLE 13.8 (*Continued*)

Floral Calendar of *Apis florea* in Andhra Pradesh and Telangana

September	October	November	December
Argemone mexicana (H)	*Moringa oleifera* (T)	*Psidium guajava* (T)	*Antigonon leptopus* (C)
Pedalium murex (H)	*Psidium guajava* (T)	*Argemone mexicana* (H)	*Portulaca oleracea* (H)
Antigonon leptopus (C)	*Argemone mexicana* (H)	*Antigonon leptopus* (C)	*Punica granatum* (T)
Portulaca oleracea (H)	*Pedalium murex* (H)	*Portulaca oleracea* (H)	*Zizyphus mauritiana* (T)
Punica granatum (T)	*Antigonon leptopus* (C)	*Punica granatum* (T)	*Zizyphus oenoplia* (S)
Zizyphus mauritiana (T)	*Portulaca oleracea* (H)	*Zizyphus mauritiana* (T)	*Bruguiera gymnorrhiza* (T)
Zizyphus oenoplia (S)	*Punica granatum* (T)	*Zizyphus oenoplia* (S)	*Bruguiera cylindrica* (T)
Bruguiera gymnorrhiza (T)	*Zizyphus mauritiana* (T)	*Bruguiera gymnorrhiza* (T)	*Ceriops decandra* (T)
Bruguiera cylindrical (T)	*Zizyphus oenoplia* (S)	*Bruguiera cylindrica* (T)	*Ceriops tagal* (T)
Ceriops decandra (T)	*Bruguiera gymnorrhiza* (T)	*Ceriops decandra* (T)	*Rosa centifolia* (S)
Rosa centifolia (S)	*Bruguiera cylindrica* (T)	*Ceriops tagal* (T)	*Rosa* sp. (S)
Rosa sp. (S)	*Ceriops decandra* (T)	*Rosa centifolia* (S)	*Hamelia patens* (S)
Borreria hispida (H)	*Rosa centifolia* (S)	*Rosa* sp. (S)	*Mitragyna parviflora* (T)
Hamelia patens (S)	*Rosa* sp. (S)	*Borreria hispida* (H)	*Tarenna asiatica* (S)
Mitragyna parviflora (T)	*Borreria hispida* (H)	*Hamelia patens* (S)	*Santalum album* (T)
Tarenna asiatica (S)	*Hamelia patens* (S)	*Mitragyna parviflora* (T)	*Sapindus emarginatus* (T)
Santalum album (T)	*Mitragyna parviflora* (T)	*Tarenna asiatica* (S)	*Solanum trilobatum* (C)
Allophylus serratus (S)	*Tarenna asiatica* (S)	*Santalum album* (T)	*Sterculia urens* (T)
Cardiospermum canescens (C)	*Datura metel* (H)	*Sapindus emarginatus* (T)	*Lantana camara* (S)
Cardiospermum halicacabum (C)	*Solanum melongena* (H)	*Datura metel* (H)	*Achras zapota* (T)
Solanum melongena (H)	*Solanum trilobatum (C)*	*Solanum melongena* (H)	*Dalbergia latifolia* (T)
Sonneratia apetala (T)	*Helicteres isora* (T)	*Solanum trilobatum* (C)	*Tecomaria capensis* (T)
Solanum trilobatum (C)	*Duranta erecta* (S)	*Lantana camara* (S)	*Glycosmosis pentaphylla* (T)
Helicteres isora (T)	*Lantana camara* (S)	*Achras zapota* (T)	*Tagetes erecta* (S)
Duranta erecta (S)	*Vitex altissima* (T)	*Dalbergia latifolia* (T)	*Eucalyptus globulus* (T)
Lantana camara (S)	*Tribulus terrestris* (H)	*Tecomaria capensis* (T)	*Cestrum diurnum* (S)
Vitex altissima (T)	*Achras zapota* (T)	*Glycosmosis pentaphylla* (T)	*Corallocarpus epigaeus* (H)
Tribulus terrestris (H)	*Dalbergia latifolia* (T)	*Tagetes erecta* (S)	*Muntingia calabura* (T)
Achras zapota (T)	*Tecomaria capensis* (T)	*Eucalyptus globulus* (T)	*Dahlia* sp. (H)
Dalbergia latifolia (T)	*Glycosmosis pentaphylla* (T)	*Cestrum diurnum* (S)	*Zinnia* sp. (H)
Tecomaria capensis (T)	*Eucalyptus globulus* (T)	*Corallocarpus epigaeus* (H)	*Capsicum annuum* (H)
Prosopis juliflora (T)	*Cestrum diurnum* (S)	*Muntingia calabura* (T)	*Achyranthes aspera* (H)
Glycosmosis pentaphylla (T)	*Muntingia calabura* (T)	*Dahlia* sp. (H)	*Syndrella nodiflora* (H)
Eucalyptus globulus (T)	*Dahlia* sp. (H)	*Zinnia* sp. (H)	
Muntingia calabura (T)	*Zinnia* sp. (H)	*Capsicum annuum* (H)	
Dahlia sp. (H)	*Capsicum annuum* (H)	*Achyranthes aspera* (H)	
Zinnia sp. (H)	*Achyranthes aspera* (H)	*Syndrella nodiflora* (H)	
Capsicum annuum (H)	*Syndrella nodiflora* (H)		
Achyranthes aspera (H)			
Syndrella nodiflora (H)			

Discussion

In this study, sample size of 20 and one numerical morphological characters of *A. florea* have been used for identifying morphoclusters in Andhra Pradesh and Telangana. These characters have been reduced to 16 by taking the ratios of certain characters such as head length and width, metatarsus length and width, forewing length and width, 3rd sternite length and width, and wax mirror length and width. The other measured characters have been taken as obtained. This consolidation has been done to enhance the accuracy in identifying morphoclusters of *A. florea*. The multivariate statistical analysis of the morphometric data of 16 morphological characters included hierarchial cluster analysis and factor analysis. The cluster analysis performed for the means of the 16 morphological characteristics of all agroclimatic zones indicated two distinct morphoclusters – the first cluster represented Zones I and II followed by Zone VII, while the second represented Zones III and IV. The Zone V populations share morphological characters with both the clusters. The isolated Zone VI populations are attributed to scarce rainfall and limited floral diversity.

The analysis of the morphometric data indicated only one morphocluster in factor scores 1–6. The factor score 1 clustering indicated that the ratio of head length and width, proboscis length, dorsal ocelli distance, antennal length, the ratio of forewing length and width, distance between two wax plates on the 3rd sternite and lateral width of the 4th tomentum are more variable. Ozkan et al. (2009) stated that in Iran, *A. florea* populations vary morphologically at higher and lower latitudes. Bhandari (1983) suggested that *A. florea* populations of hotter, drier and lower latitudes are smaller than those at the higher latitudes. The present study does not show variation in the morphological characters of *A. florea* distributed in different latitude and altitudinal points.

In honeybees, the hamuli are located near the proximal end of the anterior margin of each hind wing and couple it to a fold in the forewing to form a single, functional aerofoil during flight (Saini et al. 1983). The bees fly farther in dry conditions to get forage and hence they need effective wings. In consequence, the bees develop longer forewings and produce more numbers of hamuli on the hind wings as adaptations to fly longer distances (Richards 1949). The hamuli number of *A. florea* is 11.10 in Thailand (Rinderer et al. 1995) and 10.10 to 12.93 in Asia and the Middle East (Hepburn and Radloff 2004). In this study, the hamuli number in *A. florea* ranged from 10.25 to 12.15, indicating the regular distance it flies for forage collection. Ruttner (1988) used the cubital index in morphometric studies; the numbers varied from 2.92 to 3.30 in Iran (Ozkan et al. 2009), 3 in northeastern India to Indo-China (Radloff et al. 2011) and 2.864 in Thailand (Rinderer et al. 1995). In this study, the cubital index of *A. florea* ranged from 2.5 to 3.68, which is relatable to variation in the length and width of its forewing.

The forage plants of *A. florea* indicated that 77 to 92 forage plants flower each month in all agroclimatic zones, but trees serve as important forage sources yearlong especially during March–June when most of the herbaceous shrubs and herbs disappear. The floral calendar prepared for *A. florea* is of great help in selecting the most suitable places to move hives during the dearth period (Kevan 1984).

REFERENCES

Bhandari, V.C. 1983. Biometrical studies on *Apis florea* F. and *Apis dorsata* F. of north-western India. PhD Thesis, Himachal Pradesh University, Simla, India.

Haddad, N., Fuchs, S., Hepburn, H.R. and Radloff, S.E. 2009. *Apis florea* in Jordan: Source of the founder population. *Apidologie* 40: 508–512.

Hepburn, H.R. and Radloff, S.E. 2004. The wing coupling apparatus and the morphometric analysis of honeybee populations. *S. Afr. J. Sci.* 100: 565–570.

Hepburn, H.R., Radloff, S.E., Otis, G.W., Fuchs, S., Verma, L.R., Tan, K., Chaiyawong, T., Tahmasebi, G., and Wongsiri, S. (2005) Apis florea: morphometrics, classification and biogeography. *Apidologie* 36: 359–376.

Kevan, P.G. 1984. Insect pollination of economically important plants of tropical and Subtropical Asia. *Proceedings of the FAO (United Nations) Expert Consult on Beekeeping with* Apis mellifera *in Tropical and Sub-tropical Asia*, Bangkok, Thailand, pp. 203–210.

Kshirsagar, K.K. 1983. Morphometric studies on the Indian hive bee *Apis cerana indica* F. I-morphometric characters useful in identification of intraspecific taxa. *Proceedings of the 2nd International Conference on Apiculture in Tropical Climates*, New Delhi, India, pp. 254–261.

Ozkan, A., Gharlcko, M.M., Ozden, B. and Kandemir, I. 2009. Multivariate morphometric study on *Apis florea* distributed in Iran. *Turk. J. Zool.* 33: 93–102.

Radloff, S.E., Hepburn, H.R. and Engel, M.S. 2011. Chapter 1. The Asian species of *Apis*. In: *Honeybees of Asia*, H.R. Hepburn and S.E. Radloff (Eds.), Springer-Verlag, Berlin, Germany.

Richards, O.W. 1949. The significance of the number of wing hooks in bees and wasps. *Proc. R. Ent. Soc. Lond.* 24: 7–9.

Rinderer, T.E., Oldroyd, B.P., Wongsiri, S., Sylvester, H.A., De Guzman, L.I., Stelzer, J.A. and Riggio, R.M. 1995. A morphological comparison of the dwarf honeybees of southeastern Thailand and Palawan, Philippines. *Apidologie* 26: 387–394.

Roubik, D.W. 1995. *Pollination of Cultivated Plants in the Tropics*. FAO Agricultural Services Bulletin 118. FAO, Rome, Italy.

Ruttner, F. 1988. *Biogeography and Taxonomy of Honeybees*. Springer-Verlag, Berlin, Germany.

Ruttner, F., Mossadegh, M.S. and Kauhausen, D. 1995. Distribution and variation in size of *Apis florea* F. in Iran. *Apidologie* 26: 477–486.

Saini, M.S., Aggarwal, R. and Dhillon, S.S. 1983. Functional morphology of the wing coupling apparatus and its bearing on the taxonomy of genus *Apis* Linnaeus. *Uttar Pradesh J. Zool.* 3: 10–14.

Singh, M.P., Verma, L.R. and Daly, H.V. 1990. Morphometric analysis of the Indian honeybee in the northeast Himalayan Region. *J. Apic. Res.* 29: 3–14.

Tahmasebi, G., Ebadi, R., Tajabadi, N., Akhoundi, M. and Faraji, S. 2002. The effects of geographical and climatic conditions on the morphological variation and separation of Iranian small honeybee (*Apis florea* F.) populations. *J. Sci. Technol. Agric. Nat. Res.* 6: 169–176.

Tan, K., Radloff, S.E., Hepburn, H.R., Yang, M., Zhang, L. and Fan, X. 2007. Environmentally-induced developmental effects on morphometric characters of workers in *Apis cerana* colonies. *Apidologie* 38: 289–295.

Verma, L.R., Mattu, V.K. and Daly, H.V. 1994. Multivariate morphometrics of the Indian honeybee in the northwest Himalayan region. *Apidologie* 15: 203–223.

14

Distribution and Diversity of Apis florea *in Kerala and Its Role in Agriculture Production*

S. Devanesan

CONTENTS

Introduction

First identified in the late eighteenth century, *Apis florea* is unique for its morphology, foraging behaviour and defensive mechanisms such as making a piping noise. *A. florea* has open nests and small colonies (Plate 14.1), which makes it more susceptible to predation than cavity nesters with large numbers of defensive workers. Apart from its small size, simple exposed nests and simplified dance language, the life cycle and behaviour of this species are fairly similar to other species of *Apis*. The bees are excellent pollinators, which gives them an important ecological role in the places they inhabit. These honeybees are important pollinators to various agricultural and horticultural crops of Kerala.

The dwarf/little honeybee *A. florea* is one of the two species of wild honeybees found in Kerala, the other species being the rock bee *Apis dorsata* F. Dwarf bees are the smallest of the true honeybees found in plains up to an altitude of 300–500 m above sea level. (Rarely found at altitudes above 1500 m, the bee is absent north of the Himalayas.) They nest in dense bushes, hedge plants, small trees, corners of hay stalks, generally preferring shaded sites. It is also seen in corners of buildings, wells and on rock cliffs Mishra (1995).

PLATE 14.1

The honeybees are migratory in nature and frequently change their place whenever the colony is disturbed. They often migrate between plains and nearby low hills depending on the availability of a food source. If they are building a new nest near the old one, they salvage the wax from the old nest. Other species of honeybee do not return to the old nest to recoup wax. This behaviour is observed only in this species.

Phylogenetic analysis by Alexander (1991) indicated that (1) the small species (group of *A. florea* Fabricius) constitute the basal branch, i.e., sister group to the rest, and that (2) the giant species (group of *A. dorsata* Fabricius) are the next branch, i.e., sister to (3) the medium-sized species (group of *Apis mellifera* Linnaeus). Engel and Schultz (1997) used an architectural character and numerous larval characters, in addition to adult characters and DNA sequence data, and they reached the same conclusion when total evidence was considered. The members of group (3) construct combs in cavities; therefore, construction of combs in the open, as in other two groups, must be ancestral.

Habitat and Nesting of *A. florea*

A colony usually builds a single, exposed comb. Generally, the bees build nests in the open on tree branches or shrubs (Plate 14.2), but not without camouflage. Most nests are hung from slender branches of trees or shrubs covered with relatively dense foliage, usually from 1 to 8 m above the ground. The comb is broad at the top and serves as a landing place for the foragers. The nest comb of cells is primarily made of wax secreted by the sternal wax glands of workers. The comb is always covered by a curtain of worker bees (Plate 14.3) clinging to each other, often three or four bees deep (~10 mm). About three-quarters of the colony's worker population are employed in forming this living protective curtain of bees. A comb has 13.2 cells per cm^2, and the nest is 20 cm in diameter and contains approximately 3600 cells on each side. The upper part of the comb expands by adjoining honey storage cells to form a crest that surrounds the branch or other object from which the comb is suspended. Adjacent to the rows of honey storage cells is the section of comb, which the workers use for storing pollen. Beneath this band of pollen storage cells is the area where the worker brood is reared (Makinson et al. (2015) Plate 14.4).

PLATE 14.2

PLATE 14.3

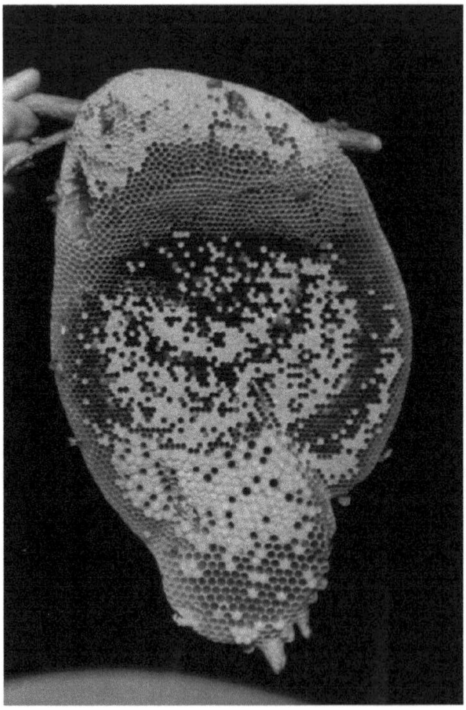

PLATE 14.4

Identifying Features

A. florea is called the dwarf honeybee due to its small size compared to other honeybees. As its name implies, the dwarf honeybee is the smallest species of honeybee, both in the body size of its workers and in the size of its nest. Workers are red-brown in colour (Plates 14.5a and b).

Drones (Plate 14.6) carry a thumb-like bifurcation called the basitarsus, which is located two-thirds along the length of the tibia. The fimbriate lobe of *A. florea* workers has three protrusions, and they sting using two stylet barbs. *A. florea* and its sister species *A. andreniformis* Smith together comprise the subgenus *Micrapis*, and they are the most primitive of the living species of *Apis*, reflected in their small colony size and simple nest construction.

The distinctiveness of the two species *A. florea* and *A. andreniformis* was established unequivocally in the 1990s. *A. florea* is red, and old workers always have a red first abdomen (younger workers are paler in colour, as is the case in giant honeybees); *A. andreniformis* is generally darker in color, and the first segment of abdomen is completely black in old bees.

Distinguishing characteristics of the dwarf honeybees are as follows: **Workers**: Body length is 7–10 mm. Overall colouration is red-brown. Proboscis length is 3.11–3.37 mm, Forewing length is 6.17–6.74 mm, forewing width is 2.12–2.32 mm, hind wing length is 3.17–4.83 mm and hind wing width 1.36 ± 0.04 mm. **Comb dimensions:** Vertical length (from support bottom) is 12.0 ± 3 cm, horizontal width is 16.9 ± 5.3 cm and branch diameter is 0.8 ± 0.07 cm, **Worker cell dimensions:** Depth of cell is 0.93 ± 0.07 cm; width of 10 cells is 2.98 ± 0.15 cm. **Drone cell dimensions:** Depth of cell is 1.33 ± 0.07 cm; width of 10 cells 4.88 ± 0.21 cm. **Queen cell dimensions:** Depth of cell is 1.41 ± 0.15 cm; internal diameter of cell is 0.47 ± 0.09 cm (Viraktamath et al 2013.).

The results of morphometric characters (15) studied from the worker bee samples of *A. florea* collected from different districts of Kerala are given below (Table 14.1):

(a)

(b)

PLATE 14.5

PLATE 14.6

TABLE 14.1

Mean and Standard Deviation of Morphometric Characters of *A. florea* in Kerala

L:W	PL	DBO	DOOD	AL	FL	TL	ML:W	FWL:W	CI	HAM	3rd TL	3rd SL: SW	WPL: W	DBWP
2.47 ± 0.23	2.88 ± 0.17	0.37 ± 0.01	0.27 ± 0.01	2.69 ± 0.07	1.60 ± 0.04	1.99 ± 0.11	2.32 ± 0.18	2.96 ± 0.05	3.10 ± 0.34	11.46 ± 0.74	1.29 ± 0.03	0.71 ± 0.01	0.58 ± 0.04	0.27 ± 0.03

L:W, Ratio of Head length and Width; PL, Proboscis Length; DBO, Distance between Two Dorsal Ocelli; DOOD, Dorsal Ocello-ocular Distance; AL, Antennal Length; FL, Femur Length; TL, Tibia Length; ML:MW, Ratio of Metatarsus Length and Width; FWL:FWW, Ratio of Forewing Length and Width; CI a:b, Ratio of Cubital Index (a:b); HAM, No of Hamuli; TL, Tergite Length; 3rd SL: SW, Ratio of 3rd Sternite Length and Width; 3rd LWM:WWM, Ratio of Wax Mirror Length and Width; DBWP, Distance between Two Wax Plates on 3rd Sternite (Plates 14.7–14.15).

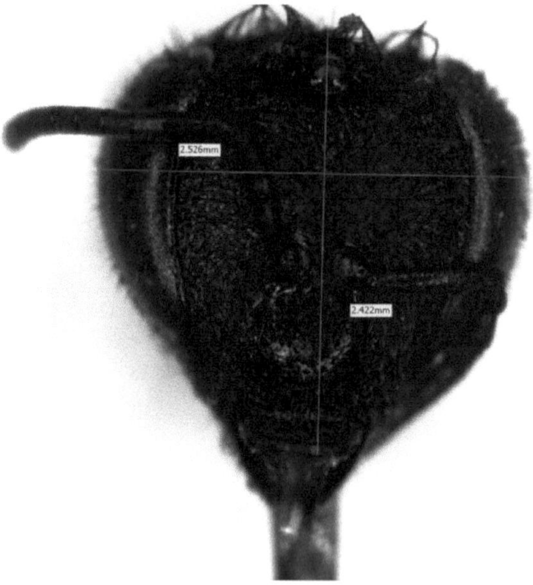

PLATE 14.7

PLATE 14.8

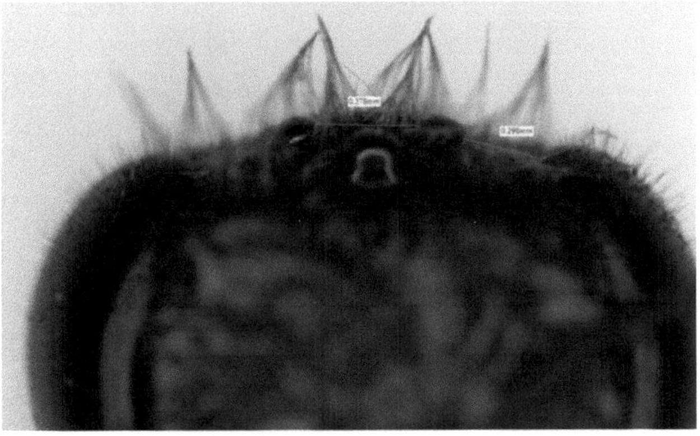

PLATE 14.9

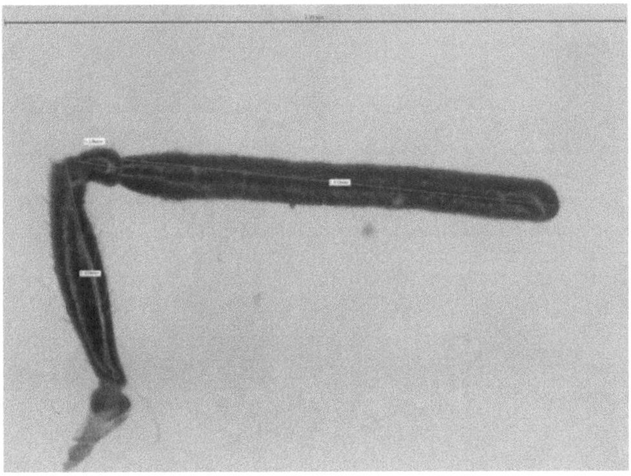

PLATE 14.10

PLATE 14.11

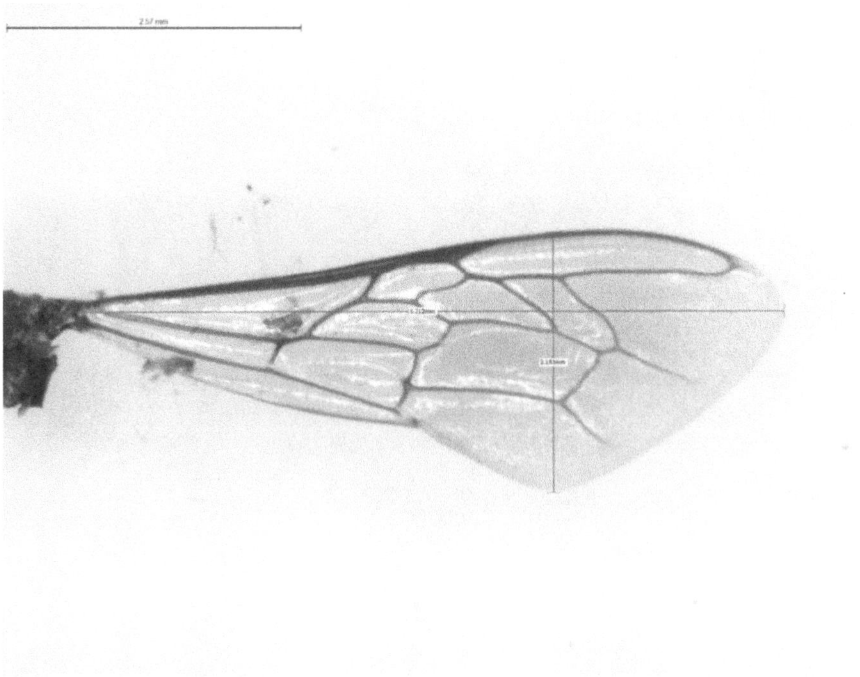

PLATE 14.12

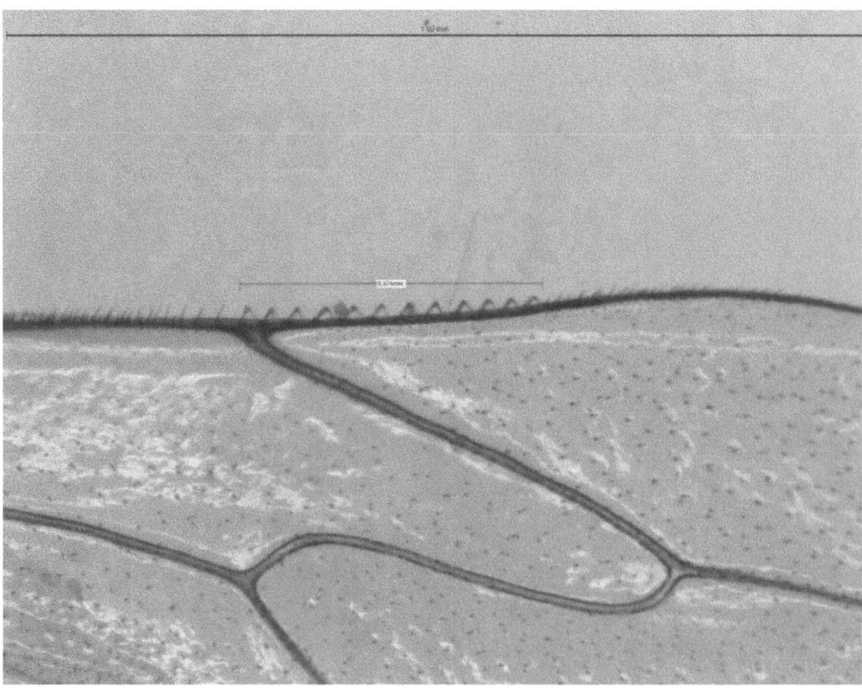

PLATE 14.13

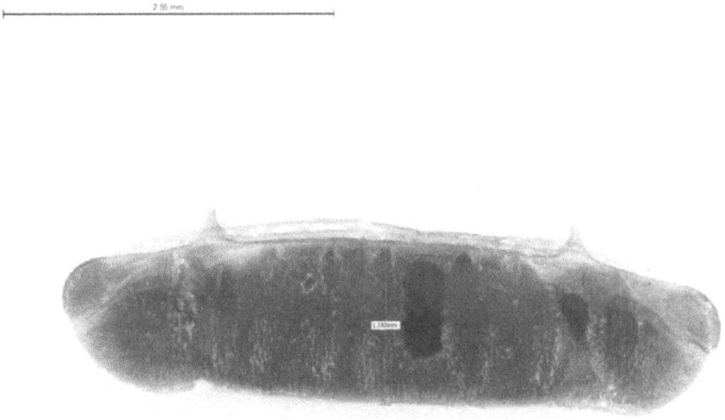

PLATE 14.14

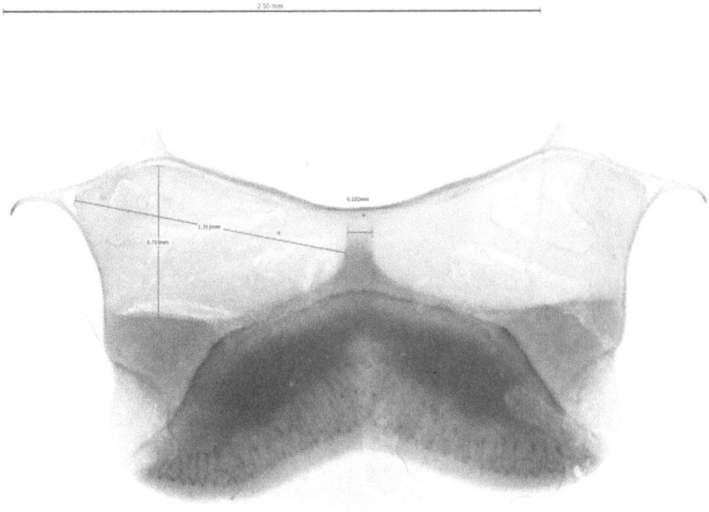

PLATE 14.15

Castes and Life Cycle

An *A. florea* colony consists of three castes, viz., a queen, hundreds of workers and few drones as in other honeybees. The queen bee lays eggs singly in a cell in a wax honeycomb, produced and shaped by the worker bees. She can lay both fertilized and unfertilized eggs. Drones develop from unfertilized eggs, while females (queens and worker bees) develop from fertilized eggs.

Larvae are initially fed with royal jelly produced by worker bees, later switching to honey and pollen. The exception is the larva fed solely on royal jelly and reared in larger cells that will develop into a queen bee. Food for larvae is provided progressively; cells are not closed until the larva has finished feeding. The larva undergoes several instars before pupating. Brood cells for workers (Plate 14.16) and storage cells for honey or pollen are alike, hexagonal; brood cells for males (Plate 14.17) are similar but larger. Queen-producing cells (Plate 14.18) tend to hang individually from brood combs of worker cells. Development from egg to emerging bee varies among queens, workers and drones. Queens emerge from their cells in 15–16 days, workers in 21 days and drones in 24 days (Plates 14.19–14.21).

Young worker bees (Plate 14.22) clean the hive and feed the larvae. When their royal jelly-producing glands begin to atrophy, they begin to produce wax and build comb cells. They progress to other tasks within the colony as they become older, such as receiving nectar and pollen from foragers and guarding the hive. Later, workers take their first orientation flights and finally leave the hive, and typically spend the remainder of their lives as foragers.

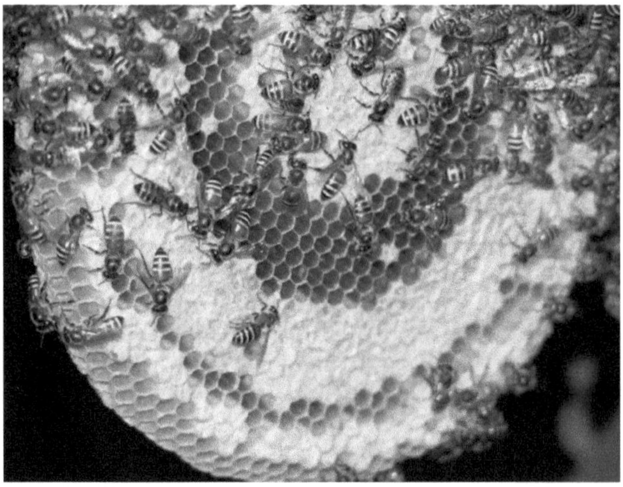

PLATE 14.16

PLATE 14.17

PLATE 14.18

PLATE 14.19

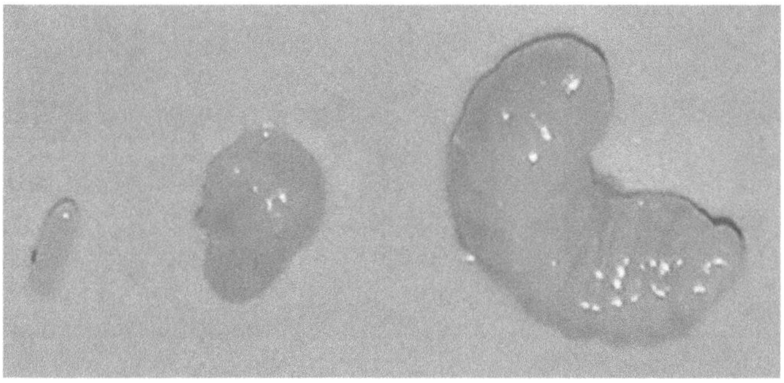

PLATE 14.20

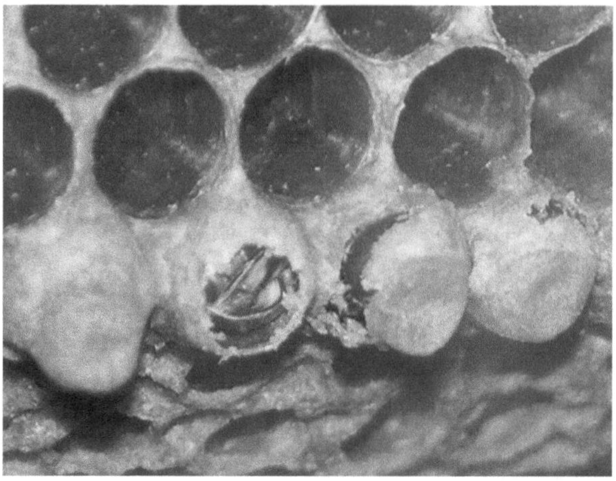

PLATE 14.21

PLATE 14.22

Breeding and Lifespan

When a queen (Plate 14.23) mates, the sperm is transferred into the spermatheca directly. Specifically, *A. florea* females produce no mucus as a sign of mating. *A. florea* males use a clasper organ at the hindleg to hold the queen's leg during mating. Anatomical studies support the hypothesis that *A. florea* bees mate via direct sperm transfer. *A. florea* queens typically mate with 3–4 males and hence use the more energetically efficient method of sperm transfer.

In a single patriline, more workers mean the colony will produce more drones. High mating frequency in *A. florea* has been evolutionarily selected for as observed in other members of the genus. The average lifespan of a drone is 15.6 days, ranging from 6 to 41 days and that of worker bees ranged from 48 to 56 days. The queen has a much longer lifespan consistent with the longevity of the colony.

PLATE 14.23

Worker Policing

In the presence of a multiply mated queen, eggs laid by workers are removed in a process called worker policing. *A. florea* colonies have workers that police to ensure that workers' eggs are destroyed at twice the rate of queens' eggs using oophagy. The queens' eggs are marked so as to avoid destruction by worker bees. Worker policing is thus a result of genetic conflicts of interest among workers used non-exclusively by honeybees and wasps. With worker policing, workers control the production of males by other workers in favour of reproduction by the queen. But *A. florea* is multiply mated, which reduces relatedness of workers to the eggs laid by other workers. If they do not share the same father, then relatedness of a worker to her half-sister's son is only 0.125, half that of her relatedness to her brothers. So in this case, each worker wants to lay her own eggs but her sisters prevent it and worker policing occurs. As an evolutionarily stable strategy (ESS), workers thus compromise by raising the eggs laid by the queen only and removing any eggs laid by other workers. Interestingly, worker policing is therefore more beneficial to workers than rearing their own eggs because they are better able to ensure the survival of their genetic material through worker policing as an ESS. This may explain why workers have inactive ovaries in queenright colonies of this multiply mated species.

Queenless colonies, therefore, face a reproductive dilemma. Once a colony becomes queenless, some workers activate their ovaries within just four days and lay eggs in a last effort to ensure gene survival before the colony disappears via absconding behaviour. In queenless colonies, worker policing decreases as all workers lay male eggs. However, the dilemma occurs when other workers parasitize a queenless colony and thus worker policing may be reinstalled to some extent in this case. Workers that contribute resources and/or time to a queenless colony may succumb to the concorde fallacy, in which investing in a colony with high rates of parasitism diminishes reproductive success and the benefits of remaining within that colony. Therefore, queenless colonies are observed to abscond within three weeks of losing a queen. This further supports the notion that *A. florea* colonies prefer to have a queenright colony and use worker policing as an ESS. Queenless *A. florea* colonies have been observed to merge with nearby queenright *A. florea* colonies, suggesting workers are attracted to queen bee pheromones. *A. florea* colonies demonstrate aggressive behaviours when competing for territory. Additionally, they compensate for their small size and limited flight range with more aggressive behaviour since their flight range is also limited.

Genetic Diversity and Division of Labour

Even in a single nest, there is high genetic diversity among the *A. florea* bees. Since honeybee queens are polygamous, strong genetic variability exists. The tendency of this species to perform certain tasks is dependent on this variation. For example, fanning of the nest is done by a specific set of workers when the nest reaches a specific temperature threshold.

Among the members of the colony there is division of labour and specialization in the performance of biological functions. The younger individuals of *A. florea* work within the nests by performing maintenance while older individuals are responsible for protection and foraging. Some specific tasks such as fanning of the comb to maintain a stable temperature are done by a specific set of workers, which is genetically motivated. This is completed by patrilines because diverse patrilines respond to a specific temperature threshold. When a queen mates with multiple unrelated males, genotypically diverse workers emerge in the colony. These colonies may be more fit than genetically uniform colonies perhaps because they are able to tolerate a wider range of environmental conditions. In such cases, *A. florea* may use genetically based task specialization, in which highly related daughters of a haploid male have genetic predispositions to undertake certain tasks. This could explain why polyandry has evolved and been maintained.

Migration and Swarming

A. florea is highly migratory, but long distance migrations are unknown. Disturbances cause the colonies to desert the comb (Plate 14.24), leaving behind honey, brood and pollen stores.

The honeybees are gentle in temperament; however, they do sting when disturbed. When disturbed, the worker bees show a 'shimmering' movement, with the individual bees shaking their abdomens from side to side in a synchronous manner; at the same time, a hissing sound is released. If the colony is further disturbed, the worker bees raise their abdomens and take off from the curtain to attack the intruder. If they are building a new nest near the old one, they salvage the wax from the old nest. Other species of honeybee do not return to the old nest to salvage wax, perhaps because of the risk of contaminating pathogens. This behaviour is only observed in this species. Even within the species

PLATE 14.24

this behaviour differs; the colonies that migrate less than 200 m are involved in the salvaging of the wax, but the colonies that migrate long distances do not return for the old wax (Hepburn et al 2010).

The distribution area of *A. florea* is generally confined to warm climates (Otis 1991). *A. florea* bees reach a decision about a new nesting site via dancing. They decide on a site when the largest number of individuals dances in the direction of the new site (Oldroyd et al 2008). The workers use an auditory signal of piping to indicate a decision about a new site, with the top of the swarm lifting off into the air first, followed by the bottom of the swarm and then the middle reaching the air last, but all within a short span of time – one minute of the swarm's initial liftoff. *A. florea* swarms travel to the new site as a group and leave to a new site if it is later discovered to be unsuitable. This makes searching for new sites a much faster process for *A. florea* but not necessarily more accurate.

Piping

Defending a large social honeybee colony containing several thousands of workers requires simultaneous action of many individuals, which ideally involves communication between workers, enabling coordinated action and a fast response. *A. florea* colonies employ a defensive strategy called 'piping', in which one individual emits an initial warning signal ('piping'), which is followed 0.3–0.7 seconds later by a general response from a large number of bees ('hissing'). With a fundamental frequency of 384 ± 31 Hz that lasts 0.82 ± 0.35 seconds, piping is audible to the human ear. Hissing is a noisy, broadband signal that is also audible to the human ear and produced by slight movements of the bees' wings. Individuals close to the piping bee are the first to hiss, which spreads rapidly to neighbours until an impressive coordinated crescendo is produced by the entire colony. Piping and hissing are accompanied by workers ceasing activity such as foraging, dancing and departures from the colony (Duangphakdee, 2005).

Colony Cycle

The annual colony cycle of honeybees involves migration, swarming and absconding. *A. florea* migrates seasonally from one habitat to another. This might increase colony fitness, as the honeybees search for new territories, resources or a reduction in natural enemies. Once a colony has outgrown its hive space, it will reproduce via swarming. During warmer seasons, ambient temperatures allow honeybees to forage actively, and they will reproduce frequently. During the colder seasons, colonies diminish in size because they depend on reserved food stores. Before swarming, the colony will build queen cells in order for virgin queens to rear young queens. Before the new queens emerge, the colony's workers search for a new nesting site. Afterwards, the bees will choose who stays and who goes. Similar to absconding behaviour, the old queen will swarm while the new queens prepare for emergence. Prior to the swarming season, drone-brood cells are added, adjoining the lower rows of the worker-brood cells. When a colony loses its queen, emergency queen cells are built from normal cells containing young worker larvae.

During the season when there is an ample supply of nectar and honey, populous colonies of the dwarf honeybee send out multiple reproductive swarms. In addition, colonies of this bee have a high degree of mobility. Disturbance by natural enemies, exposure to inclement weather and scarcity of forage are among the major causes of colonies absconding. The entire colony, accompanied by a laying queen, can fly several metres away to regroup and later abscond. Some absconding colonies are able to survive to build their new combs in a nearby area.

Attempts were made by the Vellayani Centre to hive and domesticate the *A. florea* colonies during the brood-rearing season (November) by transferring the colonies in a special hive designed in which a wire net provided for facilitating the foraging bees (Plates 14.25 and 14.26). The colony was maintained at ambient temperature and protected from direct sun and rain. Diluted honey (3:1) was fed artificially in a drip system (Plate 14.27). Even though the food reserves (nectar and pollen) were sufficient, the colonies deserted during March.

PLATE 14.25

PLATE 14.26

PLATE 14.27

Communication and Recruitment to Crops

As social bees, *A. florea* colonies require a mechanism of communication, especially when conveying important spatial information about foraging, including direction and distance. Communication among members is one of the most important biological attributes of the honeybee colony. All species employ two principal modes of social communication: pheromones and dance 'language'.

Pheromone communication among the members of a colony plays an essential role in regulating its social life. Apart from the queen pheromones and the alarm pheromone, scent-gland pheromones, known as nasomov pheromones, are used extensively by workers in indicating the hive location and during the process of swarming.

In addition to chemical communication through pheromones, workers of all honeybee species are able to attract their nest mates to a crop by dancing, with scouts announcing the discovery of a new home site or, more often, food source in this manner. The distance from the hive and the direction of the food source are conveyed through a dance 'language', which has two important basic forms: the round dance and the tail-wagging dance.

The round dance indicates that food is near the hive and does not give a direction, but the tail-wagging dance gives more details. The distance is indicated by the speed with which the dancer completes its dance cycle, while the direction is shown by the angle of deflection from the vertical made by the bee's abdomen while in movement, which is equal to the angle formed by the lines between the food source, the hive entrance and the sun. If the line of movement is upward, it signifies that the foragers should fly toward the sun, and vice versa. Further, if the dance indicates the presence of nectar, the scout will pause and give a small quantity of the nectar to its fellow workers, while if the source consists of pollen, the scout will allow the others to inspect or perceive the odour of the pollen. Finally, the intensity of the dance indicates the richness of the food source: a vigorous dance, indicating a rich source, will attract numerous recruits, while a slower dance, indicating a poorer source, will on the contrary attract fewer recruits. *A. florea* neither prioritizes nesting site nor food patch information by changing the precision of the dance.

Foraging

A. florea adopts the open-comb branch site as an information centre for foraging. The branch platforms serves as a location for bee dancing that conveys important foraging information to other bees. They are commonly seen foraging near branches and small bushes. They mainly forage for pollen and nectar as food sources as well as water and safety from predators. Their frequent foraging and long migration range display their high degree of mobility. They tend to build combs at lower elevations, away from direct sunlight and on the peripheral side of plant branches. *A. florea* forages openly, unlike *A. cerana*, which mostly forages in parts of the hive that are not visible. *A. florea* bees tend to have longer foraging durations when compared to other bees in this genus, perhaps because they require smaller territories at lower altitudes and build open combs on branches or twigs. Their small size necessitates more time to forage for pollen or nectar because they have a limit on how much food they can carry. *A. florea* adults consume primarily honey and nectar but also feed the brood on pollen.

Under normal conditions, worker bees begin to forage when they are about 2–3 weeks old. Foraging is the last chore in the life of a worker. Once the worker begins foraging, it continues in this activity for the rest of its life. Foragers are sometimes called 'field bees'. Part of the colony's stored honey is invested in foraging activity; before taking off on a flight, a field bee consumes a certain amount of honey to ensure that it will have a sufficient energy supply for the round-trip journey. To obtain a full load of nectar and/or pollen in a single trip, the bees may have to visit several hundred flowers. The amount of energy a bee expends, related to the amount of food it collects, is determined largely by such factors as the species of forage, floral density per unit area, the distance from the hive and weather conditions. For the collection of nectar and pollen, there is no specialization or division of labour among foragers. There are, however, both qualitative and quantitative differences among flowering plant species as regards nectar and pollen production. Not all plant species possess nectaries (glands secreting nectar), and for a forager to collect nectar, the nectaries must be accessible by the bee's proboscis or tongue. Nectaries may be located on many parts of the blossom, base of the stamen and stigma, petals and sepals. Moreover, some plant species in Kerala have extra floral nectaries (Rubber plant – *Hevea brasiliensis* and Maruthi – *Terminalia paniculata*) that may be visited by bees.

A forager may appear to prefer the nectar of one flower species over another. This is because it is to the forager's advantage to visit flowers producing greater quantities of nectar with a higher sugar concentration. Further, the sugar concentration in the nectar of a given plant species may vary from one place to another, or from one time of day to another or even from one plant to another in the same species, coinciding with the pollination requirement.

Workers of all honeybee species carry nectar internally, part of their alimentary canal being modified to form a 'honey sac' or 'honey stomach' to accommodate their nectar load. On returning to the hive, the forager regurgitates the nectar to one or more house bees, by which it will be converted into honey. The house bees add to the nectar the enzyme invertase, through whose action the sucrose sugar in the nectar is split into fructose and glucose, the sugars predominant in honey. Using their proboscises, the house bees expose the nectar as a thin film, thus increasing its surface area and ensuring the more rapid evaporation of the water it contains.

The entire body of a worker bee, particularly its thorax, is densely covered with fine, branched hairs, on which pollen grains are caught when the bee works on a flower. The worker bee sometimes uses its mandibles to chew off the anthers, or it deliberately rolls over the anthers to acquire the pollen. The tibiae of the bee's hind legs are equipped with rows of short setae, which it uses to scrape the pollen from its body and to form it into pellets, sometimes regurgitating a slight quantity of nectar to provide moisture and adhesiveness for this purpose (Plates 14.28–14.33). The pellets, attached to 'pollen baskets' on the bee's rear tibiae, are carried back to the hive, where the load is deposited by itself in a pollen storage cell. Whereas cells containing ripe honey are capped, pollen storage cells are not: the bees tightly pack pollen to about two-thirds of the capacity of the cell and coat the top surface of the pollen in each cell with honey; this protects the pollen against spoiling.

In addition to collecting nectar and pollen as the colony's food, field bees collect plant gum (propolis) and also water. Propolis, which is exuded by certain plants, often to protect wounds on their surface,

PLATE 14.28

PLATE 14.29

PLATE 14.30

PLATE 14.31

PLATE 14.32

PLATE 14.33

is rich in tannin and displays antibiotic activity. It is an adhesive material, which the bees use in comb construction, to coat the interior of the hive, and to seal cracks. In collecting propolis, a field bee uses its mandibles to bite the substance from the plant surface and carries it back on her rear legs to the hive, where the house bees, in turn, use their mandibles to remove it from the forager.

The honeybee colony needs water for two purposes only: to cool the hive and to dilute the honey fed to the larvae. Like nectar, water is collected by the field bee through its proboscis and is carried back to the hive in its honey stomach, being regurgitated to the house bees on arrival. During the heat of the day, some foragers may switch from nectar to water collection, or they may prefer to collect nectars with a low sugar concentration, whose water concentration is correspondingly higher.

Temperature Regulation

Honeybees, like all other insects, are unable to control their body temperature internally according to changes in the ambient temperature. Although the individual bee cannot control its body temperature, a populous honeybee colony can regulate the interior temperature of the hive, particularly within the area surrounding the developing brood. In normal colonies, the brood-nest temperature is maintained at a remarkably constant 32°C–36°C. By fanning their wings, evaporating the water film at the proboscises of the workers, and dispersing drops of water in empty cells, a honeybee colony can reduce its temperature considerably. When the external temperature is low, on the contrary, the bees reduce heat losses by clustering together, and the lower the temperature, the more compact the cluster. In addition, in order to generate more body heat, the worker bees will consume more food, especially honey: more heat is released as a result of the increased rate of food metabolism.

Collection of Honey

A. florea is a poor honey yielder, and a comb yields 200–900 g of honey that have a thin consistency. Honey hunters take away the whole comb and thus destroy a large number of colonies. The honey produced by this species is believed to have special medicinal qualities. The medicinal value, if any, must be attributed to the nectar of the medicinal plants foraged by the workers. *A. florea* honey fetches a high price. In comparison with other honeybee species, the amount of honey that *A. florea* workers will store in their nests is small. For harvesting honey, the rural people transfer the nests with bees from their natural sites to a distant place and then, using twine and two short twigs, the nest is clamped and attached to a small branch of a tree. The upper part of the comb, containing the honey, is cut out, and the honey is squeezed out from it. A period of about six to eight weeks is required for the bees to repair the comb and replenish it with honey, and then it is harvested again. Most colonies will abscond either shortly after their transfer to the new site or after the first or second harvest.

Natural Enemies

Since *A. florea* colonies use open combs, they must protect their comb from the outside, specifically at the site of a branch. They employ two lines of defence. The first is against any falling object such as leaves or other debris. *A. florea* bees use the head-pushing technique to move obstructions. The second line of defence is against ants and other arthropods. The bees accomplish this by creating sticky or repellent barriers.

A. florea bees show highly specific social defence mechanisms when they sense predators nearby. For instance, they typically display hissing and shimmering behaviour, in addition to nesting amidst dense foliage to camouflage themselves from potential predators. A more poignant example is the specific behavioural response they exhibit against their predominant predator, the *Oecophylla smaragdina* weaver ant. To ward off ant attacks, the workers coat both ends of the nest support with sticky strips of propolis, or 'plant gum', from 2.5 to 4 cm wide. *A. florea* is the only honeybee that uses this defensive technique. The reflective substance on the branch either side of the nest is propolis, which acts as a sticky and chemically repellent barrier to protect the nest from ants. When these ants are in close proximity, the bees produce and deposit sticky barriers to obstruct their path. The guard bees take cover at this sticky zone and start to alert other bees using specific hissing sounds with the goal of preventing full-blown raids by the ants. Over time, more bees are recruited to contribute to this sticky zone barrier, which reinforces their defence.

Human Interactions

A. florea colonies have two main interactions with humans: hunting and tourism. In Asia, there is widespread hunting of these bees because they are harmless to humans, and they also have social and cultural significance to the two main religions in Asia: Hinduism and Buddhism. In Hindu culture, the honey from such bees represents 'the blendedness of everything' and is served with other ingredients in a variety of traditional ceremonies. In Buddhist culture, gifting honey to fellow monks is considered equal to giving alms, one of the most expressive ways that one can reward a good deed. It is worth noting that honey-hunting sites are considered tourist sites in places such as Nepal. Increased deforestation due to industrialization has enhanced bee migration to areas that are less populated; therefore, human interaction is mainly limited to fairly rural regions.

Role of *A. florea* in Agriculture

In Kerala, the major agricultural and horticultural crops are seen visited by *A. florea*. The observations recorded over years on the foraging of *A. florea* showed that the plants are visited either for pollen/nectar or both, which in turn benefited the cross-pollination and enhanced productivity in the following crops (Table 14.2).

TABLE 14.2

Foraging Plants of *Apis florea* in Kerala

S. No.	Common Name	Scientific Name	Family	Period in Bloom	Nectar/Pollen
Plantation/Cash Crops					
1	Rubber	*Hevea brasiliensis*	Euphorbiaceae	Jan–Apr	N
2	Coconut	*Cocos nucifera*	Palmaceae	Jan–Dec	N+P
3	Areca nut	*Areca catechu*	Arecaceae	Jan–Mar	N+P
4	Cashew	*Anacardium occidentale*	Anacardiaceae	Feb–Apr	N+P
5	Coffee	*Coffea arabica*	Rubiaceae	Feb	N
6	Coffee	*Coffe arobusta*	Rubiaceae	Feb	N
7	Tea	*Camellia sinensis*	Theaceae	Jan–Jun	N+P
Medicinal Plants					
8	Touch-me-not	*Mimosa pudica*	Mimosaceae	Jan–Dec	N+P
9	Javanese wool plant	*Aerva lanata*	Amaranthaceae	Jan–Dec	N
10	Nagadandi	*Baliospermum monatanum*	Euphorbiaceae	Sept–Oct	N
11	Ixora	*Ixora coccinea*	Rubiaceae	Jan–Dec	N+P
Vegetables					
12	Sponge gourd	*Luffa cylindrica*	Cucurbitaceae	Jan–Dec	P
13	Drumstick	*Moringa oleifera*	Moringaceae	Feb–Apr	N+P
14	Chinese cabbage	*Brassica pekinensis*	Bracicaceae	Oct–Nov	N
15	Dolichos	*Dolichos lablab*	Fabaceae	Feb–Mar	N
16	Radish	*Raphanus sativus*	Brassicaceae	Feb–Mar	N+P
17	Bilimbi	*Averrhoa bilimbi*	Oxalidaceae	Jan–Dec	P
18	Onion	*Allium cepa*	Amaryllidaceae	Jul–Dec	P
19	Violet brinjal	*Solanum melongena*	Solanaceae	Jan–Dec	N+P
20	Green brinjal	*Solanum melongena*	Solanaceae	Jan–Dec	N+P
21	Sweet gourd	*Momordica chinensis*	Cucurbitaceae	Sept–Oct	P
22	Bitter gourd	*Momordica charantia*	Cucurbitaceae	Mar–Sept	N+P
23	Bottle gourd	*Lagenaria vulgaris*	Cucurbitaceae	Jun–Nov	N+P
24	Cucumber	*Cucumis sativus*	Cucurbitaceae	Mar–May	P
25	Green chilly	*Capsicum annum*	Solanaceae	Jan–Dec	N+P
26	Tomato	*Lycopersicon esculentum*	Solanaceae	Nov–Jan	N+P
27	Lady's finger	*Abelmoschus esculentus*	Malvaceae	Mar–Sept	N+P

(Continued)

TABLE 14.2 (*Continued*)

Foraging Plants of *Apis florea* in Kerala

S. No.	Common Name	Scientific Name	Family	Period in Bloom	Nectar/Pollen
Fruit Plants					
28	Banana	*Musa* spp.	Musaceae	Jan–Dec	N+P
29	Guava	*Psidium guajava*	Myrtaceae	Apr–Jul	N+P
30	Lemon	*Citrus* spp.	Rutaceae	Dec–Mar	N+P
31	Litchi	*Litchi chinensis*	Sapindaceae	Feb–Mar	N+P
32	Papaya	*Carica papaya*	Caricaceae	Jan–Dec	N+P
33	Passion fruit	*Passi flora* spp.	Passifloraceae	Jan–Jun	N+P
34	West Indian cherry	*Malpighia punicifolia*	Malpighiaceae	Jan–Feb	N
35	Jamun	*Eugenia jambolana*	Myrtaceae	Mar–Apr	N+P
Grains/Millets					
36	Bajra	*Pennisetum typhoides*	Poaceae	Feb–Mar	P
37	Maize	*Zea maize*	Poaceae	May–Jun	P
Oil Seed					
38	Sunflower	*Helianthus annus*	Compositae	Jul–Sept	N+P
39	Gingely	*Sesamum indicum*	Pedaliaceae	Mar–Apr	N+P
Spices/Condiments					
40	Burmese	*Burmese coriander*	Apiaceae	Jul	N+P
41	Tamarind	*Tamarindus indica*	Fabaceae	May–Jul	N
42	Cinnamon	*Cinnamom zeylanicum*	Lauraceae	Jan–Feb	P
43	Mustard	*Brassica juncea*	Umbelliferaceae	Jun–Aug	N
44	Cardamom	*Elettaria cardamomi*	Zingiberaceae	Feb–Oct	N
Tuber Crops					
45	Tapioca	*Manihot esculenta*	Euphorbaceae	Mar–Apr	N+P
Fibre Crops					
46	Cotton	*Gossypium hirsutum*	Malvaceae	Aug–Sept	N+P
47	Jute	*Corchorus olitorius*	Tiliaceae	Aug–Oct	P
48	Cotton tree	*Bombax malabaricum*	Malvaceae	Jan–Feb	N+P
Ornamental Plants					
49	Poinsettia	*Euphorbia pulcherrima*	Euphorbiaceae	Dec–May	N+P
50	Rose	*Rosa* spp.	Rosaceae	Oct–May	N+P
51	Balsam	*Impatiens balsamina*	Balsaminaceae	Jan–Dec	N
52	Duranta	*Duranta goldiana*	Verbenaceae	Jan–Dec	N+P
53	Hamelia	*Hamelia patens*	Rubiaceae	Jan–Dec	N
54	Marigold	*Tagetes erecta*	Asteraceae	Jan–Dec	N
55	Peacock plant	*Caesalpinia pulcherrima*	Fabaceae	Dec–Mar	N+P
56	Golden dew drop	*Duranta plumer I*	Verbenaceae	Jan–Dec	N+P
57	Canna	*Canna indica*	Cannaceae	Jan–Dec	N
58	Golden rod	*Solidago Canadensis*	Asteraceae	Jan–Dec	P
59	Ball lily	*Haemanthus cinnabarinus*	Amaryllidaceae	May–Jun	P
60	Bird of Paradise	*Heliconia rostrata*	Heliconiaceae	Jan–Dec	N
61	Gladiolus	*Gladiolus grandiflorus*	Iridaceae	Jan–Dec	N
62	Shoe flower	*Hibiscus* sp.	Malvaceae	Jan–Dec	N
63	Dahlia	*Dahlia* sp.	Asteraceae	Jan–Dec	N
64	Torch ginger	*Etlinge relatior*	Zingiberaceae	Jun–Dec	N
65	Powder puff	*Calliandra surinamensis*	Fabaceae	Jan–May	P

(*Continued*)

TABLE 14.2 (*Continued*)

Foraging Plants of *Apis florea* in Kerala

S. No.	Common Name	Scientific Name	Family	Period in Bloom	Nectar/Pollen
66	False eranthemum	*Pseuderanthemum reticulatum*	Acanthaceae	Jan–Dec	N
67	Cock's comb	*Celosia argentea* var. *plumose*	Amaranthaceae	Jan–Dec	N
68	Water leaf	*Talium triangulare*	Potulaceae	Jan–Dec	N
69	Water lilly	*Nymphea stellata*	Pymphaceae	Jan–Dec	P
70	Sage	*Salvia splendens*	Lamiaceae	Jun–Nov	P
71	Anthurium	*Anthurium andreanum*	Araceae	Jan–Dec	P
72	Sesbania	*Sesbania rostrata*	Fabaceae	Mar–Apr	N
73	Mandaram	*Bauhinia acuminate*	Fabaceae	Jan–Dec	N+P
74	Cosmos	*Cosmos sulphurous*	Asteraceae	Jan–Dec	N+P
75	False daisy	*Eclipta alba*	Asteraceae	Jan–Dec	P
76	Tube rose	*Polianthes tuberose*	Asparagaceae	Jul–Aug	N+P
77	Heliconia	*Heliconia* sp.	Heliconiaceae	Jan–Dec	N
78	Zinnia	*Zinnia elegens*	Asteraceae	Jan–Dec	N+P
79	Portulaca	*Portulaca* sp.	Portulacaceae	Jan–Dec	N+P
80	Croton	*Codiaeum* sp.	Euphorbaceae	Feb–Mar	N+P
81	Red hot cat's tail	*Acalypha hispida*	Euphorbiaceae	Jan–Dec	P
82	Oyster plant	*Rhoeodis colour*	Commelinaceae	Jan–Dec	P
83	Mexican flame	*Senecio confuses*	Asteraceae	Aug–Oct	P
84	Bottle brush	*Callistemon citrinus*	Myrtaceae	Jan–Dec	N+P
85	Cock's comb	*Celosia cristata*	Amaranthaceae	Jan–Dec	N
86	Cypress vine	*Ipomoea quamoclit*	Convolvulaceae	Mar–May	N
87	Cuphea	*Cupheahys sopifolia*	Lythraceae	Jan–Dec	N
88	Morning glory	*Ipomea* sp.	Convolvulaceae	Jul–Oct	N
89	Golden shower	*Cassia fistula*	Fabaceae	Apr–May	N
90	Chempakam	*Plumeria* sp.	Apocynaceae	Mar–May	N
91	Nerium	*Nerium oleander*	Apocynaceae	Jan–Dec	N
92	Golden trumpet	*Allaman danerifolia*	Apocynaceae	Jan–Dec	N
93	Yellow oleander	*Thevetia peruviana*	Apocynaceae	Aug–Dec	N
94	Desert rose	*Adenium* sp.	Apocynaceae	Mar–May	N
Weed Plants					
95	Rough mexican clover	*Richardia scabra*	Rubiaceae	Jan–Dec	N
96	Coral creeper	*Antigonon leptopus*	Polygonaceae	Jan–Dec	N+P
97	Crotolaria	*Crotalaria mucronata*	Fabaceae	Feb–Nov	N
98	Angel's trumpet	*Datura* sp.	Solanaceae	Feb–May	N
Wild/Forest Plants					
99	Sera rubber	*Manihot glaziovii*	Euphorbiaceae	Apr–Jun	N+P
100	Glyricidia	*Gliricidia sepium*	Fabaceae	Jan–Apr	N
101	Birds cherry	*Muntingia calabura*	Muntingaceae	Jan–Dec	N+P
102	Cannon ball tree	*Couroupita guinensis*	Lecythidaceae	Jan–May	N+P

Kerala, 'Gods own country', is having immense potential for beekeeping due to diversified flora and honeybee species including the dwarf bee *A. florea*. Due to urbanization, interference of the public destroying the natural domicile and other ecological factors, there has been a decline in the dwarf bee population. Conservation and augmentation of the species should be ensured by creating awareness among the public about the use of this species for pollination and yield enhancement. Destruction of the species by honey hunters and irrational use of pesticides should be prevented by law. As *A. florea* is contributing a major share in the pollination of many crops in Kerala, it should be protected from extinction so as to achieve ecological stability and food security.

REFERENCES

Alexander, B.A. (1991). Phylogenetic analysis of the genus *Apis*. *Annals of the Entomological Society of America* 84: 137–149.

Duangphakdee, O. (2005). Reinforcing a barrier – A specific social defense of the dwarf honeybee (*Apis florea*) released by the weaver ant (*Oecophylla smaragdina*). *Apidologie* 36(4): 505–511. doi:10.1051/apido:2005036.

Engel, M.S. and Schultz, T.R. (1997). Phylogeny and behavior in honey bees. *Annals of the Entomological Societies of America* 90: 43–53.

Hepburn, R., Duangphakdee, O., Phiancharoen, M. and Radloff, S. (2010). Comb wax salvage by the red dwarf honeybee, *Apis florea* F. *Journal of Insect Behavior* 23(2): 159–164.

Makinson, J.C., Oldroyd, B.P., Schaerf, T.M., Wattanachaiyingcharoen, W. and Beekman, M. (2015). Moving home: Nest-site selection in the red dwarf honeybee (*Apis florea*). *Behavioral Ecology and Sociobiology* 65(5): 945–958.

Mishra, R.C. (1995). *Honey Bees and Their Management in India*. ICAR, New Delhi, India, pp. 116–148.

Oldroyd, B.P., Rosalyn, S.G., Naïla, E., Wandee, W. and Madeleine, B. (2008). Nest site selection in the open nesting honey bee *Apis florea*. *Behavioral Ecology and Sociobiology* 62(10): 1643–1653.

Otis, G.W. (1991). Revised distribution of three recently recognized species of Honey bees in Asia. *Honey Bee Science* 15: 167–170.

Viraktamath, S., Rajankar, B., Lazar, K.V., Devanesan, S., Vastrad, A.S., Muthuraman, M., Raju, A.J.S. and Abrol, D.P. (2013). Morphometry of little bee *Apis florea* Fabricius. In: Viraktamath, S., Fakrudin, B., Vastrad, A.S. and Mohankumar, S. (Eds), *Monograph on Morphometry and Phylogeography of Honey Bees and Stingless Bees in India*. DBT, New Delhi, India, pp. 14–25.

15

Management and Conservation of Apis florea

M. R. Srinivasan, P. A. Saravanan and R. Aruna

CONTENTS

Introduction

Apis florea, commonly known as little bee or dwarf honeybee, is a wild honeybee and has characteristic behaviour of construction of open nests with single combs on tree twigs and branches. Two species of dwarf honeybees, viz., red dwarf honeybee *A. florea* and black dwarf honeybee *A. andreniformis*, are reported in different parts of Asia (Booncham et al. 1995; Hepburn and Radloff, 2011).

Both species are found to be similar in size, body length, nest and type of habitats but have morphological and genetic variations. The drones of *A. florea* have longer hind legs with bifurcated basitarsus whereas *A. andreniformis* bees have smaller hind legs with bifurcated basitarsus (Ruttner, 1988). The structure of the endophallus differs in both the species (Lavrekhin, 1935; Wongsiri et al., 1990; Koeniger, 1991); in worker bees, the jugal-vannal ratio of the hind wing is greater in *A. florea* than in *A. andreniformis*. Moreover, the worker bees of *A. florea* have black hairs on the hind tibia and on the dorso lateral surface of the hind basitarsus whereas the hairs are white in *A. andreniformis* (Smith, 1858), and the cubital index is less in *A. florea* than in *A. andreniformis*. Abdominal tergite 2 of *A. florea* is not deeply punctate, but it is *vice versa* in *A. andreniformis*. The marginal setae on the hind tibiae of *A. florea* are usually white and those of *A. andreniformis* dark-brown to blackish in sclerotized individuals (Hepburn et al., 2005). The combs of the two species vary considerably (Rinderer et al., 1996). *A. florea* builds comparatively larger combs than *A. andreniformis*.

Dwarf honeybees can be easily domesticated in low-cost hives for sustainable honey harvest and pollination of several field and orchard crops. The nests can be cut from their nesting habitats and easily shifted to crops fields for planned pollination. Being an excellent pollinator (Free, 1981) and much cherished honey producer, the honeybees' economic products have promoted the livelihood of several bee hunters (Akratanakul, 1976). The economic importance of *A. andreniformis* has not been documented, but the species has lot of significance in areas where it is commonly found (Wongsiri et al., 1996).

Present Status of *A. florea*

A. florea is a native bee of Asia, but the occurrence of both the species is mainly concentrated in tropical and subtropical parts of Asia (Otis, 1996). The distributions of these species are not exactly known, but *A. andreniformis* is widespread in Thailand. Besides that, it is also found in India, China, Burma, Malaysia, the Philippines and Laos (Akratanakul, 1976; Wongsiri et al., 1990; Otis, 1991; Guzman et al., 1992).

In India, *A. florea* colonies are abundant in the forest and agricultural belts of the Kutch region in Gujarat and parts of Rajasthan. The tribals in these regions consume the brood after extracting honey from the nests. Like rock bees, the colony densities of dwarf bees can be very high in places where floral resources are abundant. In Tamil Nadu, the populations of *A. florea* colonies are high in Ramanathapuram, Sivagangai and Tirunelveli districts where the mesquite (*Prosopis juliflora*) was established as an energy plantation a few decades ago. This tree is highly preferred by the bees for their nesting as well as for the its rich source of nectar and pollen.

Conservation Concerns

The value of wild honeybees is enormous and requires very significant attention of land managers and agricultural policymakers. Honeybees are considered a keystone species, i.e., those species that perform a role critical in the maintenance of habitats, and whose absence has severe impacts on the viability of other species (Buchmann and Nabhan, 1996). Beyond their role as pollinators, wild honeybees have important economic, ecological and social values. Commonly, the wild bees are hunted for their honey and brood, that provides the essential household income for many tribal people of our country (Oldroyd and Nanork, 2009). Honey hunting and its associated rituals are crucial aspects of the traditional way of life for many people (Nath et al., 1994; Anderson, 2002; Crane, 1999). Honey and honeybees also have spiritual connections with some of the main religious movements in Hinduism and Buddhism. Wild bee colonies also serve as an important source of the diet of some extremely rare bird species that are completely dependent on bees for their existence.

Factors Influencing the Conservation of Dwarf Honeybees

Seasonal Variations

Little bees are vulnerable to both habitat loss and to human influence by direct exploitation for honey and wax. Since *A. florea* will readily nest in disturbed areas and even in urbanized locations, there was no evidence for a decline in the density of little bees due to deforestation. Dutton and Free (1979) and Free (1981) regarded dearth as the stimulus that propels absconding by *A. florea* in Oman, possibly exacerbated by unusually high temperatures and abetted by pests and predators. They reported that *A. florea* abscond/migrate during unfavourable conditions to cultivated lands near villages and then return to the mountains at the end of summer. Some colonies remain permanently at the village sites. Similar absconding behaviour were observed in Bangladesh (Nash and Murrell, 1981) and Thailand (Akratanakul, 1977). In Oman, Whitcombe (1984) plotted the seasonal cycles of migration of *A. florea* in relation to temperature regimes and floral cycles. Absconding or migration occurs in May to July and wanes in September.

Akratanakul (1977) noticed that absconding/migration in *A. florea* was driven by two main stimuli: resource depletion and serious disturbances by various pests and predators (humans, ants and wax moths). Even though supplementary feeding and pollen were still abundant, a combination of weaver ant and wax moth infestations apparently precipitated the bees' absconding.

Swarming and Migration

The situation in *A. florea* is somewhat different: in some regions, migration is apparently driven by resource depletion (Thakar and Tonapi, 1962) and in others as a response to changes in ambient temperature in the microenvironment of the nest (Tirgari, 1971).

Seasonal absconding/migration by *A. florea* contrasts sharply with reproductive swarming in this species. Dutton and Free (1979) and Free (1981) observed in Oman that brood rearing begins in February and peaks in April, with drones being produced towards the end of February; a season peak of brood production occurs in autumn. Little brood is produced in the hot summer months. *A. florea* swarms during the ascendant phase of colony growth during March and April and sometimes has a second peak in September–October. The old queen leaves with the first swarm and, sometimes, as many as eight swarms in a year emanate from the same colony (Tirgari, 1971).

Brood rearing starts again at the end of August and continues until the end of November. Drones are reared in mid-March and queen cells produced. No brood is produced during December–January. Analysis of the weather patterns for both Oman and Iran suggests that the two critical periods for swarming in spring and then sometimes in autumn could be associated with changes in environmental temperatures and/or changes in day length. Swarming occurs from mid-April to May. Interestingly, some 24% of swarms were attracted to deserted combs for which old combs' wax could well be an attractant.

As to distance swarmed, Tirgari et al. (1969) observed that three colonies of *A. florea* moved between 5 and 400 m between nest sites in Iran. Whitcombe (1984) observed that seven absconding colonies only moved short distances of 4–65 m; reproductive swarming colonies moved further, 25–800 m.

Destruction of Habitat

The primary cause of habitat change in many countries harbouring the little bees is human population growth. Asia is home to some of the densest human populations, and the population is growing in most regions between 1% and 2.5% per year. Combined effects of forest clearing, urbanization, expansion of agriculture and inappropriate hunting were some of the prime reasons of habitat destruction. Patterns of forest clearing tend to result in small patches of a once continuous habitat. Distributed habitats are characterized by reduced fruit and seed set, and a loss of pollinator diversity (Aizen and Feinsinger, 1994; Cascante et al., 2002; Wu and Kuang,1987). Under such circumstances, honeybees will have additional importance as pollinators (Bond, 1994; Klein et al., 2007).

The dwarf bees are not cavity nest dwellers and are successfully preadapted to prosper in human-altered environments, especially the agricultural ecosystems where other species may not thrive due to lack of nest sites. *A. florea* can easily build nests even in disturbed habitats. The distribution range has extended into the Middle East and Africa (Mogga and Ruttner, 1988; Ruttner et al., 1995). *A. andreniformis* could not successfully colonize heavily disturbed areas away from native vegetation, i.e., rainforests, because these bees do not find suitable shady nest sites or places to establish their drone congregation areas.

Variety of anthropogenic environmental changes can potentially affect honeybee populations. Threats to wild honeybees include forest clearing, hunting, the destruction of nesting sites, natural enemies and diseases. Consequently, the management of wild honeybees is very imminent so that their populations can be maintained in perpetuity, and measures should be adopted to ensure that wild honeybee populations are not overexploited.

Problem of Bee Enemies

Generally, the wild *Apis* are tolerant to most of the pests and diseases. The dwarf bee, *A. florea*, thrives very well in hot, dry climates. It builds a single comb in open air and is not very defensive. A band of sticky resin is deposited around the branch that supports the bees' nests to ward off ants. These bands are about 2.8 ± 2.1 cm wide with a range of 0.5–10.05 cm and trap any small animals attempting to cross into a colony (Seeley et al., 1982). Repair and re-enforcement of the sticky bands by *A. florea* is strongly correlated with intrusions of weaver ants into the colony (Duangphakdee et al., 2005). These bees cannot tolerate even minor interference. They adopt a defence strategy of flight instead of fight, and they desert their nest when the colony is disturbed.

The parasitic mite *Euvarroa* affects the dwarf honey bees (Delfinado and Baker, 1974). These mites parasitize on a specific host viz., *E. sinhai* is restricted to *A. florea*, whereas *E. wongsirii* affects *A. andreniformis* (Lekprayoon and Tangkanasing, 1991; Lekprayoon and Tangkanasing, 1994; Morin and Otis, 1993). *E. sinhai* is confirmed with similar reports from Thailand, and this mite species was also found only in drone cells and adults of *A. florea* in Sri Lanka (Akratanakul, 1976). Another mite species, *Neocypholaelaps indica*, is a new parasitic mite on *A. florea* that has been reported as a pollen feeder and phoretic on bees (Koeniger et al., 1983). Both the species of brood mite are reported to be obligate parasites on the dwarf honeybees. These foreign mites were aggressively cleaned and killed by defence behaviour of *A. florea* and *A. andreniformis* (Koeniger et al., 1989; Lekprayoon et al., 1994).

Vertebrates Enemies

Several vertebrate animals such as sloth bear, rodents and monkeys could pose a threat to honeybees. The macaque monkeys (*Macaca mulata*) in groups could shake the branch that harboured an *A. florea* nest until the bees deserted it; later, the monkeys consumed the honey and brood (Seeley et al., 1982). Many insectivorous mammals, particularly tree shrews (*Tupaia glis*), eat entire nests of *A. florea* (Seeley et al., 1982).

Effect of Chemical Insecticides

The dwarf bees, *A. florea* and *A. andreniformis*, often nest in orchards. Commercial orchards are regularly sprayed with insecticides, and it is doubtful that colonies nesting in orchard trees survive this assault. Colonies of all species may suffer losses of field bees when foraging on crops that are sprayed. But usually, colonies that are not directly sprayed can survive such losses.

The chemicals sprayed in the field and on the colonies will kill the individual and can also kill the whole colony (Biesmeijer et al., 2006). The foragers exposed to the chemicals will be killed by the toxicity of the chemical (Desneux et al., 2007). The honeybees are attracted to commercial fruit crops, mainly longan (*Dimocarpus longan*), litchi (*Litchi chinensis*) and citrus, which are major honey producers (Crane et al., 1984). Other orchard trees such as mangosteen (*Garcinia mangostana*) and rambutan (*Nephphelium lappaceum*) are suitable nesting sites for dwarf bees (Oldroyd and Wongsiri, 2006). But these orchards are regularly sprayed with insecticides, which kill all colonies nesting in the tree canopy. Spraying during flowering may also affect colonies nesting outside the crop but foraging in the crop.

Management of Dwarf Bee Colonies

Absconding is a natural phenomenon in dwarf bees that may occur in any time of the season (Hepburn, 2004). Generally, the colonies tend to abscond especially during a dearth period comparatively over a long distance. Most of the little bee colonies undergo at least one migration per year in search of flora. The colony can also abscond at any time if disturbed instead of defending against the intruders. During absconding, the bees abandon their nest and leave behind the brood, food stores and the comb. Sometimes, the workers return to take back honey, pollen and wax from the old nest. Absconding swarms form an interim cluster closer to original nesting sites before reaching their newly selected nesting destination. Some of the management measures to prevent or delay the process of absconding in dwarf bees involves thorough understanding of the following:

- Acute shortage of pasturage sources
- Invasion of ants
- Exposure to smoke
- Harvesting the total honey from the crumb region of the nest

- Removal of sheltering leaves in nesting trees, which leads to exposure of the nest to sunlight
- Blackened comb
- Any kind of natural or anthropological disturbances to the colony

Little bees are passive nest defenders. They mostly rely on retreat strategies such as hiding and absconding. The bees cannot tolerate even quite minor disturbances. The nest-supporting twig has two rings of sticky resin that encircle the branch on either side of the comb. This will deter the ants and act as ant barriers. The worker bees periodically check and add more quantity of resin (Muthuraman et al., 2014). The support area provides space for honey storage followed by pollen and brood area (Ruttner, 1985, 1988). The sticky bands of *A. florea* and *A. andreniformis* have clear-cut repellent properties against the weaver ants, *Oecophylla smaragdina* (Duangphakdee et al., 2005).

Domestication of Dwarf Honeybees

Dwarf bees are rarely domesticated. The absconding habitat mainly deters the attempts to domesticate these bees. Little bee colonies never stay in the same place for a long time. In some areas, bee keepers make a 'clip' of split bamboo. After a wild colony is obtained, the honey section of the comb is removed for consumption. The remaining brood comb is placed in the clip and resuspended near the original location in the hope that the bees will reoccupy it. Occasionally, the clip will be moved to a central apiary, but the bees often abscond. In Oman, where *A. florea* colonies nest in rock crevices, bee hunters may help the bees' welfare by erecting protective barriers against sun or wind (Dutton and Simpson, 1977).

Hiving of Feral Colonies Using Cardboard Box Hive

The cardboard box hive is simple in design and made of cardboard in TNAU, Coimbatore. The hive is very well suited for the purpose of transportation and maintenance of feral colonies of *A. florea* (Muthuraman et al., 2012). Two 'U'-shaped cuts are made on each side along the side walls of the cardboard box that serve as rabbets. The width of the cuts depends on the thickness of the twig to which the comb is attached. Similarly, the depth of the cuts also varies with the height of the honey crest so as to accommodate the honey crest conveniently within the hive even after closing the cardboard box using the top lid. Small holes of 2 mm diameter are made on the sides and the front wall of the cardboard box for ventilation. The perforations permit only air movement and not the bees. A rectangular opening is cut on the front wall of the hive along the perforations to facilitate the landing and takeoff by foraging bees from their nest after shifting the colonies to the chosen location.

Arboreal nesting colonies are mainly hived. The twig with the comb is gently cut using either a secateur or a sharp knife with least disturbance to the comb and bees. The twig with the comb thus removed is kept inside the cardboard box hive in the U-shaped rabbets. Allow 15–30 minutes for settling of bees on the comb, then the top lid of the hive is closed and made bee-proof before transport. The cardboard hive serves the twin purposes of colony transport and comfortable lodging for the bees. In addition, the hive also provides additional protection to the colony against heat and rain. Colony removal, hiving and shifting are mainly done during the evening time to avoid the loss of field bees. Care should be taken to minimize the jerks during transport to avoid the breakage of combs. The front lid on the front wall of the hive is removed after reaching the destination (Muthuraman et al., 2012).

The following are the precautions to be followed during the hiving process and subsequent transport of hived colonies:

Do's

- Hive the colonies that build the nests on tree branches encircling twigs.
- Hive the colonies especially during nectar flow season.
- Hive the nest either during early morning or late evening.
- Choose the appropriate cutting tool according to the size of the twig. A thin twig is easily cut with secateurs, while a hacksaw blade is suited for cutting a thick twig.

- Cut the twig with extreme care so as to avoid detachment of comb from twig.
- Use minimum quantity of smoke with a small bunch of coconut husk and calm bees before cutting the twig.
- Delay the shifting of the nest until all the bees are settled back on the comb.
- Transport the hived colonies with minimum jerk.

Dont's

- Hive the colonies with either white- or yellow-coloured comb.
- Hive the colonies that are queenless.
- Hive the colonies that are ready to swarm.
- Hive the colonies that are about to abscond.
- Hive the colonies where the combs are not totally covered by bees.
- Hive the colonies with patchy brood.
- Use excess smoke while removing the nest along with the twig.
- Catch the queen with fingers, as handling the queen may inflict injury to her.
- Expose the naked comb to direct sunlight that can cause melting of combs and comb detachment from the twig after settling of bees.
- Alter the angle of the twig with respect to the ground while replacing the nest inside the hive.

Habitat Conservation

Densities of *A. florea* can be very high in places where floral resources are rich. Recent bee faunal and floral surveys conducted in Tamil Nadu have clearly revealed the scope for keeping *A. florea* colonies in places where banana is cultivated extensively and also in tank beds where *Acacia nilotica* and *A. planifrons* trees are grown. The nest densities of *A. florea* are also found to be high in energy plantations established with *Prosopis juliflora*. In addition, it is a major source of nectar and pollen for the bees.

Sustainable Honey Harvesting Methods

The dwarf bees are gentle in temperament but sting when irritated with their barbed sting to ward off intruders. The workers produce decenyl acetate, a more persistent enemy marker pheromone that triggers the stinging behaviour. The bee curtain deters aggression by small enemies. Spraying of water using a hand sprayer or use of mild smoke can pacify the bees and control the defending attitude.

The honey hunters usually drive away the bees from the nest by simply shaking the nest branch or using a puff of tobacco smoke. The bees desert the nest immediately and settle on the nearby branch. Then, the deserted nest attached to the branch is cut off and removed. The honey crest is sliced off from the twig and squeezed to extract the honey. The brood nest is either consumed or thrown away. This kind of destructive honey harvest can lead to absconding of colonies and also wastage of valuable bees wax.

Sustainable honey harvest consists of removing only part of the honey comb, leaving some amount of honey for the bees. Since the stored honey surplus is not harvested totally, the bees usually do not abscond. During the honey flow season, it is possible to harvest honey twice or thrice from each nest by this method without harming the bees.

Conclusion

Dwarf bees have the main role in plant pollination and biodiversity. In order to prevent extinction, these bees have to be protected from destruction of colonies. Educating the public and creating awareness about the importance of these keystone species will help in conservation. Factors responsible for the loss of honeybees such as habitat destruction and indiscriminate pesticides usage must be checked to protect the bees. Bee-friendly plants can be grown wherever possible to provide forage to honeybees. This will be the way forward in conserving the dwarf bees and preventing their extinction from this biosphere.

REFERENCES

Aizen, M.A., and Fiensinger, P. 1994. Habitat fragmentation, native insect pollinators, and feral honeybees in Argentine 'chaco serrano'. *Ecological Applications* 4: 378–392.

Akratanakul, P. 1976. Honey bees in Thailand. *American Bee Journal* 116(3): 120–126.

Akratanakul, P. 1977. The natural history of the dwarf honey bee, Apis florea F in Thailand. Thesis, Cornell University.

Anderson, C. 2002. Self-organization in relation to several similar concepts: Are the boundaries to self-organization indistinct? *The Biological Bulletin* 202: 247–255.

Biesmeijer, J.C., Roberts, S.P.M., Reemer, M., Ohlemuller, R., Edwards, M., Peeters, T., Schaffers, A.P. et al. 2006. Parallel declines in pollinators and insect-pollinated plants in Britain and the Netherlands. *Science* 313: 351–354.

Bond, W.J. 1994. Do mutualisms matter? Assessing the impact of pollinator and disperser disruption on plant extinction. *Philosophical Transactions of the Royal Society B* 344: 83–93.

Booncham, U., Wongsiri, S. and Thirakupt, K. 1995. Niche differences between *A. florea* and *A. andreniformis* in dry-evergreen forest, Thailand. *International Conference on Tropical Bees and the Environment*, Pedu Lake, Kedah, Malaysia, 11–15 March, p. 33.

Buchmann, S.L. and Nabhan, G.P. 1996. *The Forgotten Pollinators*. Island Press, Washington, DC, p. 292.

Cascante, A., Quesada, M., Lobo, J.J., and Fuchs, E.A. 2002. Effects of dry tropical forest fragmentation on the reproductive success and genetic structure of the tree Samanea saman. Conservation Biology 16: 137–147.

Crane, E. 1999. *The World History of Beekeeping and Honey Hunting*. Routledge, New York, p. 720.

Crane, E., Walker, P. and Day, R. 1984. *Directory of Important World Honey Sources*. International Bee Research Association, London, UK, p. 384.

Delfinado, M.D. and Baker, E.W. 1974. Varroidae, a new family of mites on honey bees (Mesostigmata: Acarina). *Journal of the Washington Academy of Science* 64(1): 4–10.

Desneux, N., Decourtye, A. and Delpuech, J.M. 2007. The sublethal effects of pesticides on beneficial arthropods. *Annual Review of Entomology* 52: 81–106.

Duangphakdee, O., Koeniger, N., Koeniger, G., Wongsiri, S., and Deowanish, S. 2005. Reinforcing a barrier – a specific social defense of the dwarf honeybee (*Apis florea*) released by the weaver ant (*Oecophylla smaragdina*). *Apidologie* 36: 505–512.

Dutton, R.W., and Simpson, J. 1977. Producing honey with Apis florea in Oman. Bee World 58: 71–76.

Free, J.B. 1981. Biology and behaviour of the honeybee *Apis florea*, and possibilities for beekeeping. *Bee World* 62(2): 46–59.

Guzman, L.I.D., Forbes, M., Cervancia, C., Rinderer, T.E. and Somera, S.J.R. 1992. *Apis andreniformis* Smith in Palawan, Philippines. *Journal of Apicultural Research* 31(2): 111.

Hepburn, H.R. 2004. Absconding, migration and swarming in honeybees: An ecological and evolutionary perspective. *Proceedings of the IUSSI Conference on Social Insects*, St. Petersburg, Russia.

Hepburn, H.R. and Radloff, S.E. 2011. Biogeography of the dwarf honeybees, *Apis andreniformis* and *Apis florea*. *Apidologie* 42: 293–300.

Hepburn, H.R., Radloff, S.E., Otis, W.G., Stefan, F., Verma, L.R., Tan, K., Chaiyawong, T., Tahmasebi, G., Ebadi, R. and Wongsiri, S. 2005. *Apis florea*: Morphometrics, classification and biogeography. *Apidologie* 36(3): 359–376.

Klein, A.M., Vaissiere, B.E., Cane, J.H., Steffan, D.I., Cunningham, S.A., Kremen, C. and Tscharntke, T. 2007. Importance of pollinators in changing landscapes for world crops. *Proceedings of the Royal Society B: Biological Sciences* 274: 303–313.

Koeniger, G. 1991. Diversity in *Apis* mating systems. In: Smith, D.R. (Ed.), *Diversity in the Genus* Apis, Westview Press, Boulder, CO, pp. 199–212.

Koeniger, N., Koeniger, G. and Baker, M.D. 1983. Observations on mites of the Asian Honeybee species (*Apis cerana, Apis dorsata, Apis florea*). *Apidologie* 14(3): 197–204.

Koeniger, N., Koeniger, G. and Wongsiri, S. 1989. Mating and sperm transfer in *Apis florea*. *Apidologie* 20(5): 413–418.
Lavrekhin, F.A. 1935. Contribution to the study of variation of the honey bee. 2. A comparative biometric characteristic of the sexual appendages of the drones belonging to different forms of the genus *Apis*. *Zoologicheskii Zhurnal* 14: 655–663 (in Russian).
Lekprayoon, C. and Tangkanasing, P. 1991. *Euvarroa wongsirii*, a new species of bee mite from Thailand. *International Journal of Acarology* 17(4): 255–258.
Lekprayoon, C. and Tangkanasing, P. 1994. Comparative morphology of *Euvarroa sinhai* and *Euvarroa wongsirii*: Parasites of *Apis florea* and *Apis andreniformis*. In: Connor, L.J., Rinderer, T., Sylvester, H.A. and Wongsiri, S. (Eds.), *Asian Apiculture*, Wicwas Press, Cheshire, CT, pp. 427–433.
Lekprayoon, C. and Wongsiri, S. 1991. Queen production with *Apis florea* in Thailand. *Proceedings of the First Asia-Pacific Conference of Entomology*, Chiang Mai, Thailand, 1989, pp. 621–627.
Lekprayoon, C., Wongsiri, S., Jayasvasti, S. and Leepitakrat, S. 1994. Survival of the parasitic honeybee mites *Tropilaelaps clareae* on adult *Apis florea* and *Apis andreniformis* workers. *Proceedings: Acarology IX*, Ohio.
Lord, W.G. and Nagi, S.K. 1987. *Apis florea* discovered in Africa. *Bee World* 68: 39–40.
Maa, T. 1953. An inquiry into the systematics of the tribus Apidini or honey bees (Hymenoptera). *Treubia* 21: 525–640.
Mogga, J. and Ruttner, F. 1988. *Apis florea* in Africa: Source of the founder population. *Bee World* 69(3): 100–103.
Morin, C.E. and Otis, G.W. 1993. Observations on the morphology and biology of *Euvarroa wongsirii* (Mesostigmata: Varroidae), a parasite of *Apis andreniformis* (Hymenoptera: Apidae). *International Journal of Acarology* 19(2): 167–172.
Muthuraman, M., Kaliyamoorthy, S., Ganapathy, N., Thangavel, G.K., Priyadarshini, P., and Bharathidasan, K. 2012. A simple technique to hive little bee colonies. In: Centre for Advance Faculty Training on Advances in Bio Ecology and Management of Insect Pollinators of Crops at Department of Entomology, CCS Haryana Agricultural University, Hisar. 140–142p.
Muthuraman, M., Kaliamoorthy, S., Ganapathy, N., and Saravanan, P.A. 2014. Little bee. TNAU Offset Printing press, Coimbatore, India, 37pp.
Nash, W.T., and Murrell, D.C. 1981. Beekeeping in Bangladesh. *American Bee Journal* 121: 352–356.
Nath, S., Roy, P., Leo, R., and John, M. 1994. Honeyhunters and beekeepers of Tamil Nadu. A survey document. Keystone, Pondicherry, p. 86.
Oldroyd, B.P. and Nanork, P. 2009. Conservation of Asian bees. *Apidologie* 40: 296–312.
Oldroyd, B.P. and Wongsiri, S. 2006. *Asian Honey Bees: Biology, Conservation and Human Interactions*. Harvard University Press, Cambridge, MA, p. 340.
Otis, G.W. 1991. Revised distribution of three recently recognized species of honey bees in Asia. *Honeybee Science* 15(4): 167–170.
Otis, G.W. 1996. Distributions of recently recognized species of honey bees (Hymenoptera: Apidae; *Apis*) in Asia. *Journal of the Kansas Entomological Society* 69: 311–333.
Richards, A.J. 2001. Does low biodiversity resulting from modern agricultural practice affect crop pollination and yield? *Annals of Botany* 88: 165–172.
Rinderer, T.E., Wongsiri, S., Kuang, B.Y., Liu, J.S., Oldroyd, B.P., Sylvester, H.A. and Guzman, L.I.D. 1996. Comparative nest architecture of the dwarf honey bees. *Journal of Apicultural Research* 35: 19–26.
Ruttner, F. 1985. Graded geographic variability in honeybees and environment. *Pszczelnicze Zeszyty Naukowe* 29: 81–92.
Ruttner, F. 1988. *Biogeography and Taxonomy of Honeybees*. Springer-Verlag, Berlin, Germany, 284 pp.
Ruttner, F., Mossadegh, M.S. and Kauhausen, K.D. 1995. Distribution and variation of size of *Apis florea* in Iran. *Apidologie* 26: 477–486.
Seeley, T.D., Seeley, R.H., Akratanakul, P. 1982. Colony defense strategies of honeybees in Thailand. *Ecological Monographs* 52: 43–63.
Smith, F. 1858. Catalogue of the hymenopterous insects collected at Sarawak, Borneo. *Proceedings of the Linnean Society of London* 2: 42–130.
Thakar, S.S., and Tonapi, K.V. 1962. Nesting behaviour of Indian honeybees II. Nesting habits and comb cell differentiation in *Apis florea* Fab. *Indian Bee Journal* 24: 27–31.
Tirgari, S. 1971. On the biology and manipulation of Apis (Micrapis) florea F in Iran. In: *Proceedings of the 23rd International Beekeeping Congress*, Moscow, pp. 330–332.

Tirgari, S., Polous, H., Foroughi, H., Fakhari, F., and Nabavit, M.A. 1969. On the occurrence and biology of little bee (*Micrapis florea* F) and the first account of its migration (Hymenoptera: Apidae). In: *Proceedings of the National Congress Entomology and Pest Control (Iran)*, p. 4 (in Persian).

Whitcombe, R.P. 1984. The biology of *Apis* spp in Oman with special reference to *Apis florea* Fab. Thesis, University of Durham.

Williams, I.H. 1996. Aspects of bee diversity and crop pollination in the European Union. In: Matheson, A., Buchman, S.L., O'Tool, C., Westridge, P. and Williams, I.H. (Eds.), *The Conservation of Bees*, Academic Press, London, UK.

Wongsiri, S., Lekprayoon, C., Thapa, R., Thirakupt, K., Rinderer, T.E., Sylvester, H.A., Oldroyd, B.P. and Booncham, U. 1996. Comparative biology of *Apis andreniformis* and *Apis florea* in Thailand. *Bee World* 78(1): 23–35.

Wongsiri, S., Limbipichai, K., Tangkanasing, P., Mardan, M., Rinderer, T.E., Sylvester, H.A., Koeniger, G. and Otis, G. 1990. Evidence of reproductive isolation confirms that *Apis andreniformis* (Smith, 1858) is a separate species from sympatric *Apis florea* (Fabricius, 1787). *Apidologie* 21(1): 47–52.

Wu, Y.R. and Kuang, B. 1987. Two species of small honeybee – A study of the genus *Micrapis*. *Bee World* 68(3): 153–155.

16

Diversity and Distribution of Dwarf Honeybees in Nepal and Their Role in Natural and Agricultural Systems

Ratna Thapa

CONTENTS

Introduction

Nepal has the top 10 tallest mountains in the world, including Mount Everest (8848 m), which lies in the central Himalayas. The country has 23 in the lowlands, 52 in the mid-hills, 38 in the mountains and 5 in areas common to more than one of the above zones. Thc widc mountain ranges and their deep river valleys exhibit wide ranges of precipitation, humidity and temperature, which provide unique habitats and diverse forests within a small geographical area.

Nepal has a diverse climate ranging from tropical to arctic within a distance of approximately 180 km from the south to the north of the Himalayas. Fifty-nine types of vegetation have been identified from Nepal.

Diversity

Nepal is rich in honeybee diversity. Four out of nine species of honeybee species – *Apis laboriosa*, *Apis dorsata*, *Apis florea* and *Apis cerana* – are native to Nepal and existed from the Terai belts to the foothills of the Himalayas. *A. laboriosa*, the Himalayan cliff bee or the largest honeybee in the world, builds a single colony in the open spaces along rivers (Otis, 1996). It is commonly known as the black Himalayan giant honeybee. It is found at high altitudes ranging from 1200 to 3700 m, and it forages up to 4100 m (Roubik et al., 1985; Thapa, 2001; Woyke et al., 2012). This species plays a crucial role in the conservation of the Himalayan ecosystem by pollinating wildflowers in the entire Himalayan region (Thapa, 2001). *A. dorsata*, a giant yellow honeybee, builds a single, large comb on the tallest trees, windows, balconies of buildings and water tanks. *A. dorsata* colonies seasonally migrate between 190 and 1200 m and return to the same old nesting sites (Thapa, 2001; Thapa and Wongsiri, 2011). *A. florea*, known as a dwarf honeybee, builds a single comb in the bushes (Wongsiri et al., 1996; Thapa et al., 2000). It is found from 350 to 1200 m. It is an efficient pollinator of fruit and natural flora (Wongsiri et al., 1996). *A. florea* honey is traditionally used to cure snake-bite wounds. *A. cerana*, the Asian hive bee, is kept in traditional hives such as wall and log hives, and modern Newton hives. There are three subspecies of *A. cerana* viz. *A. c. indica* in plain areas, *A. c. himalaya* in valleys and *A. cerana* in high areas up to 1500 m that are coexisting honeybees in Nepal (Thapa et al., 2000).

It is still unknown the number of *A. florea* colonies that are annually harvested for their brood and honey. *A. florea* honey is considered to have high medicinal properties, but the honey product of *A. florea* is very low and not available in local markets.

Materials and Methods

From March 2015 to 2018, I conducted a series of surveys to investigate the distribution pattern and behavioural aspects of *A. florea* in Nepal. During that time, several trips were made seasonally to several parts of Nepal to find the precise geographical distribution of *A. florea*. Observations were made on established colonies, foragers on flowers and questionnaires to the local beekeepers. The elevation of nesting sites of *A. florea* in each district was obtained from the local meteorological station in Nepal.

The Red Dwarf Honeybee: *A. florea*

The red dwarf honeybee, *A. florea* Fabricius (1787), is an indigenous honeybee species in tropical Asia, the lowlands of Asia and the Middle East (Otis, 1991). *A. florea*, an open-nesting dwarf honeybee, is a native honeybee in Nepal, where it is the dominant honeybee species (Oldroyd and Wongsiri, 2006). Like all *Apis* species, it is also eusocial.

Nesting Site

A. florea nests are found in more disturbed areas such as urban and agricultural areas. The colonies build a small single comb, which is apically attached to small branches of bushes (Figure 16.1a, b). The brood area is below the supporting branch, and the honey storage or crown is above and around the supporting branch. *A. florea* stores pollen just below the crown. The drone cells are found at the lower margins of the nest.

Nest Structure

A. florea nesting height is about 1–15 m (mean = 2.22 ± 1.7 cm, n = 17). *A. florea* nests are large (H = 12.00 ± 3.3 cm, n = 41; W = 10.19 ± 3.0 cm, n = 14; Wongsiri et al., 1996; Figure 16.1c). The honey storage capacity of an *A. florea* nest has been reported to be 1000 g or more (Mutto, 1956; Tigari, 1971). *A. florea* workers form a very compact protective curtain. During the curtain formation, the curtain formation bees hid their heads under the curtain to provide effective protection from cold temperature. The bees hid their heads under the curtain to provide effective protection from cold temperature in response to climatic changes.

FIGURE 16.1 *A. florea* nests. (a) A huge nest of *A. florea* at the base of small bushes at the level of 18 inches from the ground, (b) *A. florea* nest under the coverage of brood leaves bush, found in 2015 at Khotang district, eastern part of Nepal, (c) Brood comb of *A. florea* after harvesting honey showing four successive brood cycles.

Rain Proof Curtain Formation

During favourable climatic conditions, the curtain bees of *A. florea* formed loose curtains. During the rain, the curtain bees moved to the crown of the nest and formed an umbrella shape by spreading wings upon wings, and the curtains formation bees oriented their heads upwards and spread wings in an inverted Λ shape. The curtain bees keep their heads under the bodies of the upper bees, and raise their abdomens at about 15 degrees against the comb surfaces. That way, the lower curtain bees conceal their heads to minimize physical damage from the raindrops (Figure 16.2a, b; Thapa, 2011). Therefore, during a heavy rain, the crown of *A. florea* stays intact (Figure 16.2c).

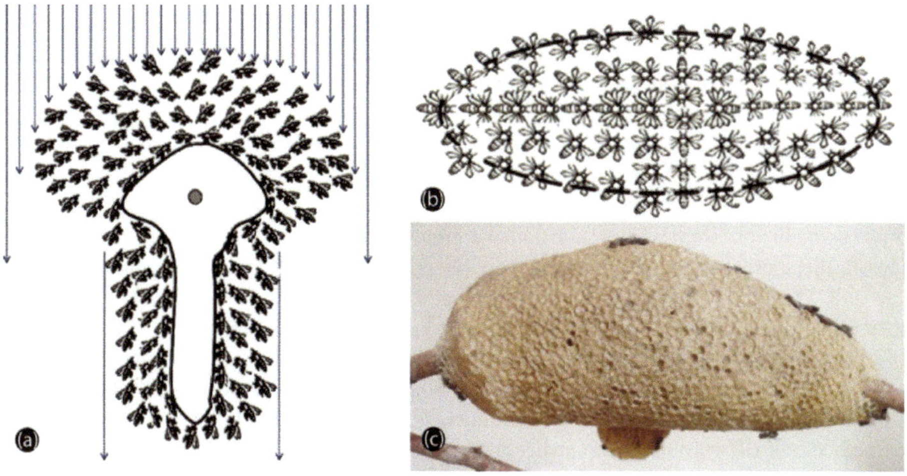

FIGURE 16.2 (a) Rainproof curtain formation of workers of *A. florea* from later view – note arrow indicates the raindrops, (b) Rainproof curtain form from top view, and (c) Crown or honey storage area of the comb after rain stops.

FIGURE 16.3 A single comb of *A. florea* nest built on an air-conditioner pipe consists of a bulbous waxen crown. (a) One hour after *A. florea* absconding, the workers collect the wax from the crown of the nest (indicated by arrows); (b) Day two, workers are still collecting the wax from the crown; (c) Day three, no workers are collecting from the abandoned nest.

Wax Salvaging Behaviour

Nest desertation by absconding or migrating colonies is a common feature of tropical honeybees (Hepburn and Radloff, 1998; Oldroyd and Wongsiri, 2006). Wax salvaging behaviour has been reported only in one *Apis* species, *A. florea* (Akaratnakul, 1977; Dutton and Free, 1979; Booncham et al., 1995; Wongsiri et al., 1996). After a short distance absconding, *A. florea* workers return to the old nest and collect wax from the abandoned comb to reuse (Booncham et al., 1995; Wongsiri et al., 1996). I observed that forgers of *A. florea* collected wax only from the crown or honey storing areas (Figure 16.3a, b), but I did not observe the workers collecting wax from the other part of the comb (Figure 16.3c). This is because the brood areas of the comb are used for brood rearing, so during the brood cycles, it contains larvae faeces, and the silky cover of pupae (Hepburn, 1986), which makes it more difficult to harvest pure wax from brood cells.

Why does only the *A. florea* species salvage the wax from the crown? This is probably because *A. florea* has a small absconding range less than 100 m. If the absconding range is more than 100 m, the foragers do not return to salvage wax (Seeley, 1985). Pirk et al. (2011) nicely explain that the energetic trade-off, which could be either between the energy of wax recovered or energy invested to replace the wax, is the reason for *A. florea* salvaging the wax.

Distribution

A. florea occurs in both tropical and subtropical regions of Asia (Otis, 1991). The distribution of *A. florea* overlaps in the south and eastern parts of Asia (Hepburn et al., 2005). It is common in central India (Soman and Chawda, 1996) and is widely distributed in Thailand (Wongsiri et al., 1996), but it has not been found in the southern Malay Peninsula and Indonesia (Figure 16.4a).

A. florea colonies are occurring in tropical and subtropical areas of Nepal, including THE Terai region. Most of the *A. florea* colonies are occurring in the southern part of Nepal (Figure 16.4b, c). In 2015, a new distribution of *A. florea* was found in Khotang (27°12′N; 86°47′E, 2000 m) the eastern part of Nepal, where the bees coexist with native honeybees, *A. cerana* and *A. laboriosa*. The distribution of *A. florea* is beyond the previous report (Ruttner et al., 1985; Wongsiri et al., 1996). This indicates that *A. florea* is a very successful colonizer, even where the conditions are less favourable, such as in the cold mountainous areas of Nepal. Its highly colonizing abilities suggest that *A. florea* is capable of adapting its nesting

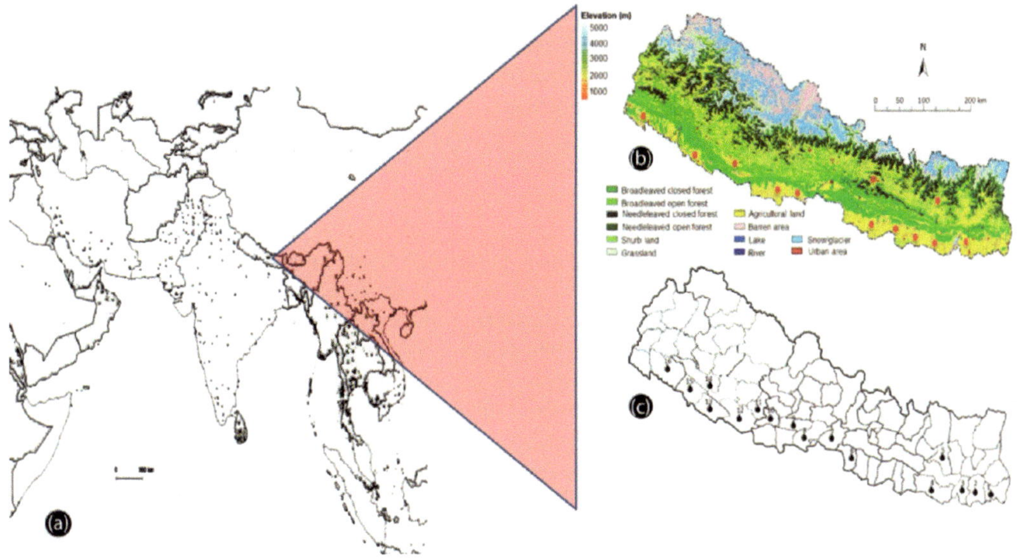

FIGURE 16.4 (a) Map of *A. florea* distribution in Asia (adapted from Hepburn et al., 2005) – full references available at http://www.edpsciences.org/apido; (b) Distribution of forest in Nepal, and elevation, lower tropical (<below 300 m), upper tropical (300–1000 m), subtropical (1000–3000 m), temperate (2000–3000 m), subalpine (3000–4000 m); (c) Areas of distribution of *A. florea* in Nepal. The symbols (♠) show newly recorded *A. florea* nest sites.

behaviour and foraging strategies in cold habitats. Most of the *A. florea* colonies were observed in village gardens, and mango and litchi orchards in the Terai regions of Nepal.

Hepburn et al. (2005) reported *A. florea* colonies from Tansen (Palpa district, Figure 16.4a). Now, I have found established colonies of *A. florea* in 16 different districts of Nepal (Table 16.1) and its coexistence with other *Apis* species – *A. laboriosa*, *A. dorsata*, *A. cerana* and *A. mellifera*.

TABLE 16.1

Geographical Distribution of *A. florea* in Nepal

Site Number	Districts	Geographic Locality	Coordinates (Latitude:Longitude)	Attitude (m)	No. Colony Observed
1	Jhapa	Terai belt	27°91′ N; 84°19′ E	506	5
2	Morong	Terai belt	26°40′ N; 87°30′ E	350	2
3	Sunsari	Terai belt	26°36′ N; 87°05′ E	349	6
4	Saptari	Terai belt	26°35′ N; 86°45′ E	350	3
5	Khotang	High mountain	27°12′ N; 86°47′ E	2,620	1
6	Birgunj	Terai	27°0′ N; 84°52′ E	92	2
7	Chitwan	Terai belt	27°35′ N; 84°30′ E	2,000	13
8	Nawalparashi	Terai belt	27°32′ N; 83°24′ E	1,937	9
9	Tansen	Inner Terai	27°52′ N; 83°3′ E	1,350	3
10	Arghakhachi	Inner Terai	27°54′ N; 83°09′ E	2,575	7
11	Pyuthan	Mountainous area	28°05′ N; 82°50′ E	1,500	1
12	Dang	Inner Terai valley	28°0′ N; 82°16′ E	1000	5
13	Banke	Terai belt	28°0′ N; 83°14′ E	350	1
14	Surket	Inner Terai Valley	28°36′ N; 81°38′ E	700	4
15	Bardiya	Terai belt	28°23′ N; 81°30′ E	374	3
16	Kailali	Terai belt	28°41′ N; 80°52′ E	1,957	5

Notes: The numbering used in this table is based on questionnaire survey from localities.
This site number corresponding with the numbers in Figure 16.4c.

During my survey, I did not observe *A. florea* colonies in Tansen in the 2017. This is probably due to a heavy application of insecticides in mustard flowers and vegetables. In areas of heavy pesticide uses, my surveys show that *A. florea* bees, including other species of *Apis*, and pollinators are drastically declining, and caused severe damaged in fruit production in Nepal.

Their Role in Natural and Agricultural Systems

A. florea is an economically important *Apis* species in Nepal. It stores small amounts of honey. Therefore, it has drawn less attention of beekeepers, who did not know the free pollinating values of *A. florea* by fruits, and vegetable growers. *A. florea* is an excellent pollinator, and it is easy to maintain in orchards (Wongsiri et al., 1996). The most important contribution of *A. florea* is its valuable pollination of many fruit plants in tropical to temperate ecosystems of Nepal. Therefore, the expansion of this species in the upper parts of Nepal may have positive impacts in adding to pollination associations for better Himalayan ecosystem functioning, and improving the productivity of highland cultivated crops. The expansion of this species in new habitats has to be monitored, an understanding of pollinator ecology needs to be investigated and potential positive, and negative, socioeconomic and ecological impacts in highland agricultural systems of Nepal need to be addressed.

REFERENCES

Akaratnakul P. (1977). The natural history of the dwarf honey bee, *Apis florea* F. in Thailand. Thesis, Cornell University, Ithaca, NY, pp. 1–91.

Booncham U., Wongsiri S., Thirakupt K. (1995). Niche differences between *A. florea* and *A. andreniformis* in dry-evergreen forest, Thailand. *International Conference on Tropical Bees and Environment*, Pedu Lake, Malaysia, p. 33.

Dutton R., Free J.B. (1979). The present status of beekeeping in Oman. *Bee World*, 60(4): 176–185.

Hepburn H.R. (1986). *Honey Bees and Wax: An Experimental Natural History.* Springer, Berlin, Germany.

Hepburn H.R., Radloff S.E. (1998). *Honey Bees of Africa.* Springer, Berlin, Germany.

Hepburn H.R., Sarah E.R., Otis G.W., Fuchs S., Verma L.R., Tan K., Chaiyawong T., Tahmasebi G., Ebadi R., Wongsiri S. (2005). *Apis florea*: Morphometrics, classification, and biogeography. *Apidologie*, 36: 359–376.

Mutto R.N. (1956). Facts About beekeeping in India. *Bee World*, 37: 125–133, 154–157.

Oldroyd B.P., Wongsiri S. (2006). *Asian Honey Bees: Biology, Conservation, and Human Interactions.* Harvard University Press, Cambridge, MA, p. 340.

Otis G.D. (1991). Revised distribution of three recently recognized species of honey bees in Asia. *Honeybee Science*, 15(4): 167–170.

Otis G.W. (1996). Distributions of recently recognized species of honey bees (Hymenoptera: Apidae; *Apis*) in Asia. *Journal of the Kansas Entomological Society*, 69(4): 311–333.

Pirk C.W.W., Kendall L.C., Daungphakdee O., Radloff S.E., Hepburn R. (2011). Economic of comb wax salvage by red honey bee, *Apis florea. Journal of Comparative Physiology, Biochemical, Systemic, and Environmental Physiology*, 181(3): 353–359.

Roubik D.W., Sakagami S.F., Kudo I. (1985). A Note on distribution and nesting of the Himalayan honey bee, *Apis laboriosa* Smith (Hymenoptera: Apidae). *Journal of the Kansas Entomological Society*, 58(4): 746–749.

Ruttner F., Pourasghar D., Kauhausen D. (1985). Die Honigbienen des RAN.1. *Apis florea* Fabricius. *Apidologie*, 16(2): 119–138.

Seeley T.D. (1985). *Honey Bee Ecology: A Study of Adaptation in Social Life.* Princeton University Press, Princeton, NJ.

Soman A.G., Chawda S.S. (1996). A contribution to the biology and behavior of the dwarf honey bee, *Apis florea* F and its economic importance in Kutch, Gujarat, India. *Indian Bee Journal*, 58: 81–88.

Thapa R. (2001). The Himalayan giant honey bee and its role in ecotourism development in Nepal. *Bee World*, 82(3): 139–141. doi:10.1080/0005772X.2001.11099516.

Thapa R. (2011). Rain proofing behaviour of *Apis florea* and *Apis andreniformis*. *Bee World*, 88(3): 53–55. doi:10.1080/0005772x.

Thapa, R., Shrestha, R., Manandhar, D. N., Bista, S., Kafle, B. (2000). Beekeeping in Nepal. In: Matsuka, M., Verma, L.R., Wongsiri, S., Shrestha, K.K., Uma Partap (Eds.), *Asian Bees and Beekeeping: Progress of Research and Development. Proceedings of the Fourth International Conference*, 23–28 March 1998, Nepal Kathmandu. Oxford and IBH Publishing Co. Pvt. Ltd., New Delhi, pp. 409–413.

Thapa R., Wongsiri S. (2011). Wax specks as remnants cues for recognition of old nest sites of immigrant colonies of *Apis dorsata* F. in northern Thailand. *Bee World*, 88: 79–81. doi:10.1080/0005772X.2011.11417432.

Tigari S. (1971). Biology and behavioral characteristics of the Iranian honey bee (*Apis florea*). *Proceeding of the International Beekeeping Congress*, 23: 344–345.

Wongsiri S., Lakprayoon A., Thapa R., Thirokupt K., Rinderer T.E., Sylvester A., Oldroyd B.P., Boonchai U. (1996). Comparative biology of *Apis andreniformis* and *Apis florea*. *Bee World*, 77(4): 23–25.

Woyke J., Wilde, J., Wilde M. (2012). Swarming and migration of *Apis dorsata* and *Apis laboriosa* Honey bees in India, Nepal and Bhutan. *Journal of Apicultural Science*, 56(1): 81–91.

17

Potential of Apis florea *as a Relevant Insect for Providing Pollination and Ecological Services in Thailand*

Preecha Rod-im and Orawan Duangphakdee

CONTENTS

Foraging Behaviour

The red dwarf honeybee, *Apis florea* Fabricius, is one of the two smallest honeybee species of Southern and South East Asia (Ruttner, 1988). The colonies construct a small, open-air nest and have become adept at surviving in various types of habitats. The red dwarf honeybee is a highly adaptive species, confined to warm climates (Seeley et al., 1991); its distribution ranges from tropical to subtropical in southern Asia, the Philippines to southern China, in Laos (east) and India (west) (Oldroyd and Wongsiri, 2006). To date, the distribution of *A. florea* extends westwards to the Middle East including Iran and Iraq, and the Arabian Peninsula, Oman, Yemen and central Saudi Arabia (Hepburn et al., 2005; Hepburn and Radloff, 2011).

One of the important traits that allows *A. florea* to be able to adapt to various habitats is its catholic taste – its ability to forage on the wide range of flora. The size of the colony should also be taken into account. As one of the smaller honeybees, a strong colony comprises between 5000 and 10,000 individuals, which is fourfold less than *Apis cerana*, sixfold less than *Apis mellifera* and tenfold less than *Apis dorsata*. The relatively low number of individuals in a colony enables *A. florea* to fill an ecological niche where floral resources are limited or do not produce sufficient resources to support the other larger honeybee species.

Apis florea is known to continue foraging on the same floral species longer than other bee species. In a study conducted in Thailand of insects visiting Makiang flowers (*Cleistocalyx paniata* var. *paniata*), eight insect species were identified, the three most abundant of which were honeybees – *A. cerana, A. mellifera* and *A. florea* (Sirinun, 2002).

These three honeybee species began actively foraging 6:00–7:00 a.m. *Apis cerana* and *A. mellifera* showed high peak foraging activity until 9:00 a.m. However, the number of *A. florea* foragers increased gradually as time passed, their apex occurring between 2:00 p.m. and 4:00 p.m. Generally, honeybee species that forage on the same resource will compete against each other. *Apis mellifera* outcompetes *A. cerana* foraging on many crops (Partap, 2000), and removal of the *A. mellifera* colonies will significantly increase the number of *A. cerana* foragers on those crops. *Apis florea* shares pollen partitioning with *A. mellifera* on *Acacia seyal*, date palm and onions by collecting at different times. Bee morphology also plays a predominant role in determining the resource partitioning of the bees. Oldroyd et al. (1992) and Duangphakdee (2000) showed a significant correlation between *A. florea* body size and tongue length and flower morphology. *Apis florea* was frequently as successful as *A. cerana* and *A. dorsata* at the foraging

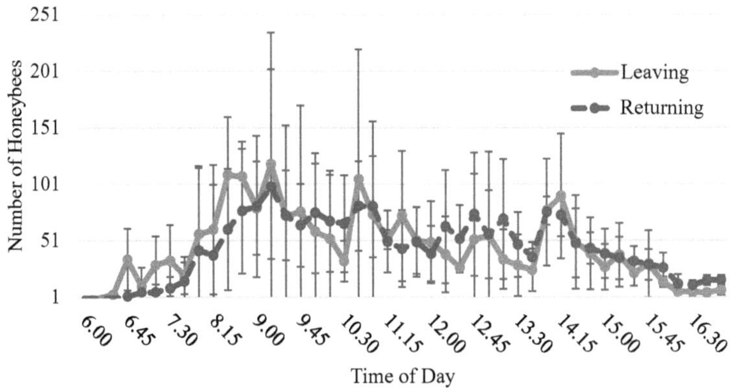

FIGURE 17.1 The mean frequency distributions of flight activities; the number of *Apis florea* foragers leaving and returning to the nest ($n = 7$). (Some data acquired from Rod-im, P. et al., *J. Apic. Res.*, 54, 246–254, 2015a).

stations, attracting a number of foragers, maintaining its dominance and inducing the departure of the other bee species (Abrol, 2011). *Apis florea* workers tend to become active around 6.15 a.m., with foragers returning to the nest approximately an hour later (Figure 17.1). The bees collect pollen throughout the day without any noticeable interruption (El-Shafie et al., 2002; Abrol, 2011). Flight activity peaked between 08:30 and 10:00 a.m. after which the bees continued foraging but at a lower rate. However, foraging activities differ in duration and by the number of visits according to the different flower species. A significant effect was noticed on foraging behaviour according to resources available and the quality of the reward. *Apis florea* foragers visited the *A. leptopus* flowers at 7:45 a.m. and gradually reached peak foraging between 12:15 p.m. and 16:00 p.m. (Figure 17.2) On the same day, and over a similar distance, *A. florea* foraging activity began slightly later, at 8:30 a.m., on *C. sulphurous* flowers (Figure 17.2). The number of foragers per flower showed multiple peaks occurred, the highest of which was 10:00–11:30 a.m. Far fewer foragers visited the *C. sulphurous* site. *A. florea* foraging activities on these two species of honeybee

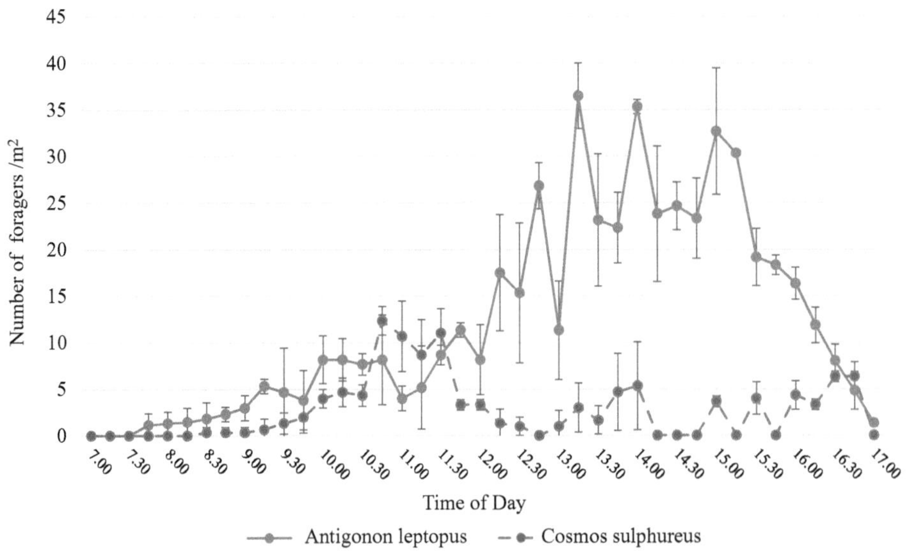

FIGURE 17.2 The mean frequency distributions of the number of *Apis florea* foragers visiting *Antigonon leptopus* and *Cosmos sulphureus* flowers.

flora showed no correlation to a noon lull. Even though the honeybee dancing behaviour reflected the noon lulls (Oldroyd et al., 1992; Wongsiri et al., 1998; Duangphakdee et al., 2009), the flow pattern of *A. florea* flight activity was consistent, with no obvious variance during the azimuth at noon for the duration of the experiment. The number of bees foraging on *A. leptopus* was significantly higher than those on *C. sulphurous* during the same period of time and over the same distance (Rod-im et al., 2015a). On *Archontophoenix alexandrea* inflorescences, a good pollen resource for honeybees, the larger species, *A. dorsata* and *A. cerana*, foraged early when the pollen was most abundant, yet at relatively low numbers. The smaller-sized *A. florea* and *A. andreniformis* were present in larger numbers and continued to forage after the pollen had been somewhat depleted by the larger species. *Apis florea* also took over the partition of the stingless bees, even though those species appear before the *Apis* species (Oldroyd et al., 1992).

Absconding and Swarming Behaviour

Absconding is one of the most important adaptive characteristics in tropical honeybees. This trait is generally restricted to the movement of one colony, or a few colonies, away from the maternal nest site to another, which is usually caused by local environmental disruptions. The striking differences between the extent of resource investment in reproductive swarming compared with absconding and migration are ascribable to the evolution of the various honeybee species in regions with different climates and associated seasonality of flowering in tropical and temperate regions (Hepburn, 2011). Absconding, migration and swarming in honeybees are strategies related to the phenology of flowering across the ranges of temperate and tropical ecosystems. Reproductive swarming almost always occurs in the ascendant phase of colony growth, driven by pollen/nectar flows (Seeley, 1985; Crane, 1993; Hepburn and Radloff, 1998), while migration (= absconding) is a resource-related, seasonal movement of tropical honeybee colonies, usually without the production of queens and drones. Absconding/migration maximize the new areas and provide a spatial refueling cycle clearly driven by r-selection, which is an evolutionary alternative to the K-selection (Verhulst, 1838; Pianka, 1970) hoarding characteristic of temperate bees (Hepburn, 2006). The primary stimuli for prepared absconding or migration are chronic resource depletion, deterioration of the microclimate of the nest or continuous disturbance by pests or predators. Nest abandonment by absconding/migration is a ubiquitous behaviour in tropical African and Asian honeybees, including *A. florea* (Hepburn, 2011). There have been numerous reports on this behaviour in the red dwarf honeybees, *A. florea*, which are subject to severe predation pressure. *Apis florea* must select for greater mobility through absconding and migration, which is also facilitated by climates and perennial flowering (Hepburn and Hepburn, 2011). Various aspects of absconding behaviour by *A. florea* including nest-site selection (Oldroyd et al., 2008), colony density and aggregations (Rinderer et al., 2002; Wattanachaiyingcharoen et al., 2008), precision of the waggle dance (Beekman et al., 2008; Duangphakdee et al., 2009) and the salvage of wax from abandoned combs (Hepburn et al., 2010) have thus far been documented. Dispersal is a critical determinant in population dynamics, and how this behaviour is adaptive and responds to changes in selection pressure remains highly relevant. Dispersal behaviour has long been thought to be at least partially under genetic control (Hepburn, 2006) and has therefore been generally treated as an adaptation generated by natural selection (Hamilton, 1996). As with any evolving life history trait, dispersal will be shaped by its cost-benefit ratio and genetic constraints. The question of *A. florea* colony density was raised by Rinderer et al. (2002), who reported that their study area contained clear aggregations of *A. andreniformis* and *A. florea*, and a possible aggregation of *A. cerana*. They recorded and plotted the distribution of a mixed population of *A. andreniformis, A. cerana, A. dorsata* and *A. florea* in orchards near Chanthaburi, Thailand, over a brief time frame of three weeks in March. They observed no emigration of *A. florea*, and the possible immigration of a very few new colonies. Nonetheless, *A. florea* is well known for its absconding/migration propensity (Akratanakul, 1977; Wongsiri et al., 1997). Wattanachaiyingcharoen et al. (2008) also conducted surveys of an area of woodland in Phitsanulok Province, Thailand, with 15 colonies of *A. florea*. The colonies had a highly aggregated spatial distribution (Standardized Morisita's Index of Dispersion = 0.59). Microsatellite analysis based on five loci showed that none of the colonies were related as mother-daughter colonies, suggesting that unrelated colonies tend to nest near existing colonies. Wattanachaiyingcharoen et al. (2008) studied an area of 4.16 km^2 divided into quadrats of 0.16 km^2 and found three clusters of aggregated colonies (5, 5 and 4 colonies), situated in very distant quadrats. This works out to a density of about 30 colonies/km^2 and capacity of 12.2 colonies month/km^2,19–42 colonies with a mean of 32.77 which, with an effective area for nesting of 2.8 km^2, equates to a range of 7.1 to 14.3 colonies km^2/month and an average carrying capacity of 12.2 colonies km^2/month.

Demographic data consisting of observations on 124 immigrant, 105 emigrant and 302 sedentary colonies were obtained in Ratchaburi, Western Province of Thailand. The data were analysed to interpret the dynamic balance of *A. florea* populations with regards to colony density, population stability, capacity for expansion or decline over an extended period of time (Duangphakdee personal obs.). The length of stay for the 302 sedentary colonies averaged 3.8 ± 2.8 months and varied between 1 and 13 months (Figure 17.3). The study found a Standardized Morisita's Index of Dispersion 0.558–0.602, which was calculated and closely coincides with the data of Wattanachaiyingcharoen et al. (2008).

The adaptive significance (if any) of nest aggregation in *Apis* remains unclear. Nesting in aggregations is likely to be associated with at least some costs. The risks of honeybees nesting in aggregations could well be reasonably high because doing so clearly holds potential disadvantages. For example, individual colonies increase the risk of acquiring pests and diseases from other colonies because of the clumped nesting distribution, using overlapping foraging ranges and the tendency for bees to drift between nests must certainly be increased by aggregation. Also, the propensity of 'robbing' weaker colonies is well known in *A. mellifera*, and it probably occurs to some degree in the other species as well. Some colonies are weak because they are diseased; robbing will both spread disease and further debilitate neighbouring colonies, if not completely destroy them. Finally, nest aggregations may enhance predators' nest-finding abilities, which could lead to return visits to the aggregation areas to prey on the remaining nests (Seeley et al., 1982). However, the fact that nest aggregations occur in at least four of the nine known species of honeybees suggests that there are likely to be adaptive benefits of aggregation that outweigh the obvious potential costs. Briefly, the suggested benefits are: (i) swarms may seek to join aggregations because of the potential for cooperative nest defence among colonies (Seeley et al., 1982); this seems unlikely because, at least in *A. dorsata*, adjacent colonies rarely join defensive swarms when a colony is attacked by a mammalian or avian predator (Kastberger and Sharma, 2000); and (ii) swarms might seek to join aggregations as a means of improving the mating success of offspring derived from queens and drones (Oldroyd et al., 2000).

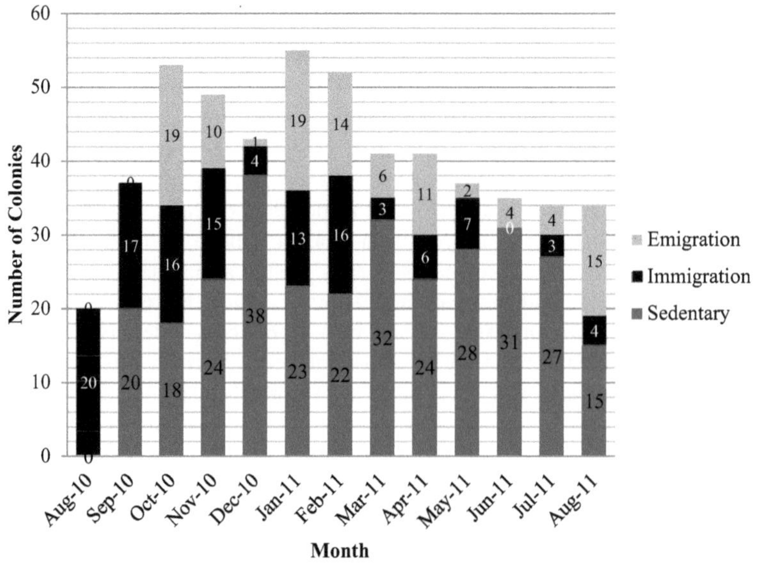

FIGURE 17.3 The seasonal phenology and number of emigrating, immigrating and sedentary *Apis florea* colonies. (N.B. Sedentary colonies were present throughout, from September 2010 to August 2011.)

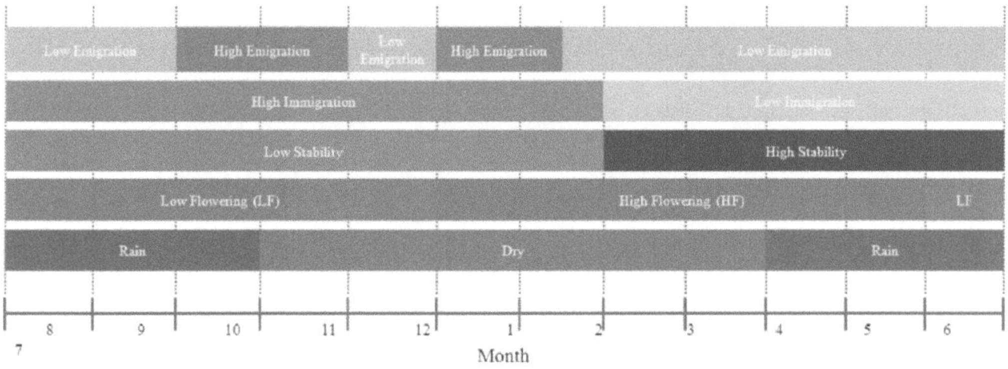

FIGURE 17.4 The seasonal phenology of (from top downwards) emigration, immigration, population stability, flowering intensity and rainfall on Ratchaburi Campus, King Mongkut's University of Technology, Chombueng, Ratchaburi, Thailand (13° 59′ N, 99° 51′ E, Altitude 86 m) between August 2010 and August 2011. (Courtesy by R. Hepburn and O. Duangphakdee.)

Oldroyd and other authors (2000) conducted a survey of *A. florea* colony density in an area of Muang District, Phitsanulok Province, Thailand. The survey took place over a period of four months, June–September 2004, during the rainy season when *A. florea* populations are at their lowest. The study area comprised numerous shrubs, trees and some buildings. *Apis florea* is not known for its high colony stability (Duangphakdee, personal observation). During the dry season, colonies are at their most stable, flowering trees are abundant and both immigration and emigration events are reduced during this time and only increase with the onset of the rains (Figure 17.3). The question as to whether immigration and emigration constitute 'absconding' is moot. We now know that roughly four months is required for an *A. florea* colony to settle at a new nest site, mature and reproductively swarm (Rod-im et al. 2015b). Seasonal fluctuations of *A. florea* are hypothesized (Hepburn, Personal observation) to be influenced by relative humidity, temperature, number of plant species in flower and total floral abundance (Figure 17.4).

Beekeeping Potential of *Apis florea*

The economic use of *A. florea* in Thailand suggests that there is a possibility for a unique beekeeping system that enhances sustainability at both economic and environmental levels. The dwarf honeybees usually construct a single comb. Onset of comb building starts immediately within the first hour after settling at the new nest site (Figure 17.5); the comb reaches maturity in fewer than 20 days. Abandonment of the nest, as documented by Hepburn et al. (2014), could be 4–13 months later. Changes in the development of an *A. florea* vertical, single comb nest once the swarm has settled is shown over a period of 17 weeks in Figure 17.6.

Rod-im et al. (2015b) documented the changes in *A. florea* nests where the different stages of construction were studied over time. The sixth day after settling, the bees had constructed a suitably sized comb, and eggs and young larvae were observed in some cells. On the eighth day, the nest had already been partitioned into an area for honey (comb crown was constructed above the foundation twig), with an underlying section for pollen storage, below which both capped and uncapped brood cells occurred. On the twentieth day, adults had emerged from the oldest patch of brood, extensive capping of brood cells continued and the nest had reached the fully mature stage.

The dimensional growth curves of the nests for the three *A. florea* colonies showed daily changes in comb length, width and area, from the inception of the nest to their completion and maturity (Figure 17.6). Initially, the lengths of the nests increased by about 4.2 cm/week, thus, doubling in length over the first three weeks. Over the subsequent 12 weeks, the nests continued to lengthen by about 8.4 cm/week, with a final growth surge reaching roughly 23 cm at full maturity at four months. The widths of the *A. florea* single

FIGURE 17.5 Comb development in *Apis florea* after 24 h of settling at a new nesting site (Courtesy by P. Rod-im).

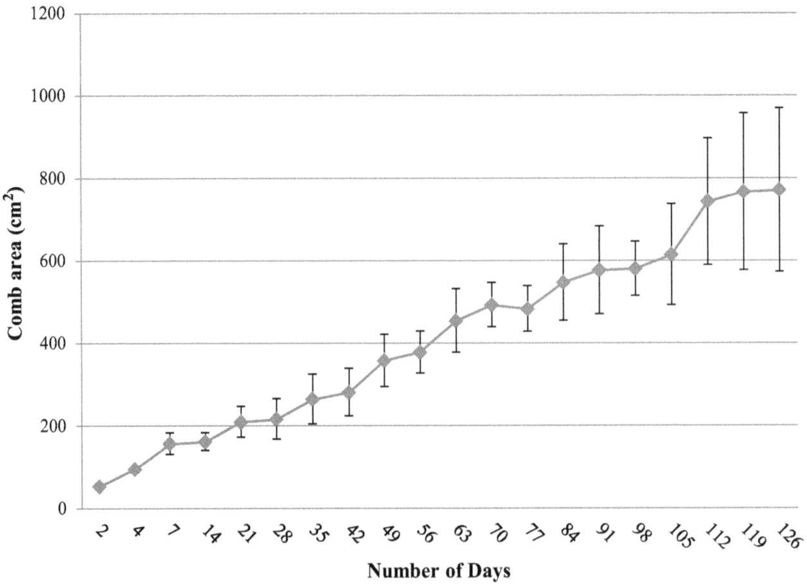

FIGURE 17.6 Changes in *Apis florea* comb area from the settling date until absconding. (From Rod-im, P. et al., Developmental cycles of the red dwarf honeybee, *Apis florea*, *Proceedings of the International Conference on Agricultural, Ecological and Medical Sciences* (AEMS-2015), pp. 96–99, 2015b.)

comb nests increased at about 3 cm/week over the first three weeks, from about 7 cm at their inception, to reach about 15 cm. In the next phase of continued expansion, the width of the nests increased to about 20 cm or 1 cm/week, and slowly continued to increase at about 0.4 cm/week until reaching some 23 cm at maturity. Over the four months required to reach maturity, the combs showed the same rate of expansion in length and width, culminating in a final structure that was close to circular, apart from the portion of drone comb that protrudes downward from the bottom of the comb. The development of the crown comb would seem to be both structurally and biologically somewhat independent. Obviously, there is not necessarily a relationship between changes in the crown comb and the subtending brood comb. Nonetheless, a substantial amount of honey crown comb construction was already apparent on the fourth day, and by the eighth day the honey had been capped. It remained as such until, on the 23rd day, worker bees opened some cells and removed the honey. Over the same period, there were changes in the length and width of the crown comb. One week after settling, the colonies had gradually expanded the length of the crown comb by about 2 cm. The crown comb continued to thicken slowly at about 2 cm/week until fully mature.

Apis florea are traditionally maintained by beekeepers to produce honey for sale in many regions of Asia (e.g., Pakistan, India, Thailand, Vietnam and Laos). An *A. florea* beekeeper searches for feral nests in natural localities. To harvest the honey, the beekeeper will cut a twig of approximately the same size and diameter as the crown twig. The twig is slid into the comb, between the honey crown and the brood, and substituted for the crown twig, ensuring the nest is still supported. The bees then build a new crown comb around the substitute twig and begin storing honey. This method of harvesting honey is sustainable: (i) it does not lead to the colony abandoning its nest and absconding; (ii) there is no destruction of the brood area; and (iii) the colony does not lose time and energy looking for a new nesting site and rebuilding its nest but is able to continue collecting resources and pollinating.

Rod-im (2015b) conducted further research that supports the sustainable and practical approach to beekeeping with *A. florea*. Throughout the 4-month nesting cycle, *A. florea* continually stored honey, capped and uncapped the honey cells in conjunction with the simultaneous production of brood (Figure 17.7). The more the honey crown is harvested, the higher the production of honey; the bees ensure there are enough stores for the growing colony. Observations of naturally occurring *A. florea* colonies showed a small-sized colony, with a comb area <270 cm^3 stored 203.75 ± 26.88 g of honey, with a range between 170 and 230 g; a medium-sized comb, <270–510 cm^3, stored 270 ± 77.89 g, ranging between 160 and 330 g; and a large-sized comb stored 338.75 ± 103.47 g, ranging between 240 and 480 g.

Rod-im (2014) established whether the colony would accept a piece of bamboo, cut in half horizontally, as a substitute foundation twig. The bees accepted the substitute and built crown comb around the bamboo within 4–7 days. As in earlier publications on colony development, Rod-im et al. (2015b) showed that by day eight, the colony had begun storing honey. This technique allows the beekeeper to harvest the honeycomb with the least amount of disturbance to the colony. The colony then rebuilds the comb crown, which is less destructive and time consuming than if the whole nest had to be rebuilt and the stores replenished (Figure 17.8). As a result, an increase of the number of honey harvest events occurred (Figure 17.9). This resulted in a honey yield of approximately 3.3 to 4.5 times higher than the average yield of the first harvest from a wild colony (Rod-im, 2015a, 2015b). The results were substantiated and are of ecological significance when compared to related data on the immigration, emigration and stability of nests over a full year (cf. section above). The yield from the wild colony was 270 ± 89.92 (Rod-im, 2014, 2015b). The consecutive removal of the *A. florea* honeycomb over the developmental cycle yielded

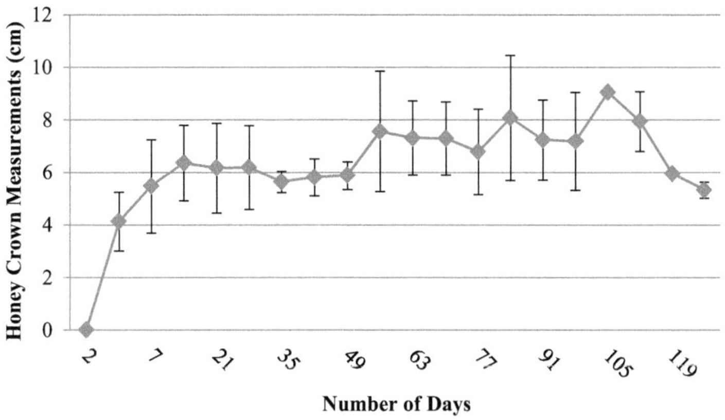

FIGURE 17.7 Measurements of *Apis florea* honey crown comb from the date of settling until absconding. (From Rod-im, P. et al., Developmental cycles of the red dwarf honeybee, *Apis florea, Proceedings of the International Conference on Agricultural, Ecological and Medical Sciences* (AEMS-2015), pp. 96–99, 2015b.)

FIGURE 17.8 *Apis florea* comb sold along the roadside in western Thailand.

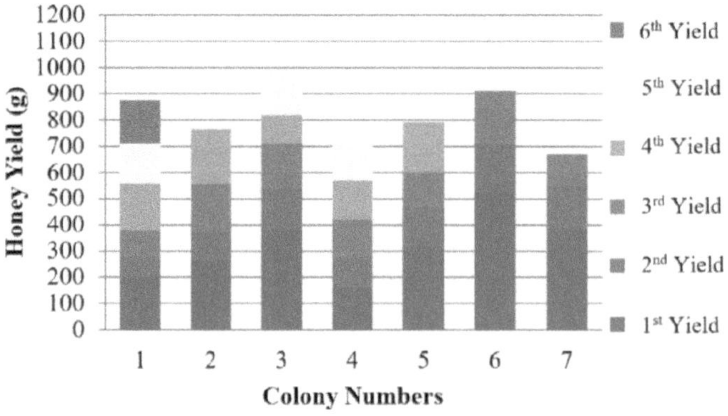

FIGURE 17.9 Honey yielded from the sustainable harvesting of the honey crown experiment conducted on seven *Apis florea* colonies. When 80% of the honey cells were capped, honey from the crown was harvested leaving the rest of the comb untouched. The experiment ran continuously until the colonies absconded. (From Rod-im, P., Possibility of Commercial Beekeeping Development of the Red Dwarf Honeybees, *Apis florea* for Honey Production in Comparison to the Traditional Honey Hunting, MSc Thesis, School of Bioresources and Technology, King Mongkut's University of Technology Thonburi, Bangkok, Thailand, 2014.)

1,229.56 ± 230.26 g honey (Figure 17.9), which is 4.5 times higher than the honey collected from wild honeybee combs by the conventional honey hunting method. The quick development, tendency to utilize a wide range of floral resources and the medium size of the colony are contributing factors to the high economic potential of *A. florea* for beekeeping in urban areas.

Apis florea as a Pollinator in Ecosystems

The majority of flowering plants are pollinated by insects and other animals. The estimation, based on 352,000 species of angiosperms, is that 67% to 96% of flowering plants need pollination. The dependence of plants on animals increases from temperate to tropical zones from 78% to 94%, respectively (Figure 17.10). Based on an estimation of 352,000 plant species of angiosperms Kier et al. (2005) and

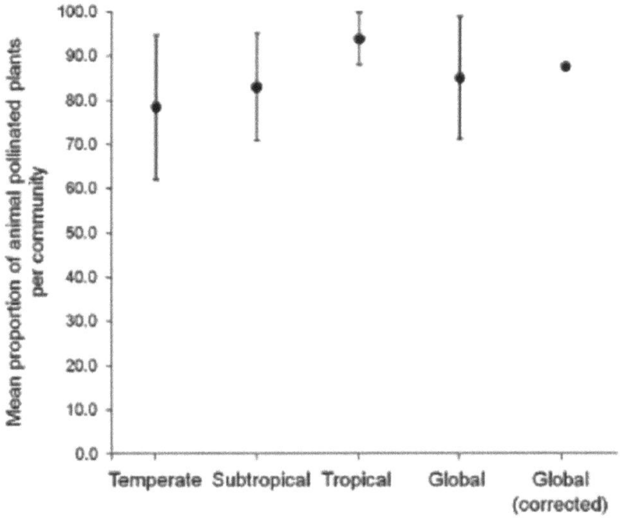

FIGURE 17.10 Mean (SD) proportion of animal pollinated plant species found in terrestrial communities at different latitudes. The global average represents the mean of all studies used in the analysis. The global (corrected) average accounts for the relative distribution of plant diversity in the different zones. Data for the three latitudinal zones were analysed with a one-way ANOVA. (From Ollerton, J. et al., *Oikos*, 120, 321–326, 2011.)

Ollerton et al. (2011) calculated that 50% of the species are distributed in the tropics, 27.7% in the subtropical and 21.8% in the temperate regions. Around 85% of flowering plants, or 299,200 species, are biotically pollinated (Figure 17.10).

Apis florea plays a vital role in the pollination of cultivated crops and plants growing in its natural habitats in Thailand. As an endemic species, the honeybees have co-evolved with the native flora, which has led to specific associations between the plant species and the bees. Because *A. florea* colonies are able to adapt to various habitats, they are regarded as one of the most suitable bee species to coexist with humans. *Apis florea* plays a critical role in maintaining the vital ecological process of cross-pollination in agricultural, disturbed and fragmented areas (Roubik et al., 2005). *Apis florea* migrates in response to the availability or scarcity of flowering plants. Neupane and Thapa (2005) reported that foraging activity is greatly influenced by weather conditions and the availability of nectar and floral resources. In relation to the behaviour of a short flowering period in tropical plants, the honeybees have developed the ability to establish and expand the colony rapidly in response to available resources to ensure survival (Itioka et al., 2001).

The sharp decline in the number of native honeybee colonies in Thailand, due to the extension of agricultural areas and urbanization, has been reported over the past four decades. This may be attributable to extensive deforestation between 1976 and 1989 due to the increase of urbanization and the construction of roads. Although no systematic study has been conducted, the number of the larger bee species, such as *A. dorsata* and *A. cerana*, has declined more rapidly than that of *A. florea* (pers. obs.). The giant honeybees, *A. dorsata*, migrate annually and form large colony aggregation areas; Ruttner (1988) reported more than 100 colonies in a single large tree. Large aggregations, such as this, require a substantial amount of floral resources, and it is this reward that drives the annual *A. dorsata* migrations (Oldroyd et al., 2000).

In a study conducted at the Thai Department of Agriculture's Chanthaburi Tropical Fruit Research Centre (CTFRC), 95 colonies of honeybees/km^2 were recorded. Of these, the dwarf bees, *A. florea* and *A. andreniformis*, were the most commonly occurring native species – 12 colonies of each, 5 colonies of *A. cerana* and 2 colonies of *A. dorsata*. The mean distance from an *A. florea* colony to the next nearest *A. florea* colony was 187 m, and the mean distance from that colony to the next nearest colony was 105 m (Rinderer et al., 2002). Liow et al. (2001) reported that the size and quality of habitats significantly impact the abundance of bee species to adjacent agricultural landscapes. In the presence of small-sized, fragmented patches of different habitats as a result of urbanization, the abundance of *A. florea* colonies signifies its role as a key pollinator species in maintaining plant diversity in these disjointed areas. Moreover, because of *A. florea*'s short foraging range, it may be least affected by pesticide usage and the destruction of natural habitats (Kremen et al., 2002). *Apis florea* has already been reported as one of the most efficient pollinators of various crops in Pakistan, e.g., canola, *Brassica napus* (Ali et al., 2011); onion, *Allium cepa* (Saeed et al., 2008; Sajjad et al., 2008); fodder, *Sesbania sesban* (Sajjad et al. 2009); fodder alfalfa, *Medicago sativa* (Ahmad, 1976); and bitter gourd, *Momordica charantia* (Saeed et al., 2012).

In a study of bee diversity and pollen sources of Apidae in Thung Salaeng Luang, the third largest National Park in Thailand, which comprises four forest types (deciduous dipterocarp, deciduous with bamboo, seasonal evergreen and dipterocarp-pine forests), it was found that *A. florea* colonies were the main pollinators of 46 of the 62 plant species recorded in the Park (Maxwell, 2004). Rod-im (2015a, 2015b), in a similar study conducted in the Tanassirim mountain range of western Thailand, found over 35 species of wild plants that were utilized by *A. florea* as a source of pollen and nectar. The morphology of this bee plays an important part in pollination because of its smaller size and widespread distribution across mainland Asia, occurring at elevations ranging from sea level to approximately 2000 m (Hepburn and Radloff, 2011). *Apis florea* has been found to be a regular visitor and pollinator in many crops, i.e., *Jatropha curcas* (Inson and Malaipan, 2011), hog plum, lychee, mango, pomelo, tomato, and it is reportedly an important visitor to cucurbits (Grewal and Sidhu, 1978). The same is true of crops being propagated for seed; *A. florea* was the most frequent visitor to cauliflower and fennel (81 %) and radish. Some observations from the 1940s recorded *A. florea* as the only species of honeybee found in sarson and toria (Rahman, 1940). *Apis mellifera* pollination of onion requires 12–15 hives/ha, as foragers tend to avoid the flowers. However, *A. florea* appears to be the most frequent *Apis* visitor on onion and was recommended for further consideration as a pollinator (Crane, 1991). *Apis florea* also plays a leading role in pollination of macadamias in northern Thailand (Kongpitak et al., 2004) and the nocturnally dehiscent king palm (*Archontophoenix alexandrae*) (Oldroyd et al., 1992), and it is a potential pollinator of *Muntingia calabura* (Hawkeswood and Sommung, 2016).

Pollination by wild animals is a key ecosystem service that benefits humans. A large proportion of the world's major crops are pollinated by animals (Richards, 2001; Ghazoul, 2005) or comes indirectly from animals fed on staple crops. Klein et al. (2007) collected primary data from 200 major food-producing countries and found 87 species of plants are dependent on animal pollination, while 28 species are not. If production volume is considered, 60% of global production does not depend on animal pollination whereas 35% does and 5% is unknown/not evaluated (Figure 17.11). Pollination is essential for 13 crops, of high importance for 30 crops, moderately important for 27, marginally important for 21 and 7 do not rely on pollination. The remaining nine crops remained unknown (Figure 17.12).

The decline of pollinator species can lead to a serious reduction in the quantity and number of plant species, biodiversity and essential nutrients in ecosystems, thus affecting humans and other living creatures. According to Williams (1994), 70% of tropical crops are likely to have at least one variety that is improved by animal pollination. Studies on honeybees have mainly focused on *A. mellifera* and *A. cerana*, the most economically valuable pollinators of crops grown in monocultures worldwide (McGregor, 1976; Watanabe, 1994; Roubik, 2002). In Thailand, *A. mellifera* colonies are introduced species and often do not visit some of the native plants. The indigenous bees, therefore, have an important role in ensuring pollination. Compared to other wild bees, *A. florea* colonies are adaptable, versatile, cheap, effective and a convenient choice as pollinators. Implementation of bee pollination must be locally managed, taking into consideration farming methods, crops cultivated, honeybee traits and local customs. Additionally, the strong turnover in the quality and quantity of floral resources is known to influence the behaviour and foraging strategy of pollinator species (Aluri and Subba Reddi, 1994). High humidity, heavy rainfall, wind, and low temperatures have a negative effect on visits to sunflower inflorescence. Different floral nectar compositions are preferred by different species (Abrol, 2011). *Apis florea* prefers flowers with low caloric rewards (Sihag and Rathi, 1992), whereas *A. dorsata* prefers flowers with high caloric rewards (Sajjad et al., 2017). However, this could be a result of resource partitioning by honeybees. In a competition, *A. florea* will lose the battle against *A. dorsata* and *A. cerana*. Although it has been reported that *A. florea* prefers flowers with low caloric rewards, this might be a disputed claim as *A. florea* is often found in high resource-rich flowers such as pumpkin and king palm.

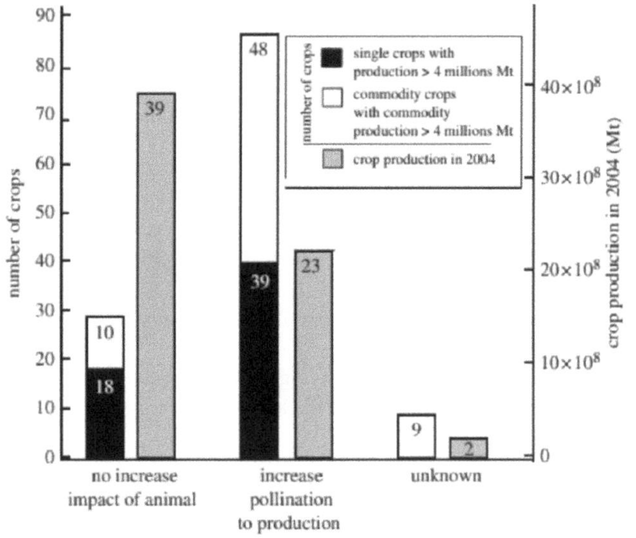

FIGURE 17.11 Relative importance of animal pollination in leading global crops and global production volume; 35% rely on pollinators, 60% do not rely on pollinators and 5% were unknown. (From Klein, A.M. et al., *Proc. Royal Soc. B*, 274, 303–313, 2007.)

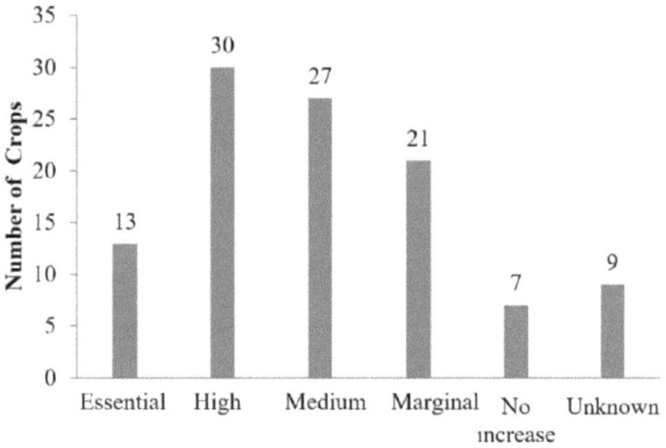

FIGURE 17.12 Level of dependence on animal-mediated pollination in 156 species of major global crops whose production was greater than 4,000,000 metric tons (Mt). (From Klein, A.M. et al., *Proc. Royal Soc. B*, 274, 303–313, 2007.)

The economic benefits of bees and other insect pollination have been discussed at both local and global levels. In considering the maintenance of honeybee populations, the conservation and management of resources within agricultural and surrounding natural landscapes need to be addressed (Zhang et al., 2007; Johannsmeier and Mostert, 2001), as well as the suitability of nesting habitats. The information on host plant species, temporal dynamics in providing floral resources, foraging preference, behavioural interaction between honeybees and flowers, and effect of foraging competition on crops between pollinators will serve as baseline studies for an ecosystem service. This would result in new beekeeping strategies being developed, providing a livelihood for local peoples (Devy and Davidar, 2006) and the conservation of natural habitats (Kovács-Hostyánszki et al., 2017).

REFERENCES

Abrol, D. P. (2011) Foraging. In: H.R. Hepburn and S.E. Radolff (Eds.), *Honeybees of Asia*. Springer, Berlin, Germany. pp. 257–292.

Akratanakul, P. (1977) The natural history of the dwarf honey bee, *Apis florea* F in Thailand. Thesis, Cornell University.

Ali, M., Saeed, S., Sajjad, A. and Whittington, A. (2011) In search of the best native pollinators for canola (*Brassica napus* L.) production in Punjab, Pakistan. *Journal of Applied Entomology and Zoology* 46: 353–361.

Aluri, J.S.R. and Subba Reddi, C. (1994) Pollination ecology and mating system of the weedy mint *Leonotis nepetaefolia* R. Br. in India. *Proceedings of the Indian Natural Science Academy B* 60: 255–268.

Ahmad, R. (1976) A note on insect pollinators of alfalfa in Pakistan. *The New Zealand Entomologist* 6: 190–191.

Beekman, M., Gloag, R.S., Evan, N., Wattanachaiyingchareon, W. and Oldroyd, B. (2008) Dance precision of *Apis florea* – clues to the evolution of the honeybee dance language. *Behavioural Ecology and Sociobiology* 62: 1259–1265.

Crane, E. (1991) *Apis* species of tropical Asia as pollinators and some rearing methods for them. *Acta Horticulture* 288: 29–48.

Crane, E. (1993) Current status of research on Asian honey bees. In: L.J. Connor, T.E. Rinderer, H.A. Sylvester and S. Wongsiri (Eds.), *Asian Apiculture*. Wicwas Press, Cheshire, CT. pp. 19–41.

Devy, M.S. and Davidar, P. (2006) Pollination systems of trees in Kakachi, a mid-elevation wet evergreen forest in the Western Ghats, India. *American Journal of Botany* 90: 650–657.

Duangphakdee, O. (2000) The compatibility of *Apis florea* Fabricius, 1787 and floral resource morphology in Chulalongkorn University. Senior Project, Faculty of Sciences, Chulalongkorn University, Bangkok, Thailand.

Duangphakdee, O., Radloff, S.E., Pirk, C.W.W. and Hepburn, H.R. (2009) Sun angle time windows for absconding by the dwarf honeybee, *Apis florea*. *Journal of Insect Physiology* 55: 1009–1012.

El-Shafie, H.A.F., Mogga, J.B.B. and Basedow, T. (2002) Studies on the possible competition for pollen between the honey bee, *Apis mellifera sudanensis*, and the imported dwarf honey bee *Apis florea* (Hym., Apidae) in North-Khartoum (Sudan). *Journal of Applied Entomology* 126: 557–562.

Ghazoul, J. (2005) Buzziness as usual? Questioning the global pollination crisis. *Trends in Ecological Evolution* 20: 367–373.

Grewal, G. S. and Sidhu, G. (1978) Insect pollination of some cucurbits in Punjab. *Indian Agricultural Journal of Science* 48: 79–83.

Hamilton, W.D. (1996) *Narrow Roads of Gene Land*. Vol. 1: Evolution of Social Behaviour. Oxford University Press, Oxford.

Hawkeswood, T.J. and Sommung, B. (2016) Pollination of *Muntingia calabura* L. (Muntingiaceae) by native bees in Bangkok, Thailand. *Calodema*. 421: 1–6.

Hepburn, H.R. (2006) Absconding, migration, and swarming in honeybees: An ecological and evolutionary perspective. In: V.E. Kipyatkov (Ed.), *Life Cycles in Social Insects, Behavior Ecology and Evolution*. St. Petersburg University Press, St. Petersburg. pp. 121–136.

Hepburn, H.R. (2011) Absconding, migration and swarming. In: H.R. Hepburn and S.E. Radloff (Eds.), *Honeybees of Asia*. Springer, Berlin, Germany. pp. 133–158.

Hepburn, H.R., Duangphakdee, O., Phiancharoen, M. and Radloff, S. (2010) Comb wax salvage by the red dwarf honeybee, *Apis florea* F. *Journal of Insect Behavior*. 23: 159–164.

Hepburn, H.R., Pirk, C.W.W. and Duangphakdee, O. (2014) *Honeybee Nests: Composition, Structure, Function*. Springer Verlag, Berlin, Germany.

Hepburn, H.R. & Hepburn, C. (2011) Bibliography of the Asian species of honeybees. In: H.R. Hepburn and S.E. Radloff (Eds) *Honeybees of Asia*. Springer, Berlin, Germany. pp. 473–657.

Hepburn, H.R. and Radloff, S.E. (1998) *Honeybees of Africa*. Springer, Berlin, Germany.

Hepburn, H.R. and Radloff, S.E. (2011) (Eds.) *Honeybees of Asia*. Springer, Berlin, Germany.

Hepburn, H.R., Radloff, S.E., Otis, G.W., Fuchs, S., Verma, L.R., Ken, T., Chaiyawong, T., Tahmasebi, G., Ebadi, R. and Wongsiri, S. (2005) *Apis florea*: Morphometrics, classification and biogeography. *Apidologie* 36: 359–376.

Inson, C. and Malaipan, S. (2011) Diversity of Bees (Hymenoptera: Apoidea) as insect pollinators on Physic nuts (Euphorbiaceae: *Jatropha curcas* L.) in Thailand. *Thai Journal* of *Agricultural Science* 44: 263–269.

Itioka, T., Inoue, T., Kaliang, H., Kato, M., Nagmitsu, T., Momose, K., Sakai, S. et al. (2001) Six year population fluctuation of the giant honey bee *Apis dorsata* (Hymenotera: Apidae) in a tropical lowland dipterocarp forest in Sarawak. *Annals of the Entomological Society of America* 94: 545–549.

Johannsmeier, M.F. and Mostert, A.J.N. (2001) South African nectar and pollen flora. In: M.F. Johannsmeier (Ed.), *Beekeeping in South Africa*, 3rd Edition. Plant Protection Research Institute, Handbook No. 14, Agricultural Research Council, Pretoria.

Kastberger, G. and Sharma, D.K. (2000) The predator-prey interaction between blue-bearded bee-eaters (*Nyctyornis athertoni*) and giant honeybees (*Apis dorsata*). *Apidologie* 31: 727–736.

Kier, G., Mutke, J., Dinerstein, E., Ricketts, T.H., Küper, W., Kreft, H. and Barthlott, W. (2005) Global patterns of plant diversity and floristic knowledge. *Journal of Biogeography* 32: 1107–1116.

Klein, A.M., Vaissière, B. E., Cane, J. H., Steffan-Dewenter, I., Cunningham, S. A., Kremen, C. and Tscharntke, T. (2007) Importance of pollinators in changing landscapes for world crops. *Proceedings of the Royal Society B* 274: 303–313.

Kongpitak, P., Matsuka, M. and Wongsiri, S. (2004). Macadamia and their pollinators in Thailand. In: *7th Asian Apicultural Association Conference and 10th Beenet Symposia and Technofora*, Los Baños, Philippines, February 23–27, 2004.

Kovács-Hostyánszki, A., Espíndola, A., Vanbergen, A. J., Settele, J., Kremen, C., and Dicks, L. V. (2017) Ecological intensification to mitigate impacts of conventional intensive land use on pollinators and pollination. *Ecology Letters* 20(5): 673–689.

Kremen, C., Williams, N.M. and Thorp, R.W. (2002) Crop pollination from native bees at risk from agricultural intensification. *Proceedings of the National Academy of Sciences USA* 99: 16812–16816.

Liow, L.H., Sodhi, N.S. and Elmquist, T. (2001) Bee diversity along a disturbance gradient in tropical lowland forests of south-east Asia. *Journal of Applied Ecology* 38: 180–192.

McGregor, S. E. (1976) *Insect Pollination of Cultivated Crop-Plants*. U.S.D.A. Agriculture Handbook No. 496: 93–98.

Maxwell, J.F. (2004) A synopsis of the vegetation of Thailand. *Natural History Journal of Chulalongkorn University* 4(2): 19–29.

Neupane, K.R. and Thapa, R.B. (2005) Pollen collection and brood production by honeybees (*Apis mellifera* L.) under Chitwan condition of Nepal. *Journal of the Institute of Agriculture and Animal Science* 26: 143–148.

Oldroyd, B.P., Gloag, R.S., Even, N., Wattanachaiyingcharoen, W. and Beekman, M. (2008) Nest-site selection in the open-nesting honey bee *Apis florea*. *Behavioral Ecology and Sociobiology* 62: 1643–1653.

Oldroyd, B.P., Osborne, K.E. and Mardan, M.B. (2000) Colony relatedness in aggregations of *Apis dorsata* Fabricius (Hymenoptera: Apidae). *Insectes Sociax* 47: 94–95.

Oldroyd, B.P, Rinderer, T. E. and Wongsiri, S. (1992) Pollen resource partitioning by *A. dorsata, A. cerana, A. andreniformis* and *A. florea* in Thailand. *Journal of Apicultural Research* 31: 3–7.

Oldroyd, B.P. and Wongsiri, S. (2006) *Asian Honey Bees. Biology, Conservation and Human Interactions*. Harvard University Press, Cambridge, MA.

Ollerton, J., Winfree, R. and Tarrant, S. (2011) How many flowering plants are pollinated by animals? *Oikos* 120: 321–326.

Partap, U. (2000) Foraging behaviour of *Apis cerana* on sweet orange (*Citrus sinensis* var. Red Junar) and its impact on fruit production. In: M. Matsuka, L.R. Verma, S. Wongsiri, K.K. Shrestha and U. Partap (Eds.), *Asian Bees and Beekeeping: Progress of Research and Development*. Oxford and IBH Publishing, New Delhi, India. pp. 174–177.

Pianka, E.R. (1970) On r- and K-selection. *American Naturalist*. 104: 592–597.

Rahman, K.A. (1940) Insect pollinators of toria (*Brassica napus* Linn. var. *Dichotoma prain*), and sarson (*B.campestris* Linn. var. *Sarson prain*) at Lyallpur. *Indian Journal of Agricultural Sciences* 10: 422–447.

Richards, A. J. (2001) Does low biodiversity resulting from modern agricultural practice affect crop pollination and yield? *Annals of Botany* 88: 165–172.

Rinderer, T.E., Oldroyd, B.P., de Guzman, L.I., Wattanachaiyingcharon, W. and Wongsiri, S. (2002) Spatial distribution of the dwarf honey bees in an agroecosystem in southeastern Thailand. *Apidologie* 33: 539–543.

Rod-im, P. (2014) Possibility of Commercial Beekeeping Development of the Red Dwarf Honeybees, *Apis florea* for Honey Production in Comparison to the Traditional Honey Hunting. MSc Thesis, School of Bioresources and Technology, King Mongkut's University of Technology Thonburi, Bangkok, Thailand.

Rod-im, P., Duangphakdee, O., Radloff, S., Pirk, C.W.W, Hepburn, C. and Hepburn, R. (2015a) Azimuth-dependent waggle dances; flight and foraging activities of the red dwarf honeybee, *Apis florea* Fabricius (1787). *Journal of Apicultural Research* 54(3): 246–254.

Rod-im, P., Hepburn, H.R. and Duangphakdee, O. (2015b) Developmental cycles of the red dwarf honeybee, *Apis florea. Proceedings of the International Conference on Agricultural, Ecological and Medical Sciences* (AEMS-2015). pp 96–99.

Roubik, R. D., Sakai, S. and Hamid, A.A. (2005) *Pollination Ecology and the Rainforest: Sarawak Studies.* Springer, New York.

Roubik, D. W. (2002) The value of bees to the coffee harvest. *Nature* 417: 708.

Ruttner, F. (1988) *Biogeography and Taxonomy of Honeybees.* Springer-Verlag, Berlin, Germany.

Saeed, S., Malik, S.A., Dad, K., Sajjad, A. and Ali, M. (2012) In search of the best native pollinators for bitter gourd (*Momordica charantia* L.) pollination in Multan, Pakistan. *Pakistan Journal of Zoology* 44: 1633–1641.

Saeed, S., Sajjad, A., Kwon, O. and Kwon, Y.J. (2008) Fidelity of Hymenoptera and Diptera pollinators in onion (*Allium cepa* L.) pollination. *Entomological Research* 38: 276–280.

Sajjad, A., Ali, M. and Saeed, S. (2017) Yearlong association of *Apis dorsata* and *Apis florea* with flowering plants: Planted forest vs. agricultural landscape. *Sociobiology* 64(1): 18–25.

Sajjad, A., Saeed, S. and Masood, A. (2008) Pollinator community of onion (*Allium cepa* L.) and its role in crop reproductive success. *Pakistan Journal of Zoology* 40: 451–456.

Sajjad, A., Saeed, S., Muhammad, W. and Arif, M.J. (2009) Role of insects in cross-pollination and yield attributing components of Sesbania sesban. *International Journal of Agriculture and Biology* 11: 77–80.

Seeley, T.D. (1985) *Honeybee Ecology: A Study of Adaptation in Social Life.* Princeton University Press, Princeton, NJ.

Seeley, T.D, Camazine, S. and Sneyd, J. (1991) Collective decision-making in honeybees: How colonies choose among nectar sources. *Behavioral Ecology and Sociobiology* 28: 277–290.

Seeley, T.D., Seeley, R.H. and Akratanakul, P. (1982) Colony defense strategies of honeybees in Thailand. *Ecological Monographs* 52: 43–63.

Sihag, R.C. and Rathi, A. (1992) Foraging modes and foraging rates of different bee pollination of pigeon pea (*Cajanus cajan* (L.) Millsp). *Proceedings of the International Symposium on Pollination in the Tropics.* pp. 93–95.

Sirinun, A. 2002. Nectar and pollen resource partitioning by honey bees and other insects in Makiang. Rajamangala University of Technology Conference, Pathum Thani (Thailand). 22–27.

Verhulst, P.F. (1838) Notice sur la loi que la population pusuit dans son accroissement. *Correspondance Mathématique et Physique* 10: 113–121. [In French].

Watanabe, M. E. (1994) Pollination worries rise as honey bees decline. *Science* 265: 1170.

Wattanachaiyingcharoen, W., Wongsiri, S. and Oldroyd, B.P. (2008) Aggregations of unrelated *Apis florea* colonies. *Apidologie* 39: 532–536.

Williams, I. H. (1994) The dependences of crop production within the European Union on pollination by honey bees. *Agricultural Zoology Reviews* 6: 229–257.

Wongsiri, S., Lekprayoon, C., Thapa, R., Thirakupt, K., Rinderer, T.E., Sylvester, H.A., Oldroyd, B.P. and Booncham, U. (1997) Comparative biology of *Apis andreniformis* and *Apis florea* in Thailand. *Bee World* 78: 23–35.

Wongsiri, S., Thapa, R. and Kongpitak, P. (1998) Longan: A major honey plant in Thailand. *Bee World* 79 (1): 23–28.

Zhang, W., Ricketts, T.H., Kremen, C., Carney, K. and Swinton, S.M. (2007) Ecosystem services and dis-services to agriculture. *Ecological Economics* 64: 253–260.

18

The Dance Language of Apis florea

Orawan Duangphakdee, Preecha Rod-im and Christian Pirk

CONTENTS

A. florea Dance as a Primitive to Their Ancestor as well as Simplicity on Their Communication

Since von Frisch, social bees have been investigated to understand how they share information on resources and how they recruit other members to these new resources. The most basal and simple way of communicating can be found in bumblebees (Dornhaus and Chittka 1999) to more advanced communication in stingless bees, and the most sophisticated way is found in honeybees (Nieh 2004). The information conveyed ranges from a rough distance using a non-directional dance for nearby sources to a dance encoding for precise details including distance, direction and quality of the target. To exaggerate the information, honeybees encode the information in proximity with pheromones and the scents of flowers (Reinhard and Srinivasan 2009).

Stingless bees (Meliponini), a sister taxa of *Apini*, which includes all 11 *Apis* species, (Rasmussen and Cameron 2009) also use different ways of information transfers about food sources and do not include a non-symbolic language as can be found in honeybees. Therefore, it was first perceived as being more basal sine it appeared that no information transfer regarding distance and direction took place (Lindauer 1957), but subsequent studies showed that stingless bees communicate the direction and distance, however less accurate than the *Apis* dances (Jarau et al. 2000).

Although basically similar in their social organization, like reproductive division of labour and age-related polyethism and the way they exploit resources (Koeniger and Vorwohl 1979), the Asian honeybee species differ markedly in body size, colony size and nesting behaviour (Dyer and Seeley 1991). Being endemic to Asia and tropical and subtropical forests (Hepburn et al. 2014), their food resources are more patchy and widely spread unlike in temperate climates with distinct seasonal mass flowering patterns. Among *Apis*, *A. florea* traits are believed to resemble most the ancestral state (Beekman et al. 2005), and it has been suggested that the species' origin laid in selecting nest sites and moving between sites rather than foraging for resource, which evolved at a later stage (Beekman and Lew 2007).

Distance Encoded in the Dance Symbol

Two types of dance have distinguished the 'round dance' that is performed if the target source is near to the nest and does not encode information on direction and distance. The round dance starts with a run in a line and return into the starting point to make a new dance cycle. The reverse point makes almost a circle move, frequently only three-quarters or halves of the circle, and there are often one or two complete circles for this dance (von Frisch 1967). Second, the 'waggle dance' is performed if the target source is farther away from the nest encoded with information both for distance and direction to the target patch. In the waggle dance, the dancer starts in a straight run and returns in a semicircle to the starting point, then again runs through the same straight line section and returns in a semicircle in the opposite direction, performing a figure of eight (von Frisch 1967). On the straight part of the run, the bee buzzes its wings and vibrates its abdomen – the so-called 'waggle'. The angle of the axis of this waggle phase is relative to the vertical and represents the angle of the goal to the sun's azimuth (Duangphakdee et al. 2011).

Distance, meanwhile, is positively correlated with the duration of the waggle. The longer the bee waggles, the farther away the food. If the target resource is only a short distance away from the nest, the waggling phase becomes quicker or 'more lively' as von Frisch originally described it. Since the 1920s, the view that honeybees have two discrete recruitment dances has been widely accepted. It was hypothesized that it is a special case of the waggle dance and is a result of the round dance referring to a target that is close by /near by and therefore a rushed version during which the waggle phase is omitted (Jensen et al. 1997). Indeed, later studies showed that both round and waggle dances convey information on distance and direction and that honeybees basically use the same dance symbols that gradually change from round to sickle shape to finally a waggle dance shape when increasing the distance from close to far (Gardner et al. 2008). The transition point of change from round dance to waggle dance is highly variable within the subspecies and between species, and the variability is highly influenced by genetic and environmental factors (Oldroyd and Wongsiri 2006). The general shape of the course and the waggle phase seem to be identical in both cavity-nesting and giant honeybees. Both dance in the vertical so that the bees move their bodies in parallel to the combs and the direction encoded in relation to the gravity. In the waggle run of *A. florea*, the bee does move differently on a horizontal plan and therefore the direction, e.g., the angle to the sun, does not have to be related to gravity but can point directly to the source. The waggling of *A. andreniformis* is similar to that of *A. florea* (Duangphakdee pers. obs.). The distance at which bees change from round to waggle dance is affected by their genetics and ecological factors. The transition from the round to the waggle dance in *A. mellifera* is around 80 m, 5 m in *A. cerana*, 2.5 m in *A. dorsata* and less than 1 m in *A. florea* (Towne 1985). Studies of *A. florea* in Thailand found that the direction component in the dance is still distinctly detectable even at less than 0.5 m [Duangphakdee et al. 2011] although the dance is partly sickle shaped (Figure 18.1).

However, a real round dance might occur in *A. florea* during absconding and swarming events, similar to the round dance during foraging described for *A. mellifera* by von Frisch. This dance was first noted by Oldroyd et al. (2008) and later confirmed by Duangphakdee et al. (2012). During absconding events, a significant number of round dances with either limited or no distance and direction information can be observed in both dwarf honeybee species, *A. florea* and *A. andreniformis.* Instead of dancing a figure of eight, they seem to dance in random circles, sometime waggling their abdomens, but without any specific duration and direction (Figure 18.2). The round dance is quite specific and limited to swarming events during which it occurs about half to four times more frequently than waggle dances and persisted until the last half hour before lift-off (Oldroyd et al. 2008). Does it serve the same function as in foraging dances of other species? The hypothesis has been suggested, but no experiments have shown it to be so (Oldroyd et al. 2008). Alternatively, the sudden ceasing of the round dance during swarming could have a similar function such as the piping in *A. mellifera* (Seeley and Tautz 2001) serving as a signal for lift-off.

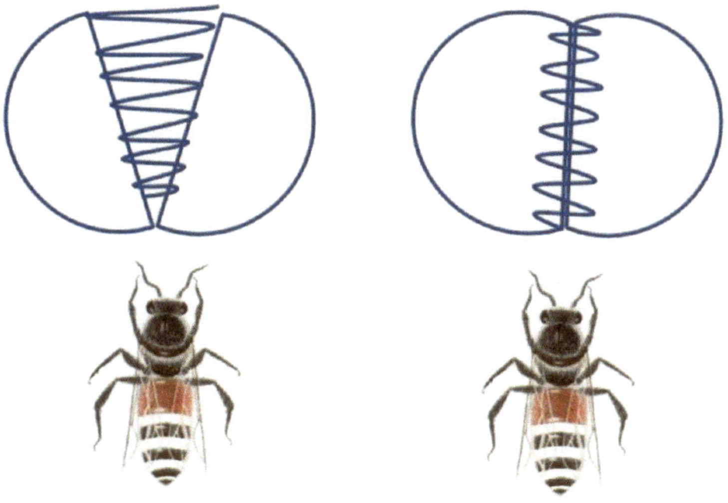

(a) (b)

FIGURE 18.1 Conceptual idea of Waggle dances in the figure of eight pattern: (a) feeder at 1 metre and (b) feeder at 10 metres.

FIGURE 18.2 The round dance in swarms of the dwarf honeybees (dance circuits frequently a b overlap. (From Duangphakdee O. et al., *Honeybees of Asia*, Springer, Berlin, Germany, pp. 313–332, 2011.)

Migration Dances

Nest-site selection in honeybees is a prime example of a distributed decision-making process. Unlike foraging – where individuals or small groups of individuals make decisions on multiple goals – finding a nest site focuses the decision process on one ultimate goal, which relies on the highly synchronized self-organization of the colony (Oldroyd et al. 2008) to be successful. Within swarm clusters, several thousands of individual bees, which have all kinds of different tasks, self-organize to fulfil the obligations towards one single goal (Gordon 2016).

All species of honeybee go through a nest-site selection process as part of their adaptive behaviour and life history. Different reasons cause bees to swarm such as diminishing resources, habitat changes and threats by natural enemies and diseases, or a mature colony produces a daughter colony; all these events could cause scout bees to start searching for a new apposite nest (Hepburn 2006). Honeybees will search for new nest sites during absconding, migration and reproductive swarming. The reproductive swarm occurs when a colony reaches maturity, starting when a queen departs from the colony with a large number of workers to establish a colony of her own. Migration is defined as a seasonal movement of many colonies of almost a whole population in the same region to another region. Absconding is usually restricted to the movement of a whole colony away from its maternal nest site to another place, usually caused by local environmental perturbations (Hepburn 2006). It has been suggested that the symbolic dance in honeybees has evolved in the context of nest-site selection (Beekman et al. 2008) and only later was adapted as well to communicate the information of resources in the context of foraging.

In *A. florea*, it is unclear if seasonal migration occurs; however, colonies move to a new nest site in response to insufficient food resources within the proximity of their current nest site.

While the process of nest-site selection in *A. mellifera* has been comprehensively studied (Lindauer 1957; Seeley et al. 2006), the processes involved in Asian honeybees is less well studied and understood.

Even though Asian species use the same system to encode the basic information in their dance communication, like *A. mellifera* does, there are significant differences in their dance related to behaviour and habitat variation.

In *A. mellifera* (Schneider and McNally 1994), *A. cerana* (Sasaki et al. 1990) and *A. dorsata* (Koeniger and Koeniger 1980), a long abdomen-protracting horizontal waggle dance, a so-called migration or absconding dance is performed during migration, most likely to stimulate a colony to prepare to move. While the underlying mechanism is still not fully understood, this dance does not necessarily have to point to the final nest-site direction; it normally only indicates a general direction of travel and signals a long-distance move during absconding. During migration, the swarms normally settle 2–30 metres away from the original nest site, and then it can take a few hours to weeks until the process of identifying and agreeing on a new nest site is finalized and the swarm will take off for the new nest destination. The absconding/migration dances had long been thought to be similar to the foraging dance, until in 1980, when Koeniger and Koeniger first demonstrated that there are fundamental differences (1980) between the dances that clearly separate the migration dance from the foraging dance.

Indeed, the migration dance includes a recruitment process and signals conveying a decision-making process. In contrast to other honeybee species, the dwarf honeybees, *A. florea* and *A. andreniformis*, conduct a final nest-site selection process right up on the mother nest. While the details differ among species, deciding either on the original nest or swarm cluster on temporary sites, scout bees explore the surroundings for potential nest sites and report on the location of potential sites using the waggle dance (Lindauer 1956). In *A. florea*, scouts of the daughter swarm will start to perform waggle dances for nest sites, which is normally initiated 3–7 days before departing from the maternal nest. Via a combination of positive and negative feedback of the site-visiting scout, the number of inferior potential sites is abandoned whereas superior nest sites are enumerated, remarked under consideration and archived until a final decision is made (Seeley et al. 2012).

In *A. florea* and *A. andreniformis*, the first migration dance can be observed 1–4 days before the colony flies off (Oldroyd et al. 2008; Duangphakdee et al. 2012) and reaches a quorum before lift-off (Figure 18.3). The process of finding new nest sites and recruiting other scouts to it and then making a decision for a specific candidate nest site can be achieved within 2–24 hours. In *A. florea*, foraging activity will decline and cease about an hour before the colony takes off (Duangphakdee et al. 2012), allowing the colony to proceed with nest-site selection and reach a consensus decision without the disruption of other hive activities. Nevertheless, the colony will wait until the last brood has emerged and is ready to join the swarm. Recently, a few dances and acts have been reported that are associated with the nest-site selection process (Beekman et al. 2008; Duangphakdee et al. 2012). The waggle dance is typically performed to share distance, direction and quality of the nest site (Lindauer 1956). While the other species have sometimes shown an incomplete cycle by wagging their abdomens without the returning run in a semicircle (Koeniger and Koeniger 1980; Sasaki et al. 1990), *Apis florea* often shows a complete cycle of waggle dance for its nest-site selection process. The nest-site selection process depends on the voting by the foragers, which accounts for approximately 30% of

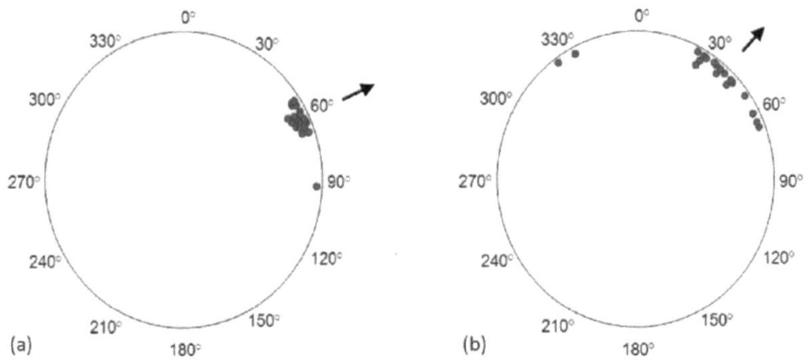

FIGURE 18.3 A schematic drawing of the last 20 waggle dances advertising a nest site before liftoff of (a) *A. florea*: new nest site located at 70 m from the original nest. The bees in this example dance with a divergence angle of 37° and (b) *A. andreniformis*: new nest site located at 68 m from the original nest. (From Duangphakdee O. et al., *Insectes Sociaux* 59, 571–577, 2012.)

individuals within the colony. The other 70% still carry on with other tasks such as foraging and nursing. When the voting process reaches a quorum, a subset of bees starts buzzing and running around, interfering and disturbing other workers in their tasks, provoking them to abandon their tasks and to get ready. The absconding dances in *A. dorsata* may start with 2–3 dancers and can increase to 5–15 dancers after 24 h. These dances seem to be more likely to trigger the colony to move and not to recruit more dancers (Koeniger and Koeniger 1980). This is different to the migration dances seen in the cavity-nesting bees, *A. mellifera* and *A. cerana*, where most of the dances observed during reproductive swarm events (Oldroyd et al. 2008) clearly attracted recruits, which is also the case during absconding in the dwarf bees, *A. florea* and *A. andreniformis* (Duangphakdee et al. 2011). As the waggle phase contains information about distance, the longer the waggle phase, the longer the distance, while shorter ones refer to nearby targets. *Apis florea* do not migrate far, either because of the shorter flying distance capacity and/or the small colony size, which results in the fact that a waggle phase in *A. florea* has never been observed for more than 3 s. This is in contrast to *A. dorsata*, which waggles up to 8 s and can even waggle for longer to indicate a long-distance flight, excluding any information on the distance. In *A. cerana*, waggles ranging from 8 seconds to more than 60 seconds can be observed (Punchihewa et al. 1985). To be noted, these observed differences among the species might well be a result of behavioural differences – if, for example, only a nearby resting place or a next new nesting site is communicated (Sasaki et al. 1990).

Dance Ultimatum

Both Western and Eastern cavity-nesting honeybees are likely to have a limited number of potential nest sites in their proximity that fulfil the requirements in terms of volume and cavity height, size, position and direction of the entrance and evidence of previous occupancy (Seeley and Morse 1978). The nest-site selection dance is performed within the nest, and sometimes the quorum can be reached before leaving the mother nest. In other cases, it seems that a new nesting site has not yet been specified and the swarm then gathers under the branch of a nearby tree not far from the mother nest; the nesting-site decision process will gradually continue while the swarm is clustering in the open. The decision process in swarms of the open-nesting *A. dorsata* is characterized by worker bees dancing repeated waggle phases without the turning phases to trigger all members of the swarm to leave the mother colony. The bees seem to reached a consensus decision before becoming airborne (Koeniger and Koeniger 1980), similar to *A. mellifera* (Seeley et al. 2006). Since *A. dorsata* is migrating long distances (Robinson 2012), the decision process has to be repeated several times during the migration.

The dwarf honeybees, *A. florea* and *A. andreniformis*, as open-nesting species, seem to have more suitable nesting sites within their proximity of the old nest site. Building combs on trees or shrubs with dense canopy allow them to utilize more potential nesting sites in contrast to cavity-nesting species. Their dance quorum suggested a broad area, probably a tree canopy and not particular branches or twigs. Also, the decision process does not need to include all scouts, as one can see during emergency swarming when many scouts still performed dances indicating nest sites in several different directions shortly before lift-off. Since they have so much more potential nest sites in their surroundings, they could trade accuracy against time/energy spent for reaching a quorum. Indeed, it has been suggested that less accuracy might be the primitive trait (Preece and Beekman 2014). As long as the bees can find the right target, even by being less accurate they would spend less time and energy and therefore could represent the optimal compromise or solution. Like the recent finding in *A. mellifera* (Seeley et al. 2006), there is no consensus in dancing for a specific nest site before reaching the quorum in *A. florea* (Oldroyd et al. 2008). Nevertheless, in *A. florea* during absconding events, the scouts dance with a relatively high angular deviation up to 45° with a target distance less than 100 m, and colonies distinctly reach a quorum instead of consensus in dancing before they are airborne (Duangphakdee et al. 2012) (Figure 18.4). One-third of total colony members, generally workers on the crown, are responsible for the selection process of where to go, and when the decision has been made, several of the bees from the crown will start to perform a zigzag run while producing an audio signal, or 'worker piping', to stimulate other bees to warm up for flight (Seeley and Tautz 2001; Seeley et al. 2003). During most of the nest-site selection process, *A. florea* does show normal foraging activity, which usually drops at least one hour before colony liftoff. This allows the dancers, communicating the new nest site, to intensively advertise those nest sites to which they have selected. By maintaining key activities, like foraging, during the consensus process, this allows the colonies to be prepared if the take-off process cannot be finalized (Duangphakdee et al. 2012) and the colony has to stay put. The piping bees stimulate every member of the swarm to warm up thoracic muscles in preparation to become airborne. This warming-up period takes approximately 30 minutes, and the achievement of final consensus happens side by side with other workers packing up honey, pollen as well as beeswax before the swarm takes off. In case the bees do not have sufficient time to actually made a consensus decision, the colony will probably still take off even if dancing scouts are still advertising different nest sites (Duangphakdee per. Obs).

Previous studies propose that the dance in honeybees follows the tuned-error hypothesis (Tanner and Visscher 2010) that the precision of the dance, represented by the divergent angle of the dance, has tuned to produce an optimal position of the target (Figure 18.5). In *A. mellifera*, dancing for nest sites is more precise than food sources, such as flower patches (Weidenmüller and Seeley 1999). Later studies have shown that the bee dance precision is more dependent on physical adjustment than tuned-error hypothesis (Towne 1985; Tanner and Visscher 2006) and that the different level of precision found in nest-site selection dances compared to foraging dances is a result of the different dancing substrates or surfaces.

A. florea dance precision is similar for nest sites and communicating food sources because they have almost identical target characters as it does not make a huge difference if one is communicating a tree canopy for nest site or a flower patch. It has been suggested that the bees' dancing is actually shaped by tuned-error hypothesis (Beekman et al. 2008; Figure 18.6); however, the similarity in the precision of dances encoding for nest sites and food sources could also be due to the fact that the bees dance on

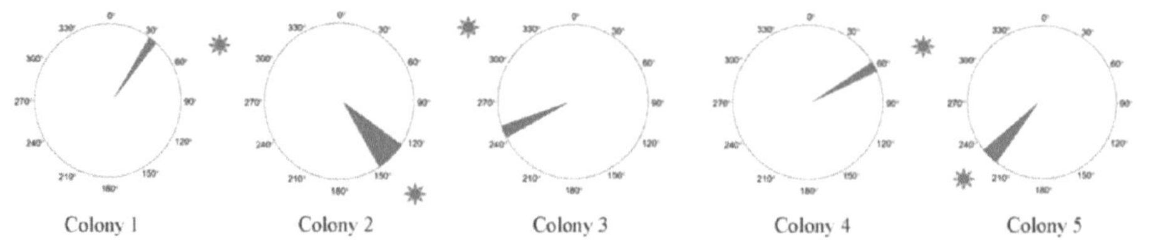

FIGURE 18.4 The mean angles and angular accuracy of the five colonies as advertised by the dances. ☼ Represent final nest site direction, sector represent 95% confidence interval of mean angle present by the dances during the last 3 min period before liftoff.

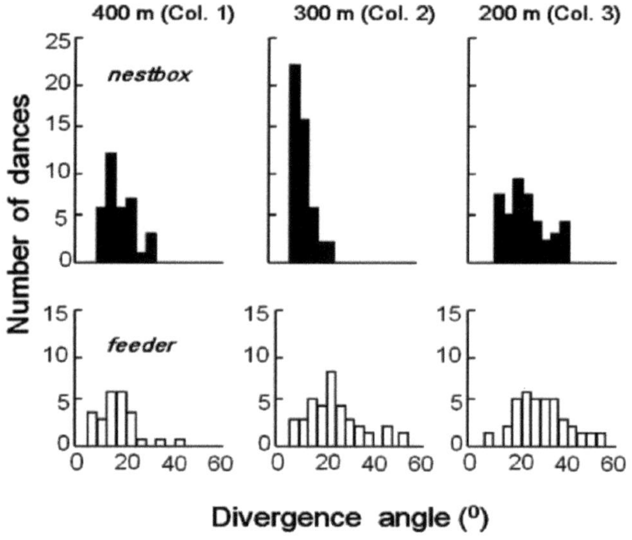

FIGURE 18.5 Distribution of divergence of angles in the dances of *A. mellifera* indicating either a nest box (filled bar) or a feeder (open bars) at difference distance. (From Weidenmüller, A. and Seeley, T., *Behav. Ecol. Sociobiol.*, 46, 190–199, 1999.)

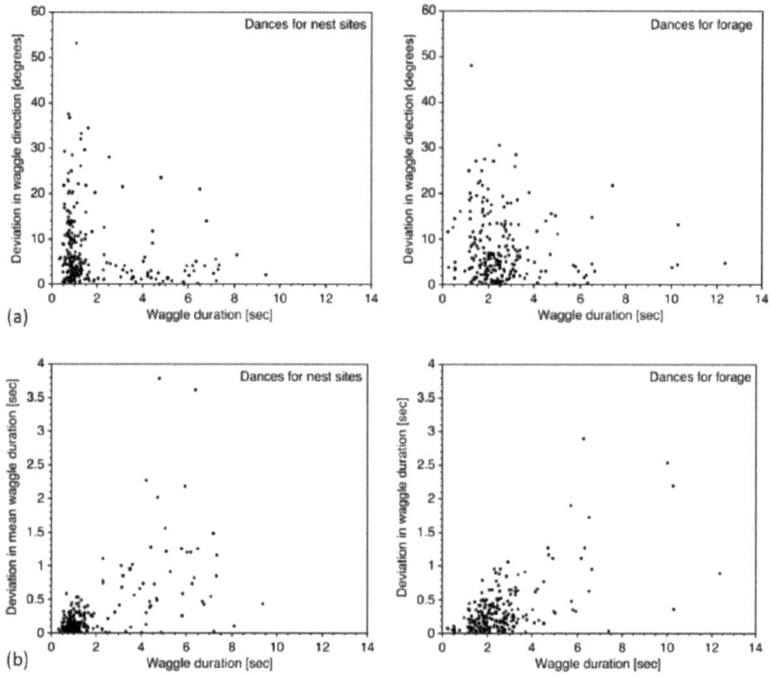

FIGURE 18.6 Angular and time deviation of the dance of *A. florea* during nest-site selection and foraging. (a) change in deviation in mean waggle direction with increasing average waggle duration (b) change in deviation in mean waggle duration with increasing average waggle duration. (From Beekman, M. et al., *Behav. Ecol. Sociobiol.*, 62, 1259–1265, 2008.)

exactly the same surfaces, a curtain of bees as the argument made by Tanner and Visscher (2006). Another fundamental different is that individual *A. florea* dances show no decay, e.g., a reduction of the number of dance circuits during the dance process as one can observe in *A. mellifera*. Since the rate of the decay is an indicator for site/food quality (Seeley 2003), it has been concluded that *A. florea* dances do not communicate the difference between low- and high-quality nest sites (Makinson et al. 2011).

According to the tuned-error hypothesis, the variability in dancing the correct angle in *A. florea* during nest-site selection is similar to dances indicating food sources, so that dance precision for nest sites and food sources is essentially the same irrespective of context. Worker bees tune the divergence angle of the dances in that way so that they produce an optimal scatter of recruits across the potential resources (Tanner and Visscher 2006). To be noted, *A. florea* also favoured the dance surface hypothesis because the bees dance on the same substrate either for nest site or food sources. The time necessary to reach a quorum in *A. florea* can be varied from as short as one hour to many days. *Apis andreniformis* seems to reach a decision faster than *A. florea* (Duangphakdee et al. 2011).

The quorum in decision-making of *A. florea* still has been perplexing. The reports are that *A. florea* does not seem to use a quorum for the final decision (Makinson et al. 2011) at the chosen nest site.

Makinson made an interesting observation that *A. florea* scouts do not really distinguish between food source and nest site in the way they are dancing (Makinson et al., 2011). This explains that if scouts communicate the quality of nest site using the same communication channels as they did for food resources, the perceived quality of the site remains constant through time (Beekman and Oldroyd 2018).

Since a potential nest site is mainly characterized by a suitable branch, potential new nest sites are rather abundant and less limiting as in *A. mellifera* or *A. cerana*; therefore, *A. florea* swarms using a less precise vectorial consensus compared to cavity-nesting bees would still be adaptive (Schaerf et al. 2011). Indeed, all scouts participate in guiding the swarm; therefore, *A. florea* does not have to reach a quorum of the chosen nest site – it is enough for some scouts to agree (Beekman and Oldroyd 2018). The final decision where to build the new nest is done when the swarms have reached the proximity of the tree they were dancing for (Makinson et al. 2011; Duangphakdee et al. 2012). Besides having ample suitable nest sites available and therefore being able to be less precise, another factor might be playing a role. Based on simulations, it has been suggested that small swarms do not have the workforce to sample all suitable nest sites in the environment, hence a precise quorum is unnecessary (Schaerf et al. 2013). Even though *A. florea* applies the same basic decision rules as *A. mellifera*, it would be sufficient for members to select a good nest site and not necessarily the best. The energy and time saved by sampling only a subset of the suitable nest sites in the surrounding area could outweigh the disadvantages of settling for a suboptimal new nest site. Moreover, since *A. florea* is the only species that commonly salvages wax and other resource for the old nest site (REF), this could further minimize potential negative effects of selecting a new nest site that is not the best. Natural selection would favour a system that has to be precise but would not waste too much energy on being perfectly precise; therefore, a messy decision-making process is adequate enough to move to a suitable and qualitatively reasonable new nest site.

Dancing for Food

Foraging in social insects like honeybees is a trade-off between how fast the individual collects honey versus the dying of the flower during the blooming periods. In tropical regions, the flowering time and the resulting rewards are the limiting factors, and the level of competition is high because many insect species try to utilize these resources (Jarau and Hrncir 2009). Many social insects have elaborate and effective foraging strategies and recruitment systems for their success in foraging efficiency (Jarau and Hrncir 2009). One of the most effective strategies is to occupy and conquer. Honeybees, due to the size of their colonies and their ability to fly, can spot patchy resources and outcompete others by recruiting more foragers to a lucrative food source (Nagamitsu and Inoue 1997). Unlike ants, they might not go to the extreme to defend a food source, but the key to their success is the effective communication system, by means of the bee dance, that allows for a trade-off between effort and usefulness of the resource (Preece and Beekman 2014). The foraging in honeybees starts and is led by experienced

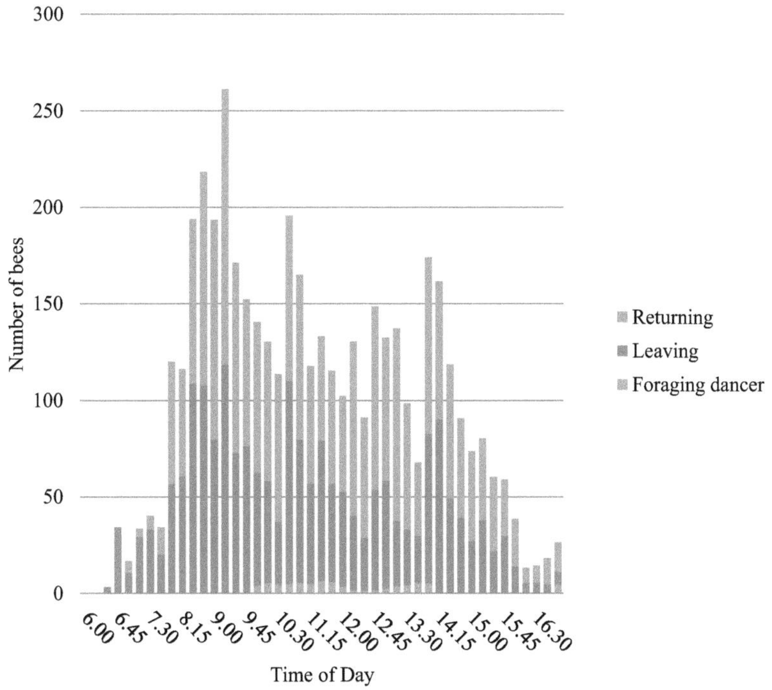

FIGURE 18.7 The mean frequency distributions of flight activities; the number of *Apis florea* foraging dancer, foragers leaving and returning to the nest ($n = 5$) over the course of the day.

foragers starting the day and searching for sufficient nectar and pollen food sources. If she commits the source reflected the high quality of foods, she will perform a dance advertising to the site. The medium size of *A. florea* colony with approximately 5,000–8,000 individuals, foragers start the day with less than 80–100 bees and reach peak between 9:00 and 10:00 hr with 400–500 bees joining the foraging activity (Rod-im et al. 2015; Figure 18.7).

Dance Floors

The dwarf honeybees are the only group that dances normally on a horizontal plane compared to the cavity-nesting and giant honeybees. Because of the lack of a horizontal surface, bees perform dances on the vertical plane, in particular on the surface of the curtain of the bees at the mouth area of the comb (Morse and Laigo 1969; Dyer and Seeley 1991). Those species, which dance on a vertical plan, translate the position of source relative to the sun angle into the angle relative to vertical (gravity axis) during their waggle phase (Lindauer 1956).

Dwarf bees dance on a relatively flat surface of the crown comb in the horizontal plane. Therefore, dancers can point directly to food sources without the need to refer to gravity as observed in the cavity-nesting or giant honeybees. Since the crown is a relatively limited area of about 40–60 mm^2 of the comb surface and is often vaulted, some dancers dance beyond the crown by sloping down to the vertical surface of the comb. The dance then takes place in the crowded dancing floor near the target (Koeniger et al. 1982; Figure 18.8). Lindauer (1956) demonstrated that *A. florea* normally dances on a horizontal plain (Duangphakdee et al. 2011). The experiment to rotate the comb so that the crown was upside down, and thus no longer horizontal, led dancing to be significantly disrupted. However, the bees will discover

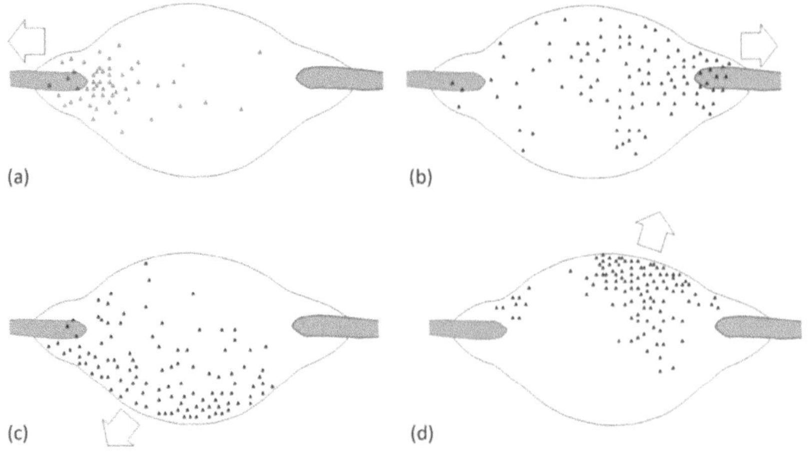

FIGURE 18.8 Dance map of *A. florea* where triangles represent the centres of the waggle phases (open triangles represent foraging dances and closed triangles represent migration dances. (a–d) Representing the dance distribution in top of the nest crown in different direction of target. (From Duangphakdee O. et al., The dance language, in *Honeybees of Asia*, Springer, Berlin, Germany, pp. 313–332, 2011.)

a horizontal floor for dancing automatically. For instance, if the crown is upside down, the bees simply dance on the lower edge of the comb that was recently turned upward. If the crown is covered by other objects, either the dance becomes disoriented, or bees perform dances on the horizontal plane above the objects that cover it. Nevertheless, if they are unable to locate a horizontal platform, bees will dance in such a manner that the straight run points to the goal, as, for instance, when bees are compelled to dance on a tiny vertical platform without sufficient room to return to waggle dancing, they will dance directly into the air pointing to the target.

A swarm cluster of the dwarf bees successfully dances in a horizontal plane. Unlike other cavity-nesting and giant honeybees, the dancers of the dwarf bee on the top of the cluster orient directly towards the goal without referring to gravity (Dyer 1994). In dwarf honeybees, however, the dances for nest sites are not limited to the top part of the comb, whereas foraging dances are spread only over the top of the crown. Interestingly, the waggle dances for nest sites are usually performed away from the foraging dance areas to indicate the direction of the target (Figure 18.9; Duangphakdee et al. 2011).

FIGURE 18.9 *A. florea* dancer (white dot) on the curtain member bees as a dance floor.

Dance Dialects

The distance towards a source or nest site is encoded by the duration of waggle phase during the dance. The relationship between distance and waggle time appears to have population and species-specific characteristics that can be visualized by plotting the relationship of waggles and distances and the different slopes; these are called dialects (von Frisch 1967). As the dance is strongly related to the survival of the bees, these differences in the dancing or these dialects are a result of evolutionary processes and represent environmentally shaped adaptive traits (Gould 1975; Dyer and Seeley 1991). This explains that the slope of the dialect correlated to the flight range of the bees represents its maximum typical flight range shaped by the environmental complexity each population or species is experiencing. Indeed, the result that honeybees measure distance based on the optical signal perceived during the flight relates to complexity of environmental visual stimulation during the bees' travel. Therefore, encoding in the waggle dance, the same bees flying over the different levels of landscape complexity should be considered as possible differences in slope dialects (Esch et al. 2001)

Dialect studies on three species of honeybees – *A. florea*, *A. dorsata* and *A. cerana* – at two different locations showed different results. The dialect curves between these three species were different when tested in Sri Lanka (Punchihewa et al. 1985), but they were virtually identical in their slope in Thailand (Dyer and Seeley 1991). Because the dialect seems to be related to flight range, the steepness of dialect curve should be inverse related to body size, and small bees such as *A. florea* should show a steeper slope compared to *A. cerana* and *A. dorsata*.

According to a previous study (Lindauer 1956), the flight range of *A. florea* is around 350 m. In a study site in Thailand, by contrast, the longest waggle phase that appeared in the dance of *A. florea* (≈30 sec) was about three times longer than what was observed in *A. cerana* (Dyer and Seeley 1991). Other studies in many parts of the world established the fact that *A. florea* does adapt its optimum flight range and dance signal according to the habitats it encounters (Figure 18.10). Dialects are highly similar among populations of the same species. *A. florea* from Sri Lanka (Lindauer 1956) showed slopes that were similar to those populations studied by Punchihewa et al. (1985) in Thailand. However, the slopes often differ among geographical races and environmental variety. The flight range of *A. florea* in Thailand is also reported at about 700 meters, twice longer than the previous studies showed (Dyer and Seeley 1991). It seems that *A. florea* is able to optimally adjust its foraging ability according to environmental

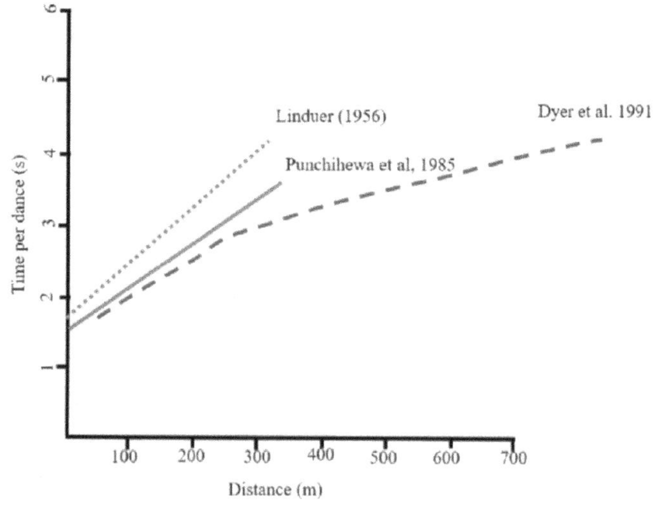

FIGURE 18.10 Relationship between foraging distance and dance time for *A. florea* in different ecotypes. (Redrawn from Lindauer, M., *Bee World*, 38, 521–557, 1956; Punchihewa, R. et al., *J. Apic. Res.*, 24, 168–175, 1985; Dyer, F.C. and Seeley, T.D., *Behav. Ecol. Sociobiol.*, 28, 227–233, 1991.)

conditions. At the greater distance, a small error of dance angle can have greater consequences; therefore, one would expect that if the variability of the dancing angle is a result of the tuned-error hypothesis, the steepness of the slope correlates with the level of dance precision.

As a consequence of the environmental and genetic factors, to apply single standard calibration curve for interspecies and intraspecies is not viable, even for the same colony. However, under the same ecotypes, the dialects in the distance signal should reflect the level of species (Duangphakdee et al. 2011).

In *A. mellifera*, the dance error actually decreases with greater foraging distance as more precise data are need at the further distance. Also, more precise data are needed in cases of communicating a new nest site compared to flower patches. Towne (1985) found the divergence of dance angle of 75° at 100 m but only 15° at 700 m. In *A. florea*, the precision of dance presenting nest site and food resources was similar in both cases, foraging or nest site, and it referred to the tree canopy as the final goal (Beekman et al. 2008). To be noted, the flower patches in tropical surroundings would typically be small – for example, a single tree – whereas in comparison in temperate zones a feasible flower patch would be represented by a forest or an orchard. At an equivalent distance, the tropical species will communicate more precise dance information than temperate species (Towne and Gould 1988).

This raises the possibility that the link between ecology and the distance code is subtler than previously supposed, if indeed such a link exists (Dyer and Seeley 1991), and again the optic flow hypothesis might be considered. Differences in foraging range also show that changing dance tempo reflects changing food distances. For *A. cerana* and *A. florea* foraging at artificial feeders, the dance tempo was observed to decline more rapidly than has been reported for *A. mellifera* as distance to the food source increased. In *A. dorsata*, the change in dance tempo with increasing distance was much slower and similar to that reported for *A. mellifera* (Punchihewa et al. 1985). *Apis florea* seems to be less accurate than the cavity-nesting bees, which supports the notion that it represents the most primitive form of the dance found in the common ancestor of *Apis* (Beekman et al. 2008; Dyer & Seeley, 1989; Makinson et al. 2011). In conclusion, ecological conditions do affect the distributed decision-making in closely related species. Even though the trait is exposited as primordial, considering from the optimum theory, the capacities of information processing are not inferior to *A. cerana* and *A. mellifera* (Dyer 1987). *Apis florea* has the similar cognitive capacities to reach the same goal even with less accurate information compared to other *Apis* species. However, the *A. florea* dance might be substantial as a primitive trait but considering from the optimal hypothesis, it is an exceptional species. *A. florea* has the ability to adjust the accuracy of its information by 'tuning' its dance to increase or reduce error according to distances of targets, types and quantity of the targets (Dyer, 2002; Gardner et al. 2007; Towne & Gould, 1988; Weidenmüller & Seeley, 1999). Indeed, even it is not yet confirmed by experiment in *A. florea*, the multisignals should also be considered. The odour and visual cues are reported to enhance a final stage of the reward-searching flight (Riley et al. 2005).

Adaptive Enhancement of the Dances

Within genus *Apis*, at the beginning of the day *Apis florea* is recruiting slowly for food sources, but it literally gains the forager numbers during the day and becomes the most active species, which can maintain activity for as long until the resource is totally depleted. However, *A. florea* officially has a 'lunch break' at noon. Between 12:00 h and 13:00 h, bees largely avoid flying or when the angle of the sun is between 55° and 65°. Given the difficulties of taking an accurate reading of the sun at angles 6° of the sun's zenith (resulting in a 1 h loss around noon), and the 2 h required to reach consensus over the final direction to be flown, the bees are simply left with two time windows – morning and afternoon – in which to abscond and, indeed, some 90% of the absconding events observed of the red dwarf honeybees fall into these two time windows (Figure 18.11). The noonday lull is not associated with high temperatures for any given day since absconding is not inhibited by high temperatures (Duangphakdee et al. 2009).

As in other honeybee species, *A. florea* uses sun and azimuth as a compass for the dances. In *A. mellifera*, *A. cerana* and *A. dorsata*, which dance in the vertical, the position of the sun relative to the food sources encode in their dance angle relative to gravity. Under the circumstances, as a result, these species compensate their dance to the changing of the sun and sun azimuth over the course of the date

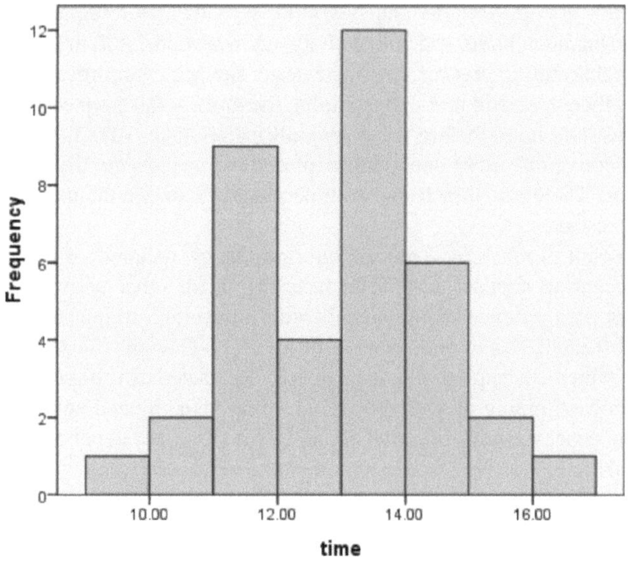

FIGURE 18.11 Frequency distribution for local clock time of absconding by *Apis florea* at Chom Bueng, Thailand (13° 59′ N, 99° 51′ E) during 2007–2009 (n = 37). (From Duangphakdee, O. et al., *J. Insect Physiol.*, 55, 1009–1012, 2009.)

and times (Duangphakdee et al. 2009). Considering that *A. florea's* dance floor is horizontal, it does not refer its dance to the gravity. Since the bees do not need to adopt an angle relative to gravity, they can point directly to the goal. *Apis florea* always points its dance toward the target and does not adjust its dance relative to the changing of azimuth (Dyer 1987). Even the sun position is not encoded in the dance as in other species. Koeniger et al. (1982) conducted an experiment demonstrating that the *A. florea* dance becomes misshapen without direct sight of the sun. The further experiment confirmed how important the direct sight of the sun is for the bees' dances. Workers adjusted their dancing according to the changing of the position of a surrogate sun made by the aid of a mirror (Koeniger et al. 1982). It may be possible that other objects or landmarks (Capaldi and Dyer 1995) – e.g., big trees, buildings – could also be encoded in *A. mellifera* dances, but there are no reports along these lines in *A. florea.* Another study showed the importance of the sun for orientation in *A. florea* (Duangphakdee et al. 2009, 2012). The dance and foraging activities are bimodal with respect to the time of the day with no activity during noon. The dancing for new nest sites was bimodal with the highest peak between 10:00–11:00 and 13:00–14:00.

The inactive in dancing and foraging of *A. florea* during noon was first considered to a result of high temperatures rather than the angle of the sun. However, a study on foraging and dancing behaviour showed that the highest temperature of the day was recorded between 14:00 and 14:30 and therefore cannot explain the break in foraging activity. Later then, the azimuth was considered to have an impact, and it was shown that dances and communication became disorientated when the azimuth is around noontime, resulting in the bimodal distribution of foraging (Duangphakdee et al. 2009). However, dancing can still be observed during noontime, but the dance lacks any information on direction. The disorientation of the dance during noontime can be explained by the relatively small difference in the angles of the sun and zenith. At noontime, due to the small angle of the sun to the zenith, the azimuth rapidly changes, causing bees to become confused in terms of compensating the sun position encoded in their dance. Similar results were found in *A. mellifer*; the dancers become disorientated when the sun is within 3° of the zenith (Lindauer 1957). This problem of distinguishing the position of the sun and the zenith is particularly prominent in the tropics; *A. florea* demonstrated that its difficulties start already when the sun is at angles ±6° of the sun's zenith. The most active of dance-related activity was also shown when the angle of the sun was between 55° and 65° (Duangphakdee et al. 2009). This disorientation related to sun azimuth occurs in both foraging and swarming activities. In foraging activity, the bees simply lull their activity during noon hours and gain activity again around 13:00 hr. However, it becomes more complicated during absconding or swarming. The nest's site selection process takes time, to reach the decision takes time, then to recruit nest mates joining the process takes time (Seeley 2003), all of which can add up to at least two hours before the colony ultimately takes off to the new nest site. The bees will decide before noon or only after due to the error of dancing closer to the time when the sun is close to the zenith position. Therefore, they have two time windows, one in the morning and the other in the afternoon, to move nest sites.

Lift-off is not a result of reaching a precise quorum but by a change of behaviour. Some *A. florea* workers will produce stop signals at 475 Hz, resulting in all other activity to stop (Towne 1985). The alerted workers than produce piping signals with an average frequency of 384 ± 31 Hz and an average duration of 0.82±0.35 s to elicit a wave of hissing behaviour (Sarma et al. 2002). In contrast to *A. mellifera*, in which the piping signal relates to the fact that a quorum is reached during the nest-site decision process, piping in *A. florea* includes both dancing and non-dancing bees throughout the decision-making process, indicating that signal to lift off is not depending on reaching a quorum (Makinson et al. 2011).

REFERENCES

Beekman M., L. Doyen, B.P. Oldroyd. (2005) Increase in dance imprecision with decreasing foraging distance in the honey bee *Apis mellifera* L. is partly explained by physical constraints. *Journal of Comparative Physiology A* **191**(12): 1107–1113.

Beekman M., R.S. Gloag, N. Even, W. Wattanachaiyingchareon, B.P. Oldroyd, et al. (2008) Dance precision of *Apis florea*—clues to the evolution of the honeybee dance language? *Behavioral Ecology and Sociobiology* **62**(8): 1259–1265.

Beekman M., J.B. Lew. (2007) Foraging in honeybees—when does it pay to dance? *Behavioral Ecology* **19**(2): 255–261.

Beekman M., B.P. Oldroyd. (2018) Different bees, different needs: How nest-site requirements have shaped the decision-making processes in homeless honeybees (*Apis* spp.). *Philosophical Transactions of the Royal Society B: Biological Sciences* **373**(1746): 20170010.

Capaldi E., F.C. Dyer. (1995) Landmarks and dance orientation in the honeybee *Apis mellifera. Naturwissenschaften* **82**(5): 245–247.

Dornhaus A., L. Chittka. (1999) Evolutionary origins of bee dances. *Nature* **401**(6748): 38–38.

Duangphakdee O., H. Hepburn, S. Radloff, C. Pirk, P. Rod-im, et al. (2012) Waggle dances in absconding colonies of the red dwarf honeybee, *Apis florea. Insectes Sociaux* **59**(4): 571–577.

Duangphakdee O., H. Hepburn, J. Tautz. (2011) The dance language. *Honeybees of Asia.* Springer, Berlin, Germany, pp. 313–332.

Duangphakdee O., S.E. Radloff, C.W. Pirk, R. Hepburn. (2009) Sun angle time windows for absconding by the dwarf honeybee, *Apis florea. Journal of Insect Physiology* **55**(11): 1009–1012.

Dyer F. (1987) New Perspectives on the dance orientation of the Asian honeybees. *Neurobiology and Behavior of Honeybees.* Springer, Berlin, Germany, pp. 54–65.

Dyer, F. (2002) The biology of the dance language. *Annual Review of Entomology* **47**: 917–949.

Dyer, F., T.D. Seeley. (1989) On the evolution of the dance language. *American Naturalist* **133**: 580–590.

Dyer, F.C., T.D. Seeley. (1994) Colony migration in the tropical honey bee *Apis dorsata F. Insectes Sociaux* **41**: 129–140.

Dyer F.C., T.D. Seeley. (1991) Dance dialects and foraging range in three Asian honey bee species. *Behavioral Ecology and Sociobiology* **28**(4): 227–233.

Esch H.E., S. Zhang, M.V. Srinivasan, J. Tautz. (2001) Honeybee dances communicate distances measured by optic flow. *Nature* **411**(6837): 581.

Gardner, K.E., T.D. Seeley, N.W. Calderone. (2007) Hypotheses on the adaptiveness or non-adaptiveness of the directional imprecision in the Honey Bee's waggle dance (Hymenoptera: Apidae: *Apis mellifera*). *Entomologia Generalis* **29**: 285–298.

Gardner K.E., T.D. Seeley, N.W. Calderone. (2008) Do honeybees have two discrete dances to advertise food sources? *Animal Behaviour* **75**(4): 1291–1300.

Gordon D.M. (2016) From division of labor to the collective behavior of social insects. *Journal of Behavioral Ecology and Sociobiology* **70**(7): 1101–1108.

Gould J.L. (1975) Honey bee recruitment: The dance-language controversy. *Science* **189**(4204): 685–693.

Hepburn H. (2006) Absconding, migration and swarming in honeybees: An ecological and evolutionary perspective. In: V.E. Kipyatkov. (Ed.), *Life Cycles in Social Insects–Behaviour, Ecology Evolution*, St. Petersburg University Press, St. Petersburg, pp. 121–136.

Hepburn R., C. Pirk, O. Duangphakdee. (2014) *Honeybee Nests: Composition, Structure, Function.* Springer-Verlag, Berlin, Heidelberg. doi:10.1007/978-3-642-54328-9.

Jarau S., M. Hrncir. (2009) *Food Exploitation by Social Insects: Ecological, Behavioral, and Theoretical Approaches.* CRC Press, Boca Raton, FL.

Jarau S., M. Hrncir, R. Zucchi, F.G. Barth. (2000) Recruitment behavior in stingless bees, *Melipona scutellaris* and *M. quadrifasciata.* I. Foraging at food sources differing in direction and distance. *Apidologie* **31**(1): 81–91.

Jensen I.L., A. Michelsen, M. Lindauer. (1997) On the directional indications in the round dances of honeybees. *Naturwissenschaften* **84**(10): 452–454.

Koeniger N., G. Koeniger, R. Punchihewa, M. Fabritius, M. Fabritius. (1982) Observations and experiments on dance communication in *Apis florea* in Sri Lanka. *Journal of Apicultural Research* **21**(1): 45–52.

Koeniger N., G. Koeniger. (1980) Observations and experiments on migration and dance communication of *Apsis dorsata* in Sri Lanka. *Journal of Apicultural Research* **19**(1): 21–34.

Koeniger N., G. Vorwohl. (1979) Competition for food among four sympatric species of Apini in Sri Lanka (*Apis dorsata, Apis cerana, Apis florea* and *Trigona iridipennis*). *Journal of Apicultural Research* **18**(2): 95–109.

Lindauer M. (1956) On understanding Asian honeybees [In German]. *Bee World* **38**: 521–557.

Lindauer M. (1957) Communication among the honeybees and stingless bees of India. *Bee World* **38**(1): 3–14.

Makinson J.C., B.P. Oldroyd, T.M. Schaerf, W. Wattanachaiyingcharoen, M. Beekman, et al. (2011) Moving home: Nest-site selection in the red dwarf honeybee (*Apis florea*). *Behavioral Ecology and Sociobiology* **65**(5): 945–958.

Morse R.A., F. Laigo. (1969) Apis dorsata *in the Philippines (including an annotated bibliography).* Philippine Association of Entomologists, Los Baños, Philippines.

Nagamitsu T., T. Inoue. (1997) Aggressive foraging of social bees as a mechanism of floral resource partitioning in an Asian tropical rainforest. *Oecologia* **110**(3): 432–439.

Nieh J. (2004) Recruitment communication in stingless bees (Hymenoptera, Apidae, Meliponini). *Apidologie* **35**:159–182 doi: 10.1051/apido:2004007.

Oldroyd B.P., R.S. Gloag, N. Even, W. Wattanachaiyingcharoen, M. Beekman, et al. (2008) Nest site selection in the open-nesting honeybee *Apis florea. Behavioral Ecology and Sociobiology* **62**(10): 1643–1653.

Oldroyd B.P., S. Wongsiri. (2006) *Asian Honey Bees: Biology, Conservation, and Human Interactions.* Harvard University Press, Cambridge, MA.

Preece K., M. Beekman. (2014) Honeybee waggle dance error: Adaption or constraint? Unravelling the complex dance language of honeybees. *Animal Behaviour* **94**: 19–26.

Punchihewa R., N. Koeniger, P. Kevan, R. Gadawski. (1985) Observations on the dance communication and natural foraging ranges of *Apis cerana, Apis dorsata* and *Apis florea* in Sri Lanka. *Journal of Apicultural Research* **24**(3): 168–175.

Rasmussen C., S.A. Cameron. (2009) Global stingless bee phylogeny supports ancient divergence, vicariance, and long distance dispersal. *Biological Journal of the Linnean Society* **99**(1): 206–232.

Reinhard J., M. V. Srinivasan. (2009) The role of scents in honey bee foraging and recruitment. In: S. Jarau, M. Hrncir (Eds.), *Food exploitation by Social Insects: An Ecological, Behavioral, and Theoretical Approach.* CRC-Press, Boca Raton, FL, pp. 165–182.

Riley J.R., U. Greggers, A.D. Smith, D.R. Reynolds, R. Menzel. (2005) The flight paths of honeybees recruited by the waggle dance. *Nature* **435**(7039): 205.

Robinson W.S. (2012) Migrating giant honey bees (*Apis dorsata*) congregate annually at stopover site in Thailand. *PLoS One* **7**(9): e44976.

Rod-im P., O. Duangphakdee, S. Radloff, C.W. Pirk, C. Hepburn, et al. (2015) Azimuth-dependent waggle dances; flight and foraging activities of the red dwarf honeybee, *Apis florea* Fabricius (1787). *Journal of Apicultural Research* **54**(3): 246–254.

Sarma M.S., S. Fuchs, C. Werber, J. Tautz. (2002) Worker piping triggers hissing for coordinated colony defence in the dwarf honeybee *Apis florea. Zoology* **105**(3): 215–223.
Sasaki M. (1990) Absconding dance in *Apis cerana japonica*: A slow and long tail-waggling motivates the whole colony to translocate. *Proceedings of 11th International Congress*, IUSSI, Bangalore, India, pp. 125–126.
Schaerf T.M., J.C. Makinson, M.R. Myerscough, M. Beekman. (2013) Do small swarms have an advantage when house hunting? The effect of swarm size on nest-site selection by *Apis mellifera* **10**(87): 20130533.
Schaerf T.M., M.R. Myerscough, J.C. Makinson, M. Beekman. (2011) Inaccurate and unverified information in decision making: A model for the nest site selection process of *Apis florea. Animal Behaviour* **82**(5): 995–1013.
Schneider S., L. McNally. (1994) Waggle dance behavior associated with seasonal absconding in colonies of the African honey bee, *Apis mellifera scutellata. Insectes Sociaux* **41**(2): 115–127.
Seeley T.D. (2003) Consensus building during nest-site selection in honey bee swarms: The expiration of dissent. *Behavioral Ecology Sociobiology* **53**(6): 417–424.
Seeley T.D., M. Kleinhenz, B. Bujok, J. Tautz. (2003) Thorough warm-up before take-off in honey bee swarms. *Naturwissenschaften* **90**(6): 256–260.
Seeley T.D., R.A. Morse. (1978) Nest site selection by the honey bee, *Apis mellifera. Insectes Sociaux* **25**(4): 323–337.
Seeley T.D., J. Tautz. (2001) Worker piping in honey bee swarms and its role in preparing for liftoff. *Journal of Comparative Physiology A* **187**(8): 667–676.
Seeley T.D., P.K. Visscher, K.M. Passino. (2006) Group Decision Making in Honey Bee Swarms: When 10,000 bees go house hunting, how do they cooperatively choose their new nesting site? *American Scientist* **94**(3): 220–229.
Seeley T.D., P.K. Visscher, T. Schlegel, P.M. Hogan, N.R. Franks, et al. (2012) Stop signals provide cross inhibition in collective decision-making by honeybee swarms. *Science* **335**(6064): 108–111.
Tanner D.A., K. Visscher. (2006) Do honey bees tune error in their dances in nectar-foraging and house-hunting? *Behavioral Ecology and Sociobiology* **59**(4): 571–576.
Tanner D.A., P.K. Visscher. (2010) Adaptation or constraint? Reference-dependent scatter in honey bee dances. *Behavioral Ecology and Sociobiology* **64**(7): 1081–1086.
Towne W.F. (1985) *The Spatial Precision and Mechanisms of the Dance Communication of Honey Bees: Experimental and Comparative Studies.* Princeton University, Princeton, NJ.
Towne W.F., J.L. Gould. (1988) The spatial precision of the honey bees' dance communication. *Journal of Insect Behavior* **1**(2): 129–155.
von Frisch, K. (1967). *The Dance Language and Orientation of Bees.* Harvard University Press, Cambridge.
Weidenmüller A., T. Seeley. (1999) Imprecision in waggle dances of the honeybee (*Apis mellifera*) for nearby food sources: Error or adaptation? [Elektronische Ressource]. Behavioral Ecology and Sociobiology 46: 190 -199. https://doi.org/10.1007/s002650050609.

19

Diversity and Distribution of Dwarf Honeybees in Iran and Their Role in Natural and Agricultural Systems

Ayça Özkan Koca and İrfan Kandemir

CONTENTS

Introduction

The dwarf honeybee, *Apis florea*, is one of several species in the genus *Apis*. *Apis florea* is naturally distributed in suitable habitats in the Palearctic and Oriental zoogeographic regions. Like other bee species, dwarf honeybees also play a great role in providing ecosystem services as pollinators in their hot distribution areas. However, the research on *Apis florea* is not even comparable to the western honeybee, *Apis mellifera* and its subspecies. Thus, the research on dwarf honeybee is scarce, and it should be a target organism in the future by scientists and be studied thoroughly. Here we compiled a review on research mostly done on *Apis florea* in Iran. As it will be seen in the following parts, research on biology, morphometry, geometric morphometry, caryology and also a limited number of molecular studies are brought together for researchers who are interested in doing a study on dwarf honeybees.

Taxonomic Status of *Apis florea*

The genus *Apis* has been separated into 11 distinct species based on morphological and behavioural characteristics and also on geographic distribution (Engel 1999). Members of genus *Apis* are classified according to their wing and body size and divided into three groups: large, medium and dwarf honeybees. *Apis florea* species belong to the third group. Fore wing length (in the worker bees) in the dwarf honeybees (*A. florea* and *A. andreniformis*) species group has been evaluated to be between 5 and 7 mm. In this group, the male bees have an inner lobe on their hind legs. The dwarf honeybees, *Apis andreniformis* and *Apis florea*, constructed a single-comb nest and are sister species with a partially sympatric distribution in southern Asia. It is difficult to distinguish between two species because the worker bees are morphologically similar and the species are partially sympatric (Otis 1996). Discrimination of *A. florea* and *A. andreniformis* as biological species has now been based on the morphology of the drone genitalia (Wu and Kuang 1986, 1987; Ruttner 1988; Chen 1993; Wongsiri et al. 1996), differences in nest structure (Dung et al. 1996; Rinderer et al. 1996), chemical profiles of beeswax composition (Aichholz and Lorbeer 2000), morphometrics (Jayavasti and Wongsiri 1992; Rinderer et al. 1995; Rattanawannee et al. 2007), allozyme polymorphism (Nunamaker et al. 1984; Li et al. 1986; Gan et al. 1991), mtDNA sequence divergences (Smith 1991; Nanork et al. 2001), flight mechanics (Radloff et al. 2001), differences in timing of mating flights (Rinderer et al. 1993; Otis et al. 2001), sexual selection (Baer 2005), ecological niche differences (Oldroyd et al. 1992; Booncham 1995; Rinderer et al. 2002; Takahashi 2006) and reproductive isolation between the two species (Koeniger and Koeniger 1991, 2000, 2001; Otis 1991; Dung et al. 1996).

Biological Status of *Apis florea* in Iran

The dwarf honeybee (*Apis florea*, Fabricius) is an open-nesting eusocial honeybee species and also a primary pollinator that plays a great role in pollination in Iran and South East Asia to the Far East, especially on the southern coastline extending from Afghanistan to Iraq. *A. florea* could be defined as one-third of a honeybee, *Apis mellifera* in size, with an orange-coloured abdomen and distinctive black-and-white bands at the tip. Body length of colony members is as follows: workers are 8 mm, drones are 12 mm and the queen is about twice the size of workers, that is 14 mm (Moradi and Kandemir 2005). The whole colony lives on a single comb. *A. florea* usually builds a single-comb nest, but double-comb nests were also reported (Whitcombe 1984). It is known that *A. florea* builds a single nest in diverse places; colonies are able to build a wax comb 16 to 25 cm wide and 17 to 40 cm deep (Mossadegh 2014). Hepburn and Radloff (2011) stated that *A. florea* built hives in South East Asia in different places such as on high tree branches, on the walls of buildings and on the edge of the roofs of buildings; while in the warm and dry regions of west Asia, they built hives in the buildings and caves. As in many distribution regions, Moradi and Kandemir (2005) observed that *A. florea* built a single comb in various places in Iran, while they have built comb on cliffs or in the caves in the mountainous area of warm regions in Iran (Figure 19.1).

FIGURE 19.1 *Apis florea* colony site in a rock cave.

The nests of *A. florea* are usually exposed to sunlight and often have one surface of the comb exposed to direct sunlight for several hours a day. Dwarf honeybees built their combs in a west/east direction at the east side of trees and built in the spring and summer, and at the south side in winter (Mossadegh 1990a). Whitcombe (1984) in Oman stated that 'the nests occurred more frequently in the southeast quadrant of trees than they did in any of the other quadrants'. Moradi and Kandemir (2005) observed the hives in both tree branches and cavities during their trip in Iran and also stated that all the combs were placed on the southern part of the tree; similarly, all the nests in caves were open to the south, as the bees need solar energy to heat up their bodies after a cool night.

A. florea can tolerate high temperatures (40°C); the distribution ranges from the east to west coastal zones (Hashemi 2004). Abrol and Kapil (1987) mentioned that *A. florea* can forage at the extremely high temperature of 43°C and utilizes highly concentrated nectar of 68.0%. In the southern and the tropical provinces of Iran, where it is not possible to keep *A. m. meda*, the beekeepers move their colonies to northern parts of Iran, and the dwarf honeybees play a vital role and have a great importance in pollinating the cross-pollinating plants due to their suitability to warm weather. *A. florea* is one of the most important pollinators on the southern coastline in Iran. When the temperature reaches 45°C–50°C in hot summers, other pollinators are not active, but *A. florea* is active in these regions (Mossadegh and Bandpay 1993). For example, when the temperature in Shuster is around 40°C, only dwarf honeybees have pollination activity as a pollinator (Moradi and Kandemir 2005). Thermoregulation and cold resistance were at constant temperatures within the range of +5°C to −5°C (Mossadegh 2014).

To benefit from honey, in many parts of Asia, beekeeping with *A. florea* is being practiced. Kandemir and Moradi (2004) mentioned that there is not much known about the beekeeping with *Apis florea* in Iran. *A. florea* pollination activity starts in the spring and ends in late fall. Depending on the size of the colonies, the honey yield ranges from 1 to 6 kg. According to a honey hunter observation in Shuster, *A. florea* colonies can produce as much as 4 kg of honey (1–5 per a colony, depending on colony size) in a single comb. The dwarf honeybee's honey has been believed to be a medicine for pain, stomach ache, heartburn, constipation, etc. by local people (Kandemir and Moradi 2004).

On the tree, the comb is almost always parallel with the main branch to which it is attached. The brood area is situated below the branch. Honey is stored in cells around and above the branch. In the combs of both *A. florea* and *A. andreniformis*, pollen is stored below the honey, just above the brood. Drone cells are found at the lower margin of the comb below worker brood cells, and swarm queen cells are found vertically from the bottom of the brood nest (Wongsiri et al. 1996). Moradi and Kandemir (2005) observed a fairly big colony in their trip throughout Iran provinces and gave some numerical values about the colony: In the colony there was 12 queen cells, more than 500 drone cells and more than 5000 brood cells. On the same comb, they calculated the percentages of the regions used for different types of tasks. According to their observation, the largest area was the brood-rearing area (approximately 65% of the total comb). The second largest one was the honey storage area (approximately 25% of the total comb). The third area consisted of the drone-rearing area and the queen cells, which comprised less than 10% of the total comb area. Because honeybee queens are polygamous, there is high genetic diversity among the *A. florea* bees even in a single nest. This variation allows species to perform certain tasks (Jones et al. 2007).

The combined rainfall and temperature changes from east to west affected the swarming and migration biology of *A. florea* (Hepburn et al. 2005; Hepburn 2006). *A. florea* swarms during the ascendant phase of colony growth during March and April and sometimes has a second peak in September–October. Little brood is produced in the hot summer months. The old queen leaves with the first swarm and, sometimes, as many as eight swarms a year radiate from the same colony (Tirgari 1971). A similar cycle occurs in Iran where brood rearing begins in April and peaks in May when queens are produced and swarming occurs (Mossadegh 2014). Likewise, there may be a second phase of swarming in autumn. As to distance swarmed, Tirgari et al. (1969) observed that three colonies of *A. florea* moved between 5 and 400 m between nest sites in Iran. Whitcombe (1984) observed that seven absconding colonies only moved short distances of 4–65 m; reproductive swarming colonies were further, 25–800 m.

The mite *Euvarroa sinhai* Delfinado and Baker (Acarina: Mesostigmata) is a common parasite of drone broods of the honeybee *A. florea* in many regions of Asia including Iran (Delfinado and Baker 1974; Koeniger et al. 1983; Mossadegh 1991). Apart from *E. sinhai*, other mites and some fungi have been reported from this honeybee species (Mossadegh 2014). Two acarid mites, *Tyrophagus putrescentiae* (Schrank) and *Caloglyphus oudemansi* (Zak.), were identified from two different areas (Ahwaz and Dezful) in southwest Iran (Mossadegh 1998). *Galleria mellonella* (L.) and *Plodia interpunctella* (Hub.) are the other two important pests in Iran. *G. mellonella* consumes wax and often does considerable damage to combs of traditional and modern beehives, while *P. interpunctella* feeds on cereals, dried fruits, meal, dead insects and many other stored products (Mossadegh 1979, Mossadegh and Birjandi 1982).

Much is known of the microbial pathogens associated with *A. mellifera* but, as yet, almost nothing is known of their association with *A. florea*. Black queen cell virus (BQCV) is associated with *A. mellifera* queen cells, which causes the development of dark brown to black cell walls, and was seen in drone cells of *A .florea* colonies (Mossadegh 1990b, 1992). Using serology, BQCV was detected in diseased drone brood of *A. florea* in Iran (Allen and Ball 1996). Stonebrood is a widespread and potentially destructive disease of the dwarf honeybee in southwest Iran. Of the fungi studied, *Aspergillus flavus* appeared most often associated with drone pupae. *A. niger* and *A. fumigates* were also major causes of the disease in dwarf honeybees (Mossadegh 2014).

Pollination: Honey and Pollen Sources

Dwarf honeybees are distributed in the southern half of Iran; they are found in the subtropical to tropical climates in Asia and Africa and have an important role in the pollination of plant species in these regions. Mossadegh (2014) reported that *A. florea* is a valuable pollinator of many crops during hot summers (45°C–55°C) in south and southwest Iran as well as other parts of its distribution regions.

The foraging ranges of pollinators provide information regarding their intrinsic capabilities in gathering nectar and pollen resources that largely depend on their body size and energy needs (Abrol 1986; Abrol and Kapil 1987). *A. florea*, with relatively low-energy requirements and poor thermoregulatory capabilities, visit flowers in the late morning hours when energy expenditure due to prevailing higher temperature is minimal. Abrol (2006) found that *A. florea* visited fewer flowers/umbel during the earlier and later parts of the day, with a maximum of flowers/min visited between 12:00 and 02:00. Abrol (1988) reported that *A. florea* workers forage up to 150 m from their nesting sites. Bees generally visit flowers within a radius of 1–3 km.

Asian honeybees have been reported to visit wild species in a wide range of plant families. A few families are clearly overrepresented in these reports, most strikingly the Euphorbiaceae, which may often fail to pollinate because male and female flowers are separate and only pollen is collected. Other major families where many species are visited by honeybees include Anacardiaceae, Apocynaceae, Aquifoliaceae, Arecaceae (palms), Asteraceae, Burseraceae, Dipterocarpaceae, Elaeocarpaceae, Fabaceae (legumes), Lamiaceae, Malvaceae (in the broad sense, including Tiliaceae and Sterculiaceae), Myrtaceae, Pentaphylacaceae (including Eurya), Rosaceae, Rubiaceae, Rutaceae, Sapindaceae, Sapotaceae and Verbenaceae (Mossadegh 2014). Moradi and Kandemir (2005) reported that they had found pollen belonging to the families of Myrtaceae, Chenopodiaceae, Rosaceae, Poaceae (Gramineae), Ericaceae, Rhamnaceae in the analysis of *A. florea* honey collected from Iran.

The value of dwarf honeybee honey is high due to its scarcity and medicinal properties. The taste and the texture of the honey are similar to *Apis mellifera* honey. Moradi and Kandemir (2005) conducted chemical composition analysis of *A. florea* honey and also compared it with the values of flower and pine honey types (Sunay et al. 2003) in Turkey; The moisture (%) content was found to be lower (15.1) than the other two types of honey (Flower honey = 18.31; Pine honey = 17.62). In addition, one other parameter was the amount of diastase, which was much lower (6.5) than that of reference honeys (Flower honey = 18.36; Pine honey = 17.65). Moradi and Kandemir (2005) reported that this result may be related to the plant sources and the ability of dwarf honeybees to produce diastase enzyme. Moreover, Moradi and Kandemir (2005) found that the HMF (mg/kg) and total disaccharides (%) are much higher (80.64 and 16.58, respectively) than the other two honey references (HMF; Flower honey = 5.07 and Pine honey = 4.68, Disaccharides; Flower honey = 5.37 and Pine honey = 8.85, respectively).

Distribution Region of Dwarf Honeybees (*Apis florea*) in Iran

The dwarf honeybee (*Apis florea*, Fabricius) is an open-nesting eusocial honeybee species, which naturally occupies rainforests, savannas, subtropical steppes and semidesert habitats and is widely distributed in the subtropical to tropical climates from South East Asia to the Far East (Hepburn et al. 2005). It is not only confined to this area, but it also has continuously been expanding westward from its origin. It is found in Iran, Pakistan and Afghanistan (Otis 1996); it has very limited distribution in coastal areas of southern Iraq, Oman, and Yemen (Wongsiri et al. 1996); it has established colonies in Central Saudi Arabia (Hepburn et al. 2005; Hepburn and Radloff 2011) around Aqaba, Jordan (Haddad et al. 2008, 2009; Moritz et al. 2010) and Eilat in Israel (Moritz et al. 2010). Throughout this expansion along the Arabian Peninsula and into Africa, the species appears to have successfully colonized and become well adapted to the hot arid conditions. It has been recently found in Sudan (Mogga and Ruttner 1988) and is gradually expanding its territory to the whole of Sudan (Moritz et al. 2010) and to neighbouring countries in Djibouti (Steiner 2017) and northeastern Egypt (Pauly and Zewdu 2013; Zewdu et al. 2016; Shebl 2017).

Today, *A. florea* populations are continuously expanding westward along the Arabian Peninsula into Africa. In warm and dry regions in Africa, *A. florea* has built strong colonies and established founder populations. Moreover, the rapid and uncontrolled spread of *A. florea* colonies continues to threaten agriculture with invasion and disease to both urban and rural areas and affects native biodiversity (Haddad et al. 2009; Moritz et al. 2010). Also, *A. florea* colonies may spread from Iran and Iraq to Turkey and southern European countries in the future. Eventually, *A. florea* threatens both agriculture and native biodiversity of these new territories. The expansion of the dwarf honeybee populations on the continent may have positive impacts in adding to pollination of plants for better ecosystem functioning and improving the productivity of cultivated crops. The expansion of the species in the new habitats has to be monitored and needs to be investigated to deepen our understanding of pollinator ecology and to evaluate potential positive and negative socioeconomic and ecological impacts in the new ecosystems (Archer et al. 2014).

Iran is a unique country where three different honeybee species (*Apis mellifera meda*, *Apis florea* and *Apis cerana*) were distributed natively (Crane 1993). *A. m. meda* is distributed across Iran and comes into contact with *A. m. anatoliaca* in southeastern Turkey (Ruttner et al. 1985). However, there is not enough information in the literature on *A. cerana* in Iran except the distribution given by Crane (1993). In addition, Kandemir and Moradi (2004) reported that *A. cerana* was distributed especially at the border to Pakistan. As dwarf honeybee (*A. florea*) is called in Persian 'Zambur riz', *A. florea*, is one of the other honeybee species distributed in Iran. *A. florea* is distributed in the southern regions (including the southwest, south and southeast) of the country (Tahmasebi et al. 2002). The distribution of this species starts from Qasr-e Shirin in Kermanshah and spreads to Ilam, Lorestan, Khuzestan, Bushehr, Fars, Hormozgan, Kerman, Sistan and Baluchestan, and Kohkilooyeh and Boyer Ahmad (Mossadegh 1993; Ruttner et al. 1995; Özkan et al. 2009, Figure 19.2). Moradi and Kandemir (2005) reported that the distribution may extend north to the Kurdistan states. Hashemi (2004) stated that the western distribution starts from Qhasr-e-Shirin in Kurdistan, and Sehristani (1997) mentioned the presence of this species in Kirman-Jiroft in Iran.

FIGURE 19.2 Distribution region of *A. florea* in the southern part of Iran. Vertical lines: area of *A. florea* populations in Iran; horizontal lines: reported area in Iraq. (From Ruttner, F. et al., *Apidologie*, 26, 477–486, 1995.)

Researches on *Apis florea* Distributed in Iran

Apis florea has been distributed in a large area in Asia, and its populations have been spreading to new regions in Asia and Africa. Different approaches have been applied to assess its geographical and genetic variations in this wide range from natural distribution regions to the invasion regions. In the invaded regions, morphometric analyses (Ruttner 1988; Haddad et al. 2009; Zewdu et al. 2016) and microsatellite markers (Moritz et al. 2010) have been used to evaluate the variability of the invasive populations. The morphometric variation of *A. florea* in naturally distributed areas was studied extensively (Mossadegh 1993; Rinderer et al. 1995; Ruttner et al. 1995; Chaiyawong 2001; Tahmasebi et al. 2002; Chaiyawong et al. 2004; Özkan et al. 2009; Al-Kahtani and Taha 2014). Several morphometric studies on regional or country bases have appeared through the years. Based on multivariate statistical analysis of morphometric characters, Ruttner (1988) emphasized that *A. florea* was divided in three geographic types: one found in Sri Lanka and south India; one distributed in Iran, Oman and Pakistan; and the third found in Thailand. Geographic variation and clinical type of variability exist in *A. florea* populations: larger bees found in the north and smaller ones in the south. In addition, *A. florea* was separated into two indistinct morphoclusters in Iran (Tahmasebi et al. 2002), and morphoclusters were changed clinically with latitude in a much wider geography extending from Vietnam and southeastern China to Iran and Oman (~7000 km) (Hepburn et al. 2005).

Morphometric Studies on Iranian Dwarf Honeybees

Attempts have been made to explain the geographic variation of *A. florea* through the use of multivariate statistical analysis of various morphometric characters on the worker bee body: fore wing, hind leg, abdomen, wing angles, and the number of hamuli, proboscis and antenna (Ruttner et al. 1995; Chaiyawong et al. 2004; Hepburn et al. 2005; Özkan et al. 2009; Parichehreh et al. 2013). Ruttner et al. (1985) studied populations of *A. florea* from the southern part of Iran based on morphometric data and the collected morphometric data in conjunction with other populations of dwarf honeybees in the world, and comparing the obtained results and information and the separation of the populations of dwarf honeybees using principal component analysis (PCA), indicated that the dwarf honeybees of Iran are clustered in a different group. Ruttner (1988) firstly demonstrated that geographic variation and clinical type of variability (geocline) exists in *A. florea* populations, meaning that larger bees were found in the north and smaller bees in the south.

Besides geography-related variability, climate-related variation (ecocline) was also reported for *A. florea*. Ruttner et al. (1995) used 15 colonies of *A. florea* from the Kermanshah province in the northwest to east Baluchistan in the southeast, covering a distance of about 2000 km for morphological studies. The results indicated that the measured morphometric characters in the populations of *A. florea* in mountainous areas (with high altitude) are larger than the populations of the coastal bees, leading to a positive correlation between the size/geographical altitude.

In another important study, Tahmasebi et al. (2002) investigated 40 colonies of *A. florea* belonging to nine states (Ilam, Kermanshah, Khuzestan, Fars, Kohgiluyeh and Boyer Ahmad, Bushehr, Hormozgan, Kerman and Sistan and Baluchestan) from the distribution regions of *A. florea* in Iran based on multivariate statistical analysis of 12 standard morphological characters. Tahmasebi et al. (2002) also reported geographic variation in *A. florea* based on analysis of morphometric variables. The colonies of *A. florea* were divided into two groups in the PCA: one group included a western group of larger bees from southwest regions of the country found at higher latitudes, and the another group included an eastern group of smaller bees from southeast regions of the country found at lower latitudes. Hepburn et al. (2005) collected morphometric data from previous studies related to *A. florea* in all parts of Asia, and the previous gaps in the distribution of this species were completed in their research. In this study, *A. florea* colonies from Iran were clustered with the Pakistan colonies, and this group was separated from the other two clusters (Figure 19.3). Consequently, *A. florea* was separated into two indistinct morphoclusters in Iran (Tahmasebi et al. 2002), and morphoclusters were changed clinically with latitude in a much wider geography extending from Vietnam and southeastern China to Iran and Oman (Hepburn et al. 2005).

Özkan et al. (2009) also studied about 115 colonies of *A. florea* from four states (Ilam, Khuzestan, Bushehr and Hormozgan) on the coastal north-south diagonal in Iran (extending over eight parallels, from Hormuzgan (26°) on the south to Ilam (33°) on the north) based on multivariate statistical analysis of 29 morphological characters, 7 of which were new characters for honeybees. For the first time, new size and angle characters from the fore wings and hind wings were utilized in the morphometric analysis of *A. florea* populations. These characters included fore wing distance D (WDL), fore wing distance C (WCL) (Nazzi 1992), hind wing length (HWL), hind wing width (HWW) and hind wing angles (HW1, HW2, HW3, HW4, HW5). Also, this study was the most comprehensive study in terms of colony number and the number of morphometric characters used. Canonical variates analysis (CVA) supported morphological differentiation in *A. florea* colonies. The results of CVA indicated that the colonies from Ilam, Khuzestan and Bushehr showed an overlapped cluster, while the colonies from Hormuzgan state formed

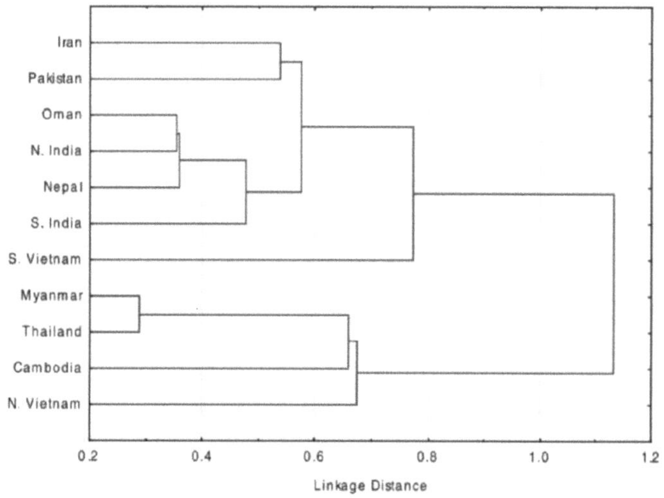

FIGURE 19.3 Hierarchical clustering dendrogram of *A. florea* in Asia, derived from single linkage clustering on seven morphometric characters. (From Hepburn, H.R. et al., *Apidologie*, 36, 359–376, 2005.)

an apparent non-overlapping cluster with the other states. Overall, based on the measured morphological features, the dwarf honeybees belonging to colonies of lower latitudes (Hormozgan) are smaller in comparison to colonies of bees belonging to higher latitudes.

Consequently, geography-related variation was pronounced in *Apis* by Ruttner, who stated that a clinical type of geographical variability exists in *A. florea*, meaning that larger bees were found in the north and smaller ones were found in the south. Tahmasebi et al. (2002) also emphasized that a western group of larger bees was found at a higher latitude (29°–34°) and an eastern group of smaller bees was found at a lower latitude (<29°) in PCA in Iran. Based on morphometric variables, geography-related findings were supported by Özkan et al. (2009). The researchers generally agreed that dwarf honeybees have larger sizes in northern, higher and colder areas, which is consistent with Bergmann's rule as in other animals. This rule is one of the best-known generalizations in animals that the body size increases with latitude (Bergmann 1847).

Geometric-Morphometric Studies on Iranian Dwarf Honeybees

Morphometry is a common method for the identification and classification of distinct species and subspecies of bees. In standard morphometric approach, wing shape has been studied by using angles and distances (Ruttner 1988). Within honeybees (Apidae: Apis), especially *A. mellifera*, this approach based on angles and distances has been extensively applied to recognize the groups, subspecies and species. From the late 1980s, geometric morphometrics, a relatively new field dealing with the study of shape – deriving character states that are found in landmarks (Rohlf and Marcus 1993) – was developed. In recent years, the landmark-based geometric morphometric (GM) method based on wing shape has been applied to recognize and discriminate the groups in *Apis* (Rattanawannee et al. 2010), especially *A. mellifera* (Francoy et al. 2008; Tofilski 2008; Kandemir et al. 2011; Özkan Koca 2012; Barour and Baylac 2016; Nawrocka et al. 2018), *A. dorsata* (Rattanawannee et al. 2012), *A. cerena* (Zhou et al. 2016), as well as stingless bees (Francoy et al. 2009, 2011) and others (Villemant et al. 2007; Francoy et al. 2012). Two-dimensional shape analysis of fore wing has frequently been used for lower and higher taxonomic levels in order to find geographic variations, taxonomic status or population structures in Apidae. The landmark-based method utilized two- or three-dimensional coordinates of biological shapes, thus eliminating non-shape variation and allowing different ways of statistical analysis. In addition, outline-based methods are used to digitize points along an outline, fit the points with a mathematical function (usually some form of a Fourier analysis) and compare curves by using the coefficients of the functions as shape variables in multivariate statistical analyses (Adams et al. 2004). In Apidae, the outline-based method was firstly applied to distinguish species of Euglossine bees (Hymenoptera, Apidae, Euglossini) using fore wing cells (Francoy et al. 2012). In *Apis*, the features from a single wing cell were used to distinguish different racial groups of *Apis mellifera* (Hymenoptera, Apidae) (Francoy et al. 2006).

Kandemir et al. (2009) and Özkan Koca et al. (2018) conducted GM studies based on the landmark method and outline method [by using elliptic Fourier analysis (EFA)], respectively, on dwarf honeybee samples collected from Hormuzgan, Bushehr and Khuzestan along the coast of the Persian Gulf and from an inland state, Ilam. In these studies, a new methodology was applied to investigate wing (fore and hind) and wing cell shape variation in *A. florea* populations distributed in four states in Iran and also to find out geographic variation with respect to wing and wing cell shapes. As a result, Özkan Koca et al. (2018) found the most powerful cell on the wings by comparing the discriminatory power of different cell shapes in distinguishing *A. florea* populations in Iran, and they also found the best method to distinguish *A. florea* populations by comparing three different methods: standard morphometry, landmark-based geometric morphometry and outline-based geometric morphometry.

Firstly, Kandemir et al. (2009) studied the population structure of the Iranian dwarf honeybees (same data in Özkan et al. 2009) using the GM method based on the shape of the wing and identified a total of 20 landmarks on fore wing and a total of five landmarks on hind wing. They found that the populations from different states were significantly ($P < 0.001$) different for both fore wing and hind wing. Discriminant function analysis (DFA) resulted in clear separation of groups when fore wing shape was used. The colonies from Bushehr and Khuzestan showed overlapping groups, while the colonies from Ilam and Hormuzgan state formed a non-overlapping cluster; also, samples from Hormuzgan state formed a distant and non-overlapping cluster. DFA results based on GM support the results obtained from morphological research. Therefore, one can easily conclude that for honeybees, landmark based-GM can be used as a useful tool in population structure studies. Spatial autocorrelation analysis resulted in a total of 26 significant correlograms, and most of these were clinical-type autocorrelograms showing gradual changes from a southeast to a northwest direction.

Secondly, Özkan Koca et al. (2018) investigated cell shape differentiation among *A. florea* populations distributed in four states (Hormuzgan, Bushehr, Khuzestan and Ilam) of Iran (same data in Özkan et al. 2009; Kandemir et al. 2009) using elliptic Fourier analysis (EFA-outline-based GM analysis) on all fore wing cells. When EFA was performed on all closed cells on the fore wings of *A. florea* colonies, researchers determined a significant shape differentiation among *A. florea* populations in all fore wing cells analyzed. The differences between *A. florea* populations were tested with MANOVA based on each wing cell, and the result was found to be statistically significant (Wilks' $\lambda = 0.000$, $p < 0.05$). The research also demonstrated that geographical location plays a key role for the morphological differentiation as in previous studies. DFA based on harmonics of each cell resulted in two clusters (main distinct geographic groups), as in previous morphometric studies: southern *A. florea* populations from lower latitude (Hormuzgan state) and the rest of the populations from higher latitude (Ilam, Khuzestan, Bushehr state) with different differential discrimination values (1st submarginal cell = 75.4%, 2nd submarginal cell = 86.0%, 1st medial cell and 2nd medial cell = 86.8%, marginal cell = 88.6%, 3rd submarginal cell = 90.4%, 2nd cubital cell = 92.1%). The shape analyses of all cells showed that the most efficient cells to discriminate populations were 2nd cubital, 3rd submarginal and marginal cells respectively. Although DFA of each cell shape data displayed similar results, the results obtained from the 3rd submarginal and the 2nd cubital cells displayed better clustering among *A. florea* populations. In both analyses, all colonies classified to their original group with a high probability, and a cross-validation test confirmed that all *A. florea* colonies were reassigned to their original group with over 85% probability. This research revealed that EFA has been functional and effective for distinguishing *A. florea* populations based on the wing cell shapes and verified that two different populations were available.

Özkan Koca et al. (2018) also compared the colony assignments with respect to *A. florea* groups based on standard morphometric characters, shape analysis of fore wing and shape analysis of each wing cell. In three methods (except the shape analysis of the 2nd submarginal cell) used to distinguish populations, all the colonies from Hormuzgan were correctly classified to their assigned groups. The other colonies are classified in preassigned groups with high classification percentages. Standard morphometric and landmark method correctly classified population groups to 95% and 95.7% to their preassigned groups. In addition, 2nd cubital cell and 3rd submarginal cell, the best distinguishing cell, classified the groups as 92.1% and 90.4%, respectively. Accordingly, the outline-based GM applied in the last study has been as successful as the other morphometric approaches such as standard morphometric and landmark-based GM in previous studies (Kandemir et al. 2009; Özkan et al. 2009).

When the square root of the area of each fore wing cell was used as a size estimator in the analysis, fore wing cell size showed geographical variation. The size of cells has a positive correlation with latitude. The square root of the area of the most distinguishing cells belonging to four main populations decreased from north to south. Dot density diagrams based on the size of two cells and latitude showed a clinical type of geographical variability in populations, meaning that bees with large wing cells were found at a higher latitude (in the north) and smaller ones were found at a lower latitude (in the south).

In the study on dwarf honeybee populations distributed in Iran, first, Tahmasebi et al. (2002) demonstrated that colonies from a lower latitude formed a non-overlapping distant cluster based on multivariate analysis of standard morphometrical characters. Later, Özkan et al. (2009) and Kandemir et al. (2009) confirmed the results of Tahmasebi et al. (2002). Finally, a study on the shape of fore wing cells verified that two different populations were available in Iran: southern *A. florea* populations from a lower latitude (Hormuzgan state) and the rest of the populations from a higher latitude (Ilam, Khuzestan, Bushehr state) (Özkan Koca et al. 2018).

Chromosomal Study on Iranian Dwarf Honeybees

Fahrenhorst (1977) reported that the four *Apis* species – *A. florea, A. dorsata, A. cerana* and *A. mellifera* – uniformly had n = 16 chromosomes. Asadi et al. (2010) investigated the karyotype status of *A. florea* using C-banding analysis of the haploid chromosomes. C-banding pattern indicated heterochromatic components in the centromere regions of all chromosomes. Total chromosome length (TCL) indicated that the longest metaphase chromosome of this species is about 2.76 mm, and the shortest one is equal to 0.86 mm. According to the study, the 16 chromosomes of species are classified as two groups: 4 metacentric chromosomes (1, 4, 7 and 11) and 12 chromosomes (including submetacentric chromosomes, subtelocentric and 2 acrocentric chromosomes).

Molecular Studies on Iranian Dwarf Honeybees

In the genetic diversity research realized in dwarf honeybees, morphological studies are given priority rather than studies at the molecular level. There are only two studies at the molecular level on *A. florea* populations in a limited region in Iran. In the first study, Terzi et al. (2014) investigated the nuclear polymorphism within and among *A. florea* populations in three states (Ilam, Khuzestan and Bushehr) from the western part of Iran by using RAPD-PCR analysis. A total of 158 colonies from these three states of Iran were evaluated. Of 25 RAPD primers tested, 10 were identified with a total of 115 fragments. This study realized that the populations have high levels of genetic variation. On the tree based on Nei genetic distance, *A. floera* populations from three states grouped as one big cluster. In addition, populations in each state showed higher similarities than the populations of other states.

The second study was applied to determine genetic diversity within and between the population of *A. florea* in southern Iran by using six microsatellite markers (A88, A107, A7, B124, A113 and A35), which were reported by other researchers on other species of bees (Asadi et al. 2016). Worker bees used in the study were collected from 48 nests (colonies) in four different regions covering the species distribution in southern Iran: 1-Minab, 2-Bandar Abbas, 3-Bandar Khami and 4-Qeshm Island. Sixty-seven alleles were found for six microsatellite loci, and six loci had different numbers of alleles in the sampled colonies ranging from 7 (locus A107) to 3 (loci A7, A35). Samples from Minab showed higher levels of polymorphism, with an average of 3.3 alleles per locus. In contrast, colonies from Qeshm Island had the lowest average alleles per locus (average, 2.2). Gene diversity in *A. florea* ranged from 0.491 to 0.595. Asadi et al. (2016) emphasized that this range probably reflects the spreading of nests in a large region with a varied climate. Phylogenetic relationships among the populations showed two distinct clusters: populations of Bandar Abbas, Bandar Khamir and Qeshm Island were located in the first group. The second group included dwarf honeybee population of Minab. As it is seen from the scarcity of molecular studies, molecular studies do not fully reflect the genetic diversity of the dwarf honeybees in Iran. Further research on all populations using all sorts of molecular markers is needed to develop a better understanding of genetic variation in Iran.

Conclusion

Although there exist tools to study biology, behaviour, biochemistry, pest and diseases, and genetics in the same way that *Apis mellifera* has been studied, the potential of *A. florea* as an organism – as documented in the above-mentioned research – has not been thoroughly utilized. In addition, the potential for the involvement of *A. florea* as an economically valuable organism for providing ecosystem services is mostly unexploited. As this review brought up-to-date research on *A. florea*, especially in Iran, it is apparent that further research is needed in all aspects of the organism for a complete study.

REFERENCES

Abrol, D.P. (1986). Ecophysiological adaptations between pollinating bees and their flowers. *Environment and Ecology*, 4, 161–162.

Abrol, D.P. (1988). Ecology and behaviour of three bee species pollinating loquat (*Eriobotrya japonica Lindley*). *Proceeding of the Indian National Science Academy B*, 54, 161–163.

Abrol, D.P. (2006). Diversity of pollinating insects visiting litchi flowers (*Litchi chinensis* Sonn.) and path analysis of environmental factors influencing foraging behaviour of four honeybee species. *Journal of Apicultural Research*, 45(4), 180–187.

Abrol, D.P., Kapil, R.P. (1987). Nectar dilution pattern of bees in semi-arid environments. *Current Science*, 56, 681.

Adams, D.C., Rohlf, F.J., Slice, D.E. (2004). Geometric morphometrics: Ten years of progress following the 'revolution'. *Italian Journal of Zoology*, 71, 5–16.

Aichholz, R., Lorbeer, E. (2000). Investigation of combwax of honeybees with high-temperature gas chromatography and high-temperature gas chromatography-chemical ionization mass spectrometry. II. High-temperature gas chromatography-chemical ionization mass spectrometry. *Journal of Chromatography A*, 883, 75–88.

Al-Kahtani, S.N., Taha, E.K.A. (2014). Morphometric studies on dwarf honey bee *Apis florea* F. workers in Saudi Arabia. *Journal of Apicultural Science*, 58(1), 127–134.

Allen, M., Ball, B. (1996). The incidence and world distribution of honeybee viruses. *Bee World*, 77, 141–162.

Archer, C.R., Pirk, C.W.W., Caralheiro, L.G., Nicolson, S.W. (2014). Economic and ecological implications of geographic biasin pollinator ecology in the light of pollinator declines. *Oikos*, 123, 401–407.

Asadi, N., Ghafari, S.M., Gharahdaghi, A.A., Tahmasebi, G.H., Khederzadeh, S. (2010). Karyotype and C-banding analyses of haploid male chromosomes of *Apis florea* F. *African Journal of Biotechnology*, 9(27), 4471–4474.

Asadi, N., Rahimi, A., Ghaheri, M., Kahrizi, D., Bagheri Dehbaghi, M., Khederzadeh, S., Banabazi, M.H. et al. (2016), Genetic diversity of the dwarf honeybee (*Apis florea* Fabricius, 1787) populations based on microsatellite markers N. *Cellular and Molecular Biology*, 62(12), 51–55.

Baer, B. (2005). Sexual selection in *Apis* bees. *Apidologie*, 36(2), 187–200.

Barour, C., Baylac, M. (2016). Geometric morphometric discrimination of the three African honeybee subspecies *Apis mellifera intermissa*, *A. m. sahariensis* and *A. m. capensis* (Hymenoptera, Apidae): Fore wing and hind wing landmark configurations. *Journal of Hymenoptera Research*, 52, 61–70.

Bergmann, C. (1847). Ueber die verha ¨ltnisse der wa ¨rmeo ¨konomie der thiere zu ihrer grösse. *Göttinger Studien*, 3, 595–708.

Booncham, U. (1995). Niche differences between *A. florea* and *A. andreniform* is in dry-evergreen forest, Thailand, PhD Thesis, University of Chulalongkorn, Chulalongkorn (in Thai).

Chaiyawong, T. (2001). Morphometric study of *Apis florea* in Thailand, MSc. Thesis, Chulalongkorn University, Bangkok, Thailand, 108 p.

Chaiyawong, T., Deowanish, S., Wongsiri, S., Sylvester, H.A., Rinderer, T.E., De Guzman, L. (2004). Multivariate morphometric study of *Apis florea* in Thailand. *Journal of Apicultural Research*, 43(3), 123–127.

Chen, Y. (1993). *Chinese Apiculture*. Agricultural Publishing House, Beijing, China.

Crane, E. (1993). Current status of research on Asian honey bees. In: Conner, L.J., Rinderer, T., Sylvester, H.A., Wongsiri, S. (ed.) *Asian Apiculture*, Wicwas press, Cheshire, UK pp. 19–41.

Delfinado, M.D., Baker, E.W. (1974). Varroidae, a new family of mites on honey bees (Mesostigmata: Acarina). *Journal of Washington Academy Sciences*, 64, 4–10.

Dung, P.X., Thai, P.H., Trung, L.Q. (1996). Some comparative characteristics between *Apis florea* and *Apis andreniformis* in Minh Hai Province. *Proceedings of the 3th Asian Apiculture Association International Conference*. Hanoi, Vietnam, 62 p.

Engel, M.S. (1999). The taxonomy of recent and fossil honey bees (Hymenoptera: Apida); *Apis*. *Journal Hymenopteran Research*, 8, 165–196.

Fahrenhorst, H. (1977), Nachweis übereinstimmender Chromosomen-Zahlen (n = 16) bei allen 4 *Apis*-Arten. *Apidologie*, 8(1), 89–100.

Francoy, T.M., Faria Franco, F., Roubik, D.W. (2012). Integrated landmark and outline-based morphometric methods efficiently distinguish species of Euglossa (Hymenoptera, Apidae, *Euglossini*). *Apidologie*, 43(6), 609–617.

Francoy, T.M., Grassi, M.L., Imperatriz-Fonseca, V.L., May-Itzá, W.J., Quezada-Euán, J.J.G. (2011). Geometric morphometrics of the wing as a tool for assigning genetic lineages and geographic origin to *Melipona beecheii* (Hymenoptera: Meliponini). *Apidologie*, 42(4), 499–507.

Francoy, T.M., Prado, P.P.R., Gonçalves, L.S., Costa, L.D., De Jong, D. (2006). Morphometric differences in a single wing cell can discriminate Apis mellifera racial types. *Apidologie*, 37(1), 91–97.

Francoy, T.M., Silva, R.A.O., Nunes-Silva, P., Menezes, C., Imperatriz-Fonseca, V.L. (2009). Gender identification in five genera of stingless bees (Apidae, Meliponini) based on wing morphology. *Genetics and Molecular Research*, 8, 207–214.

Francoy, T.M., Wittmann, D., Drauschke, M., Muler, S., Steinhage, V., Bezerra-Laure, M.A.F., De Jong, D., Gonçalves, L.S. (2008). Identification of Africanized honey bees through wing morphometrics: Two fast and efficient procedures. *Apidologie*, 39(5), 488–494.
Gan, Y.Y., Otis, G.W., Mardan, M.B., Tan, S.G. (1991). Allozyme diversity in Asian *Apis*. In: Smith, D.R. (ed.) *Diversity in the Genus Apis*. Westview, Boulder, CO, pp. 131–176.
Haddad, N., De Miranda, J.R., Bataena, A. (2008). Discovery of *Apis florea* in Aqaba, Jordan. *Journal of Apicultural Research*, 47(2), 173–174.
Haddad, N., Fuchs, S., Hepburn, H.R., Radloff, S.E. (2009). *Apis florea* in Jordan: Source of the founder population. *Apidologie*, 40(4), 508–512.
Hashemi, M. (2004). *The Complete Guide to Beekeeping*. 2nd Edition, 730 p.
Hepburn, H.R. (2006). Absconding, migration and swarming in honeybees: An ecological and evolutionary perspective. In: Kipyatkov, V.E. (ed.) *Life Cycles in Social Insects – Behaviour, Ecology and Evolution*. St. Petersburg University Press, St. Petersburg, pp. 121–136.
Hepburn, H.R., Radloff, S.E. (2011). *Honeybees of Asia*. Springer, New York, 669 p.
Hepburn, H.R., Radloff, S.E., Otis, G.W., Fuchs, S., Verma, L.R., Ken, T., Chaiyawong, T., Tahmasebi, G., Ebadi, R., Wongsiri, S. (2005). *Apis florea*: Morphometrics, classification and biogeography. *Apidologie*, 36(3), 359–376.
Jayavasti, S., Wongsiri, S. (1992). Scanning electron microscopy analysis of honeybees (*Apis florea*, *Apis dorsata*, *Apis cerana*, *Apis mellifera*, *Apis andreniformis* and *Apis koschevnikovi*) stings. *Recent Advances in Toxinology Research*, 2, 193–204.
Jones, C., Piyamas, N, Benjamin, P.O. (2007). The role of genetic diversity in nest cooling in a wild honey bee, *Apis florea*. *Journal of Comparative Physiology*, 5, 159–165.
Kandemir, I., Moradi, M.G. (2004). Persian honey bees and beekeeping observations in Iran. *American Bee Journal*, 144(9), 99–703.
Kandemir, I., Moradi, M.G., Özden, B., Özkan, A. (2009). Wing geometry as a tool for studying the population structure of dwarf honey bees (*Apis florea* Fabricius 1876) in Iran. *Journal of Apicultural Research*, 48, 238–246.
Kandemir, I., Özkan, A., Fuchs, S. (2011). Reevaluation of honey bee (*Apis mellifera*) microtaxonomy: A geometric morphometric approach. *Apidologie*, 42(5), 618–627.
Koeniger, N., Koeniger, G. (1991). An evolutionary approach to mating behaviour and drone copulatory organs in *Apis*. *Apidologie*, 22(6), 581–590.
Koeniger, N., Koeniger, G. (2000). Reproductive isolation among species of the genus *Apis*. *Apidologie*, 31(2), 313–339.
Koeniger, N., Koeniger, G. (2001). Behavioural mating barriers among sympatric species of the genus, *Apis*. *Proceedings of the 7th International Conference on Tropical Bees*, Chiang Mai, Thailand, pp. 80–85.
Koeniger, N., Koeniger, G., Delfinado-Baker, M. (1983). Observations onmites of the Asian honey bee species. *Apidologie*, 14(3), 197–204.
Li, S., Meng, Y.P., Chang, J.T., Li, J.H., He, S.Y., Kuang, B.Y. (1986). A comparative study of esterase isozymes in 6 species of *Apis* and 9 genera of Apoidea. *Journal of Apicultural Research*, 25(3), 129–133.
Mogga, G.B., Ruttner, F. (1988). *Apis Florea* in Africa: Source of the founder population. *Bee World*, 69(3), 100–103.
Moradi, M.G., Kandemir, I. (2005). Observations on *Apis florea* "the Dwarf Honey Bee" in Iran. *American Bee Journal*, 145, 498–502.
Moritz, R.F.A., Haddad, N., Bataieneh, A., Shalmon, B., Hefetz, A. (2010). Invasion of the dwarf honey bee *Apis florea* into the near East. *Biological Invasions*, 12, 1093–1099.
Mossadegh, M.S. (1979). The Indian-meal moth *Plodia interpunctella* (Hub.). *Journal of Scientific Agriculture*, 6, 52–68.
Mossadegh, M.S. (1990a). Nesting behavior of *Apis florea* F. (Hymenoptera: Apidae) in Khuzestan, Iran. *11th International Congress*, IUSSI, India, pp. 669–670.
Mossadegh, M.S. (1990b). Some honey bee virus diseases in Iran. *The Scientific Journal of Agriculture*, 13 (13): 64–72.
Mossadegh, M.S. (1991). Geographical distribution, levels of infestation and population density of the mite *Euvarroa sinhai* Delfinado and Baker (Acarina: Mesostigmata) in *Apis florea* F colonies in Iran. *Apidologie*, 22(2), 127–134.
Mossadegh, M.S. (1992). Effect of the black queen cell virus on honeybee drone of *Apis florea* F. (Hym.: Apidae). *Fifth International conference on Apiculture in Tropical Climate, Trinidad and Tobago*.

Mossadegh, M.S. (1993). New geographical distribution line of *Apis florea* in Iran. In: Conner, L.J., Rinderer, T., Sylvester, H.A., Wongsiri, S. (ed.) *Asian Apiculture*, Wicwas press, Cheshire, UK, pp. 64–66.

Mossadegh, M.S. (1998). Mites (Acari) associated with *Apis florea* F. in Iran. *Proceeding of the 13th Iranian Plant Protection Congress*, Karaj, Iran, 242 p.

Mossadegh, M.S. (2014). *Know the Dwarf Honey Bee Apis (Micrapis) florea F. (Hymenoptera: Apidae)*. published by Shafid Chamran University of Ahvaz; Iran, 73 p. (English); 437 p. (Persian).

Mossadegh, M.S., Bandpay, V. (1993). Plants visited by *Apis florea* F. in Iran. In: Conner, L.J., Rinderer, T., Sylvester, H.A., Wongsiri, S. (ed.) *Asian Apiculture*, Wicwas press, Cheshire, UK, pp. 369–371.

Mossadegh, M.S., Birjandi, A.K. (1982). Wax moth. *Ministry of Agriculture, Extension Service*, 6, 1–9.

Nanork, P., Deowanish, S., Wongsiri, S. (2001). Mitochondrial DNA variability of dwarf honeybee (*Apis florea* Fabricius (1787)) in Thailand by using PCR-RFLP technique. *Proceedings of the 7th International Conference on Tropical Bees*, Chiang Mai, Thailand, pp. 341–345.

Nawrocka, A., Kandemir, İ., Fuchs, S., Tofilski, A. (2018). Computer software for identification of honey bee subspecies and evolutionary lineages. *Apidologie*, 49(2), 172–184.

Nazzi, F. (1992). Morphometric analysis of honeybees from an area of racial hybridization in north-eastern Italy. *Apidologie*, 23(2), 89–96.

Nunamaker, R.A., Wilson, W.T., Ahmad, R. (1984). Malate dehydrogenase and non-specific esterase isoenzymes of *Apis florea*, *A. dorsata*, and *A. cerana* Biogeography of *A. andreniformis* and *A. florea* as detected by isoelectric focusing. *Journal of Kansas Entomology Society*, 57, 591–595.

Oldroyd, B.P., Rinderer, T.E., Wongsiri, S. (1992). Pollen resource partitioning by *Apis dorsata*, *A. cerana*, *A. andreniformis* and *A. florea* in Thailand. *Journal of Apicultural Research*, 31(1), 3–7.

Otis, G.W. (1991). A review of the diversity of species within *Apis*. In: Smith, D.R. (ed.) *Diversity in the Genus Apis*, pp. 29–49. Westview, Boulder, CO.

Otis, G.W. (1996). Distribution of recently recognized species of honeybees (Hymenoptera: Apidae: *Apis*) in Asia. *Journal of Kansas Entomological Society*, 69, 311–333.

Otis, G.W., Hadisoesilo, S., Sugiyani, Cooper, A.M. (2001). Using social bees as indicators of habitat quality in Sulawesi, Indonesia. *Proceedings of the 7th International Conference on Tropical Bees*, Chiang Mai, Thailand, pp. 313–317.

Özkan, A., Gharleko, M.M., Özden, B., Kandemir, I. (2009). Multivariate morphometric study of Apis florea distributed in Iran. *Turkish Journal of Zoology*, 33, 93–102.

Özkan Koca, A. (2012). Analysis of *Apis mellifera* L. (Hymenoptera: Apidae) subspecies distributed in the Middle East by using geometric morphometric methods, PhD Thesis, Ankara University, Ankara (in Turkish).

Özkan Koca, A., Moradi, M.G., Deliklitaş, Ö., Uçan, A., Kandemir, İ. (2018). Discrimination of dwarf honey bee (*Apis florea*, Fabricius 1876) populations in Iran using elliptic Fourier wing cell shape analysis. *Journal of Apicultural Research*, 57, 195–202.

Parichehreh, S.H., Farshineh Adl, M.B., Fallahzadeh, M. (2013). Study and comparison of morphological characteristics of dwarf honey bees, *Apis florea* F. (Hymenoptera, Apidae) in Iran. *Journal of Entomological Research*, 5(4), 315–330.

Pauly, A., Zewdu, A.H. (2013). Apini and Meliponini from Ethiopia (Hymenoptera: Apoidea: Apidae: Apinae). *Belgian Journal of Entomology*, 16, 1–36.

Radloff, S.E., Hepburn, H.R., Otis, G.W. (2001). Flight machinery of honey bees (Hymenoptera: Apidae, *Apis*). *Proceedings of the 7th International Conference on Tropical Climates*, Chiang Mai, Thailand, pp. 157–168.

Rattanawannee, A., Chanchao, C., Wongsiri, S. (2007). Morphometric and genetic variation of small dwarf honeybees *Apis andreniformis* Smith, 1858 in Thailand. *Insectes Sociaux*, 14, 451–460.

Rattanawannee, A., Chanchao, C., Wongsiri, S. (2010). Gender and species identification of four native honey bees (Apidae: Apis) in Thailand based on wing morphometric analysis. *Annals of the Entomological Society of America*, 103 (6), 965–970.

Rattanawannee, A., Chanchao, C., Wongsiri, S. (2012). Geometric morphometric analysis of giant honeybee (*Apis dorsata* Fabricius, 1973) populations in Thailand. *Journal of Asia-Pacific Entomology*, 15(4), 611–618.

Rinderer, T.E., Oldroyd, B.P., de Guzman, L., Wattanachaiyingchareon, W., Wongsiri, S. (2002). Spatial distribution of the dwarf honey bees in an agroecosystem in southeastern Thailand. *Apidologie*, 33(6), 539–543.

Rinderer, T.E., Oldroyd, B.P., Wongsiri, S., Sylvester, H.A., de Guzman, L.I., Potichot, S., Sheppard, W.S., Buchmann, S.L. (1993). Time of drone flight in four honeybee species in south-eastern Thailand. *Journal of Apicultural Research*, 32, 27–33.

Rinderer, T.E., Oldroyd, B.P., Wongsiri, S., Sylvester, H.A., de Guzman, L., Stelzer, J.A., Riggio, R.M. (1995). A morphological comparison of the dwarf honey bees of southeastern Thailand and Palawan, Philippines. *Apidologie*, 26(5), 387–394.

Rinderer, T.E., Wongsiri, S., Kuang, B.Y., Liu, J.S., Oldroyd, B.P., Sylvester, H.A., de Guzman, L.I. (1996). Comparative nest architecture of the dwarf honey bees. *Journal of Apicultural Research*, 35(1), 19–26.

Rohlf, F.J., Marcus, L.F. (1993). A revolution in morphometrics. *Trends in Ecology and Evolution*, 8, 129–132.

Ruttner, F. (1988). *Biogeography and Taxonomy of Honeybees*. Springer-Verlag, Berlin, Germany, 300 p.

Ruttner, F., Mossadegh, M.S., Kauhausenkeller, D. (1995). Distribution and variation of size of *Apis florea* in Iran. *Apidologie*, 26(6), 477–486.

Ruttner, F, Pourasghar, D, Kauhausen, D. (1985). Die Honigbienen des Iran. I. *Apis florea* F. *Apidologie*, 16(2): 119–125.

Sehristani, N. (1997). *Honey Bees and Beekeeping*. Sipehr Press, Tehran, Iran, 455 p.

Shebl, M.A. (2017). Discovery of *Apis florea* colonies in northeastern Egypt. *African Entomology*, 25(1), 248–249.

Smith, D.R. (1991). Mitochondrial DNA and honey bee biogeography. In: Smith, D.R. (ed.) *Diversity in the Genus Apis*, pp. 131–176. Westview, Boulder, CO.

Steiner, W.E. (2017). A scientific note on the arrival of the dwarf honey bee, *Apis florea* (Hymenoptera: Apidae), in Djibouti. *Apidologie*, 48(5), 657–659.

Sunay, A.E., Altiparmak, O., Dogaroglu, M., Gokcen, J. (2003). Honey production in Turkey and in the world, honey trading and faced problems. *2nd Marmara Beekeeping Congress Proceedings*, Yalova, Turkey pp. 151–183.

Tahmasebi, G., Ebadi, R., Tajabadi, N., Akhondi, N., Faraji, S. (2002). The effect of geographical and climatological conditions on the morphometrical variation and separation of Iranian small honeybee (*Apis florea* F.) populations. *Journal of Science and Technology of Agriculture and Natural Resources*, 6, 176–185.

Takahashi, J. (2006). Taxonomy and phylogeny of the genus *Apis*, 2: ecology of the dwarf honeybee *Apis florea* and *A. andreniformis*. *Honeybee Science*, 27, 23–32.

Terzi, B., Özkan Koca, A., Gharkheloo, M.M., Sözen, M., Kandemir, İ. (2014). Genetic variation of dwarf honeybee (*Apis florea* Fabricius) populations distributed in the western part of Iran based on RAPD analysis. *Uludag Bee Journal*, 14(1), 4–14.

Tirgari, S. (1971). On the biology and manipulation of *Apis (Micrapis) florea* F. in Iran. *Proceedings of 23rd International Beekeepers Congress*, pp. 330–332.

Tirgari, S., Polous, H., Foroughi, H., Fakhari, F., Nabavit, M.A. (1969). On the occurrence and biology of little bee (*Micrapis florea* F.) and the first account of its migration (Hymenoptera: Apidae). *Proceedings of National Congress On Entomology and Pest Control* (Iran), 4 p (in Persian).

Tofilski, A. (2008). Using geometric morphometrics and standard morphometry to discriminate three honey bee subspecies. *Apidologie*, 39(5), 558–563.

Villemant, C., Simbolotti, G., Kenis, M. (2007). Discrimination of *Eubazus* (Hymenoptera, Braconidae) sibling species using geometric morphometrics analysis of wing venation. *Systematic Entomology*, 32, 625–634.

Whitcombe, R.P. (1984). Bedouin bee. *Saudi Aramco World*, 35, 34–40.

Wongsiri, S., Thapa, R., Oldroyd, B.P., Burgett, M.D. (1996). A magic bee tree: Home of *Apis dorsata*. *American Bee Journal*, 136, 196–199.

Wu, Y., Kuang, B. (1987). Two species of small honeybee – a study of the genus *Micrapis*. *Bee World*, 68, 153-155.

Wu, Y.R., Kuang, B.(1986). A study of the genus *Micrapis* (Apidae). *Zoological Research*, 7, 99–102.

Zewdu, A., Desalegn, B., Amssalu, B., Gebreamlak, B., Tolera, K. (2016). Assessment of alien honeybee species (*Apis florea*) in North West and Northern Ethiopia. *Greener Journal of Agricultural Sciences*, 6(3), 93–101.

Zhou, S., Zhu, X., Xu, X., Zheng, X., Zhou, B. (2016). Assessing of geometric morphometric analyses in microtaxonomy of the *Apis cerana* Fabricius (Hymenoptera: Apidae) within China. *Journal of the Kansas Entomological Society*, 89, 297–305.

20

Apis florea Invasion in Sudan: Present Status and Future Implication

Mohammed M. Ibrahim

CONTENTS

Introduction

The dwarf honeybee *Apis florea* Fab. (Order: Hymenoptera, genus: *Apis*) was described in 1787 by Fabricus. *A. florea* is a small wild honeybee species originally found in tropical and subtropical areas of South East Asia (Ruttner, 1988). *A. florea* covers about 9 million square kilometres, with a good invasive potential to occupied areas. The little bee is steadily expanding westwards, both naturally and inadvertently via global transportation (Mogga and Ruttner, 1988; Hepburn et al., 2005) from its origin area and is found to be very well adapted to warm climate conditions where it does quite well (Ruttner, 1988; Suwannapong et al., 2011). Recently, the species of *A. florea* has become one of great concern for apiculturists and due to its a wider distribution including Southeast China, India, Burma, Laos, Vietnam, Malaysia, Indonesia and the Philippines (Suwannapong et al., 2011). Also, the species has been detected in the Middle East including Iran, Jordan and Egypt, but it has very limited distribution in the coastal area of Southern Iraq and has established abundant populations on the Arabian Peninsula in Oman, Yemen and central Saudi Arabia (Hepburn et al., 2005; Hepburn and Radloff, 2011). In 1985, Mogga and Ruttner reported the prevalence of *A. florea* in a garden and forest of Khartoum in Sudan not far from the airport. They proposed that *A. florea* may be arriving by boat to the port of Khartoum (Ruttner, 1988). Later in 1987, 20 additional colonies had been found at about a 12 km distance from the detected area in an irrigated garden in bushes and trees 0.4 to 8.0 m above ground. Subsequent bee populations have become established in all areas where they are able to survive, specifically around the Nile River sites. The fact that over 80% of the reports were of *A. florea* confirms the ability of the species to survive under harsh climates, with limited forage within the vicinity of central Khartoum, and revealed a wide range of nesting sites. Now, *A. florea* is is spreading rapidly over most parts of Sudan (Omer, 2007; Mortez et al., 2010). Figure 20.1 shows the successful establishment of *A. florea* in Khartoum, and the border with Ethiopia in Al-Gadarif State represents the first recording of the species in Africa. In 2013, Pauly and Zewdu reported that *A. florea* for the first time had invaded 15 localities in Ethiopia. These confirmations make Ethiopia the second invaded country in the Africa region next to Sudan.

A. florea populations can invade areas where even native *A. mellifera* populations are densely populated with apparently no obstacle for the range of expansion (El Shafie et al., 2002). *A. florea* appears to have effectively colonized and become well adapted to the hot arid conditions of northwest and north

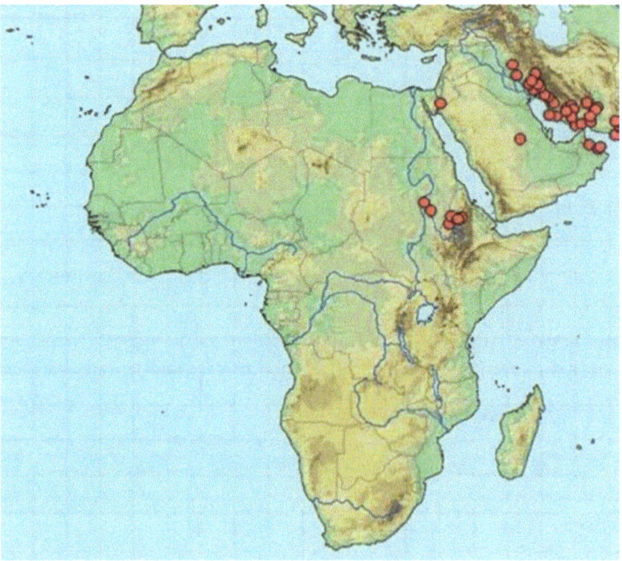

FIGURE 20.1 Map of *A. florea* distribution in Africa.

Ethiopia without being affected by the competition from local honeybees, *A. mellifera*. This natural expansion of the species may not have a pronounced negative effect at this time but may pose some threat to the local ecosystem in the future (Bezabih et al., 2014).

Nests of *A. florea*

Dwarf honeybees are characterized by their outdoor nesting with single comb (usually less than 25 cm across) built around a small branch (El Shafie et al., 2002). Typically, the bees establish their colonies in cryptic nest sites, i.e., balconies, bathrooms, car park canopies and kitchens; due to the fact that they are not very aggressive, they can easily stay undetected for a long time (El Shafie et al., 2002). The nest contains a crown above the branch for honey storage that is also used by the bees as a platform for the foragers leaving and arriving at the nest. The brood comb is suspended below the supporting branch in a single comb. The honeybees normally produce sticky propolis (in strips of 2.5 to 4 cm wide) and use it as a repellent barrier to protect the nest from attack by other insects, such as ants. The curtain of bees for a large dwarf honeybee colony is usually 3–4 bees. When they are disturbed, the bee curtain shows a movement called shimmering in which the workers shake their abdomens from side to side in a synchronous way; at the same time, they emit a sound as if whistling. If a major disturbance occurs, the bee curtain is scattered and attacks the intruder. Concerning preference of plant species to nest on, out of the four active nests, three of them were found on a small branch of dried and live *Ziziphus* spp. (El Shafie et al., 2002). Most of the nests were found hanging on twigs, partly hidden and very difficult to find. These field observations also supported the results from the interview, strengthening that *A. florea* in the assessed localities prefers small branches of *Ziziphus* for various nesting habitats (El Shafie et al., 2002). The colonies were doing well and expanding, with nests of normal size (16 × 14 cm to 46 × 36 cm) with brood and honey stores and in one case, with numerous queen cells. The following measurements were recorded from the combs of the two colonies: vertical length of the comb 279.4 mm (including 50.8 mm of honey chamber), horizontal length 215.9 mm, brood area vertically 147.3 mm and horizontally 167.6 mm, depth of honey cells 18.32 mm (14.29–22.35 mm), depth of pollen cells 12.7 mm and depth of brood cells 9.53 mm (El Shafie et al., 2002).

Morphometric of *A. florea*

The morphometrics measurements of *A. florea* were found to be similar to *A. mellifera*. El Shafie et al. (2002) reported that the proboscis length ranged between 3.250 and 3.407 mm, whereas the body length of foraging worker bees was found to be 7–10 mm. The forewing length ranged between 6.0 to 6.9 mm while the width ranged from 2.00 to 2.299 mm. The cubital index was found to be 2.91 mm. Also, the measurements of the third leg length were 5.239 mm and (3 and 4) tergites length 2.843 mm. On the basis of a principal component analysis of bees in Sudan, the bees were relatively large in size like those from countries of western Asia (Pakistan, Iran and Oman) and much larger than samples from South East Asia (Omer, 2007). This is possibly because *A. florea* was introduced from a country of western Asia, most likely from Pakistan to Africa (Omer, 2007). Human predation/nest destruction, forage shortage and wildfire are outlined as major reasons of absconding in the areas assessed. Unlike *A. mellifera*, these bees are not very aggressive and can stay undetected while coming closer and closer to their nest. This and the cryptic nature of the nests could help *A. florea* colonies to avoid nest destruction by predators including human beings.

Future Implications

Beekeeping in Sudan faces many challenges including selection of the most appropriate honeybee species and a lack of proper identification of ecotypes. Besides that, dwarf honeybees in Sudan have not received many biological, ecological and systematic studies. Fortunately, *A. florea* has been successfully introduced and is well adapted to hot, arid conditions of both urban and rural landscapes (El Shafie, 1992). The bee colonies were found in irrigated gardens near human settlements. In summer, they are found on trees shaded by leaves and buildings; in the winter season, the colonies move to sun-exposed house walls (El Shafie et al., 2002). The bees have the ability to find nest sites very easily; consequently, they spread rapidly and are apparently unaffected by competition from any local species of *A. mellifera*. Currently, the species occupies forests, savannas, subtropical, semidesert steppes and cultivated areas (El Shafie, 1992). In spite of permanent exploitation and destruction of nests by human populations to produce a tasty honey, up to some 4000 colonies have been found in the vicinity of small towns. Dwarf honeybees have not been domesticated for honey production or pollination services in Sudan, but naturally, *A. florea* provides pollination service to a small irrigated area beside the Nile (El Shafie et al., 2002). *A. florea* is found likely to visit crops grown in small-scale agriculture, such as vegetable (cucurbits and *Allium cepa*), and fruit plants (lemon tree, hog plum, lychee, mango and pomelo) in the garden of a suburban house in Khartoum around the Nile River as well. Also, *A. florea* was reported to prefer foraging on *Prosopis* spp. Flowers and contributes positively in the production and buildup of such exotic plants (Laxen, 2007). El Shafie (1992) reported that during January to March the pollen was collected from Talh (*Acacia seyal*), and in July to October it was collected from Sunut (*A. nilotica*). Also, *Eucalyptus* spp. is a good nectar source in April to June. The main pollen flow was between early December and late March. Maximum foraging activities of *A. florea* were rarely observed in early morning, whereas the minimum activities were recorded in the late afternoon. But during late morning, almost equal numbers of foragers of both species were observed. As the time proceeded from 1:00 p.m. to 2:00 p.m., a lot of *A. florea* foragers were observed, especially on *Ziziphus* flower. This could be related to the natural daily rhythms of the species (El Shafie et al., 2002). *A. florea* was collecting pollen all day without interruption, even at very high air temperatures (Mogga, 1994). In all, the spectrum of plant species exploited as main pollen sources for *A. florea* under study showed 11 common species or genera (of 22) such as *Zeamays*, *E. sativa*, *A. seyal*, *Helianthus annuus*, *Acacia nilotica*, *Maerea oblongifolia*, *Phoenix dactylifera*, *A. cepa*, *Convolvulus* spp., *Solanum* spp. and *Corchorus* spp. Additional plant species, which are visited occasionally, may be regarded as not important for the bee. El Shafie (1992) reported that the colony population reaches its peak during September to December. In the season where there is a lot of nectar, the most populated colonies swarm naturally. As remarked by the local community, colony absconding is very common with these alien honeybees. Swarming occurs in January to March, and September, and it coincides with the availability of *Acacia* pollen (El Shafie et al., 2002). During dry months, the chance of finding the nests of *A. florea* in the areas becomes less, indicating fewer colony populations. The use of open nesting of *A. florea* for crop pollination service shows interesting possibilities in some areas. However, it is obvious that the pollen harvested by bees is dependent on the plant sources around the hives (Free, 1970), which can vary between seasons and years. In spite of this, the bees seem to coexist in the Khartoum area by alternating their use of the pollen resources in the course of the day, presumably because of different temperature requirements, and not because of competition. This observation may suggest a possible explanation for our case (El Shafie et al., 2002). The larger honeybees were active earlier (i.e., at lower temperatures) than the small honeybee *A. florea*. But it cannot be excluded that the high numbers of *A. florea* may yet affect the pollen foraging of *A. m. Sudanensis* (El Shafie et al., 2002). The introduced dwarf honeybee *A. florea*, which cannot be used for the production of honey, has become an important visitor to plants in Sudan. Studies conducted by Rahman (1940) on the Indian subcontinent, the original area of distribution of *A. florea*, give hints that the daily rhythm that was observed in the Khartoum area is similar to that in India. Rahman has shown in India that *A. florea* appeared on *Brassica* spp. later in the morning than the other pollinators. Similar studies in Sudan tried to conduct an inquiry on the question of whether the introduction of *A. florea* had exerted a negative influence on beekeeping in the Khartoum area. But still no documentations exists (El Shafie et al., 2002).

Conclusion

It is important to investigate the distribution of *A. florea*, the species' current influence on beekeeping in the invaded areas and future ecological fear for the species and other pollinator bees. Since then, this alien honeybee species has become widely distributed in Sudan. Throughout this expansion, *A. florea* appears to have effectively colonized and become well adapted to the hot, arid conditions of Sudan without being affected by the competition from local honeybees, *A. mellifera* (Zewdu et al., 2016). It is likely that *A. florea* will become permanently established in settlements of the African Sahel Zone, as an integrated part of the Africa fauna. This natural expansion of the species poses some threat to local honeybees, though the majority of the respondents indicated that these bees had no negative effects on beekeeping of the areas under question. Therefore, quick expansion of *A. florea* in the new habitat has to be monitored, and more detailed research is needed to understand the species' ecological impacts on biodiversity of ecosystem and potentially possible economic advantages and disadvantages.

REFERENCES

Bezabih, G., Adgaba, N., Hepburn, H. R. and Pirk, C. W. (2014). The territorial invasion of *Apis florea*. *African Entomology*, 22: 888–890.

El Shafie, H. A., Mogga, J. B. and Basedow, T. (2002). Studies on the possible competition for pollen between the honey bee, *A. mellifera sudanensis*, and the imported dwarf honey bee *A. florea* (Hym., Apidae) in North-Khartoum (Sudan). *Journal of Applied Entomology*, 126: 557–562.

El Shafie, H. A. F. (1992). Studies on some factors affecting pollen gathering activity of honeybees *Apis mellifera* L. (Hymenoptera: Apidae) under Shambat conditions. M.Sc. Thesis, (Agric.), Khartoum University, Khartoum.

Free, J. B. (1970). *Insect Pollination of Crops*. Academic Press, London, UK.

Hepburn, H. R. and Radloff, S. E. (2011). Biogeography of the dwarf honeybees, *Apis andreniformis* and *Apis florae*. *Apidologie*, 42: 293–300.

Hepburn, H. R., Radloff, S. E., Otis, G. W., Fuchs, S., Verma, L. R., Tan, K., Chaiyawong, T., Tahmasebi, G. and Wongsiri, S. (2005). *Apis florea*: Morphometrics, classification and biogeography. *Apidologie*, 36: 359–376.

Laxen, J. (2007). Is prosopis a curse or a blessing. An ecological-economic analysis of an invasive alien tree species in Sudan. Tropical Forests Report 32. Academic dissertation. Faculty of Agriculture and Forestry of the University of Helsinki, Fabianinkatu 34, on Friday 27 April, Helsinki 2007.

Mogga, G. B. and Ruttner, F. (1988). *Apis florea* in Africa: Source of the founder population. *Bee World*, 69: 100–103.

Mogga, J. B. B. (1994). Comparative studies of the indigenous honeybee (*Apis mellifera* L.) and the recently introduced honeybee (*Apis florea* F.) in Sudan. PhD thesis, Khartoum University, Khartoum.

Mortez, R. F. A., Haddad, N., Bataieneh, A., Shalmon., B. and Hefetz, A. (2010). Invasion of the dwarf honeybee *Apisflorea* into the near East. *Biological Invasions* 12: 1093–1099.

Omer, E. A. (2007). Studies on the taxonomy of honeybees in the Sudan. Theses and Dissertations, Agriculture, Crop protection, University of Khartoum.

Pauly, A. and Zewdu, A. H. (2013). *Apini* and *Meliponini* from Ethiopia (Hymenoptera: Apoidea: Apidae: Apinae). *Belgian Journal of Entomology*, 16: 1–36.

Rahman, K. A. (1940). Insect pollinators of toria (*Brassica napus* Linn. var. *Dichotoma sarson* Prain) and (*Brassica campestris* Linn. var. *sarson* Prain) at Lyallpur. *Indian Journal of Agriculture Science*, 10: 422–427.

Ruttner, F. (1988). *Biogeography and Taxonomy of Honeybees.* Springer-Verlag, Berlin, Germany.

Suwannapong, G., Benbow, M. E. and Nieh, J. C. (2011). Biology of Thai honeybees: Natural history and threats. In: *Bees: Biology, Threats and Colonies* (Richard, M. F. ed). Nova Science Publishers, New York. pp. 1–98.

Zewdu, A., Desalegn, B., Amssalu, B., Gebreamlak, B. and Tolera, K. (2016). Assessment of alien honeybee species (*Apis florea*) in North West and Northern Ethiopia. *Greener Journal of Agricultural Sciences*, 6(3): 93–101.

21

Distribution, Nesting Trees and Genetic Diversity of Apis andreniformis *in West Sumatra*

Rika Raffiudin, Siti Salmah and Jasmi Jambak

CONTENTS

Introduction

Sumatra is home to the dwarf Indo-Malayan honeybee, *Apis andreniformis.* As shown in Figure 21.1, the characteristics of *A. andreniformis* are as follows: body length, 12–14 mm for queen, 10–11 mm for drone, and 8–9 mm for worker; black or reddish-brown scutellum; soft and flat brownish wings; black abdomen and reddish brown tergum (first, second or third part) for queen. The worker has a brown to black abdomen and has dark or black hairs on the tibia and hind basitarsus (Maa 1953; Wu and Kuang 1986). A mature colony of *A. andreniformis* contain one queen, 3511–5901 worker individuals and 0–126 drone individuals. Young colonies have one queen, 3263–6904 workers, and no drones (unpublished data).

Compared with its sister species *Apis florea*, *A. andreniformis* has almost the same body size. These two species differ in the colour of the abdomen and hairs on the hind tibia, and on the dorsolateral surface of the base of the hind legs of worker bees. In *A. andreniformis*, the worker abdomen is brown to black, and the tibia hairs on the hind legs are dark or black. In *A. florea*, the worker abdomen is brown, and the tibia hairs on the hind legs and on the surface of the dorsolateral basitarsus of hind legs are white (Wu and Kuang 1986). In addition, *A. andreniformis* is much more aggressive than *A. florea*, and it is able to chase its enemies as far as 30–40 m if disturbed (Wu and Kuang 1986).

Each *A. andreniformis* colony inhabits a single open nest, like *A. florea*. The shape of the comb widens at the top and narrows at the bottom. The honey production area is at the top of the comb (approximately 20% of the comb area). Worker bee cells are in the middle and occupy 50%–60% of the area of

FIGURE 21.1 Colony of dwarf honeybee, *Apis andreniformis*, in West Sumatra. (a) queen, (b) drone, and (c) worker.

the comb, while drone cells are located at the bottom and account for approximately 20% of the total comb area. The queen cells are located at the bottom of the comb (Wu and Kuang 1986). Each nest of *A. andreniformis* comprises a single comb that hangs from branches or shrubs.

This chapter discusses the results of previous surveys of the distribution and nesting sites of this species in West Sumatra, Indonesia, and describes some of its molecular characteristics. The population of this species has declined since 2011, indicating an urgent need for appropriate management strategies.

Distribution of *Apis andreniformis* in West Sumatra

West Sumatra (Sumatera Barat Province) is located on the western part of central Sumatra, and its northern quarter lies on the Equator. The tropical regions are generally defined as the low-latitude areas lying between the Tropics of Capricorn and Cancer, where the air temperature never drops below 18°C and the mean monthly air temperature remains constant at about 27°C–28°C throughout the year. Seasonal variation is considerably less than diurnal extremes, which normally range from 23°C to 33°C.

A. andreniformis generally lives in tropical climates and inhabits secondary forests and its surrounding areas, where it plays an important role as a pollinator. In Indonesia, the bees are distributed in Sumatra, Bangka, Java and Kalimantan, while *A. florea* is distributed in Sumatra, Java and Kalimantan (Maa 1953). Previous studies reported that *A. andreniformis* colonies were found at 0–500 m above sea level (asl) in West Sumatra (Salmah et al. 1990), and at altitudes lower than 1000 m asl in China. Thus, the distribution of *A. andreniformis* is narrower than that of *A. florea* (Wu and Kuang 1986).

Several methods were used to collect *A. andreniformis* in previous surveys, including: (1) collection of insects from flowers using a net; (2) spraying diluted honey (±40% sugar concentration) onto leaves and sweeping insect visitors from the leaves a few hours later; (3) direct collection from the nest; (4) collection of bees attracted to perspiration, moist ash or cement. Nests were observed in diverse tree species (Table 21.1). At least one nest was collected for each species to calculate the number of workers, volume of the nest, production of honey, pollen, wax, or resin and other characteristics. The conditions and altitude of the nest location were recorded using the following codes: primary forest (p); secondary forest (s); orchard area (o); coconut plantation (cc) and village or kampong (v). Habitats other than p and s are collectively called disturbed areas (orchards and villages). To determine the distribution of *A. andreniformis* in West Sumatra, workers and nests were collected and observed.

Figure 21.2 shows the distribution areas of workers and nests collected and observed between 1984 and 1986. Colonies of *A. andreniformis* were found in four locations in 1984 to 1986. Many nests of *A. andreniformis* were found in orchards and residential areas in Padang Sarai in 1984 and 1985. Nests were also found in primary forests and secondary forests in Pesisir Selatan Regency, Bukit Tapan Watas (±470 m asl), and in secondary forests in Kampar, Batu Basurek (Bukit Kincang) (±155 m asl). In Lubuk

TABLE 21.1

Distribution of *Apis andreniformis* in Several Locations in West Sumatra in Three Observation Periods

No.	Location of *Apis andreniformis* Colony	Altitude (m asl)	Vegetation type	Year 1984–1986	1993	2011–2015
	Koto Tangah District					
1	Padang Sarai	2	s, cc; v	n; w	n; w	w
	2 × 11 Enam Lingkung District					
2	Padang Lapai	60–210	v	–	n; w	–
3	Parit Melintang	<100	cc, s	–	–	w; n
4	Pakandangan	<100	cc; s	–	–	w; n
	Padang Utara District					
5	Gunung Pangilun	<100	v; s	–	–	w; n
	Batang Anai District					
6	Sungai Buluh	<100	cc; s; o	–	–	w; n
	VII Koto Sungai Sarik District					
7	Lurah Ampalu	50–300	cc; sc; o	–	–	w; n
	Sungai Geringging District					
8	Batu Gadang	200–400	cc; o	–	–	w; n
9	Lancang	>500	cc; o; sc	–	–	–
	Pesisir Selatan Regency					
10	Bukit Tapan Watas	470	p; s	w	–	–
	Kampar					
11	Batu Basurek	155	s	w	–	–
	Solok Selatan Regency					
12	Lubuk Gadang	750	p; s	w	–	–
	Batipuh District					
13	Andaleh	>1,000	sc; o	–	–	–

Note: p is the primary forest; s is the secondary forest; o is the orchard area; cc is the coconut plantation; v is the village (kampong); w is the worker; n is the nest.

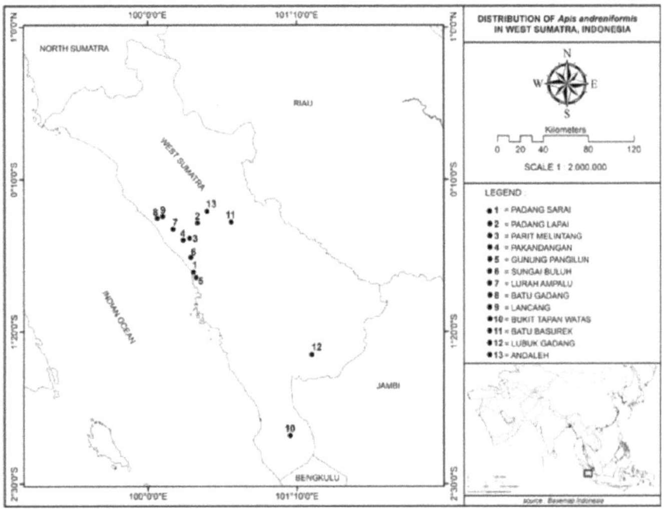

FIGURE 21.2 Distribution of *Apis andreniformis* in West Sumatra, Sumatra, Indonesia. Location numbers are referred to Table 21.1.

TABLE 21.2

Average Size of Morphological Characters of *Apis andreniformis* Workers Collected at Two Altitudes in West Sumatra (1993)

	Altitude				
	Padang Sarai (2 m)		Padang Lapai (210 m)		
Parameter (mm)	Average	Sd.	Average	Sd.	Significance
1. Length of proboscis (LP)	2.6950	0.2339	2.8850	0.0800	*
2. Breadth of thorax (BTh)	2.1850	0.1205	2.4883	0.0429	*
3. Length of forewing (LFw)	6.2083	0.1340	6.3633	0.1066	*
4. Breadth of forewing (BFw)	2.1117	0.0520	2.1417	0.0588	*
5. Length of radial cell III (RL)	2.4033	0.0524	2.4733	0.0410	*
6. Breadth of radial cell III (RD)	0.3917	0.0231	0.4033	0.0183	*
7. Length of hind wing (LHw)	4.4250	0.0898	4.5933	0.0828	*
8. Length of vein RL of hind wing	1.0773	0.0296	1.1467	0.0264	*
9. Length of vein VL of hind wing	0.7120	0.0304	0.7267	0.0259	*
10. Length of vein ML of hind wing	0.8533	0.0369	0.8573	0.0374	ns
11. Length of femur of hind leg (Fe)	1.6533	0.0450	1.6720	0.0412	ns
12. Length of tibia of hind leg (Ti)	2.0533	0.0723	2.0853	0.0522	ns
*) Cubital index (CI)	4.6535	0.6769	4.7011	0.7137	ns
Number of hamuli (NH)	9.9667	0.9279	10.5333	1.0080	*

Note: Sd. is the standard deviation, *) is the unit in the form of ratio and amount, * is the significant difference at 5% level; ns is not significantly different.

Gadang (±750 m asl), *A. andreniformis* nests were found in primary and secondary forests. In 1993, nests of *A. andreniformis* were found between Padang Sarai (±2 m asl) and Padang Lapai villages (±210 m asl) District 2 × 11 Enam Lingkung, Padang Pariaman Regency.

As shown in Table 21.2, *A. andreniformis* still existed in the lowlands (Padang Sarai) and in Padang Lapai (200 m asl) in 1993. Despite the difference of only 200 m altitude between these two sites, the population at these two sites showed significant differences in the following body and wing characteristics: Length of proboscis (LP), Breadth of thorax (BTh), front wing length (LFw), front wing width (BFw), length of cell radius III of the front wing (RL), width of cell radius III of the front wing (RD), rear wing length (LHw), length of RL vein on rear wing, length of VL vein on rear wing and number of hamuli (NH).

An analysis of the distribution of *A. andreniformis* during 2011–2015 conducted in seven locations in West Sumatra produced different results from those recorded previously. In earlier surveys, many nests of *A. andreniformis* colonies were found at the Padang Sarai site, but in 2011–2015, only workers were found. The declining nest numbers of *A. andreniformis* in Padang Sarai from 2010 onwards is due to the conversion of orchard and coconut plantations to residential areas (see Table 21.1). *A. andreniformis* has become a rare species due to habitat loss. Therefore, this dwarf honeybee now has species priority status in Indonesia (Mardiastuti et al. 2008). A priority species is one that needs more conservation efforts than other species. The common criteria and indicators of priority species are population status, habitat conditions, degree of threat and species management status (Mardiastuti et al. 2008). This species is endangered by non-eco-friendly agriculture and plantation, conflagration, and land transformation.

As shown in Figure 21.3, West Sumatra is also inhabited by three sympatric species of *Apis*: *A. andreniformis*, *Apis cerana* and *Apis dorsata*. *Apis andreniformis* is distributed mainly up to altitudes of 500 m asl, while *A. dorsata* and *A. cerana* are distributed up to altitudes of 1000 and 1500 m asl, respectively.

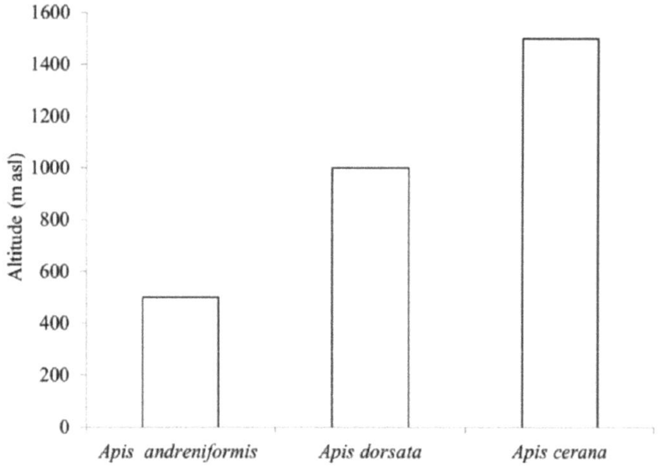

FIGURE 21.3 Distribution of *Apis* species by altitude in West Sumatra.

Nesting Trees of *Apis andreniformis* in West Sumatra

The nesting site is a key factor for survival of the wild honeybee, *A. andreniformis*. The availability of nesting sites is important for wild honeybees to survive in their habitat so that they are able to reproduce a new generation. Here, we discuss the landscape and diversity of nesting plants, the characteristics of nesting sites and habitats of *A. andreniformis* in West Sumatra, Indonesia.

Landscape of Nest Habitat of *Apis andreniformis* in West Sumatra

Landscapes with high plant species diversity and an abundance of agricultural plants provide pollen and nectar resources, and thus, are suitable habitats for bees. In West Sumatra, *A. andreniformis* is distributed in lowlands covered in shrubs and trees in mixed agricultural plantations. Most nests are in cultivated trees, and a few are in several native plant species. In Thailand, *A. andreniformis* nests are found in the orchard at the Research Centre for Tropical Fruit in South East Thailand (Rinderer et al. 2002). This habitat in shrubs and trees is similar to that of *A. florea* in Nepal (Yadav et al. 2017).

Although the nests of *A. andreniformis* are in open areas, the nesting trees in mixed agricultural plantations are under tall tree canopies in the villages of West Sumatra. The strata of the canopies vary depending on the cultivated plant species. For example, the three strata of durian, mango and cacao are the most frequent nesting sites of *A. andreniformis*. The main crop trees in lowland areas are cacao, coconut tree, areca nut and durian (Jasmi et al. 2014). In Thailand, *A. andreniformis* commonly nests in mango and mangosteen trees (Rinderer et al. 2002). In uncultivated areas, the nesting trees are overshadowed by other taller trees. For example, *A. andreniformis* nests in the shrub plants *Piper aduncum* (spiked pepper) and *Melastoma polianthum*, both of which grow under the canopy of taller trees. Mixed agricultural plantations in the lowland areas provide nesting sites and an abundance of nectar and pollen resources for foraging (Aizen et al. 1994).

In West Sumatra, 76 out of 94 plant species that are foraged by the cavity-nesting species *A. cerana* are in mixed plantations in the lowlands. Among them, 46 species are cultivated and 39 species are uncultivated. On Sumba Island, there is a higher plant diversity with 145 species, including endangered and rare plants from 52 families that are managed under the Kaliwu system. On Sumba Island, there is a higher plant diversity with 145 species, including endangered and rare plants from 52 families that are managed under the Kaliwu system (Njurumana et al. 2014). The Center of Tropical Fruits in South East Thailand is home to the sympatric bee species *A. andreniformis, A. cerana* and *A. florea*. The scout bees of *A. florea* choose the plant species for foraging (Makinson et al. 2011).

Diversity of Nesting Site Plants

Apis andreniformis nests have been widely found in secondary forests, plantations, agricultural areas and open areas in three survey periods as shown in Table 21.1. Plants that generally dominate in the lowlands are cacao, palm oil, coconut, durian, pulses and cloves, while the main plants in the highlands are coffee, cinnamon and cacao.

Species Diversity

As shown in Table 21.3 (unpublished data), nests of *A. andreniformis* have been found in 24 plant species in 19 families in West Sumatra. Out of those 24 plant species, 4 were identified as nesting sites of *A. andreniformis* from 2014 until 2018; *Phithocellobium lobatum* (djengkol/jering), *Myristica fragrans* (nutmeg), *Baccaurea racemosa* (Indonesian: *menteng/kepundung*) and cacao. The residents of Padang Pariaman tend to cultivate industrial fruit plants (BPS-Statistics of Padang Pariaman Regency 2014). Nests of *A. andreniformis* are commonly found in cultivated industrial plants, such as coconut, palm oil, cacao, cinnamon and nutmeg trees. The nests are also commonly found in cultivated fruit plants, such as avocado, mango, rambutan, *Lansium* sp., *Citrus*, durian, guava, papaya, banana, mangosteen, jackfruit, *Gnetum gnemon* (*melinjo*), pineapple and sapodilla (*Manilkara zapota*) trees.

Nest Height

The height of *A. andreniformis* nests above the ground varies among trees of different species and different stages of development, i.e., seedling, sapling and tree (Table 21.3). Nests were found in a durian sapling (9.86 cm trunk diameter) at 375 cm height, in a durian tree (19.6 cm diameter) at 845 cm height above

TABLE 21.3
Species and Characteristics of Plants Nested in by *Apis andreniformis* in West Sumatra

Family/Species	Local name	Status	Habitus	Diameter (cm)	Bark surface	Leaf (Type/Category)	HAG (cm)
I. Bombaceae							
1. *Durio zibetinus*	*Durian*	C	A	19.6	Smooth	S/M	375–845
II. Clusiaceae							
2. *Garcinia mangostana*	*Manggis*	C	A	12.8	Smooth	S/L	410–700
3. *Garcinia xanthochymus*	*Asam kandi*	C	A	12.1–18.58	Smooth	S/S	700–930
4. *Garcinia* sp.	–	W1	F	6.8	Smooth	S/M	575
III. Euphorbiaceae							
5. *Codiaeum varigatum*	*Pudiang paga*	W2	F	1.15	Smooth	S/M	185
6. *Manihot esculenta*	*Ubi gajah*	W2	F	6.8	Smooth	Co/L	330
IV. Fabaceae							
7. *Parkia speciosa*	*Patai*	C	A	28.35	Smooth	Co/S	980
V. Laoraceae							
8. *Persea americana*	*Alpokat*	C	A	11.6	Rough	S/L	315
VI. Malastomaceae							
9. *Melastoma polianthum*	*Sikaduduk*	W1	F	1.26	Smooth	S/S	100–185

(Continued)

TABLE 21.3 (*Continued*)

Species and Characteristics of Plants Nested in by *Apis andreniformis* in West Sumatra

Family/Species	Local name	Status	Habitus	Diameter (cm)	Bark surface	Leaf (Type/Category)	HAG (cm)
VII. Meliaceae							
10. *Lansium domesticum*	*Duku*	C	A	33.12	Smooth	S/L	320
11. *Swietenia macrophylla*	*Mahoni*	C	A	35.44	Rough	S/L	380
VIII. Mimosaceae							
12. *Phithocellobium lobatum*	*Jengkol*	C	A	19.74	Smooth	S/M	800–13200
IX. Moraceae							
13. *Artocarpus heterophyllus*	*Nangka*	C	A	10.68	Smooth	S/M	355
14. *Ficus* sp.	*Buku-buku*	W1	A"	9.54	Smooth	S/L	450
X. Myristicaceae							
15. *Myristica fragrans**	*Pala*	C	A	26.38	Smooth	S/M	635–1115
XI. Myrtaceae							
16. *Syzygium oleosum*	*Pucuk merah*	C	F	2.3	Smooth	S/S	290
XII. Phyllanthaceae							
17. *Baccaurea racemosa**	*Kepundung*	C	A	23.56	Smooth	S/L	500
XIII. Piperaceae							
18. *Piper aduncum*	*Sirih hutan*	W1	F	2.22	Smooth	S/L	250
XIV. Poaceae							
19. *Dinochloa* sp.	*Batuang Sariak*	W2	Ca	2.4	Smooth	S/M	300
XV. Rutaceae							
20. *Citrus aurantifolia*	*Jeruk nipis*	C	A	6.8	Smooth	S/S	185
XVI. Sapindaceae							
21. *Nephelium lappaceum*	*Rambutan*	C	A	27.7	Smooth	S/S	630
XVII. Sapotaceae							
22. *Manilkara zapota*	*Sawo*	C	A	15.84	Rough	S/M	190
XVIII. Sterculiaceae							
23. *Theobroma caca**[a]	*Cokelat*	C	A	1.4–11.14	Smooth	S/L	157–365
XIX. Verbenaceae							
24. *Vitex pinnata*	*Laban*	C	A"	3.82	Smooth	S/L	247
Total							

Note: *: nests found at all sites, [a]: nests found at least twice during observation time, C: crop, W1: wild plant in non-backyard/plantation, W2: wild plant in backyard/plantation, A: arboreus, F: frutex, A': sapling, Ca: calmus, S: simple leaf, Co: complex leaf, L: large leaf, M: medium leaf, S: small leaf, HAG: Height above ground.

the ground and in cacao trees of various sizes (stem diameter range, 4–11.14 cm) at 157–365 cm height above the ground. The lowest nest height was 1 m above the ground in a *M. polianthum* shrub, and the tallest nest height was 13.20 m above the ground in djengkol and jering trees. The reported height range of *A. andreniformis* nests in Thailand is similar (0.7–9.0 m above the ground) (Wongsiri et al. 1996) is relatively the same as in West Sumatra. Cacao trees are the most common sites of *A. andreniformis* nests, which usually hangs from branches close to the leaves. The Padang Pariaman district has 17,789 hectares of cacao plantations in 17 subdistricts (BPS-Statistics of Sumatera Barat Provinces 2017) with around 386–3061 hectares of cacao plantations per subdistrict.

A. andreniformis lives sympatrically with other Asian honeybees, such as *A. cerana* and *A. dorsata*. The nesting tree species differ among the various honeybee species. For example, *A. andreniformis* nests in bushes, bamboo, banana palms and small trees such as coffee and tea in Thailand (Wongsiri et al. 1996). The giant honeybee *A. dorsata* nests in 15 plant species in Borneo (Starr et al. 1987) but in 32 plant species (20 families) in the Nilgiri Biosphere Reserve, India (Thomas et al. 2009). In Indonesia, *A. cerana* nests in 18 plant species (15 families) and other non-plant sites polyculture plantations in West Sumatra (Jasmi et al. 2014).

Habitat Diversity

A. andreniformis nests in many plant species in West Sumatra, Indonesia (Table 21.3). Nests are found in trees (trunk diameter > 10 cm) and saplings (trunk diameter < 10 cm). The residents of West Sumatra cultivate a variety of plants in their home gardens and hence, different kind of plant habitats exist in that region. The most common plant habitats are arboreus and frutex. Restoration of plant diversity on agricultural land and in surrounding areas can improve the habitat of wild bees (Nicholls and Altieri 2012).

Characteristics of Nesting Plants of *Apis andreniformis*

Stems

Diameter of stem: The stem diameter of nesting plants of *A. andreniformis* ranges from 1.4–35.44 cm (Table 21.3). Stems with those dimensions are able to support the weight of *A. andreniformis* nests. The average length and width of an *A. andreniformis* nest is 10.03 ± 3.29 cm and 12.18 ± 3.62 cm, respectively (Wongsiri et al. 1996).

Bark surface: *A. andreniformis* builds nests in plants with a smooth bark surface (*P. aduncum*) or a rough bark surface (*Swietenia macrophylla*) (Table 21.3). There is a weak correlation between nest location and bark surface. Nine out of fifteen nesting plants species of *A. dorsata*, which is similar to *A. andreniformis*, were found in trees with smooth bark surfaces in the Nilgiri Biosphere Reserve, India (Thomas et al. 2009).

Branches and twigs: All of the nesting tree species have many branches and twigs, with abundant leaves at the end of the branches that give the plant a thick or leafy appearance. The number of branches and twigs on the plant is an important factor for nesting sites of *A. andreniformis* in West Sumatra. The nest comb wraps around the twigs close to the leaves. The dwarf honeybee *A. florea* in Nepal also builds nests on twigs (Yadav et al. 2017). There are several reasons why *A. andreniformis* builds its nests at the tips of the branches: (a) large predators cannot climb to the ends of branches; (b) the end of the branch is close to the leaves, so the leaves hide the colony from predators and (c) the end of the twig is generally close to flowers as the nectar and pollen source.

Leaves

Leaf size and number: The criterion of leaf size is estimated based on the mature comb size of *A. andreniformis*. The average length and width of an *A. andreniformis* nest comb is 10.03 ± 3.29 cm and 12.18 ± 3.62 cm, respectively (Wongsiri et al. 1996). Based on this size, leaves can be categorized as

FIGURE 21.4 Leaf size categories of plants nested in by *Apis andreniformis* in West Sumatra. (a) large leaves, (b) medium leaves and (c) small leaves.

large, medium and small (width at widest part, >10 cm, 6–9 cm, and <5 cm, respectively) (Figure 21.4a–c). The large, medium and small leaves can cover approximately 50%, 30% and 15% of an *A. andreniformis* nest, respectively. In West Sumatra, many *A. andreniformis* nests are found in plants with compound leaves (Table 21.3). Such plants produce large amounts of leaves or shady leaves. Presumably, the dense large leaves that form the canopy hide the *A. andreniformis* nest from predators. Making nests in such sheltered positions might be one of the behavioural adaptation strategies of *A. andreniformis*.

A. andreniformis is sensitive to interference and so its population is declining. This honeybee often moves from one tree to another if it is disturbed. There is no time interval pattern for this honeybee to build a nest in a new tree after absconding from another tree. The two main factors that cause this behaviour are pest interference and human activity. The moth *Galeria mellonella* is a major pest of this honeybee (Ellis et al. 2013), and human activities also contribute to the absconding of *A. andreniformis* from nests in cultivated areas. When the bees are disturbed, they become very aggressive and attack farmers. The presence of nests are often found on cultivated plants and hence, the bees abscond when they are disturbed by farmers. The persistence of colonies, therefore, depends on the activities of farmers to a large extent. Species management strategies need to be developed for *A. andreniformis*.

Genetic Variations in *Apis andreniformis* in Sumatra

A. andreniformis is one of the native bee species with declining populations in West Sumatra (Table 21.1). A molecular study was conducted to reveal the genetic variations among different nests of bees in Padang Pariaman, West Sumatra (Table 21.1).

In previous studies, similarities between *A. andreniformis* and its sister species *A. florea* have been judged based on morphology and the sequences of mitochondrial and nuclear genes. Previous molecular analyses of two mitochondrial genes (*rrnl* and *cox*2) and one nuclear gene (*itpr*) among all species of honeybees defined *A. andreniformis* and its sister species *A. florea* as the basal species in *Apis* (Raffiudin and Crozier 2007; Lo et. al 2010). Variations among populations of *A. andreniformis* in West Sumatra were further analyzed based on sequences of the *COI* gene and the nuclear *PFK* gene.

Low Frequency of Nucleotide Variations among *Apis andreniformis* Nests in West Sumatra

This is the first study on the *COI* gene in the dwarf honeybee *A. andreniformis* in Indonesia. Five colonies were collected from four locations in Padang Pariaman (Figure 21.2 no. 7). The 585-bp sequence of the *COI* gene of *A. andreniformis* from the five colonies showed the best match to the *COI* sequence from *A. andreniformis* from Nepal (Genbank Accession No. KF736157.1.) with 99% identity, 100% query cover and 0 E value.

Multiple alignments revealed a low level of genetic variation among the five colonies of *A. andreniformis* from West Sumatra. As shown in Table 21.4, the nucleotide at position 164 was A or T, and that at position 450 was C, A or T. On the basis of these nucleotide variations, the genetic distance among samples was low (0.000–0.003 within the Sumatran and non-Sumatran *A. andreniformis*) (Table 21.5).

When *A. florea* was used as the out-group, the two dwarf bee species could be differentiated based on the sequence of the *COI* gene (genetic difference, 0.064–0.067). The phylogenetic analyses using

TABLE 21.4

Nucleotide Variations in *COI* Gene of *Apis andreniformis*

	Nucleotide Sites		
No.	**Sample Codes**	**164**	**450**
1	LC427576 *A. andreniformis1* Sumatra	**T**	**C**
2	LC427580 *A. andreniformis5* Sumatra	**A**	.
3	LC427578 *A. andreniformis3* Sumatra	.	**T**
4	LC427577 *A. andreniformis2* Sumatra	.	.
5	LC427579 *A. andreniformis4* Sumatra	.	.
6	KF736157 *A. andreniformis* Nepal	.	**A**
7	KC294228 *A. andreniformis* China	.	**A**
8	EU163123 *A. andreniformis* China	.	**A**

TABLE 21.5

Nucleotide Sequence-Based p-distance of *COI* Gene in *Apis andreniformis*. GenBank Accession Numbers of *A. andreniformis* 1–5 Sumatra refer to Table 21.4

	[1]	[2]	[3]	[4]	[5]	[6]	[7]	[8]	[9]
[1]									
[2]	0.000								
[3]	0.000	0.000							
[4]	0.002	0.002	0.002						
[5]	0.002	0.002	0.002	0.000					
[6]	0.003	0.003	0.003	0.002	0.002				
[7]	0.002	0.002	0.002	0.002	0.002	0.003			
[8]	0.002	0.002	0.002	0.000	0.000	0.002	0.002		
[9]	0.064	0.064	0.064	0.065	0.065	0.067	0.065	0.065	
[10]	0.095	0.095	0.095	0.095	0.095	0.096	0.093	0.095	0.077

(Continued)

TABLE 21.5 (*Continued*)

Nucleotide Sequence-Based p-distance of *COI* Gene in *Apis andreniformis*

Note		
[1]	:	EU163123 *A. andreniformis* China
[2]	:	KC294228 *A. andreniformis* China
[3]	:	KF736157 *A. andreniformis* Nepal
[4]	:	LC427579 *A. andreniformis4* Sumatra
[5]	:	LC427576 *A. andreniformis1* Sumatra
[6]	:	LC427580 *A. andreniformis5* Sumatra
[7]	:	LC427578 *A. andreniformis3* Sumatra
[8]	:	LC427577 *A. andreniformis2* Sumatra
[9]	:	KC170303 *A. florea*
[10]	:	L06178 *A. mellifera*

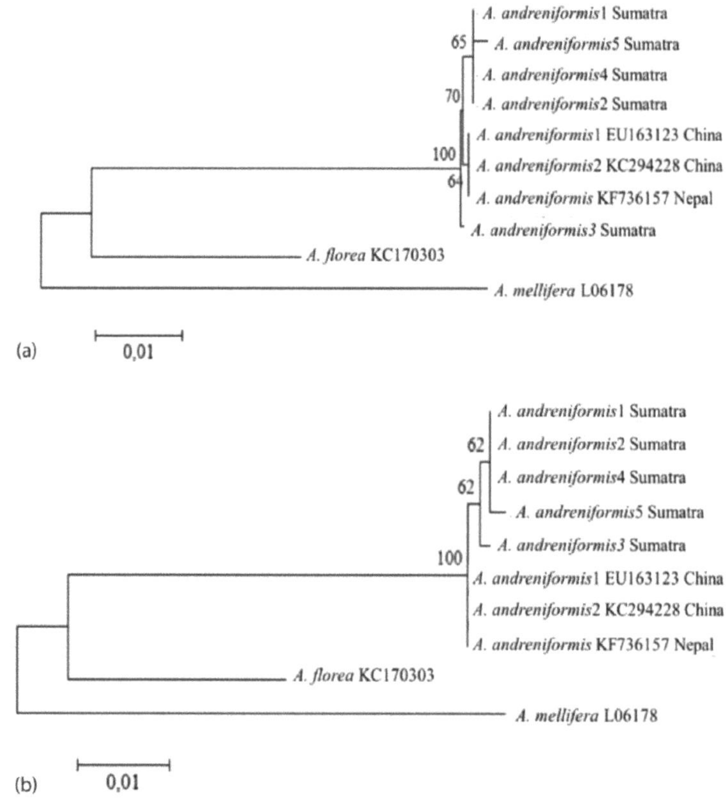

FIGURE 21.5 Phylogenetic trees constructed using nucleotide sequences of *COI* gene in *Apis andreniformis* using neighbour-joining (a) and maximum-likelihood (b) methods with 1000 bootstrap replications.

neighbour-joining (NJ) and maximum likelihood (ML) approaches with 1000 bootstrap replicates showed different topologies. The Sumatran *A. andreniformis* dwarf bees were clustered together in the ML tree but not in the NJ tree (see Figure 21.5).

Rattanawannee et al. (2007) analyzed the sequences of the mitochondrial cytochrome oxidase subunit b gene (*cytb*) and found genetic variations among *A. andreniformis* from the Thailand Peninsula, Phuket Island, and Chiangmai in northern Thailand. However, these analyses did not give a clear geographic signal.

Genetic Variations in *Apis andreniformis* 6-phosphofructokinase gene: Exon–Intron Analyses

Another functional nuclear gene of *A. andreniformis*, *PFK*, was also analyzed. The *PFK* gene, which encodes the key glycolytic enzyme 6-phosphofructokinase, was successfully amplified from *A. andreniformis* collected from West Sumatra (Figure 21.2 no, 7) (DDBJ, LC318760-61). BLAST-N algorithm-based searches of the nucleotide collection (nr/nt) database showed that the *A. andreniformis PFK* gene sequence is closely related to that of its sister species, *A. florea* (GenBank XM_012485123.1) with 100% identity, 53% query cover and 5e-94 E value (Shullia et al. 2019).

The exon–intron organization of *PFK* from *A. andreniformis* was determined by comparison with the *A. florea PFK* coding region. The *A. andreniformis PFK* sequence contains six exons (exons 2, 3, 4, 6, 7 and 8) and four introns (introns 2, 3, 6 and 7) (Shullia et al. 2019). A pairwise alignment of *PFK*s from *A. andreniformis* and *A. florea* is shown in Figure 21.6. The lengths of complete exons were the same between the *A. andreniformis* and *A. florea* sequences, while the intron lengths varied due to nucleotide insertions or deletions. As shown in Table 21.6, the exon–intron boundaries in the *A. andreniformis PFK* gene sequence followed GT–AG rules.

The sequence alignment between the *A. florea PFK* gene and the *A. andreniformis PFK* gene (DDBJ, LC318760-61) revealed 12 nucleotide variations out of 969 nucleotides in exon regions and 15 nucleotide variations of 881 nucleotides in intron regions (Table 21.7). Half of all nucleotide variations in exon regions are in exon 4, and intron 5 is the most variable intron region. This result suggests that exon 4 and intron 5 could be developed as molecular markers for honeybees, especially dwarf honeybees. The mitochondrial large subunit ribosomal RNA gene (*rrnl*) can also differentiate among dwarf honeybees based on a size polymorphism (Higgs et al. 2010). Out of the 27 substitutions between the *A. andreniformis* and *A. florea PFK* genes, 63% and 27% are transition and transversion substitutions, respectively. As shown in Table 21.8, all of the nucleotide substitutions are in the 1st and 2nd codons, thus altering the amino acid. The high number of nucleotide variations in exon regions is consistent with previous studies suggesting that carbohydrate metabolism-related genes such as *PFK* are among the most rapidly evolving genes in eusocial bees, especially honeybees (Woodard et al. 2011).

The deduced *A. andreniformis* PFK amino acid sequence has the following 19-amino-acid consensus pattern: [RK]-x(4)-[GAS]-H-x-[QL]-[QR]-[GS]-[GF]-x(5)-[DE]-[RL] (PS00433), compared with

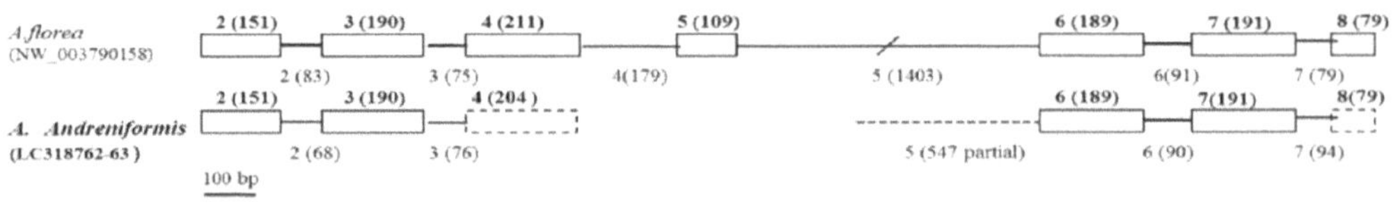

FIGURE 21.6 Schematic representation of exon and intron structures of *PFK* genes in *Apis andreniformis* and *Apis florea*. Numbers above the boxes indicate exon number; numbers in brackets indicate exon length. Intron numbers and intron lengths are shown below each line. (From Shullia et al., *Trop. Life Sci. Res.*, 30, 89–107, 2019.)

TABLE 21.6

Lengths of Introns in Partial *Apis andreniformis PFK* Gene Sequences (Shullia et al. 2019)

Gene	Intron Number	Intron Length	5′ Splice Site (Exon/Intron)	3′ Splice Site (Intron/Exon)	Intron Phase
PFK (Part A)	2	68	AAA/gtatatattatg	attttactttag/GGT	0
	3	76	GAA, G/gtaaataaaa	gtttatttag/GA, GAA	1
PFK (Part B)	5	547 ?	?	atataatttcag/GAA	?
	6	90	TTG/gttagttattat	taataataatag/GGT	0
	7	79	GGA, AA/gtatgtctt	atatttttag/G, GGA	2

Source: Shullia et al., *Trop. Life Sci. Res.*, 30, 89–107. 2019.

TABLE 21.7

Nucleotide Variations in Exon Regions of 6-phosphofructokinase (*PFK*) Gene Sequences between *Apis andreniformis* and *Apis florea*

Exon or Intron Position	**Exon 2**	**Intron 2**			**Exon 3**		**Intron 3**	**Exon 4**					
Nucleotide Position	96	192	202	214	263	375(1st)	461	511(1st)	530(2nd)	552	557(2nd)	639	702
A. andreniformis LC318760-61	A	a	t	a	T	C	a	G	A	C	C	C	G
A. florea NW 003790158	T	g	a	t	C	A	g	A	T	T	T	T	A

Exon or Intron Position	**Intron 5**									**Exon 6**		**Intron 7**	
Nucleotide Position	**816**	**913**	**926**	**962**	**969**	**1052**	**1058**	**1119**	**1163**	**1255**	**1360**	**1393**	**1726**	**1727**
A. andreniformis LC318760-61	a	c	t	t	c	t	a	a	t	C	G	C	c	g
A. florea NW 003790158	g	t	g	c	a	c	t	t	a	T	A	T	t	a

TABLE 21.8

Nucleotide and Amino Acid Variations in Exon Regions of 6-phosphofructokinase (*PFK*) Sequences between *Apis andreniformis* and *Apis florea*

Exon or Intron Position	**Exon 3**	**Exon 4**		
Nucleotide/Amino Acid Position	**375/ 95**	**511/ 115**	**530/ 121**	**557/ 130**
A. andreniformis LC318760-61	C/L	G/V	A/Y	C/A
A. florea NW 003790158	A/I	A/I	T/F	T/V

RITVLGHVQRGGNPSAFDR in the deduced *A. andreniformis* PFK amino acid sequence (Shullia et al. 2019). In the rabbit muscle PFK, the arginine (R) and histidine (H) amino acids are located in the N- and C-halves of two adjacent subunits and it defines the binding site of the substrate, fructose-6-phosphate, which is important in glycolytic pathway (Poorman et al. 1984).

In the neighbour-joining phylogenetic tree based on *PFK* nucleotide sequences, *A. andreniformis* and *A. florea* are in the basal position, separated from *A. dorsata* and the *A. mellifera* and *A. cerana* group that form a monophyletic clade (Figure 21.7). This phylogenetic tree is consistent with those constructed from the molecular sequences of five mitochondrial and nuclear genes and behavioural states

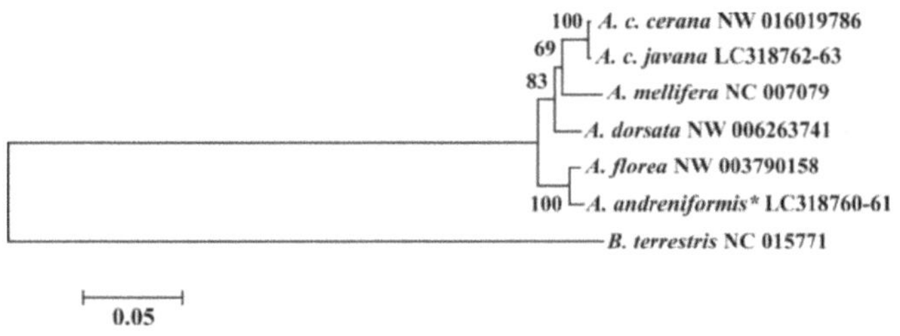

FIGURE 21.7 Phylogenetic tree constructed using nucleotide sequences of *PFK* genes from *Apis* and *Bombus* genera with neighbour-joining method and 1000 bootstrap replications. *indicates species used in this study.

(Raffiudin and Crozier 2007). Thus, both nucleotide variations in exon and intron regions of the *PFK* gene contribute to the topology of the phylogenetic tree. It also confirms the topology of previously reported phylogenetic trees based on two mitochondrial genes (*rrnL*, *cox2*) and one nuclear gene (*itpr*) (Lo et al. 2010). All of those molecular-based honeybee phylogenetic trees are consistent with a morphology-based phylogenetic tree (Alexander 1991).

REFERENCES

Aizen, Marcelo A., and Peter Feinsinger. "Habitat fragmentation, native insect pollinators, and feral honey bees in Argentine Chaco Serrano." *Ecological Applications* 4 (1994):378–392.

Alexander, Byron A. "Phylogenetic analysis of the genus *Apis* (Hymenoptera: Apidae)." *Entomological Society of America* 84 (1991):137–149.

BPS-Statistics of Padang Pariaman Regency. *Padang Pariaman in Figures.* Pariaman: BPS-Statistics of Padang Pariaman, 2014.

BPS-Statistics of Sumatera Barat Provinces. *Sumatera Barat Provinces in Figures.* Padang: BPS-Statistics of Sumatera Barat, 2017.

Ellis, James D., Jason R. Graham, and Ashley Mortensen. "Standard methods for wax moth research." *Journal of Apicultural Research* 52 (1) (2013):1–18.

Higgs, Jessica S., Marcus McHale, and Benjamin P. Oldroyd. "A scientific note on a rapid method for the molecular discrimination of *Apis andreniformis* and *A. florea*." *Apidologie* 41 (2010):96–98.

Jasmi, Siti Salmah, and Syamsuardi Dahelmi. "Nesting site *Apis cerana* Fabr. (Hymenoptera: Apidae) in two different altitude of polyculture plantations in West Sumatra." *Hayati Journal of Biosciences* 21 (3) (2014):135–143.

Lo, Nathan, Rosalyn S. Gloag, Denis Anderson, and Benjamin P. Oldroyd. "A molecular phylogeny of the genus *Apis* suggests that the Giant Honey Bee of the Philippines, *A. breviligula* Maa, and the Plains Honey Bee of southern India, *A. indica* Fabricius, are valid species." *Systematic Entomology* 35 (2010):226–233.

Maa, T. "An inquiry into the systematics of the tribus *Apidini* or honeybees (Hymenoptera)." *Treubia Reinwardtia* 21 (3) (1953):525–639.

Makinson, James C., Benjamin P. Oldroyd, Timothy M. Schaerf, Wandee Wattanachaiyingcharoen, and Madeleine Beekman. "Moving home: Nest-site selection in the red dwarf honeybee (*Apis florea*)." *Behavioural Ecology and Sociobiology* 65 (2011):945–958.

Mardiastuti, Ani., Mirza D. Kusrini, Yeni A. Mulyani, Sastrawan Manullang, and Tonny Soehartono. *Arahan strategis konservasi spesies nasional 2008–2018.* Jakarta: Direktorat Jenderal Perlindungan Hutan dan Konservasi Alam–Departemen Kehutanan RI. JICA, 2008.

Nicholls, Clara I., and Miguel A. Altieri. "Plant biodiversity enhances bees and other insect pollinators in agroecosystems. A review." *Agronomy for Sustainable* 33 (2) (2012):1–20.

Njurumana, Gerson N., Djoko Marsono, Irham, and Ronggo Sadono. "Plant Biodiversity Conservation on Kaliwu System at Sumba Island." *Jurnal Manusia dan Lingkungan* 21 (1) (2014):75–82.

Poorman, Roger A., Anne Randolph, Robert G. Kemp, and Robert L. Heinrikson. "Evolution of *PFK* gene duplication and creation of new effector sites." *Nature* 48 (1984):17–26.

Raffiudin, Rika, and Ross H. Crozier. "Phylogenetic analysis of honey bee behavioural evolution." *Molecular Phylogenetic Evolution* 43 (2007):543–552.

Rattanawannee, Atsalek, Chanpen Chanchao, and Siriwat Wongsiri. "Morphometric and genetic variation of small dwarf honeybees *Apis andreniformis* Smith, 1858 in Thailand." *Insect Science* 14 (2007):451–460.

Rinderer, Thomas E., Benjamin P. Oldroyd, Lilia I. de Guzman, Wandee Wattanachaiyingchareon, and Siriwat Wongsiri. "Spatial distribution of the dwarf honey bees in an agroeco-system in southeastern Thailand." *Apidologie* 33 (2002):539–543.

Salmah, Siti, T. Inoue, and S.F. Sakagami. "Analysis of apid bee richness (Apidae) in Central Sumatra." In *Natural History of Social Wasp and Bees in Equatorial Sumatra*, pp. 139–174. Sapporo: Hokkaido University Press, 1990.

Shullia, Nurul I., Rika Raffiudin, Berry Juliandi. "The Phosphofructokinase and Pyruvate Kinase Genes in *Apis andreniformis* and *Apis cerana* indica: Exon Intron Organisation and Evolution." *Tropical Life Sciences Research* 30 (1) (2019): 89–107.

Starr, Christhoper K., Patricia J. Schmdt, and Justin O. Schmidt. "Nest-site preferences of the giant honey bee, *Apis dorsata* (Hymenoptera: Apidae), in Borneo." *Pan-Pacific Entomologist* 63 (1) (1987):37–42.

Thomas, Sumin G., Anita Varghese, Pratim Roy, Nicola Bradbearf, Simon G. Potts, and Priya Davidar. "Characteristics of trees used as nest sites by *Apis dorsata* (Hymenoptera, Apidae) in the Nilgiri Biosphere Reserve, India." *Journal of Tropical Ecology* 25 (2009):559–562.

Wongsiri, Siriwat, C. Lekprayoon, R. Thapa, et al. "Comparative biology of *Apis andre-niformis* and *Apis florea* in Thailand." *Bee World* 77 (1996):24–35.

Woodard, S. Hollis, Brielle J. Fischman, Aarti Venkat, et al. "Genes involved in convergent evolution of eusociality in bees." *Proceedings of the National Academy Sciences* 108 (2011):7472–7477.

Wu, Yan-Ru, and Kuang Bangyu. "A study of genus *Micrapis* (Apidae)." *Zoological Research* 7 (1986):99–102.

Yadav, Sunita, Yogesh Kumar, and Babul L. Jat. "Honeybee: Diversity, castes and life cycle." In *Industrial Entomology*, pp. 5–34. Singapore: Springer Nature Singapore, 2017.

22

*Distribution, Nesting Biology and Floral Resources of Red Dwarf Honeybee (*Apis florea *Fabricius) in Nagaland, India*

Avinash Chauhan and H. K. Singh

CONTENTS

Introduction

Honeybees are social insects that are directly beneficial to humans. They play a vital role in the environment by pollinating about one-sixth of the world's flowering plant species and many agricultural crops as they forage for nectar and pollen. Besides producing honey, there are several valuable nonfood products produced by honeybees such as beeswax, bee venom, propolis, etc., that are used in cosmetics and pharmaceutical industries.

In India, a total of four species of honeybees viz. *Apis mellifera*, *A. cerana*, *A. dorsata* and *A. florea* are available along with stingless bees (Chauhan et al., 2019). Out of these, *A. mellifera* and *A. cerana* are domesticated ones, while *A. dorsata* and *A. florae* are found in the wild (Deowanish et al., 2001). Later species are migratory in nature and return to the older nests if the distance is less than 200 m (Pirk et al., 2011). Lots of studies have been made on all the honeybee species except *Apis florae* in our country (Narayanaswamy and Basarajavappa, 2013). The species is found to forage on agriculturally important crops, such as Brassica, cucurbits, maize, etc (Deowanish et al., 2001).

Nagaland is an important biodiversity hotspot having vast bee flora and fauna diversity in 16,579 sq. km geographical area. Agriculture is the main commercial activity, and beekeeping is an important integral part of the people in the state since time immemorial (Singh, 2012). Among the four species available in India, there is a rich abundance of *Apis cerana*, *Apis dorsata* and *Apis florea* (Akum et al., 2012). *Apis mellifera* is not present in Nagaland but during recent years, attempts were made to introduce it in the lowlands of the state.

Distribution

Presently in Nagaland, beekeeping is practised in a traditional manner (in simple tree trunks, logs, bamboo and underground domiciles), with *Apis cerana* and stingless bees in houses and *Apis florea* and *Apis dorsata* as wild in specific marked areas around the villages in all the 11 districts (Figure 22.1).

- Green marked areas have high *Apis cerana, Apis florae, Apis dorsata* and stingless bees
- Yellow marked areas have stingless bees, *Apis cerana* and *Apis florae*
- Red marked areas have high stingless bees and *Apis cerana* population
- Light Blue ones having *Apis cerana* and stingless bees in abundance but other bees in smaller pockets

FIGURE 22.1 Distribution of bee abundance in different district Nagaland.

However, the density of nest finding of wild bees is more in lowland areas of Nagaland (NBHM, 2018). Morphologically, the individuals of these bees are smaller in size than *Apis cerana* but bigger than stingless bees (Sharma et al., 2016).

Apis florea nests can be easily seen in the forest areas of villages where they were marked by the villagers for collection of honey (NBHM, 2018). In the Mima village of Nagaland, villagers perform different rituals that are associated with wild bee migration (Singh, 2014a). The nests of *Apis florea* are mainly found attached to bushes and trees of thick canopy. The nests are open and superficial. The bees are highly migratory in nature, and migration was observed in the months of November and December in lower areas of Nagaland while in higher regions, it was observed in October. The little bee builds a single comb nest and the nest is fixed to branches of mainly wild hedges, sweet orange, Nahor and bamboo. The comb present on the peripheral regions of the hosting plant is broad from the top, and the bee lands on this part of the comb (Basavarajappa, 2010). It is always covered by the bees. These bees seldom sting and have a very gentle temperament but are susceptible to swarming (Figure 22.2a and b).

(a) (b)

FIGURE 22.2 (a) *Apis florea* nest in wild bushes and (b) Deserted nest showing creamish white combs.

TABLE 22.1

Different Attributes of *Apis florea*

S. No	Treatment	Replication R1	R2	R3	R4	R5	Mean
1	Period of migration (Standard week)	November	November	November	December	December	–
2	Density of the nest in per unit area	2	1	2	1	1	1.40
3	Length of comb (cm)	16.5	15.0	17.5	20	18	17.4 ± 0.83
4	Width of the comb (cm)	11.5	10.5	13.5	13.5	12	12.2 ± 0.58
5	Height of the nest from ground (cm)	110	70	160	55	150	109 ± 20.88
6	Thickness of base/ support (cm)	3.0	3.0	3.5	2	4.5	3.2 ± 0.41
7	Name of the plant species	Bamboo	Wild hedge	Sweet orange	Bamboo	Nahor	–
8	Nearest water source (metre)	70	50	25	90	20	51.0 ± 13.27
9	Arrival standard week of the colony	46th	47th	47th	49th	49th	47.6 ± 0.6
9	Departure standard week of the colony	9th	16th	1st	13th	10th	9.8 ± 2.52
10	Host plant in vicinity	Sweet orange, Nahor, Litchi, Mutabulis, Pigeon pea, Mustard,					
11	Colour of comb	Creamy white	Creamy white	Creamy white	Creamy white	Creamy white	Creamy white
12	GPS data	310/ N 25° 48.292 E 093° 51.64	310/ N 25° 48.288 E 093° 51.65	310/ N 25° 48.298 E 093° 51.68	310/ N 25° 48.289 E 093° 51.65	310/ N 25° 48.288 E 093° 51.67	–

Different studies were made and surveys were conducted for the location of *Apis florea* period of arrival, measurement of comb, nesting site, distance of water source, departure and host plants around the Dimapur district of Nagaland. After a survey of two years, it was observed that a period of maximum nest abundance is in the standard week of 47.6 ± 0.6 of the year, while the bees departed in 9.8 ± 2.52 in the standard week of the next year. Measurement of combs, length and width of combs were recorded as 17.4 ± 0.83 cm and 12.2 ± 0.58 cm, respectively (Table 22.1).

Nesting Sites

The nesting sites of *Apis florea* were located in bamboo shoots and leaves, wild hedges and shrubs, branches of sweet orange in the wild and nahor trees. However, some nests can be present in thick, long grasses grown wild in Nagaland. The water sources were mostly available within 90 m of the nesting sites. The comb colour of *A. florea* was observed to be creamy white (Figure 22.2b). Bees make buzzing sounds if one approaches near their nest.

Apis florae bees usually migrate from higher hills to lower hills in November and then migrate back to higher hills in February of the next year. These bees play an important role in the conservation of biodiversity through pollination of wild as well as cultivated crops during migration. The people of Nagaland treat these pollinators with high respect, and these bees are an important part of their lives. In this state, one can easily find 2–20 colonies of domesticated bees as well as some protected nests of wild bees (*Apis florea* and *Apis dorsata*). The different flora on which *Apis florea* bees can be located under lowland conditions of Nagaland are vegetable, fruit, seed, ornamental, wild bushes and other medicinal plants Figure 22.3. The flora details are given in Table 22.2.

FIGURE 22.3 *Apis florea* foraging on ash gourd flowers.

TABLE 22.2

Flora of *Apis florea* in Lowlands of Nagaland

Common Name	Scientific Name	Family
Adhatoda	*Adhatoda zeylarica*	Acanthaceae
Aonla	*Emblica officinalis*	Euphorbiaceae
Ash Gourd	*Benincasa hispida*	Cucurbitaceae
Banana	*Musa* spp	Musaceae
Ber	*Ziziphus mauritiana*	Rhamnaceae
Bitter gourd	*Momordica charanta*	Cucurbitaceae
Bottle brush	*Callistemon lanceolatus*	Myrtaceae
Bottle Gourd	*Lagenaria vulgaris*	Cucurbitaceae
Carambola	*Averrhoa carambola*	Oxalidaceae
Cashew nut	*Anacardium occidentale*	Anacardiaceae
Castor	*Ricinus communis*	Euphorbiaceae
Cherry	*Prunus avium*	Rosacea
Chow-chow	*Sechium edule*	Cucurbitaceae
Climbing Hemp	*Mikania micrantha*	Asteraceae
Cucumber	*Cucumis sativus*	Cucurbitaceae
Eucalyptus	*Eucalyptus spp*	Myrtaceae
Firmania	*Firmania colorata*	Sterculiaceae
Goat weeds	*Ageratum conyzoides*	Asteraceae
Guava	*Psidium guajava*	Rhamnaceae
Gul Mohor	*Delonix regia*	Leguminoceae
Indian Neem	*Azadirachta indica*	Meliaceae
Jack fruit	*Artocarpus heterophyllus*	Moraceae
Justicia	*Justicia gendarussa*	Acanthaceae
Ladies' Finger	*Abelmoschus esculentus*	Malvaceae
Lemon	*Citrus spp*	Rutaceae
Litchi	*Litchi chinensis*	Sapindaceae
Mango	*Mangifera indica*	Anacardiaceae
Marigold	*Calendula officinalis*	Asteraceae

(Continued)

TABLE 22.2 (*Continued*)

Flora of *Apis florea* in Lowlands of Nagaland

Common Name	Scientific Name	Family
Mulberry	*Morus serrata*	Moraceae
Mustard	*Brassica juncea*	Brassicae
Naga tenga	*Rhus semialata*	Anacardiaceae
Needle tree	*Schima wallichi*	Theaceae
Papaya	*Carica papaya*	Caricaceae
Passion fruit	*Passiflora spp*	Passifloraceae
Pea	*Pisum sativum*	Leguminoceae
Peach	*Prunus persica*	Rosacea
Pears	*Pyrus communis*	Rosacea
Peltophorum	*Peltophorum spp*	Caesalpiniaceae
Phlogacanthus	*Phlogacanthus thysiflorus*	Acanthaceae
Phlogacanthus	*Phlogacanthus pubinervius*	Acanthaceae

Role in Pollination

The data on the scientific management of *Apis florea* for pollination are still not available in Nagaland, as these bee nests in thick forests and foraging activity on cultivated crops is observed only for few crops such as cucumber, ash gourd, tomato, brinjal, litchi, guava, pigeon pea and sweet orange (Singh, 2014; Singh and Singh 2014; Singh et al., 2014; Chauhan and Singh, 2019; Anonymous, 2019). However, emphasis on other crops pollinated by *Apis florea* is of much importance. Moreover, in Nagaland, other pollinators (halictid, megachilid, dipterans and coleopteran, etc.) are plentiful, so farmers do not keep vigilant about the role of bees as pollinators. However, the presence of *Apis florea* on blooms of cultivated and wild plants reflects their important place in nature for conservation of these plant species. Unfortunately, the honey extraction from wild pollinators is not scientifically developed due to restricted movement to hilly terrains. The local honey hunters in remote areas take out the whole comb and then squeeze it along with the brood, thus destroying the whole colony (NBHM, 2018). In recent years, the All India Coordinated Research Project on Honey Bees and Pollinators, along with team members from the Nagaland Beekeeping and Honey Mission, has given training to farmers/villagers for proper harvesting of honey without damage to the bees. With the intervention of scientific techniques, now villagers are becoming aware of the importance of bees in their lives and livelihoods. The day is not too far off when bees will become an integral part of everyone's life and thus everyone will help and contribute in conservation of these important parts of our ecosystem.

REFERENCES

Akum Z, Singh H K, Sezie K and Singh A K. 2012. Biometric and forage studies of stingless bees in Nagaland. *Indian Journal of Entomology* **74**: 343–347.

Anonymous. 2019. Biennial report of All India Coordinated Research Project on Honey Bees and Pollinators, Medziphema Nagaland.

Basavarajappa S. 2010. Nesting plants of dwarf honeybee, *Apis florea* F. under tropical conditions of Karnataka. *Animal Biology* **60**: 437–447.

Chauhan A, Singh H K, Kumaranag K M and Awomi O S. 2019. Problems and its mitigation in honey bee keeping in Nagaland. AICRP Honey Bees & Pollinators, SASRD, Nagaland University: 1–9.

Deowanish S, Wattanachaiyingcharoen W, Wongsiri S, Oldroyd B P, Leepitakrat S, Rinderer T E and H A sylvister. 2001. Biodiversity of dwarf honey bees in Thailand." *Proceedings 7th International Conference on Tropical Bees*, Chiang Mai, Thailand.

Nagaland Beekeeping and Honey Mission (NBHM). 2018. Bees our future. Published in Souvenir Nagaland Honey bee day 5th December 2018.

Narayanaswamy R and Basavarajappa S. 2013. Nesting attributes of dwarf bee, *Apis florea* F. under urban ecosystem of Manasagangotri campus, Mysore, India. *European Journal of Zoological Research* 2 (6): 6–15.

Pirk C W W, Kendall L C, Orawan D, Sarah E R and Randall H. 2011. Economics of combwax salvage by the red dwarf honey bee, *Apis florea*. *Journal of Comparative Physiology* 353–359.

Sharma H K, Chauhan A, Katna S, Rana K, Rana B S, and Thakur M. 2016. Nesting biology of dwarf bee, *Apis florae* in Himachal Pradesh. *International Journal of Farm Sciences* 6 (4):1–5.

Singh A K. 2012. Importance of insect pollinators for sustainable agriculture and second green revolution. *Agricultural Education, Research and Extension in India: Problem, Solution and Prospects*. Podder Publications: 64–68.

Singh A K. 2014a. Traditional beekeeping shows great promises for endangered indigenous bee *Apis cerana*. *Indian Journal of Traditional Knowledge* 13 (3): 582–588.

Singh A K. 2014b. Diurnal abundance of diverse insect pollinators of sweet orange. National symposium on Entomology as a science and IPM as a technology the way forward: 2014, Pasighat.

Singh A K and Singh H K. 2014. Diversity of true insect pollinators of sweet orange. National symposium on Entomology as a science and IPM as a technology the way forward: 2014, Pasighat.

Singh A K, Singh H K and Thakur R K. 2014. Role of insect pollinators on the productivity of pigeon pea (*Cajanus cajan* L.). National conference on Emerging challenges and opportunities in biotic and abiotic stress management: 2014, Hyderabad.

23

Distribution, Nesting Biology and Floral Resources of Red Dwarf Honey Bee (Apis florea *Fabricius*) *in West Bengal, India*

Ujjwal Layek and Prakash Karmakar

CONTENTS

Introduction

Honey Bees (*Apis* spp., Apidae) are the most important insects to humans, accounting for a large proportion of total global pollination services for crops (Rader et al., 2016), as well as providing bee products such as honey, pollen and wax. In India, three wild species of true honey bees (*Apis cerana* Fabricius, *Apis dorsata* Fabricius and *Apis florea* Linnaeus) can coexist. Additionally, an exotic species, *Apis mellifera* Linnaeus, has been introduced from the western countries (Thomas et al., 2002). Coexisting species differ not only morphometrically but also with respect to nesting characteristics, habitat preferences, activity patterns, use of floral resources and effectiveness as pollinators (Corlett, 2011). Phylogenetic studies suggest that the different species form separate monophyletic groups, with the small species basal and the honey bees as sisters to the stingless bees and bumble bees (Arias and Sheppard, 2005).

In this chapter we have tried to provide information about the nesting habit and floral resources of red the dwarf honey bee (*A. florea*) in West Bengal, India.

Distribution of *A. florea*

The red dwarf honey bee settled in a wide variety of habitats such as semideserts, subtropical steppes, savannas and tropical rainforests. The honey bee is distributed from south east Asia, extending to Vietnam and southeastern China, across mainland Asia, along and below the southern Himalayas, westwards to the Plateau of Iran and south into Oman and Sudan (Mogga and Ruttner, 1988; Hepburn and Hepburn, 2005). However, the major area of occurrence of this bee species is in India, Pakistan, Sri Lanka, Thailand, Malaysia and part of Indonesia and Palawan at altitudes below 1000 m (Ruttner, 1988). In India, the colonies of *A. florea* occurs mostly in arid regions, some colonies were also seen in plains and forest areas. The density of wild colonies greatly varied among the different districts of West Bengal. The 'rarh' region, characterized by hot and dry summers and cold winters, is dominated by the bee species. The number of wild colonies remains higher in human settlement areas than others (Table 23.1).

TABLE 23.1

Number of *Apis florea* Colonies per km^2 Area in West Bengal

Season	Human Settlement Area	Forest Area	Agricultural Field Area
Summer	4.2 ± 1.32	2.3 ± 1.34	0.4 ± 0.70
Monsoon	0.9 ± 1.10	0	0
Autumn	0.7 ± 0.82	0	0
Late autumn	1.1 ± 1.10	1.2 ± 1.23	0
Winter	3.8 ± 1.48	2.3 ± 1.49	0.7 ± 0.82
Spring	4.4 ± 1.35	2.4 ± 1.26	0.6 ± 0.70
Mean	2.52^a ± 2.01	1.37^b ± 1.48	0.28^c ± 0.58

Note: Values given as mean ± standard deviation. The superscript letters (a, b, and c) are the mean values in the last row indicating significant differences (determined by DMRT at 5%).

The bees' abundance also significantly varied among the seasons. Higher numbers of wild colonies were recorded during winter, spring and summer. Thereafter, the colonies migrate to other regions due to unfavourable weather conditions and dearth of floral resources. As a result, a number of colonies drastically decreased during monsoon and autumn.

Nesting Characteristics of *A. florea*

A colony builds a single, exposed comb, which is attached to a branch of shrubs, small trees or within house buildings and accessories (Figure 23.1). The majority of the studied nests (>60%) were situated within house buildings and hanging on cornices, window lintels and frames. About 34% of the studied nests were on living plant branches. The plants selected by the bee species as nesting substrate were *Citrus × aurantiifolia*, *Citrus maxima*, *Lantana camera*, *Mangifera indica*, *Millettia pinnata*, *Moringa oleifera*, *Psidium guajava* and *Punica granatum*. Height of the nest from ground level ranged from 1.31 to 10.97 m with an average value of 5.01 ± 3.10 m (Table 23.2). In general, the height of nests situated in buildings (6.69 ± 2.75 m) was higher in comparison to those situated in plant branches (2.24 ± 0.74 m). Horizontal and vertical dimensions of hives are 18.68 ± 3.89 cm and 18.92 ± 4.21 cm, respectively. Nests situated in house buildings have a slightly higher vertical dimension (20.56 ± 4.19 cm) than those situated on shrub or tree branches. The upper region of the hive serves as a honey storage area and the lower region as a brood area (Figure 23.2). Pollen is stored at the top of the brood area. Drone cells are found at the lower margin of the nest, and queen cells are found protruding vertically from the lower edges of the nest. The depth and width of drone cells are slightly greater than worker cells (Table 23.2).

Floral Resources

Bee-foraged plants were determined by melissopalynological analysis of honey and pollen samples (Layek et al., 2015a, 2015b; Layek and Karmakar, 2016). Additionally, field surveys were also performed. However, melissopalynological studies were more effective in quantifying the resource (nectar and pollen grain) contribution of bee flora. Micrographs of some pollen types obtained from palynological analysis of honey and pollen loads of *A. florea* are given in Figure 23.3. Predominant pollen types in honey samples were *Borassus flabellifer* type, *Brassica* type, *Bridelia retusa* type, *Butea monosperma* type, *Eucalyptus* type, *Lannea coromandelica* type and *Terminalia arjuna* type. Very frequent pollen types in pollen loads were *Borassus flabellifer* type, *Brassica* type, *Bridelia retusa* type, *Coriandrum sativum* type, *Eucalyptus* type, *Lannea coromandelica* type, *Phoenix sylvestris* type, *Terminalia arjuna* type and *Trianthema portulacastrum* type. Within the forage spectrum of the bee species, highly represented families were Fabaceae and Asteraceae (Table 23.3). Among the foraged plants, most taxa (69.91%) were utilized by the bee species for both nectar and pollen sources.

FIGURE 23.1 Nesting substrates of *Apis florea*. **a–f.** Alive plant branches (**a**, **b**. *Citrus* × *aurantiifolia*. **c**. *Lantana camera*. **d**. *Mangifera indica*. **e**. *Moringa oleifera*. **f**. *Psidium guajava*). **g**. Dead bamboo stick. **h**. Cart. **i–l**. Living houses. Scale bar = 10 cm.

TABLE 23.2

Nesting Characteristics of *Apis florea* in West Bengal, India

Parameter	Range	Mean	Standard Deviation
Height of nest (m)	1.31–10.97	5.01	3.10
Dimensions of nest (cm)			
Horizontal length	10.7–26.6	18.68	3.89
Vertical length	10.3–28.8	18.92	4.21
Thickness of honey storage area	4.7–6.2	5.52	0.33
Thickness of brood area	1.5–1.8	1.67	0.13
Worker cell size (cm)			
Depth of cell	0.9–1.1	0.98	0.09
Width of cell	0.3–0.4	0.33	0.05
Drone cell size (cm)			
Depth of cell	1.2–1.5	1.40	0.09
Width of cell	0.4–0.5	0.44	0.05
Queen cell size (cm)			
Depth of cell	1.3–1.6	1.45	0.13
Internal diameter	0.4–0.5	0.48	0.05

FIGURE 23.2 Hive of *Apis florea* showing different areas. **A**. Honey storage area. **B**. Brood area. **C**. Pollen cell. Scale bar = 5 cm.

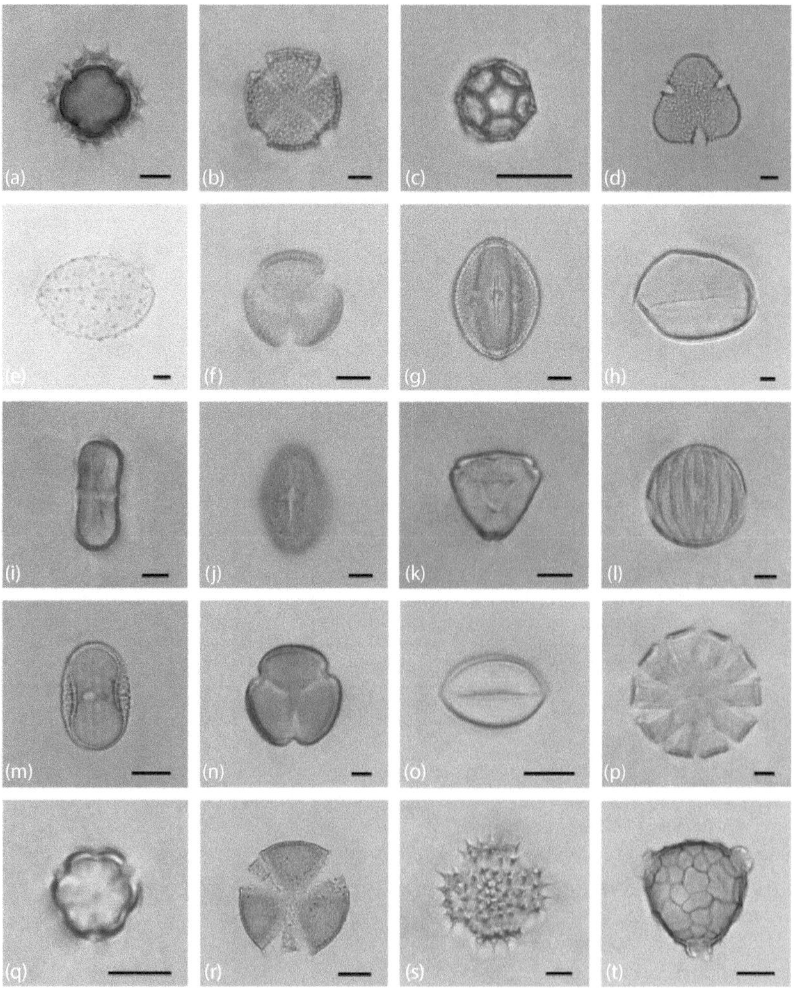

FIGURE 23.3 Photographs of some pollen types obtained from honey and pollen samples of *Apis florea*. **a**. *Acmella radicans*. **b**. *Aegle marmelos*. **c**. *Alternanthera paronychioides*. **d**. *Bombax ceiba*. **e**. *Borassus flabellifer*. **f**. *Brassica* type. **g**. *Cissus quadrangularis*. **h**. *Cocos nucifera*. **i**. *Corchorus olitorius*. **j**. *Coriandrum sativum*. **k**. *Eucalyptus* type. **l**. *Hygrophila auriculata*. **m**. *Justicia simplex*. **n**. *Leucaena leucocephala*. **o**. *Phoenix sylvestris*. **p**. *Sesamum indicum*. **q**. *Terminalia arjuna*. **r**. *Trianthema portulacastrum*. **s**. *Tridax procumbens*. **t**. *Vigna radiata*. Scale bar = 10 μm.

TABLE 23.3
List of Plant Species Foraged by *Apis florea* in West Bengal, India

			Season					
Family	**Plant Species**	**Resource**	**Summer**	**Monsoon**	**Autumn**	**Late Autumn**	**Winter**	**Spring**
Acanthaceae	*Hygrophila auriculata*	N + P	+	–	–	+	+	+
	Justicia adhatoda	N	+	+	–	–	–	–
	Justicia gendarussa	N	–	–	–	–	–	+
	Justicia simplex	N + P	–	+	+	+	–	–
Aizoaceae	*Trianthema portulacastrum*	N + P	+	++	++	–	–	–
Amaranthaceae	*Alternanthera paronychioides*	N	+	+	–	–	–	–
	Amaranthus spinosus	P	+	+	–	–	–	–
	Spinacia oleracea	P	–	–	–	–	+	+
Amaryllidaceae	*Allium cepa*	N + P	–	–	–	–	–	+
Anacardiaceae	*Lannea coromandelica*	N + P	–	–	–	–	–	+++
	Mangifera indica	N	–	–	–	–	++	+
	Semecarpus anacardium	N + P	+	–	–	–	–	–
	Spondias pinnata	N + P	–	–	–	–	++	++
Apiaceae	*Coriandrum sativum*	N + P	–	–	–	–	+++	+++
	Foeniculum vulgare	N + P	–	–	–	–	+	+
Apocynaceae	*Calotropis procera*	N	+	+	–	–	–	–
	Gymnema sylvestris	N	+	+	–	–	–	–
Arecaceae	*Borassus flabellifer*	N + P	+++	–	–	–	–	+++
	Cocos nucifera	N + P	+	++	++	+	+	+
	Phoenix sylvestris	N + P	–	–	–	–	+++	+++
Asparagaceae	*Polianthes tuberosa*	P	+	+	–	–	–	–
Asteraceae	*Acmella radicans*	N + P	–	–	–	–	++	++
	Ageratum conyzoides	N + P	–	+	+	–	–	–
	Chrysanthemum indicum	N + P	–	–	–	–	+	+
	Cyanthillium cinereum	N + P	+	–	–	–	–	+
	Eupatorium odoratum	N + P	–	–	–	–	+	+
	Helianthus annuus	N + P	–	–	–	–	+	+
	Mikania scandens	N + P	–	–	–	–	+	+
	Tagetes erecta	N + P	–	–	–	–	+	+
	Taraxacum officinale	N + P	+	–	–	–	–	+
	Tridax procumbens	N + P	+	+	+	–	–	–
Brassicaceae	*Brassica juncea*	N + P	–	–	–	–	+++	+++
	Brassica nigra	N + P	–	–	–	–	+++	+++
	Raphanus sativus	N + P	–	–	–	–	+	+
Cactaceae	*Cereus pterogonus*	N + P	–	+	–	–	–	–
Cannabaceae	*Trema orientalis*	N + P	+	+	–	–	–	–
Capparaceae	*Capparis zeylanica*	P	–	–	–	–	–	+
Cleomaceae	*Cleome viscosa*	N + P	+	+	–	–	–	–
Combretaceae	*Terminalia arjuna*	N + P	+++	–	–	–	–	–

(*Continued*)

TABLE 23.3 (*Continued*)

List of Plant Species Foraged by *Apis florea* in West Bengal, India

			Season					
Family	**Plant Species**	**Resource**	**Summer**	**Monsoon**	**Autumn**	**Late Autumn**	**Winter**	**Spring**
Commelinaceae	*Commelina benghalensis*	N	−	+	+	+	−	−
	Cyanotis axillaris	N + P	−	+	+	+	−	−
Convolvulaceae	*Evolvulus nummularius*	N + P	−	+	−	−	−	−
Cornaceae	*Alangium salviifolium*	N + P	−	−	−	−	−	++
Cucurbitaceae	*Citrullus lanatus*	N + P	+	−	−	−	−	+
	Coccinia grandis	N + P	+	+	−	−	−	−
	Cucumis sativus	N	−	+	−	−	−	−
	Cucurbita maxima	N	++	++	−	−	−	−
	Luffa cylindrica	N + P	+	+	+	−	−	−
	Momordica charantia	N + P	+	+	−	−	−	+
Dipterocarpaceae	*Shorea robusta*	N + P	−	−	−	−	−	+
Euphorbiaceae	*Croton bonplandianum*	N + P	+	+	+	−	−	−
	Ricinus communis	N + P	−	+	+	+	+	−
Fabaceae	*Acacia auriculiformis*	N + P	−	+	++	+	−	−
	Acacia nilotica	N + P	−	+	+	−	−	−
	Albizia lebbeck	N + P	+	−	−	−	−	+
	Butea monosperma	N	−	−	−	−	−	+++
	Cassia fistula	N + P	−	−	−	−	−	+
	Dalbergia sissoo	N + P	+	+	−	−	−	−
	Lablab purpureus	N	−	−	−	−	+	+
	Leucaena leucocephala	N + P	+	+	+	−	−	−
	Millettia pinnata	N	++	−	−	−	−	++
	Mimosa pudica	N + P	+	+	−	−	−	−
	Peltophorum pterocarpum	N + P	++	−	−	−	−	+
	Pithecellobium dulce	N	−	−	−	−	+	+
	Tamarindus indica	N	++	++	−	−	−	−
	Tephrosia purpurea	N	−	+	−	−	−	−
	Vigna radiata	N	+	+	−	−	−	−
Lamiaceae	*Gmelina arborea*	N + P	+	−	−	−	−	+
	Ocimum sanctum	N + P	−	+	−	−	−	−
	Vitex negundo	N	+	+	−	−	−	−
Lythraceae	*Lagerstroemia speciosa*	N + P	+	+	−	−	−	−
	Punica granatum	P	+	+	−	−	−	−
Malvaceae	*Bombax ceiba*	N + P	−	−	−	−	−	+
	Ceiba pentandra	N + P	−	−	−	−	−	+
	Corchorus capsularis	N + P	+	−	−	−	−	−
	Corchorus olitorius	N + P	+	−	−	−	−	−
	Grewia asiatica	N + P	−	−	−	−	+	−
Meliaceae	*Azadirachta indica*	N	+	−	−	−	−	+
	Melia azedarach	N	−	−	−	−	−	+

(*Continued*)

TABLE 23.3 (*Continued*)

List of Plant Species Foraged by *Apis florea* in West Bengal, India

			Season					
Family	**Plant Species**	**Resource**	**Summer**	**Monsoon**	**Autumn**	**Late Autumn**	**Winter**	**Spring**
Moringaceae	*Moringa oleifera*	N	–	–	–	–	++	++
Myrtaceae	*Eucalyptus globulus*	N + P	–	–	–	++	+++	++
	Eucalyptus tereticornis	N + P	–	–	–	++	+++	++
	Psidium guajava	N + P	+	–	–	–	+	+
	Syzygium cumini	N + P	–	–	–	–	–	++
	Syzygium jambos	N + P	–	–	–	–	+	+
	Syzygium reticulatum	N + P	++	–	–	–	–	–
Nyctaginaceae	*Boerhavia diffusa*	N + P	+	+	–	–	–	–
Oxalidaceae	*Averrhoa carambola*	N	+	+	–	–	–	–
Pedaliaceae	*Sesamum indicum*	N + P	++	–	–	–	–	–
Phyllanthaceae	*Bridelia retusa*	N + P	–	++	–	–	–	–
	Phyllanthus emblica	N + P	+	+	–	–	–	–
Poaceae	*Zea mays*	P	+	+	–	–	–	–
Polygonaceae	*Antigonon leptopus*	N	+	+	+	–	–	–
Ranunculaceae	*Nigella sativa*	N + P	–	–	–	–	+	+
Rhamnaceae	*Ziziphus mauritiana*	N + P	–	+	+	–	–	–
Rubiaceae	*Haldina cordifolia*	N + P	–	+	+	–	–	–
	Neolamarckia cadamba	N + P	–	+	–	–	–	–
	Vangueria spinosa	N	+	+	–	–	–	–
Rutaceae	*Aegle marmelos*	N	+	–	–	–	–	–
	Citrus × aurantiifolia	N + P	+	++	–	–	–	–
	Citrus maxima	N + P	–	–	–	–	–	+
	Murraya koenigii	N	–	–	–	–	–	+
	Murraya paniculata	N + P	+	+	–	–	–	+
Salicaceae	*Flacourtia jangomas*	N + P	–	–	–	–	–	+
Sapindaceae	*Litchi chinensis*	N + P	–	–	–	–	–	+
	Sapindus mukorossi	N + P	–	–	–	–	+	–
Sapotaceae	*Madhuca longifolia*	N + P	–	–	–	–	–	++
	Mimusops elengi	N + P	+	+	–	–	–	–
Solanaceae	*Solanum melongena*	P	+	–	–	–	–	–
	Solanum sisymbriifolium	P	+	–	–	–	–	–
Verbenaceae	*Phyla nodiflora*	N + P	+	++	–	–	–	–
Vitaceae	*Cissus quadrangularis*	N	–	+	–	–	–	–
Ulmaceae	*Holoptelea integrifolia*	N + P	–	–	–	–	–	++

Note: N = nectar; P = pollen; frequency: high (+++), medium (++) and low (+).

In terms of season, the number of hosted plant species was very high during spring, summer and monsoon. In autumn and late autumn, the number of resource-supplied plant taxa was very low. However, number of hosted plant species and the amount of resource contribution to the bee species were not always correlated. In monsoon and autumn, predominant and very frequent pollen types were not found. These seasons were manifested as floral dearth period for the bee species in West Bengal.

FIGURE 23.4 *Apis florea* collecting floral rewards from wild and cultivated plants. **a**. *Acmella radicans*. **b**. *Brassica nigra*. **c**. *Citrullus lanatus*. **d**. *Coriandrum sativum*. **e**. *Cucurbita maxima*. **f**. *Foeniculum vulgare*. **g**. *Mangifera indica*. **h**. *Nigella sativa*. **i**. *Phyla nodiflora*. **j**. *Syzygium reticulatum*. **k**. *Vigna radiata*. **l**. *Vitex negundo*. Scale bar = 10 mm.

Forage spectrum of the bee species comprises both wild and cultivated plant species (Figure 23.4). To collect floral rewards, the bees foraged on angiosperms with a high degree of floral fidelity. This behaviour makes them effective pollinators of several plant species. Thereby, they play an important role to conserve the local flora and increase the fruit and seed set of some crop plants such as *Allium cepa*, *Brassica* spp., *Citrus* spp., *Coriandrum sativum*, *Cucumis sativus*, *Helianthus annuus*, *Momordica charantia* and *Psidium guajava* (Abrol, 2010; Sihag, 2019). The bee-visited plants were much diversified with respect to plant habit and flowering phenology. Regarding plant habits of bee-foraged plants, most were trees (46.02%), followed by herbs (30.97%), shrubs and climbers. Most of them show a synchronous type of flowering. The bee-foraged plant showed much variation in flower shapes including brush, cruciform, dish, funnel, gullet, pea-shaped, saucer and stallate etc. However, most common were dish-, brush- and stallate-shaped flowers. The bee species are able to forage on very small-sized (viz. *Allium cepa*, *Trema orientalis*) to very large-sized flowers (viz. *Bombax ceiba*, *Cereus pterogonus*, *Cucurbita maxima*). Most of the foraged plants bear small-sized (49.56%) to medium-sized (38.94%) flowers. The colours of bee-visited flowers (as realized by humans) were also diversified. However, yellow, white and cream-coloured flowers were dominant within the bee flora.

Conclusion

The red dwarf honey bee frequently occurs in 'rarh' regions of West Bengal. The number of wild colonies was significantly higher in human settlement areas especially during winter to summer due to favourable weather conditions, availability of nesting sites and floral resources. The foraging habit of the bee species was polylectic, as the bees utilized diversified angiosperm flora and the foraging strategy remains profitable in terms of utilization of almost all the bee plants for both nectar and pollen sources. Based on the presence of wide plant habits and flower characteristics within the bee flora and a high degree of floral constancy of the bee species, we can conclude that *A. florea* is an obligatory generalist flower visitor vis-a-vis pollinator.

ACKNOWLEDGEMENTS

We are thankful to authorities of Vidyasagar University (VU) for providing necessary laboratory facilities. We are indeed thankful to the USIC section of VU as well as to Mr Dipankar Mandal for microscopy. We are also grateful to Prof. D. P. Abrol (faculty of agriculture, SKUAST) for his invitation as well as cooperation as editor.

REFERENCES

Abrol DP. 2010. Foraging behaviour of *Apis florea* F., an important pollinator of *Allium cepa* L. *Journal of Apicultural Research* 49(4): 318–325.

Arias CA, Sheppard WS. 2005. Phylogenetic relationships of honey bees (Hymenoptera: Apinae: Apini) inferred from nuclear and mitochondrial DNA sequence data. *Molecular Phylogenetics and Evolution* 37: 25–35.

Corlett RT. 2011. Honeybees in natural ecosystems. In: R. H. Hepburn and S. E. Radloff (Eds.). *Honeybees of Asia*. Berlin, Germany: Springer.

Hepburn HR, Hepburn C. 2005. Bibliography of *Apis florea*. *Apidologie* 36(3): 377–378.

Layek U, Bhakat RK, Karmakar P. 2015a. Foraging behavior of *Apis florea* Fabricius during winter and spring-summer in Bankura and Paschim Medinipur districts, West Bengal. *Global Journal of Bio-Science and Biotechnology* 4(3): 255–263.

Layek U, Karmakar P. 2016. Bee plants used as nectar sources by *Apis florea* Fabricius in Bankura and Paschim Medinipur districts, West Bengal. *Geophytology* 46(1): 1–14.

Layek U, Karmakar P, Pal PK. 2015b. Significance of *Trianthema portulacastrum* L. in sustenance of *Apis florea* colonies during dearth period. *International Journal of Biosciences* 7(1): 22–29.

Mogga J, Ruttner F. 1988. *Apis florea* in Africa: Source of the founder population. *Bee World* 69: 100–103.

Rader R, Bartomeus I, Garibaldi LA, Garratte MPD et al. 2016. Non-bee insects are important contributors to global crop pollination. *Proceedings of the National Academy of Science of the USA* 113: 146–151.

Ruttner F. 1988. *Biogeography and Taxonomy of Honey Bees*. Heidelberg, Germany: Springer-Verlag.

Sihag RC. 2019. Crop hosts and pollination potential of the red dwarf honey bee (*Apis florea* F.) in the semi-arid environment of North West India. *Journal of Applied Sciences* 19(6): 551–556.

Thomas D, Pal N, Subba RK. 2002. Bee management and productivity of Indian honeybees. *Apiacta* 3: 1–15.

Index

Note: Page numbers in italics and bold refer to figures and tables, respectively.

B

F

G

N